Tributes
Volume 29

Computational Models of Rationality
Essays Dedicated to Gabriele Kern-Isberner on the Occasion of her 60th Birthday

Volume 18
Insolubles and Consequences. Essays in Honour of Stephen Read.
Catarina Dutilh Novaes and Ole Thomassen Hjortland, eds.

Volume 19
From Quantification to Conversation. Festschrift for Robin Cooper on the occasion of his 65th birthday
Staffan Larsson and Lars Borin, eds.

Volume 20
The Goals of Cognition. Essays in Honour of Cristiano Castelfranchi
Fabio Paglieri, Luca Tummolini, Rino Falcone and Maria Miceli, eds.

Volume 21
From Knowledge Representation to Argumentation in AI, Law and Policy Making. A Festschrift in Honour of Trevor Bench-Capon on the Occasion of his 60th Birthday
Katie Atkinson, Henry Prakken and Adam Wyner, eds.

Volume 22
Foundational Adventures. Essays in Honour of Harvey M. Friedman
Neil Tennant, ed.

Volume 23
Infinity, Computability, and Metamathematics. Festschrift celebrating the 60th birthdays of Peter Koepke and Philip Welch
Stefan Geschke, Benedikt Löwe and Philipp Schlicht, eds.

Volume 24
Modestly Radical or Radically Modest. Festschrift for Jean Paul Van Bendegem on the Occasion of his 60th Birthday
Patrick Allo and Bart Van Kerkhove, eds.

Volume 25
The Facts Matter. Essays on Logic and Cognition in Honour of Rineke Verbrugge
Sujata Ghosh and Jakub Szymanik, eds.

Volume 26
Learning and Inferring. Festschrift for Alejandro C. Frery on the Occasion of his 55th Birrthday
Bruno Lopes and Talita Perciano, eds.

Volume 27
Why is this a Proof? Festschrift for Luiz Carlos Pereira
Edward Hermann Haeusler, Wagner de Campos Sanz and Bruno Lopes, eds.

Volume 28
Conceptual Clarifications. Tributes to Patrick Suppes (1922-2014)
Jean-Yves Béziau, Décio Krause and Jonas R. Becker Arenhart, eds.

Volume 29
Computational Models of Rationality. Essays Dedicated to Gabriele Kern-Isberner on the Occasion of her 60th Birthday
Christoph Beierle, Gerhard Brewka and Matthias Thimm, eds.

Tributes Series Editor
Dov Gabbay   dov.gabbay@kcl.ac.uk

Computational Models of Rationality

Essays Dedicated to Gabriele Kern-Isberner on the Occasion of her 60th Birthday

edited by

Christoph Beierle

Gerhard Brewka

and

Matthias Thimm

© Individual authors and College Publications 2016. All rights reserved.

ISBN 978-1-84890-198-8

College Publications
Scientific Director: Dov Gabbay
Managing Director: Jane Spurr

http://www.collegepublications.co.uk

Cover design by Laraine Welch

Printed by Lightning Source, Milton Keynes, UK

All rights reserved. No part of this publication may be reproduced, stored in a retrieval system or transmitted in any form, or by any means, electronic, mechanical, photocopying, recording or otherwise without prior permission, in writing, from the publisher.

Preface

Gabriele Kern-Isberner has made numerous scientific contributions to Artificial Intelligence, focussing on qualitative and quantitative approaches to knowledge representation and investigating the foundations of rational reasoning. This book is a Festschrift dedicated to her on the occasion of her 60th birthday. It contains contributions by her students, colleagues, and friends who are scientifically close to her. The articles, written by computer scientists, mathematicians, philosophers, and psychologists, address recent research in areas such as rationality and non-monotonic reasoning, problem solving and query answering, belief revision, uncertain reasoning, and argumentation. In addition, the first chapter of the Festschrift contains a summary of Gabriele Kern-Isberner's contributions compiled by the editors of this volume.

We are grateful to all authors for their contributions to this Festschrift, and we would like to thank all colleagues who served as reviewers for the articles.

This Festschrift was handed over to Gabriele in a festive colloquium that took place at the TU Dortmund University in Dortmund, Germany. We are deeply indebted to Christian Eichhorn and the other member's of Gabriele's group in Dortmund for their efforts and help in organizing this colloquium.

January 2016   Christoph Beierle
                                                            Gerhard Brewka
                                                            Matthias Thimm

Table of Contents

Christoph Beierle, Gerhard Brewka, Matthias Thimm
A Personal Glimpse on Gabriele Kern-Isberner's
Contributions to Artificial Intelligence 1

I Foundations

Benjamin Fine, Gerhard Rosenberger
i as a Quadratic Residue in the Gaussian Integers 23

Anja Moldenhauer, Gerhard Rosenberger
Cryptosystems Using Automorphisms of Finitely
Generated Free Groups.. 31

Bernhard Thalheim, Marina Tropmann-Frick
Evaluation and Capability of Models....................... 52

II Rationality and Non-monotonic Reasoning

James Delgrande, Bryan Renne
On a Minimal Logic of Default Conditionals 73

Sven Ove Hansson
Alternatives to the Ramsey Test 84

Marco Ragni
Can non-monotonic logics model human reasoning?........ 98

Wolfgang Spohn
Enumerative Induction 112

III Problem Solving and Query Answering

Joachim Biskup, Cornelia Tadros
On the Simulation Assumption for Controlled Interaction
Processing.. 133

Thomas Eiter, Christoph Redl, Peter Schüller
Problem Solving Using the HEX Family 150

Thomas Lukasiewicz, Maria Vanina Martinez, Cristian Molinaro, Livia Predoiu, Gerardo I. Simari
Ranking Answers to Datalog+/− Ontologies based on Trust and Reliability of Subjective Reports 175

IV Belief Revision

Michael Abraham, Israel Belfer, Uri Schild, Dov Gabbay
Identity Merging and Identity Revision in Talmudic Logic: An Outline Paper 195

Salem Benferhat, Amen Ajroud
On the Syntactic Representation of Multiple Iterated Belief c-Revision 210

Marcelo A. Falappa, Alejandro Garcia, Guillermo Simari
A Set of Operations for Stratified Belief Bases 223

Eduardo Fermé, Sara Gonçalves
On the Iteration of KM-Update 243

V Uncertain Reasoning

Igor Douven
On de Finetti on Iterated Conditionals..................... 265

Didier Dubois, Henri Prade
Qualitative and Semi-Quantitative Modeling of Uncertain Knowledge - A Discussion 280

Marc Finthammer, Christoph Beierle
On the Relationship Between Aggregating Semantics and FO-PCL Grounding Semantics for Relational Probabilistic Conditionals................................. 297

Elizabeth Howarth, Jeff Paris
The Finite Values Property 316

Nico Potyka
Relationships Between Semantics for Relational Probabilistic Conditional Logics........................... 332

Wilhelm Rödder, Friedhelm Kulmann and Andreas Dellnitz
A New Rationality in Network Analysis - Status of Actors in a Conditional-logical Framework................. 348

Klaus-Dieter Schewe, Flavio Ferrarotti, Loredana Tec, Qing Wang
Towards a Behavioural Theory for Random Parallel Computing 365

VI Argumentation

Leila Amgoud
On Argumentation-based Paraconsistent Logics 377

Dov Gabbay, Odinaldo Rodrigues
Further Applications of the Gabbay-Rodrigues Iteration Schema in Argumentation and Revision Theories 392

Author Index 409

A Personal Glimpse on Gabriele Kern-Isberner's Contributions to Artificial Intelligence

Christoph Beierle[1], Gerhard Brewka[2], Matthias Thimm[3]

Abstract. This article is a tribute to Gabriele Kern-Isberner on the occasion of her 60th birthday. We start out with a brief sketch of her scientific career and then provide a summary of her main research contributions. Needless to say the selection of topics we describe is highly subjective and far from covering the whole range of areas Gabriele has contributed to. We also include a description of the numerous services she provided to the community. The article concludes with some personal remarks from each of the authors.

1 Introduction

Gabriele Kern-Isberner has made numerous scientific contributions in the area of Artificial Intelligence. A major focus of her research is on qualitative and quantitative approaches to knowledge representation, dealing with default and non-monotonic logics, uncertain reasoning, probabilistic reasoning, belief revision, and argumentation, as well as multi-agent systems and knowledge discovery. Also by crossing borders to philosophy and psychology and collaborating with specialist in these fields, she has been investigating foundations of rational reasoning.

This article briefly sketches Gabriele's scientific career, provides a summary of her research contributions, and describes some of her many activities within the scientific community. From different perspectives, the three of us have known Gabriele and have been working with her for many years. As this article is a personal glimpse on her contributions, we will also add some personal comments on our collaborations with Gabriele.

2 Study, Dissertation, Family Leave, and Habilitation: From Mathematics to Computer Science

After studying mathematics and economics at the University of Dortmund and receiving a Diploma degree in 1979, Gabriele Kern-Isberner joined the University of Dortmund as a research assistant at the department of mathematics, her

---

[1] Faculty of Mathematics and Computer Science, University of Hagen, 58084 Hagen, Germany
[2] Intelligent Systems Department, Computer Science Institute, University of Leipzig, Germany
[3] Institute for Web Science and Technologies, University Koblenz-Landau, Koblenz, Germany

research focus being in the field of combinatorial group theory. In 1985, Gabriele received a Ph.D. in mathematics from the University of Dortmund. Together with her Ph.D. supervisor Gerhard Rosenberger, Gabriele has published results in mathematics even as recently as 2015 [99–101, 50].

From 1983 until 1992, Gabriele was on family leave with three children, born in 1981, 1985, and 1988. Supported by a scholarship for the re-integration of women after maternity leave, she joined the group of Wilhelm Rödder at the department of economics of the University of Hagen in 1992 where she got interested in the formal foundations of the expert system shell SPRIT [132], i.e., probablistic logics [120, 124], conditionals [1, 117, 121] and the principle of maximum entropy [133, 122, 123], and started to work in these areas [94, 128–130, 96–98].

In order to continue and extend her research on knowledge representation and reasoning and to obtain a habilitation degree in computer science, Gabriele moved to computer science and joined the group of Christoph Beierle at the University of Hagen in 1996. From 1996 until 1998 she was supported by a Lise-Meitner habilitation grant. In a series of publications [61–71, 73–76], she developed a new theory of conditionals where conditionals are viewed as agents; the following is a quote from the book [72]:

> *Conditionals are considered as agents shifting possible worlds in order to establish relationships and beliefs. This understanding of conditionals yields a rich methodological theory, which makes complex interactions between conditionals transparent and operational. Moreover, it provides a unifying and enhanced framework for knowledge representation, non-monotonic reasoning, and belief revision, and even for knowledge discovery. In separating structural from numerical aspects, the basic techniques for conditionals introduced in this book are applied both in a qualitative and in a numerical setting, elaborating fundamental lines of reasoning.* [72, p. vii]

In September 2000, Gabriele received a habilitation from the Department of Computer Science at the University of Hagen; the reviewers of her thesis in the habilitation committee were Christoph Beierle, Gerhard Brewka, and Dov Gabbay. Her habilitation thesis was published as a monograph in the Springer Lecture Notes in Artificial Intelligence series [72].

After her habilitation, Gabriele was research assistant and lecturer at the University of Hagen and leading scientist in the DFG[4] project CONDOR. In the years between 2000 and 2004, she was substitute professor of computer science at the University of Hagen, at the University of Leipzig, and at the University of Dortmund. In 2004, she joined the University of Dortmund as a professor of information engineering at the departement of computer science.

3 Scientific Contributions

In this section we try to do the impossible, namely providing – in a couple of pages – an overview of the numerous contributions Gabriele has made to

---

[4] Deutsche Forschungsgemeinschaft; German research foundation

various areas in knowledge representation and reasoning, and to Artificial Intelligence in general. We are fully aware that we can hardly scratch the surface, and that our short description is far from doing justice to the breadth and depth of Gabriele's thinking. The only excuse we have is the sheer amount of different topics covered by Gabriele. Indeed, she has made substantial contributions at least in the following areas: qualitative and quantitative approaches to knowledge representation; default and non-monotonic logics; uncertain and probabilistic reasoning; belief revision; argumentation; multi-agent systems; knowledge discovery. In addition, Gabriele has always tried to demonstrate the usefulness of her foundational results by applying and evaluating them in the context of a broad range of real world applications.

In spite of the diverse areas covered, there is a guiding theme, sort of a leitmotif, to be found in much of her work, namely the integration of ideas, view points and approaches from different fields. Much of the success of this integration hinges on a core method she developed, namely a new theory of conditionals where conditionals are viewed as agents.

3.1 Conditionals, Belief Revision, Probabilistic Reasoning, and Nonmonotonic Logics

In the focus of many works of Gabriele lies the *conditional*, i.e., a structure of the form $(B \mid A)$ modelling the (default) rule "if A than (usually, probably) B". With the help of an algebraic machinery, the behaviour of conditionals in a knowledge base can be concisely represented and used for various problems. More precisely, using constant symbols a_i^+, a_i^-, and 1, representing the states "Conditional i is verified", "Conditional i is falsified", and "Conditional i is not applicable", interpretations of a propositional logic can be completely described by products of these symbols, and, more importantly, various problems on the knowledge representation layer can be represented as algebraic operations on these products. This approach also allows to cover a broad range of models for handling uncertainty, such as ranking functions and probability theory, and to investigate the aforementioned problems in a general manner. In particular, the problem of *revision*, i.e., the addition of contradictory information to a knowledge base, and similar problems such as *iterated revision* and *update* have been handled within this framework through a series of technical papers [77, 72, 78–80, 82, 83, 91], in both qualitative and quantitive settings. Using abstract state machines, a verified implementation of belief revision in the programming language AsmL is developed in [17, 22]. Further works investigate more specialised operations for belief dynamics such as non-prioritized and multiple belief revision [46, 45], the relationship between incision and selection functions [42], the relationship belief revision and argumentation [44, 43], and problems of belief revision in logic programming [92, 111].

By employing Goguen and Burstall's notion of institution for the formalization of logical systems, different forms of semantics that have been proposed for conditionals, ranging from quantitative probability distributions to pure qualitative approaches using e.g. plausibility orderings, possibility distributions [38], or conditional objects, are formalized as institutions in [14, 16, 19, 20, 24–

27], and the precise formal relationships among these logics are established as institution morphisms.

Another important approach in many of Gabriele's works, in particular for the probabilistic setting, is the *principle of maximum entropy*, see also her recent overview paper [29]. This principle can be used for the problem of completing incomplete information in the most unbiased way and can be characterised by few simple properties, cf. [80]. Besides using this principle for problems related to belief dynamics (see above) it has also been used for the problem of *probabilistic abduction* in [36, 37]. Further works include investigations of the algorithmic challenges for computing probability distributions with maximum entropy [52, 51] and the problem of repairing inconsistent probabilistic knowledge bases [55]. Exploiting the fact that maximum entropy distributions can be computed by solving nonlinear equation systems that reflect the conditional logical structure of these distributions, in [107–109] the theory of Gröbner bases is applied to the polynomial system which is associated with a maximum entropy distribution, in order to obtain results for reasoning with maximum entropy.

3.2 Inductive Knowledge Completion, Knowledge Discovery, and Learning

The methodological theory of conditionals developed by Gabriele allows to describe the aim of knowledge discovery in a very general sense: to reveal structures of knowledge which can be seen as *structural relationships* being represented by conditionals. For developing a concrete algorithm for knowledge discovery, Gabriele proposed two key ideas. First, knowledge discovery is understood as a process which is inverse to inductive knowledge representation. So the relevance of discovered information is judged with respect to the chosen induction method. Second, the link between structural and numerical knowledge is established by an algebraic theory of conditionals, which makes it possible to consider complex interactions between rules, see e.g. [70, 72, 73, 81, 84]. By applying this theory, an algorithm that computes sets of probabilistic rules from distributions was developed and implemented in the functional programming language Haskell [90, 56, 57]. In [105, 106], this approach to knowledge discovery is transferred and successfully applied in a qualitative setting. A formalization of the inverse representation problem in the framework of institutions is given in [15, 18], and in [17], abstract state machines are used for modelling conditional knowledge discovery and belief revision. Aspects of self-learning are investigated in [131], and in [116] a hybrid learning method combining low-level, non-cognitive reinforcement learning with high-level epistemic belief revision is proposed.

3.3 Relational Conditionals and Maximum Entropy Reasoning

Most of Gabriele's works discussed so far consider propositional logic as the underlying knowledge representation formalism. More recent works also address the challenge of using first-order logic, or restrictions thereof, for the same purpose. In this setting, probabilistic conditionals have the form $(B(\boldsymbol{x}) \mid A(\boldsymbol{x}))[p]$ with first-order formulas $B(\boldsymbol{x})$ and $A(\boldsymbol{x})$ and probability p.

The initial works [119, 93] address the challenge of combining reasoning based on the principle of maximum entropy with logic programming. Here, logic programming rules are interpreted as schemas of the above form and semantics is given to a probabilistic logic program by grounding. However, grounding probabilistic conditionals may introduce conflicts. For example, the general conditional (likes(X, fred) | elephant(X))[0.1], stating that elephants usually do not like keeper Fred, and the conditional (likes(clyde, fred) | elephant(clyde))[0.9], stating that elephant Clyde usually likes Fred, are incompatible as "clyde" can also be instantiated in the first conditional, yielding different probabilities for the same statement. In order to overcome this problem, *grounding operators* [118] can be used to disallow certain instantiations wrt. their structural specificity. However, this solution is a technical one and more recent works of Gabriele are concerned with a deeper understanding how general conditionals and specific information on individuals can be brought together. The works [103, 138, 142] propose novel semantical approaches to ascribe meaning to relational conditionals by joining the interpretation of first-order logic and probabilistic conditionals in a single framework. These works also apply the principle of maximum entropy in order to complete incomplete information and investigate the whole framework wrt. several rationality postulates of non-monotonic reasoning.

The approaches investigated in this setting have also been implemented in the integrated development environment **KReator** [136, 8], which also provides support for several other approaches for combining first-order logic and probabilistic reasoning. A software system for the computation, visualization, and comparison of conditional structures for relational probabilistic knowledge bases is presented in [32, 7]. Other works provide general discussions of these hybrid approaches and give evaluation criteria for investigating their usefulness wrt. a diverse range of features [9, 87, 86]. In [126], the learning approach of solving the inverse representatation problem (cf. Section 3.2) is transferred to a relational setting. The concept of belief revision by sets of conditionals (cf. Section 3.1) is applied to first-order conditionals in [125, 127], and algorithms for computing the change operations are developed.

3.4 Qualitative Approaches and Ranking Functions

While probabilistic conditionals were at the starting point of Gabriele's work in knowledge representation and reasoning, the techniques and concepts of her theory of conditionals, e.g. the notions of c-representation and c-revision [72], also fully cover qualitative settings. Making use of ordinal conditional functions (OCFs) [134], she has investigated and developed qualitative and semi-quantitative approaches to reasoning with conditionals. This encompasses preference fusion for default reasoning [95], the intensional combination of ranking functions, inductive reasoning with conditionals, and the development and use of OCF networks [88, 89, 39–41]. Qualitative probabilistic inference with default inheritance is studied in [143]. In [30], an implementation for reasoning with ranking functions given by c-representations using abstract state machines is given, and in [31, 13, 6] c-representations and their induced inference relation are modelled by constraint satisfaction problems and computed using constrained logic programming. In [104], ranking functions for first-order conditionals are introduced, and in [85], a system Z-like approach for first-order default reasoning is developed.

3.5 Argumentation

The area of argumentation has become highly active over recent years. The basic idea is to come up with computational models of how we usually make decisions or come to conclusions, namely by (1) generating arguments for and against certain options, (2) evaluating these arguments, that is, selecting reasonable and coherent sets of of arguments, and (3) accepting the positions supported by the selected arguments. Gabriele made various substantial contributions to this area centering around the following three topics: semantical foundations and relationships to other nonmonotonic formalisms [102, 140, 141], relationships between argumentation and belief revision [49, 47, 48, 113, 58], and distributed, agent based argumentation [3, 137, 139].

Combinations with nonmonotonic formalisms: Gabriele was always interested in bridging different areas. This also applies to her work on argumentation.

The relationship between the DeLP (Defeasible Logic Programming) approach to argumentation on one hand and answer set programming on the other is investigated in [140]. The paper introduces two types of translations from DeLP to answer set programming which handle strict rules somewhat differently. Both translations result in inference notions which are slightly weaker than inference via the dialectical warrant procedure in DeLP. The investigation of the subtle differences between these notions not only clarifies the respective forms of inference, but also leads to a better understanding of the concept of warrant which is at the heart of the DeLP approach.

In [102] an interesting combination of DeLP with ideas originally developed for default reasoning is presented. More specifically, possible worlds semantics from default reasoning is used to provide examples for arguments. This allows the notion of plausibility to be carried over to DeLP. Moreover, the priority relation between DeLP arguments is derived form the tolerance partitioning of system Z. The paper very nicely illustrates how fruitful it can be to cross the borders of a particular subarea, one of the strengths to be found throughout Gabriele's work.

Argumentation and belief revision: The relationship between these two important areas is analyzed in [47]. The authors draw a "big picture", illustrating the various reasoning tasks rational agents need to perform, and highlight the different, complementary roles argumentation and belief revision play in this picture. In a nutshell, while argumentation is more appropriate for the evaluation of information, taking the information's origin into account, belief revision is useful for the dynamics of beliefs. Combinations of both areas allow reasoning processes to be modelled in a much more fine-grained manner than within each of the areas on their own. This is convincingly demonstrated on the technical level in a series of papers [49, 58, 113, 135].

Agent based and distributed argumentation: In [3] the benefits of using argumentation in multi-agent settings are worked out (see also [35]). The paper develops an agent architecture where an agent is composed of modules, each one responsible for a basic capability or reasoning task. A local argumentation theory in the module provides preferred decision options. Also the inter-module coordination is based on a local argumentation theory. A distributed, DeLP based argumentation framework which allows different agents with different opinions to argue with each other is developed in [137, 139].

In addition, Gabriele has shown how argumentation can be successfully applied to realistic application problems [12, 11]. These applications will be discussed in Section 3.7.

3.6 Agents and Multi-Agent Systems

One of the central issues in modeling rational agents is the representation of their epistemic states and the definition of adequate operations on these states which are needed to incorporate new information or to adjust to changes in the environment. These issues have been thoroughly analyzed and specific methods for handling them were proposed in [21, 23, 115, 110].

Aspects of confidentiality and secrecy in multi-agent systems are the topic of [34, 33, 112]. Assume an agent wants to protect sensitive information according to certain secrecy constraints in a secrecy policy. A secrecy constraint intuitively expresses the desire of an agent A that another agent B should not come to believe some specific information. To select its actions, A has to interpret secrecy constraints under uncertainty about the epistemic state of B. It is shown how A can be equipped with a secrecy reasoner which classifies the agent's possible actions according to their compliance with its secrecy policy.

One of the most influential agent frameworks is the BDI approach which divides an agent's mental state into beliefs, desires, and intentions. In [114, 60] it is convincingly argued that in addition also the agent's motivation has to be taken into account as this allows for flexible and proactive behavior of intelligent agents in unreliable environments. Motives characterize the agent's personality and are the driving force behind the creation of desires and the selection of goals.

3.7 Applications

With her strong background in theory and foundations, Gabriele has not only tried to connect her work to other disciplines like philosophy (e.g. [59, 143]), she has also been active in putting theories at work across her broad field of interest. In [2, 4, 5], applications of answer set programming are presented. In [11, 12], DeLP is used to develop an argumentation-based decision support system in private law.

Probabilistic reasoning and the principle of maximum entropy as well as powerful belief change operations by sets of conditionals are advocated in many of Gabriele's foundational publications. In [10, 144, 29], probabilistic maximum entropy reasoning and belief and knowledge management operations are applied in the medical domain, the analysis of clinical brain tumor data. Applications in the biomedical domain are presented [54, 53] where probabilistic relational learning is used to support bronchial carcinoma diagnosis based on ion mobility spectrometry. In current project activities, Gabriele is involved in developing advanced and demanding application scenarios in e.g. medicine and logistics.

3.8 Scientific Community

Gabriele has served the scientific community in many ways. She is member of the steering committees for the DFG SPP Priority Program *New Frameworks of Rationality* (SPP1516) and for the DFG SPP Priority Program *Intentional Forgetting in Organisations* (SPP1921), and she has played a leading role in the initiation and organization of various DFG projects. She has been organizing many workshops and conferences in the areas of artificial intelligence and knowledge representation, she is member of the steering committee of the FoIKS symposia, and reviewer, guest editor, and member of editorial board for leading international journals. At the Technische Universität Dortmund, she served as Dean of Studies and as Dean of Faculty of the department of computer science.

Besides supervising her PhD students in Dortmund, Gabriele has also been external reviewer for PhD theses at many other universities, e.g. at the universities of Leipzig, Hagen, KTH Stockholm, Mines-ParisTech Sophia Antipolis, Madeira, and Manchester. An indication of Gabriele's engagement and success in teaching is the German textbook on methods of knowledge-based systems [28] that has recently appeared already in its 5th edition.

A specific feature of Gabriele's numerous activities in the scientific community is her success in bringing together people from different scientific areas and with different backgrounds. She has organized multi-disciplinary Dagstuhl seminars, and she is collaborating with leading specialists and experts not only from different areas of computer science, but also from philosophy and psychology as well as from e.g. medicine, logistics, and the services sector, in her quest for investigating the formal foundations of rationality and in applying the emerging concepts, ideas, and theories.

4 Conclusions and Personal Remarks

In this paper we so far tried to highlight some of Gabriele's numerous outstanding scientific contributions. As mentioned earlier, we are not claiming that the picture we provide is anywhere near to being complete. We are fully aware that the selection of topics we briefly discussed here is subjective and biased by our own interests (and capabilities, for that matter). We still believe it gives at least some idea about the originality, breadth, depth and lasting influence of Gabriele's scientific contributions.

Of course, a tribute to Gabriele would be totally incomplete without some more personal remarks. We figured the best way to express our relationship to Gabriele was by including a few paragraphs, written by each of the three authors.

Christoph: In May 1995, I met Gabriele for the first time. We were sitting in my office, at that time a flat rented by the university, near Hagen's main train station, and she was presenting her plans. After obtaining a PhD in mathematics in group theory, a family leave with three children (all born almost in the same years as my own children), and returning back to work at the department of economics, she outlined her research interests in knowledge representation and reasoning that would fit better in the area of computer science. Her intention was to apply for a habilitation grant in computer science, and she asked me whether I would support this and be her habilitation supervisor. Relying on her presentation, her ideas and her determination, I agreed. Gabriele made numerous excellent scientific contributions and a remarkable career in computer science, and I have had the opportunity and privilege to work with her on many topics and in various projects and events. Gabriele is a fascinating researcher and scientist, and she has an exceptional talent to integrate seemingly different views and opinions and to motivate people and bring them together for achieving a goal. I have also always enjoyed talking with Gabriele about many things beyond our common scientific interests. So now, twenty years later and on the occasion of her 60th birthday, I am happy to say thank you, Gaby, for everything, for your enthusiasm and your energy, for your inspiring ideas, for many years of pleasant collaboration, and for your friendship. Herzlichen Glückwunsch zu Deinem Geburtstag und alles Gute!

Gerd: I first heard of Gabriele in the mid nineties when Christoph Beierle contacted me and informed me about her plans to do a habilitation, after a long maternal leave and, moreover, in an area quite different from her PhD. Needless to say I was rather sceptical at first, but accepted to become an external supervisor for her habilitation anyway. I had no clue at the time what a wise decision I had made! From that point on, she kept impressing me and still continues to do so today.

There are many ways in which Gabriele is special. The broad coverage of topics that – we hope – manifests itself in this short overview of her contributions; her originality and depth of thinking; the mathematical rigor she applies when addressing difficult problems. But maybe the most distinguishing feature is her ability to build bridges: bridges between different scientific areas, like qualitative and quantitative approaches, belief revision and argumentation, default reasoning and conditionals, to name a few; bridges between theory and applications, like the applications of advanced answer set programming technology to difficult problems in logistics she is currently working on in the context of the Hybris project; and maybe most importantly bridges between people. Her ability to connect people with different scientific backgrounds and interests, among them not only computer scientists, but also philosophers, psychologists, researchers from various application domains, and many others, is just amazing.

Gabriele, it is a pleasure to have you around in our community.

Matthias: When I started my undergraduate studies at the University of Dortmund, I already knew that I wanted to pursue an academic career in AI. However, it was only after I took Gabriele's course on "Knowledge Representation and Reasoning" during my master studies that I decided to do this in logic-based AI and knowledge representation. After this introductory course, I enrolled in all her other courses, including a focused course on belief dynamics and information fusion, as well as several seminars on various aspects of AI. I was engaged by her approach how to join the study on intelligent systems with mathematical clarity and formality. I soon took over the job as a student assistant in her group, now also teaching in the same course that raised my initial interest in the topic. After finishing my master thesis under her guidance in the area of formal argumentation, I stayed and also did my PhD thesis on probabilistic reasoning with her.

Working with Gabriele is always a pleasure. One could always expect that her critical comments indeed target the weak parts of an approach and that her positive encouragement really meant that the corresponding feature is a good thing. All I know today on how to write a scientifically accurate and also engaging paper, I learned from the discussions with her and the collaborative writing of many papers. She also showed me the benefits of being a researcher through open information exchange, giving me the opportunity to work abroad, introducing me to several experts in the field, and, most of all, the freedom to pursue my own ideas. Gabriele never restrained my curiosity and interest to work on areas not directly related to the current project. She gave me the opportunity to visit various conferences, workshops, and summer schools to present my work and meet further researchers. She helped me in doing my PhD and, because of her, I am still doing research in the very same area she introduced to me almost 10 years ago.

I thank you, Gaby, for all your guidance and encouraging words. I wish you all the best for your 60th birthday!

References

1. E. Adams. *The Logic of Conditionals*. D. Reidel, Dordrecht, 1975.
2. E. Albrecht, P. Krümpelmann, and G. Kern-Isberner. Construction of explanation graphs from extended dependency graphs for answer set programs. In M. Hanus and R. Rocha, editors, *Declarative Programming and Knowledge Management - Declarative Programming Days, KDPD 2013, Unifying INAP, WFLP, and WLP, Kiel, Germany, September 11-13, 2013, Revised Selected Papers*, volume 8439 of *LNCS*, pages 1–16. Springer, 2014.
3. L. Amgoud, A. Kakas, G. Kern-Isberner, N. Maudet, and P. Moraitis. ABA: Argumentation based agents. In H. Coelho, R. Studer, and M. Wooldridge, editors, *Proceedings 19th European Conference on Artificial Intelligence, ECAI'2010*, volume 215 of *Frontiers in Artificial Intelligence and Applications*. IOS Press, 2010.
4. C. Beierle, O. Dusso, and G. Kern-Isberner. Modelling and implementing a knowledge base for checking medical invoices with DLV. In G. Brewka, I. Niemelä, T. Schaub, and M. Truszczynski, editors, *Nonmonotonic Reasoning, Answer Set Programming and Constraints*, number 05171 in Dagstuhl Seminar Proceedings. Internationales Begegnungs- und Forschungszentrum (IBFI), Schloss Dagstuhl, Germany, 2005.
5. C. Beierle, O. Dusso, and G. Kern-Isberner. Using answer set programming for a decision support system. In C. Baral, G. Greco, N. Leone, and G. Terracina,

editors, *Proceedings 8th International Conference on Logic Programming and Non Monotonic Reasoning (LPNMR 2005)*, volume 3662 of *LNAI*. Springer-Verlag, 2005.
6. C. Beierle, C. Eichhorn, and G. Kern-Isberner. Skeptical inference based on c-representations and its characterization as a constraint satisfaction problem. In *Foundations of Information and Knowledge Systems (FoIKS 2016)*, LNCS. Springer, 2016. (to appear).
7. C. Beierle, M. Finthammer, and G. Kern-Isberner. Relational probabilistic conditionals and their instantiations under maximum entropy semantics for first-order knowledge bases. *Entropy*, 17(2):852–865, 2015.
8. C. Beierle, M. Finthammer, G. Kern-Isberner, and M. Thimm. Automated reasoning for relational probabilistic knowledge representation. In J. Giesl and R. Hähnle, editors, *Proceedings 5th International Joint Conference on Automated Reasoning IJCAR'2010*, volume 6173 of *LNCS*, pages 218–224. Springer, 2010.
9. C. Beierle, M. Finthammer, G. Kern-Isberner, and M. Thimm. Evaluation and comparison criteria for approaches to probabilistic relational knowledge representation. In J. Bach and S. Edelkamp, editors, *Proceedings 34th Annual German Conference on Artificial Intelligence, KI'2011*, volume 7006 of *Lecture Notes in Computer Science*, pages 63–74. Springer, 2011.
10. C. Beierle, M. Finthammer, N. Potyka, J. Varghese, and G. Kern-Isberner. A case study on the application of probabilistic conditional modelling and reasoning to clinical patient data in neurosurgery. In L. C. van der Gaag, editor, *Proceedings of the 12th European Conference on Symbolic and Quantitative Approaches to Reasoning with Uncertainty, ECSQARU 2013*, number 7958 in Lecture Notes in Computer Science, pages 49–60, Berlin, 2013. Springer.
11. C. Beierle, B. Freund, G. Kern-Isberner, and M. Thimm. Can Bobby demand delivery? Towards a knowledge-based system for private law. In *Proceedings 24th Workshop on Constraint Logic Programming (WLP 2010)*, pages 34–44. German University in Cairo, 2010.
12. C. Beierle, B. Freund, G. Kern-Isberner, and M. Thimm. Using defeasible logic programming for argumentation-based decision support in private law. In P. Baroni, F. Cerutti, M. Giacomin, and G. R. Simari, editors, *Computational Models of Argument. Proceedings of COMMA 2010*, pages 87–98. IOS Press, 2010.
13. C. Beierle, R. Hermsen, and G. Kern-Isberner. Observations on the minimality of ranking functions for qualitative conditional knowledge bases and their computation. In *Proceedings of the 27th International FLAIRS Conference, FLAIRS-2014*, pages 480–485, Menlo Park, CA, 2014. AAAI Press.
14. C. Beierle and G. Kern-Isberner. Using institutions for the study of qualitative and quantitative conditional logics. In *Proceedings of the 8th European Conference on Logics in Artificial Intelligence, JELIA'02*, volume LNCS 2424, pages 161–172, Berlin Heidelberg New York, 2002. Springer.
15. C. Beierle and G. Kern-Isberner. An alternative view of knowledge discovery. In *Proceedings 36th Hawaii International Conference on System Sciences, HICSS-36*. IEEE Computer Press, 2003.
16. C. Beierle and G. Kern-Isberner. A logical study on qualitative default reasoning with probabilities. In *Proceedings International Conference on Logic for Programming, Artificial Intelligence, and Reasoning, LPAR'03*, pages 376–388, 2003.
17. C. Beierle and G. Kern-Isberner. Modelling conditional knowledge discovery and belief revision by abstract state machines. In E. Boerger, A. Gargantini, and E. Riccobene, editors, *Abstract State Machines 2003 – Advances in Theory*

and Applications, Proceedings 10th International Workshop, ASM'2003, pages 186–203. Springer, LNCS 2589, 2003.
18. C. Beierle and G. Kern-Isberner. Footprints of conditionals. In *Mechanizing Mathematical Reasoning: Essays in Honor of Jörg Siekmann on the Occasion of His 60th Birthday*, number LNCS 2605 in Lecture Notes in Computer Science. Springer, 2005.
19. C. Beierle and G. Kern-Isberner. Looking at probabilistic conditionals from an institutional point of view. In *Conditionals, Information, and Inference*, number LNAI 3301 in Lecture Notes in Artificial Intelligence. Springer, Berlin New York, 2005.
20. C. Beierle and G. Kern-Isberner. A note on comparing semantics for conditionals. In *Proceedings 19th International FLAIRS Conference, FLAIRS'2006*, pages 794–799. AAAI Press, Menlo Park, California, 2006.
21. C. Beierle and G. Kern-Isberner. On the modelling of an agent's epistemic state and its dynamic changes. *Electronic Communications of the European Association of Software Science and Technology*, 12, 2008.
22. C. Beierle and G. Kern-Isberner. A verified AsmL implementation of belief revision. In E. Börger, M. Butler, J. P. Bowen, and P. Boca, editors, *Abstract State Machines, B and Z, First International Conference, ABZ 2008, London, UK, September 16-18, 2008. Proceedings*, volume 5238 of *LNCS*, pages 98–111. Springer, 2008.
23. C. Beierle and G. Kern-Isberner. A conceptual agent model based on a uniform approach to various belief operations. In B. Mertsching, M. Hund, and Z. Aziz, editors, *Proceedings 32nd Annual German Conference on Artificial Intelligence, KI-2009*, number LNAI 5803 in Lecture Notes in Artificial Intelligence, pages 273–280, Berlin Heidelberg New York, 2009. Springer.
24. C. Beierle and G. Kern-Isberner. Formal similarities and differences between qualitative conditional semantics. *International Journal of Approximate Reasoning*, 50:1333–1346, 2009.
25. C. Beierle and G. Kern-Isberner. The relationship of the logic of big-stepped probabilities to standard probabilistic logics. In S. Link and H. Prade, editors, *Proceedings Sixth International Symposium on Foundations of Information and Knowledge Systems (FoIKS 2010)*, number 5956 in Lecture Notes in Computer Science, pages 191–210. Springer-Verlag, 2010.
26. C. Beierle and G. Kern-Isberner. Conditional objects revisited: Variants and model translations. In G. M. Youngblood and P. M. McCarthy, editors, *Proceedings of the Twenty-Fifth International Florida Artificial Intelligence Research Society Conference, FLAIRS-25, Marco Island, Florida. May 23-25, 2012*, pages 561–564. AAAI Press, Menlo Park, California, 2012.
27. C. Beierle and G. Kern-Isberner. Semantical investigations into nonmonotonic and probabilistic logics. *Annals of Mathematics and Artificial Intelligence*, 65(2):123–158, 2012.
28. C. Beierle and G. Kern-Isberner. *Methoden wissensbasierter Systeme - Grundlagen, Algorithmen, Anwendungen*. Springer Vieweg, Wiesbaden, 2014. (5th, revised and extended edition).
29. C. Beierle, G. Kern-Isberner, M. Finthammer, and N. Potyka. Extending and completing probabilistic knowledge and beliefs without bias. *KI*, 29(3):255–262, 2015.
30. C. Beierle, G. Kern-Isberner, and N. Koch. A high-level implementation of a system for automated reasoning with default rules (system description). In A. Armando, P. Baumgartner, and G. Dowek, editors, *Proc. of the 4th International Joint Conference on Automated Reasoning (IJCAR-2008)*, volume 5195 of *LNCS*, pages 147–153. Springer, 2008.

31. C. Beierle, G. Kern-Isberner, and K. Soedler. A declarative approach for computing ordinal conditional functions using constraint logic programming. In H. Tompits, S. Abreu, J. Oetsch, J. Pührer, D. Seipel, M. Umeda, and A. Wolf, editors, *Applications of Declarative Programming and Knowledge Management, Revised Selected Papers from the 19th International Conference on Applications of Declarative Programming and Knowledge Management, INAP 2011, and 25th Workshop on Logic Programming, WLP 2011, Wien, Austria*, volume 7773 of *LNAI*, pages 1–18. Springer, 2013.
32. C. Beierle, S. Kuche, M. Finthammer, and G. Kern-Isberner. A software system for the computation, visualization, and comparison of conditional structures for relational probabilistic knowledge bases. In *Proc. of the Twenty-Eigth International Florida Artificial Intelligence Research Society Conference (FLAIRS 2015)*, pages 558–563, Menlo Park, CA, 2015. AAAI Press.
33. J. Biskup, G. Kern-Isberner, P. Krümpelmann, and C. Tadros. Reasoning on secrecy constraints under uncertainty to classify possible actions. In C. Beierle and C. Meghini, editors, *Proceedings Eighth International Symposium on Foundations of Information and Knowledge Systems, FoIKS-2014*, number 8367 in LNCS, pages 97–116, Switzerland, 2014. Springer International Publishing.
34. J. Biskup, G. Kern-Isberner, and M. Thimm. Towards enforcement of confidentiality in agent interactions. In M. Pagnucco and M. Thielscher, editors, *Proceedings of the 12th International Workshop on Non-Monotonic Reasoning (NMR'08)*, pages 104–112. University of New South Wales, UNSW-CSE-TR-0819, 2008.
35. J. Dix, S. O. Hansson, G. Kern-Isberner, and G. S. Simari. Belief Change and Argumentation in Multi-Agent Scenarios (Dagstuhl Seminar 13231). *Dagstuhl Reports*, 3(6):1–21, 2013.
36. D. Dubois, A. Gilio, and G. Kern-Isberner. Probabilistic abduction without priors. In *Proceedings 10th International Conference on Principles of Knowledge Representation and Reasoning, KR'2006*, pages 420–430, Menlo Park, Ca., 2006. AAAI Press.
37. D. Dubois, A. Gilio, and G. Kern-Isberner. Probabilistic abduction without priors. *International Journal of Approximate Reasoning*, 47(3):333–351, 2008.
38. D. Dubois and H. Prade. Possibility theory and its applications: Where do we stand? In J. Kacprzyk and W. Pedrycz, editors, *Handbook of Computational Intelligence*, pages 31–60. Springer, 2015.
39. C. Eichhorn and G. Kern-Isberner. LEG networks for ranking functions. In E. Fermé and J. Leite, editors, *Proceedings 14th European Conference on Logics in Artificial Intelligence, JELIA-2014*, volume 8761 of *LNAI*, pages 210–223. Springer, 2014.
40. C. Eichhorn and G. Kern-Isberner. Qualitative and semi-quantitative inductive reasoning with conditionals - technical project report. *KI*, 29(3):279–289, 2015.
41. C. Eichhorn and G. Kern-Isberner. Using inductive reasoning for completing OCF-networks. *J. Applied Logic*, 13(4):605–627, 2015.
42. M. Falappa, E. Fermé, and G. Kern-Isberner. On the logic of theory change: Relations between incision and selection functions. In *Proceedings 17th European Conference on Artificial Intelligence, ECAI'2006*, pages 402–406. IOS Press, 2006.
43. M. Falappa, A. Garcia, G. Kern-Isberner, and G. Simari. On the evolving relation between belief revision and argumentation. *Knowledge Engineering Review*, 26(1):35–43, 2011.
44. M. Falappa, A. Garcia, G. Kern-Isberner, and G. Simari. Stratified belief bases revision with argumentative inference. *Journal of Philosophical Logic*, 42(1):161–193, 2013.

45. M. Falappa, G. Kern-Isberner, M. Reis, and G. Simari. Prioritized and non-prioritized multiple change on belief bases. *Journal of Philosophical Logic*, 41(1):77–113, 2012.
46. M. Falappa, G. Kern-Isberner, and G. Simari. Explanations, belief revision and defeasible reasoning. *Artificial Intelligence*, 141(1-2):1–28, 2002.
47. M. Falappa, G. Kern-Isberner, and G. Simari. Belief revision and argumentation theory. In G. Simari and I. Rahwan, editors, *Argumentation in Artificial Intelligence*. Springer, 2009.
48. M. A. Falappa, A. J. García, G. Kern-Isberner, and G. R. Simari. On the evolving relation between belief revision and argumentation. *Knowledge Eng. Review*, 26(1):35–43, 2011.
49. M. A. Falappa, G. Kern-Isberner, and G. R. Simari. Explanations, belief revision and defeasible reasoning. *Artif. Intell.*, 141(1/2):1–28, 2002.
50. B. Fine, G. Kern-Isberner, A. Moldenhauer, and G. Rosenberger. On the generalized Hurwitz equation and the Baragar-Umeda equation. *Results in Mathematics*, 2015. (to appear).
51. M. Finthammer, C. Beierle, B. Berger, and G. Kern-Isberner. An implementation of belief change operations based on probabilistic conditional logic. In E. Erdem, F. Lin, and T. Schaub, editors, *Proceedings 10th International Conference on Logic Programming and Nonmonotonic Reasoning, LPNMR'09*, volume 5753 of *Lecture Notes in Computer Science*, pages 496–501. Springer, 2009.
52. M. Finthammer, C. Beierle, B. Berger, and G. Kern-Isberner. Probabilistic reasoning at optimum entropy with the MECoRe system. In H. Lane and H. Guesgen, editors, *Proceedings 22nd International FLAIRS Conference, FLAIRS-09*, Menlo Park, CA., 2009. AAAI Press.
53. M. Finthammer, C. Beierle, J. Fisseler, G. Kern-Isberner, and J. Baumbach. Using probabilistic relational learning to support bronchial carcinoma diagnosis based on ion mobility spectrometry. *IMS*, 50:1333–1346, 2010.
54. M. Finthammer, C. Beierle, J. Fisseler, G. Kern-Isberner, B. Möller, and J. I. Baumbach. Probabilistic relational learning for medical diagnosis based on ion mobility spectrometry. In E. Hüllermeier, R. Kruse, and F. Hoffmann, editors, *Proc. IPMU 2010*, CCIS Vol. 80, pages 365–375. Springer, 2010.
55. M. Finthammer, G. Kern-Isberner, and M. Ritterskamp. Resolving inconsistencies in probabilistic knowledge bases. In J. Hertzberg, M. Beetz, and R. Englert, editors, *30th Annual German Conference on AI, KI 2007, Osnabrück, Germany*, volume 4667 of *Lecture Notes in Computer Science*, pages 114–128. Springer, September 2007.
56. J. Fisseler, G. Kern-Isberner, and C. Beierle. Learning uncertain rules with CONDORCKD. In *Proceedings 20th International FLAIRS Conference, FLAIRS'07*. AAAI Press, Menlo Park, California, 2007.
57. J. Fisseler, G. Kern-Isberner, C. Beierle, A. Koch, and C. Müller. Algebraic knowledge discovery using Haskell. In M. Hanus, editor, *Practical Aspects of Declarative Languages, 9th International Symposium, PADL 2007*, Lecture Notes in Computer Science, pages 80–93. Springer-Verlag, 2007.
58. D. Garcia, S. Gottifredi, P. Krümpelmann, M. Thimm, G. Kern-Isberner, M. Falappa, and A. Garcia. On influence and contractions in defeasible logic programming. In J. P. Delgrande and W. Faber, editors, *Proceedings 11th International Conference on Logic Programming and Nonmonotonic Reasoning LPNMR-11*, volume 6645 of *Lecture Notes in Computer Science*, pages 199–204. Springer, 2011.
59. T. Grundmann, C. Beierle, G. Kern-Isberner, and N. Pfeifer. Wissen. In A. Stephan and S. Walter, editors, *Handbuch Kognitionswissenschaft*, pages 488–500. Metzler'sche Verlagsbuchhandlung, Stuttgart, 2013.

60. D. Hölzgen, T. Vengels, P. Krümpelmann, M. Thimm, and G. Kern-Isberner. Argonauts: a working system for motivated cooperative agents. *Annals of Mathematics and Artificial Intelligence*, 61(4):309–332, 2011.
61. G. Kern-Isberner. A conditional-logical approach to minimum cross-entropy. In *Proceedings 14th Symposium on Theoretical Aspects of Computer Science STACS'97*, pages 237 – 248, Berlin Heidelberg New York, 1997. Springer.
62. G. Kern-Isberner. A logically sound method for uncertain reasoning with quantified conditionals. In *Proceedings First International Conference on Qualitative and Quantitative Practical Reasoning, ECSQARU-FAPR'97*, pages 365 – 379, Berlin Heidelberg New York, 1997. Springer.
63. G. Kern-Isberner. The principle of minimum cross-entropy and conditional logic. In *Proceedings of the Third Dutch/German Workshop on Nonmonotonic Reasoning Techniques and their Applications, DGNMR-97*, pages 73–82, Saarbrücken, Germany, 1997. MPI for Computer Science.
64. G. Kern-Isberner. Characterizing the principle of minimum cross-entropy within a conditional-logical framework. *Artificial Intelligence*, 98:169–208, 1998.
65. G. Kern-Isberner. Nonmonotonic reasoning in probabilistics. In *Proceedings European Conference on Artificial Intelligence, ECAI'98*, pages 580 – 584, West Sussex, UK, 1998. Wiley & Sons.
66. G. Kern-Isberner. A note on conditional logics and entropy. *International Journal of Approximate Reasoning*, 19:231–246, 1998.
67. G. Kern-Isberner. Following conditional structures of knowledge. In *KI-99: Advances in Artificial Intelligence, Proceedings of the 23rd Annual German Conference on Artificial Intelligence*, pages 125–136. Springer Lecture Notes in Artificial Intelligence LNAI 1701, 1999.
68. G. Kern-Isberner. Postulates for conditional belief revision. In *Proceedings Sixteenth International Joint Conference on Artificial Intelligence, IJCAI-99*, pages 186–191. Morgan Kaufmann, 1999.
69. G. Kern-Isberner. Revising by conditional beliefs. In *Proceedings Fourth Dutch-German Workshop on Nonmonotonic Reasoning Rechniques And Their Applications, DGNMR-99*, University of Amsterdam, 1999. Institute for Logic, Language and Computation.
70. G. Kern-Isberner. Solving the inverse representation problem. In *Proceedings 14th European Conference on Artificial Intelligence, ECAI'2000*, pages 581–585, Berlin, 2000. IOS Press.
71. G. Kern-Isberner. Conditional indifference and conditional preservation. *Journal of Applied Non-Classical Logics*, 11(1-2):85–106, 2001.
72. G. Kern-Isberner. *Conditionals in nonmonotonic reasoning and belief revision*, volume 2087 of *LNAI*. Springer, Lecture Notes in Artificial Intelligence LNAI 2087, 2001.
73. G. Kern-Isberner. Discovering most informative rules from data. In *Proceedings International Conference on Intelligent Agents, Web Technologies and Internet Commerce, IAWTIC'2001*, 2001.
74. G. Kern-Isberner. Handling conditionals adequately in uncertain reasoning. In *Proceedings European Conference on Symbolic and Quantitative Approaches to Reasoning with Uncertainty, ECSQARU'01*, pages 604–615. Springer LNAI 2143, 2001.
75. G. Kern-Isberner. Representing and learning conditional information in possibility theory. In *Proceedings 7th Fuzzy Days, Dortmund, Germany*, pages 194–217. Springer LNCS 2206, 2001.
76. G. Kern-Isberner. Revising and updating probabilistic beliefs. In M.-A. Williams and H. Rott, editors, *Frontiers in belief revision*, pages 329–344. Kluwer Academic Publishers, Dordrecht, 2001.

77. G. Kern-Isberner. Handling conditionals adequately in uncertain reasoning and belief revision. *Journal of Applied Non-Classical Logics*, 12(2):215–237, 2002.
78. G. Kern-Isberner. The principle of conditional preservation in belief revision. In *Proceedings of the Second International Symposium on Foundations of Information and Knowledge Systems, FoIKS 2002*, pages 105–129. Springer LNCS 2284, 2002.
79. G. Kern-Isberner. A structural approach to default reasoning. In *Proceedings of the Eighth International Conference on Principles of Knowledge Representation and Reasoning, KR'2002*, pages 147–157, San Francisco, Ca., 2002. Morgan Kaufmann.
80. G. Kern-Isberner. A thorough axiomatization of a principle of conditional preservation in belief revision. *Annals of Mathematics and Artificial Intelligence*, 40(1-2):127–164, 2004.
81. G. Kern-Isberner. Using group theory for knowledge representation and discovery. *Contemporary Mathematics, Special Issue on Combinatorial Group Theory, Number Theory and Discrete Groups*, 421:169–186, 2006.
82. G. Kern-Isberner. A conceptual framework for (iterated) revision, update, and nonmonotonic reasoning. In *Proceedings Dagstuhl Seminar 07351 Formal Models of Belief Change in Rational Agents*, 2007.
83. G. Kern-Isberner. Linking iterated belief change operations to nonmonotonic reasoning. In G. Brewka and J. Lang, editors, *Proceedings 11th International Conference on Knowledge Representation and Reasoning, KR'2008*, pages 166–176, Menlo Park, CA, 2008. AAAI Press.
84. G. Kern-Isberner. Learning and understanding. In N. Seel, editor, *Encyclopedia of the Sciences of Learning*, pages 1805–1807. Springer Science and Business Media, 2012.
85. G. Kern-Isberner and C. Beierle. A system Z-like approach for first-order default reasoning. In *Advances in Knowledge Representation, Logic Programming, and Abstract Argumentation*, volume 9060 of *LNAI*, pages 81–95. Springer, 2015.
86. G. Kern-Isberner, C. Beierle, M. Finthammer, and M. Thimm. Probabilistic logics in expert systems: Approaches, implementations, and applications. In A. Hameurlain, S. W. Liddle, K.-D. Schewe, and X. Zhou, editors, *Proceedings 22nd International Conference on Database and Expert Systems Applications, DEXA 2011*, volume 6860 of *Lecture Notes in Computer Science*, pages 27–46. Springer, 2011.
87. G. Kern-Isberner, C. Beierle, M. Finthammer, and M. Thimm. Comparing and evaluating approaches to probabilistic reasoning: Theory, implementation, and applications. *Transactions on Large-Scale Data- and Knowledge-Centered Systems*, 6:31–75, 2012.
88. G. Kern-Isberner and C. Eichhorn. Intensional combination of rankings for OCF-networks. In C. Boonthum-Denecke and M. Youngblood, editors, *Proceedings of the 26th International FLAIRS Conference FLAIRS-2013*, pages 615–620. AAAI Press, 2013.
89. G. Kern-Isberner and C. Eichhorn. Structural inference from conditional knowledge bases. *Studia Logica, Special Issue Logic and Probability: Reasoning in Uncertain Environments*, 102(4), 2014.
90. G. Kern-Isberner and J. Fisseler. Knowledge discovery by reversing inductive knowledge representation. In *Proceedings of the Ninth International Conference on the Principles of Knowledge Representation and Reasoning, KR-2004*, pages 34–44. AAAI Press, 2004.
91. G. Kern-Isberner and D. Huvermann. Multiple iterated belief revision without independence. In *Proc. of the Twenty-Eigth International Florida Artificial In-

telligence Research Society Conference (FLAIRS 2015), pages 570–575, Menlo Park, CA, 2015. AAAI Press.

92. G. Kern-Isberner and P. Krümpelmann. A constructive approach to independent and evidence retaining belief revision by general information sets. In T. Walsh, editor, *Proceedings 22nd International Joint Conference on Artificial Intelligence, IJCAI'2011*, pages 937–942, Menlo Park, CA, 2011. AAAI Press.
93. G. Kern-Isberner and T. Lukasiewicz. Combining probabilistic logic programming with the power of maximum entropy. *Artificial Intelligence, Special Issue on Nonmonotonic Reasoning*, 157(1-2):139–202, 2004.
94. G. Kern-Isberner and H. Reidmacher. Interpreting a contingency table by rules. *International Journal of Intelligent Systems*, 11(6), 1996.
95. G. Kern-Isberner and M. Ritterskamp. Preference fusion for default reasoning beyond system Z. *Journal of Automated Reasoning, Special Issue on Uncertain Reasoning*, 45(1):3–19, 2010.
96. G. Kern-Isberner and W. Rödder. Belief revision and information fusion in a probabilistic environment. In *Proceedings 16th International FLAIRS Conference, FLAIRS'03*, pages 506–510, Menlo Park, California, 2003. AAAI Press.
97. G. Kern-Isberner and W. Rödder. Fusing probabilistic information on maximum entropy. In *Proceedings 26th German Conference on Artificial Intelligence, KI-2003*, number 2821 in LNAI, pages 407–420, Berlin Heidelberg New York, 2003. Springer.
98. G. Kern-Isberner and W. Rödder. Belief revision and information fusion on optimum entropy. *International Journal of Intelligent Systems, Special Issue on Uncertain Reasoning*, 19(9):837 – 857, 2004.
99. G. Kern-Isberner and G. Rosenberger. Über Diskretheitsbedingungen und die diophantische Gleichung $ax^2 + by^2 + cz^2 = dxyz$. *Arch. Math.*, 34:481–493, 1980.
100. G. Kern-Isberner and G. Rosenberger. Einige Bemerkungen über Untergruppen der *PSL*(2, $\mathbb{C}$). *Resultate der Mathematik*, 6:40–47, 1983.
101. G. Kern-Isberner and G. Rosenberger. A note on numbers of the form $n = x^2 + Ny^2$. *Arch. Math.*, 43:148–156, 1984.
102. G. Kern-Isberner and G. Simari. A default logical semantics for defeasible argumentation. In *Proceedings of the 24th Florida Artificial Intelligence Research Society Conference FLAIRS-24*. AAAI Press, 2011.
103. G. Kern-Isberner and M. Thimm. Novel semantical approaches to relational probabilistic conditionals. In F. Lin, U. Sattler, and M. Truszczynski, editors, *Proceedings Twelfth International Conference on the Principles of Knowledge Representation and Reasoning, KR'2010*, pages 382–391. AAAI Press, 2010.
104. G. Kern-Isberner and M. Thimm. A ranking semantics for first-order conditionals. In L. De Raedt, C. Bessiere, D. Dubois, P. Doherty, P. Frasconi, F. Heintz, and P. Lucas, editors, *Proceedings 20th European Conference on Artificial Intelligence, ECAI-2012*, number 242 in Frontiers in Artificial Intelligence and Applications, pages 456–461. IOS Press, 2012.
105. G. Kern-Isberner, M. Thimm, and M. Finthammer. Qualitative knowledge discovery. In K.-D. Schewe and B. Thalheim, editors, *Proceedings International Workshop on Semantics in Data and Knowledge Bases, SDKB 2008*, LNCS 4925, pages 77–102. Springer, 2008.
106. G. Kern-Isberner, M. Thimm, M. Finthammer, and J. Fisseler. Mining default rules from statistical data. In H. Lane and H. Guesgen, editors, *Proceedings 22nd International FLAIRS Conference, FLAIRS-09*, Menlo Park, CA., 2009. AAAI Press.

107. G. Kern-Isberner, M. Wilhelm, and C. Beierle. A novel methodology for processing probabilistic knowledge bases under maximum entropy. In *Proceedings of the 27th International Florida Artificial Intelligence Research Society Conference, FLAIRS 2014*, pages 496–501, Menlo Park, CA, 2014. AAAI Press.
108. G. Kern-Isberner, M. Wilhelm, and C. Beierle. Probabilistic knowledge representation using Gröbner basis theory. In *International Symposium on Artificial Intelligence and Mathematics, ISAIM 2014, Fort Lauderdale, FL, USA, January 6-8*, 2014.
109. G. Kern-Isberner, M. Wilhelm, and C. Beierle. Probabilistic knowledge representation using the principle of maximum entropy and Gröbner basis theory. *Annals of Mathematics and Artificial Intelligence*, 2015. (to appear).
110. P. Krümpelmann, T. Janus, and G. Kern-Isberner. Angerona - A flexible Multiagent Framework for Knowledge-based Agents. In N. Bulling, editor, *Proceedings of the 12th European Conference on Multi-Agent Systems*, Springer, 2014.
111. P. Krümpelmann and G. Kern-Isberner. Belief base change operations for answer set programming. In L. F. del Cerro, A. Herzig, and J. Mengin, editors, *Proceedings 13th European Conference on Logics in Artificial Intelligence, JELIA-2012*, number 7519 in LNCS, pages 294–306. Springer, 2012.
112. P. Krümpelmann and G. Kern-Isberner. Secrecy preserving BDI agents based on answer set programming. In M. Klusch, M. Paprzycki, and M. Thimm, editors, *Proceedings 11th German Conference on Multiagent System Technologies MATES-2013*, number 8076 in Lecture Notes in Computer Science, pages 124–137. Springer, 2013.
113. P. Krümpelmann, M. Thimm, M. Falappa, A. García, G. Kern-Isberner, and G. Simari. Selective revision by deductive argumentation. In S. Modgil, N. Oren, and F. Toni, editors, *Theory and Applications of Formal Argumentation - First International Workshop, TAFA'2011, Workshop at IJCAI'2011, Revised selected papers*, volume 7132 of *Lecture Notes in Computer Science*. Springer, 2012.
114. P. Krümpelmann, M. Thimm, G. Kern-Isberner, and R. Fritsch. Motivating agents in unreliable environments: A computational model. In F. Klügl and S. Ossowski, editors, *Multiagent System Technologies - 9th German Conference, MATES 2011, Berlin, Germany, October 6-7, 2011. Proceedings*, volume 6973 of *Lecture Notes in Computer Science*, pages 65–76. Springer, 2011.
115. P. Krümpelmann, M. Thimm, M. Ritterskamp and G. Kern-Isberner. Belief operations for motivated BDI agents. In L. Padgham, D. C. Parkes, J. P. Müller, and S. Parsons, editors, *Proceedings Seventh International Joint Conference on Autonomous Agents and Multiagent Systems, AAMAS-08*, pages 421–428, 2008.
116. T. Leopold, G. Kern-Isberner, and G. Peters. Belief revision with reinforcement learning for interactive object recognition. In *Proceedings 18th European Conference on Artificial Intelligence, ECAI'08*, 2008.
117. D. Lewis. Probabilities of conditionals and conditional probabilities. *The Philosophical Review*, 85:297–315, 1976.
118. S. Loh, M. Thimm, and G. Kern-Isberner. On the problem of grounding a relational probabilistic conditional knowledge base. In T. Meyer and E. Ternovska, editors, *Proceedings 13th International Workshop on Nonmonotonic Reasoning NMR'2010, Subworkshop on NMR and Uncertainty*, 2010.
119. T. Lukasiewicz and G. Kern-Isberner. Probabilistic logic programming under maximum entropy. In *Proceedings ECSQARU-99*, volume 1638, pages 279–292. Springer Lecture Notes in Artificial Intelligence, 1999.
120. N. Nilsson. Probabilistic logic. *Artificial Intelligence*, 28:71–87, 1986.
121. D. Nute. *Topics in Conditional Logic*. D. Reidel Publishing Company, Dordrecht, Holland, 1980.

122. J. Paris. *The uncertain reasoner's companion – A mathematical perspective.* Cambridge University Press, 1994.
123. J. Paris and A. Vencovska. In defence of the maximum entropy inference process. *International Journal of Approximate Reasoning*, 17(1):77–103, 1997.
124. J. Pearl. *Probabilistic Reasoning in Intelligent Systems: Networks of Plausible Inference.* Morgan Kaufmann, 1988.
125. N. Potyka, C. Beierle, and G. Kern-Isberner. Changes of relational probabilistic belief states and their computation under optimum entropy semantics. In I. Timm and M. Thimm, editors, *Advances in Artificial Intelligence: Proceedings of the 36th Annual German Conference on Artificial Intelligence, KI-2013*, pages 176–187. Springer, 2013.
126. N. Potyka, C. Beierle, and G. Kern-Isberner. On the problem of reversing relational inductive knowledge representation. In L. C. van der Gaag, editor, *Proceedings of the 12th European Conference on Symbolic and Quantitative Approaches to Reasoning with Uncertainty, ECSQARU 2013*, number 7958 in Lecture Notes in Computer Science, pages 485–496, Berlin, 2013. Springer.
127. N. Potyka, C. Beierle, and G. Kern-Isberner. A concept for the evolution of relational probabilistic belief states and the computation of their changes under optimum entropy semantics. *Journal of Applied Logic*, 13:414–440, 2015.
128. W. Rödder and G. Kern-Isberner. Léa Sombé und entropie-optimale Informationsverarbeitung mit der Expertensystem-Shell SPIRIT. *OR Spektrum*, 19(3):41–46, 1997.
129. W. Rödder and G. Kern-Isberner. Representation and extraction of information by probabilistic logic. *Information Systems*, 21(8):637–652, 1997.
130. W. Rödder and G. Kern-Isberner. From information to probability: an axiomatic approach. *International Journal of Intelligent Systems*, 18(4):383–403, 2003.
131. W. Rödder and G. Kern-Isberner. Selflearning or how to make a knowledge base curious about itself. In *Proceedings 26th German Conference on Artificial Intelligence, KI-2003*, number 2821 in LNAI, pages 465–474, Berlin Heidelberg New York, 2003. Springer.
132. W. Rödder and C.-H. Meyer. Coherent knowledge processing at maximum entropy by SPIRIT. In E. Horvitz and F. Jensen, editors, *Proceedings 12th Conference on Uncertainty in Artificial Intelligence*, pages 470–476, San Francisco, Ca., 1996. Morgan Kaufmann.
133. J. Shore and R. Johnson. Axiomatic derivation of the principle of maximum entropy and the principle of minimum cross-entropy. *IEEE Transactions on Information Theory*, IT-26:26–37, 1980.
134. W. Spohn. Ordinal conditional functions: a dynamic theory of epistemic states. In W. Harper and B. Skyrms, editors, *Causation in Decision, Belief Change, and Statistics, II*, pages 105–134. Kluwer Academic Publishers, 1988.
135. L. Tamargo, M. Thimm, P. Krümpelmann, A. Garcia, M. Falappa, G. Simari, and G. Kern-Isberner. Credibility-based selective revision by deductive argumentation in multi-agent systems. In E. L. Fermé, D. M. Gabbay, and G. R. Simari, editors, *Trends in Belief Revision and Argumentation Dynamics*, number 48 in Logic and Cognitive Systems. College Publications, London, UK, 2013.
136. M. Thimm, M. Finthammer, S. Loh, G. Kern-Isberner, and C. Beierle. A system for relational probabilistic reasoning on maximum entropy. In H. W. Guesgen and R. C. Murray, editors, *Proceedings 23rd International FLAIRS Conference, FLAIRS'10*, pages 116–121, Menlo Park, California, 2010. AAAI Press.
137. M. Thimm, A. Garcia, G. Kern-Isberner, and G. Simari. Using collaborations for distributed argumentation with defeasible logic programming. In M. Pagnucco and M. Thielscher, editors, *Proceedings of the 12th International Workshop on*

Non-Monotonic Reasoning (NMR'08), pages 179–188. University of New South Wales, UNSW-CSE-TR-0819, 2008.

138. M. Thimm and G. Kern-Isberner. On probabilistic inference in relational conditional logics. *Logic Journal of the IGPL, Special Issue on Relational approaches to knowledge representation and learning*, 20(5):872–908, 2012.

139. M. Thimm and G. Kern-Isberner. A distributed argumentation framework using defeasible logic programming. In P. Besnard, S. Doutre, and A. Hunter, editors, *Proceedings of the 2nd International Conference on Computational Models of Argument COMMA'08*, pages 381–392. IOS Press, 2008.

140. M. Thimm and G. Kern-Isberner. On the relationship of defeasible argumentation and answer set programming. In P. Besnard, S. Doutre, and A. Hunter, editors, *Proceedings of the 2nd International Conference on Computational Models of Argument COMMA'08*, pages 393–404. IOS Press, 2008.

141. M. Thimm and G. Kern-Isberner. On Controversiality of Arguments and Stratified Labelings. In *Proceedings of the Fifth International Conference on Computational Models of Argumentation COMMA'14*, Pitlochry, UK, September 2014.

142. M. Thimm, G. Kern-Isberner, and J. Fisseler. Relational probabilistic conditional reasoning at maximum entropy. In W. Liu, editor, *Proceedings 11th European Conference on Symbolic and Quantitative Approaches to Reasoning with Uncertainty, ECSQARU'11*, volume 6717 of *LNCS*, pages 447–458. Springer, 2011.

143. P. D. Thorn, C. Eichhorn, G. Kern-Isberner, and G. Schurz. Qualitative probabilistic inference with default inheritance. In *Proceedings of the 5th Workshop on Dynamics of Knowledge and Belief (DKB-2015) and the 4th Workshop KI & Kognition (KIK-2015) co-located with 38th German Conference on Artificial Intelligence (KI-2015)*, volume 1444 of *CEUR Workshop Proceedings*, pages 16–28. CEUR-WS.org, 2015.

144. J. Varghese, C. Beierle, N. Potyka, and G. Kern-Isberner. Using probabilistic logic and the principle of maximum entropy for the analysis of clinical brain tumor data. In P. Pereira Rodrigues, M. Pechenizkiy, J. Gama, R. Cruz-Correia, J. Liu, A. Traina, P. Lucas, and P. Soda, editors, *Proceedings of the 26th IEEE International Symposium on Computer-Based Medical Systems, CBMS 2013*, pages 401–404. IEEE Conference Publications, 2013.

Part I
Foundations

i as a Quadratic Residue in the Gaussian Integers

Benjamin Fine[1], Gerhard Rosenberger[2]

Abstract. In this note we use the structure of the Generalized Picard Group $PGL(2,\mathbb{Z}[i])$ to provide a characterization of Gaussian integers α for which i is a quadratic residue. This is a two-square theorem analogous to Fermat's two-square theorem.

1 Introduction

One version of Fermat's well-known two-square theorem says that for a natural number n, -1 is a quadratic residue modulo n if and only if there exists $a, b \in \mathbb{Z}$ with $(a,b) = 1$ and $n = a^2 + b^2$. A proof of this result using the structure of the classical modular group $M = PSL(2,\mathbb{Z})$ was given by Fine in [1]. This method was generalized in several different directions by Kern-Isberner and Rosenberger [7] and [8] and by Fine [1], [2] and [3]. In this note we apply similar techniques using the generalized Picard group $\Gamma_1 = PGL(2,\mathbb{Z}[i])$ to prove a two-square theorem for the Gaussian integers. In particular we prove that if $\alpha \in \mathbb{Z}[i]$ then i is a quadratic residue modulo α if and only if there exist relatively prime Gaussian integers a, b with $\alpha = a^2 + ib^2$ or $\alpha = -a^2 + ib^2$. In addition the technique shows that any Gaussian integer can be written as a sum of two relatively prime squares. Finally we use the group theoretical structure of $\Gamma = PSL(2,\mathbb{Z}[i])$ given in [2] to examine the group theoretical structure of $PGL(2,\mathbb{Z}[i])$.

2 Main Results

The **Gaussian integers** or **complex integers** are the ring $\mathbb{Z}[i]$. Using the standard complex norm, these form a Euclidean domain and hence a unique factorization domain.

The **Picard group** is the group $\Gamma = PSL(2,\mathbb{Z}[i])$. This consists of linear fractional transformations

$$z' = \frac{az+b}{cz+d} \text{ with } a,b,c,d \in \mathbb{Z}[i], ad-bc = 1.$$

Since -1 has a square root within $\mathbb{Z}[i]$, linear transformations with determinant -1, such as $z' = -z$, are also in Γ.

---
[1] Department of Mathematics, Fairfield University, Fairfield, Connecticut 06430, United States
[2] Fachbereich Mathematik, University of Hamburg, 20146 Hamburg, Germany

Each linear transformation corresponds to $\pm$ a matrix in $SL(2, \mathbb{Z}[i])$ that is

$$\pm \begin{pmatrix} a & b \\ c & d \end{pmatrix} \text{ with } ad - bc = 1.$$

Multiplication of linear transformations is done via matrix multiplication. For further information on these see [6] or [4].

Recall that an integer m is a **quadratic residue** modulo n if there exists a solution for $x^2 \equiv m \bmod n$. Fermat's two square theorem characterizes those integers n for which -1 is a quadratic residue modulo n. There are several versions of this result. For this note we recall the following.

Theorem 1. *Let $n \in \mathbb{N}$. Then -1 is a quadratic residue modulo n if and only if there exists $a, b \in \mathbb{Z}$ with $(a,b) = 1$ such that $n = a^2 + b^2$.*

In the standard proofs (see for example [6]) the congruence $x^2 \equiv -1$ is considered modulo primes and the difference in the situation for $p \equiv 1 \bmod 4$ and $p \equiv 3 \bmod 4$ is examined. In [1] a direct proof of this result was given using the structure of the classical modular group (see also [6]). As mentioned in the introduction, this technique was generalized by Kern-Isberner and Rosenberger ([7] and [8]) to consider integers that can be represented as $n = x^2 + Ny^2$ for various values of N. In a different direction they were extended by Fine in [3] to consider representations of integers by certain quadratic forms related to the traces of unimodular matrices. Finally in [3] Fermat's two-square theorem was shown to hold in a wide variety of rings called **sum of squares rings**.

Within the Gaussian integers $\mathbb{Z}[i]$ we have $i^2 = -1$. Hence -1 is a quadratic residue modulo any Gaussian integer α. Our main result shows that we get an analogous two-square result considering Gaussian integers α for which i is a quadratic residue. In particular:

Theorem 2. *Let $\alpha \in \mathbb{Z}[i]$. Then i is a quadratic residue modulo α if and only if there exists relatively prime Gaussian integers a, b such that $\alpha = a^2 + ib^2$ or $\alpha = -a^2 + ib^2$.*

We let $\Gamma_1 = PGL(2, \mathbb{Z}[i])$, the extended Picard group. Note that if $A \in \Gamma_1$ then $\det(A) = 1$ or $\det(A) = i$. For the proof, we first need the following lemma.

Lemma 1. *Let $A \in PGL(2, \mathbb{Z}[i])$. Then if $\operatorname{trace}(A) = 0$ and $\det(A) = i$ we have that A is conjugate within Γ_1 to*

$$X = \pm \begin{pmatrix} 0 & i \\ -1 & 0 \end{pmatrix}.$$

That is, there exists $T \in \Gamma_1$ with $T^{-1}XT = A$.

Proof. Let $A = \pm \begin{pmatrix} \alpha & \beta \\ \gamma & -\alpha \end{pmatrix}$ with $-\alpha^2 - \beta\gamma = i$. Let S be the set of conjugates of A within Γ_1 so that

$$S = \{T^{-1}AT; T \in M\}.$$

Since conjugation preserves both trace and determinant, S consists of matrices of trace zero and determinant i. Let

$$Y = \pm \begin{pmatrix} a & b \\ c & -a \end{pmatrix}$$

and

$$-a^2 - bc = i$$

be an element of S with $|a|$ minimal. This exists since the norms of Gaussian integers are rational integers and then from the well-ordering of $\mathbb{N}$. We show that a must equal zero.

Assume that $a \neq 0$. If either $b = 0$ or $c = 0$ then $a^2 = -i$. However $\sqrt{i}$ is not a Gaussian integer and therefore we may assume that $b \neq 0$ and $c \neq 0$.

We then have

$$-a^2 - bc = i \Rightarrow -bc = a^2 + i \Rightarrow |b||c| = |a^2 + i| \leq |a|^2 + 1.$$

It follows then that $b \neq 0, c \neq 0$ and either $|b| \leq |a|$ or $|c| \leq |a|$. Assume first that $|c| \leq |a|$.

By the division algorithm within $\mathbb{Z}[i]$ we have $a - qc = r$ with $r = 0$ or $|r| < |c| \leq |a|$. We may assume that $a > 0$ and $c > 0$. Then

$$0 < a - c < a.$$

Now conjugate Y by $T = \pm \begin{pmatrix} 1 & q \\ 0 & 1 \end{pmatrix}$. Then $T^{-1} = \pm \begin{pmatrix} 1 & -q \\ 0 & 1 \end{pmatrix}$ and

$$T^{-1}YT = \pm \begin{pmatrix} 1 & -q \\ 0 & 1 \end{pmatrix} \begin{pmatrix} a & b \\ c & -a \end{pmatrix} \begin{pmatrix} 1 & q \\ 0 & 1 \end{pmatrix} = \pm \begin{pmatrix} a - cq & * \\ * & cq - a \end{pmatrix}.$$

But then $0 < |a - qc| < |c| \leq |a|$ contradicting the minimality of $|a|$.

If $|b| \leq |a|$ then we have $a - qb = r$ with $r = 0$ or $|r| < |b| \leq |a|$. Now conjugate Y by $T = \pm \begin{pmatrix} 1 & 0 \\ -q & 1 \end{pmatrix}$. Then $T^{-1} = \pm \begin{pmatrix} 1 & 0 \\ q & 1 \end{pmatrix}$ and

$$T^{-1}YT = \pm \begin{pmatrix} a - qb & * \\ * & qb - a \end{pmatrix}.$$

Again $0 < |aq - b| < |a|$ contradicting the minimality of $|a|$.

Therefore in a minimal conjugate of A we must have $a = 0$ and hence $-bc = i$. It follows that $b = \pm i$ and $c = \mp 1$ or $c = \mp i$ and $b = \pm 1$. If $c = \mp i$ and $b = \pm 1$ then

$$\pm \begin{pmatrix} 0 & 1 \\ -1 & 0 \end{pmatrix} \begin{pmatrix} 0 & -i \\ -1 & 0 \end{pmatrix} \begin{pmatrix} 0 & -1 \\ 1 & 0 \end{pmatrix} = \pm \begin{pmatrix} 0 & 1 \\ i & 0 \end{pmatrix}.$$

Hence let $b = \pm i$ and $c = \mp 1$. Then

$$Y = \pm \begin{pmatrix} 0 & i \\ -1 & 0 \end{pmatrix} = X$$

completing the proof.

Using an analogous proof of the previous lemma applied to matrices of trace 0 and determinant 1 gives the following result.

Lemma 2. *Let $A \in PGL(2, \mathbb{Z}[i])$. Then if $\operatorname{trace}(A) = 0$ and $\det(A) = 1$ we have that A is conjugate within Γ_1 to*

$$X = \pm \begin{pmatrix} 0 & 1 \\ -1 & 0 \end{pmatrix}.$$

That is there exists $T \in \Gamma_1$ with $T^{-1}XT = A$.

We can now prove Theorem 2.2.

Proof. Let $\alpha \in \mathbb{Z}[i]$ and suppose that i is a quadratic residue modulo α. Then there exists $x \in \mathbb{Z}[i]$ such that $x^2 \equiv i \bmod \alpha$ so that there exists $\gamma \in \mathbb{Z}[i]$ with $x^2 - i = \alpha\gamma$ or $-x^2 - \alpha\gamma = i$. This implies that the projective matrix

$$A = \pm \begin{pmatrix} x & \alpha \\ \gamma & -x \end{pmatrix}$$

is in Γ_1 with determinant i. Since it has trace 0, then from Lemma 2.1 it follows that A is conjugate within Γ_1 to

$$X = \pm \begin{pmatrix} 0 & i \\ -1 & 0 \end{pmatrix}.$$

Now consider conjugates of X within Γ_1. Let

$$T = \pm \begin{pmatrix} a & b \\ c & d \end{pmatrix}.$$

Assume first that $\det(T) = 1$ so that

$$T^{-1} = \pm \begin{pmatrix} d & -b \\ -c & a \end{pmatrix}.$$

Then computing the conjugate of X by T we find

$$TXT^{-1} = \pm(-i) \begin{pmatrix} a & b \\ c & d \end{pmatrix} \begin{pmatrix} 0 & i \\ -1 & 0 \end{pmatrix} \begin{pmatrix} d & -b \\ -c & a \end{pmatrix}$$

$$= \pm \begin{pmatrix} * & b^2 + ia^2 \\ -(d^2 + ic^2) & * \end{pmatrix}. \tag{1}$$

Therefore any conjugate of X by a projective matrix of determinant 1 must have form (1).

It follows that if $\det(T) = 1$ the matrix A must have this form so that $\alpha = b^2 + ia^2$ or $\alpha = -b^2 - ia^2 = (ib)^2 + i(ia)^2$. Further since $ad - bc = 1$ it follows that $(a, b) = 1$.

Assume next that $\det(T) = i$ so that

$$T^{-1} = (-i)(\pm \begin{pmatrix} d & -b \\ -c & a \end{pmatrix}).$$

Then computing the conjugate of X by T we find

$$TXT^{-1} = (-i)(\pm \begin{pmatrix} a & b \\ c & d \end{pmatrix} \begin{pmatrix} 0 & i \\ -1 & 0 \end{pmatrix} \begin{pmatrix} d & -b \\ -c & a \end{pmatrix})$$

$$= \pm \begin{pmatrix} * & a^2 - ib^2 \\ -(c^2 - id^2) & * \end{pmatrix}. \quad (2)$$

Therefore any conjugate of X by a projective matrix of determinant i must have form (2).

It follows that if $\det(T) = i$ the matrix A must have this form so that $\alpha = a^2 - ib^2$ or $\alpha = (ia)^2 - i(ib)^2$. Further since $ad - bc = i$ it follows that $(a, b) = 1$.

Putting both cases together we have that if $\alpha \in \mathbb{Z}[i]$ and i is a quadratic residue modulo α then there exists $a, b \in \mathbb{Z}[i]$ with $(a, b) = 1$ and

$$\alpha = a^2 + ib^2 \text{ or } \alpha = a^2 - ib^2.$$

Conversely suppose that

$$\alpha = b^2 + ia^2 \text{ or } \alpha = b^2 - ia^2$$

with $(a, b) = 1$. We handle the first case, the other case is analogous. Since $(a, b) = 1$ there exists $x, y \in \mathbb{Z}[i]$ with $ax + by = 1$. It follows that the projective matrix

$$T = \pm \begin{pmatrix} a & b \\ -y & x \end{pmatrix}$$

is in Γ_1 with determinant 1. Hence $TXT^1 \in \Gamma_1$. Conjugating X by T and computing we get.

$$TXT^{-1} = (-i)(\pm \begin{pmatrix} a & b \\ -y & x \end{pmatrix} \begin{pmatrix} 0 & i \\ -1 & 0 \end{pmatrix} \begin{pmatrix} x & -b \\ y & a \end{pmatrix})$$

$$= \begin{pmatrix} * & b^2 + ia^2 \\ -(c^2 - id^2) & * \end{pmatrix}.$$

Therefore there exists a matrix S, with determinant i, and trace 0. Since S is conjugate to X and which has α in its upper right hand entry. That is:

$$S = \pm \begin{pmatrix} x & \alpha \\ \gamma & -x \end{pmatrix}.$$

Since the determinant is i we have

$$x^2 - \alpha\gamma = i \text{ so that } x^2 = i + \alpha\gamma.$$

Hence i is a quadratic residue modulo α, completing the theorem.

Within the Gaussian integers, $i^2 = -1$, and therefore -1 is a quadratic residue modulo α for any $\alpha \in \mathbb{Z}[i]$ with $\alpha \neq 0$ and α not a unit. Applying exactly the same proof as in Theorem 2.1, and using Lemma 2.2, we obtain further that every nonzero, nonunit Gaussian integer is either the sum of two relatively prime squares or i times the sum of two relatively prime squares.

Theorem 3. Let $\alpha \in \mathbb{Z}[i]$ with $\alpha \neq 0$ and α not a unit. Then there exists $a, b \in \mathbb{Z}[i]$ with $(a, b) = 1$ so that $\alpha = a^2 + b^2$ or $\alpha = i(a^2 + b^2)$.

Proof. Suppose that $\alpha \in \mathbb{Z}[i]$ with $\alpha \neq 0$ and α not a unit. Then -1 is a quadratic residue mod α and hence there exists $x, \gamma \in \mathbb{Z}[i]$ with $x^2 = -1 + \alpha\gamma$. Therefore the projective matrix

$$S = \pm \begin{pmatrix} x & \alpha \\ -\gamma & -x \end{pmatrix} \in \Gamma_1.$$

Since the determinant is 1, and the trace is 0, it follows from Lemma 2.2 that S must be conjugate within Γ_1 to

$$Y = \pm \begin{pmatrix} 0 & 1 \\ -1 & 0 \end{pmatrix}.$$

First conjugate Y by $T \in \Gamma_1$ with $\det(T) = 1$ so that

$$T^{-1} = \pm \begin{pmatrix} d & -b \\ -c & a \end{pmatrix}.$$

Then we find

$$TYT^{-1} = \pm \begin{pmatrix} a & b \\ c & d \end{pmatrix} \begin{pmatrix} 0 & 1 \\ -1 & 0 \end{pmatrix} \begin{pmatrix} d & -b \\ -c & a \end{pmatrix} = \pm \begin{pmatrix} * & b^2 + a^2 \\ -(d^2 + ic^2) & * \end{pmatrix}. \quad (3)$$

Therefore any conjugate of Y by a projective matrix of determinant 1 must have form (3).

It follows that if $\det(T) = 1$ the matrix A must have this form so that $\alpha = a^2 + b^2$ or $\alpha = (ia)^2 = (ib)^2$. Further since $ad - bc = 1$ it follows that $(a, b) = 1$.

Next conjugate Y by $T \in \Gamma_1$ with $\det(T) = i$ so that

$$T^{-1} = (-i)(\pm \begin{pmatrix} d & -b \\ -c & a \end{pmatrix}).$$

Then we find

$$TYT^{-1} = (-i)(\pm \begin{pmatrix} a & b \\ c & d \end{pmatrix} \begin{pmatrix} 0 & 1 \\ -1 & 0 \end{pmatrix} \begin{pmatrix} d & -b \\ -c & a \end{pmatrix})$$

$$= \pm \begin{pmatrix} * & i(b^2 + a^2) \\ -i(d^2 + ic^2) & * \end{pmatrix}. \quad (4)$$

Therefore any conjugate of Y by a projective matrix of determinant 1 must have form (4).

It follows that if $\alpha \neq 0$ and α is not a unit then the result follows; that is,

$$\alpha = a^2 + b^2 \text{ or } \alpha = i(a^2 + b^2)$$

for some $a, b \in \mathbb{Z}[i]$ with $(a, b) = 1$.

3 The Structure of $PGL(2, \mathbb{Z}[i])$

The Picard group $\Gamma = PSL(2, \mathbb{Z}[i])$ has the presentation (see [4] and [5])

$$\langle a, \ell, t, u; a^2 = \ell^2 = (a\ell)^2 = (t\ell)^2 = (u\ell)^2 = (at)^3 = (ua\ell)^3 = 1, [t,u] = 1 \rangle$$

where the transformations a, ℓ, t, u are

$$a : z' = -\frac{1}{z}, \ell : z' = -z, t : z' = z+1, u : z' = z+i.$$

Using this presentation it was shown that (again see [4]) that Γ has a very nice free product with amalgamation structure built up from finite groups. In the following theorem, S_3 is the symmetric group on 3 symbols, A_4 is the alternating group on 4 symbols, D_2 is the dihedral group of order 4 and M is the classical modular group $PSL(2, \mathbb{Z})$.

Theorem 4. *The group $\Gamma = PSL(2, \mathbb{Z}[i])$ is given group theoretically as the free product with amalgamation*

$$\Gamma = G_1 \underset{M}{\star} G_2$$

where

$$G_1 = S_3 \underset{\mathbb{Z}_3}{\star} A_4$$

$$G_2 = S_3 \underset{\mathbb{Z}_2}{\star} D_2$$

and M is the classical modular group with $M = \mathbb{Z}_2 \star \mathbb{Z}_3$.

In [3] and see also [4] this presentation was used to do an extensive examination of the group theoretical properties of Γ. For more information on this presentation see [5].

In this paper we used the extended Picard group $\Gamma_1 = PGL(2, \mathbb{Z}[i])$. Γ_1 contains Γ as a normal subgroup of index 2 and Γ_1 is Γ extended by the transformation $v : z' = iz$. In this short final section we write down a presentation for Γ_1 and from this presentation give its group theoretical structure.

Theorem 5. *The extended Picard group $\Gamma_1 = PGL(2, \mathbb{Z}[i])$ has the presentation*

$$\langle a, \ell, t, u, v; a^2 = \ell^2 = (a\ell)^2 = (t\ell)^2 = (u\ell)^2 = (at)^3 = (ua\ell)^3 = 1, [t,u] = 1,$$

$$v^4 = (va)^2 = 1, v^2 = \ell, tv = vu, uv = vt^{-1} \rangle$$

where the transformations a, ℓ, t, u, v are

$$a : z' = -\frac{1}{z}, \ell : z' = -z, t : z' = z+1, u : z' = z+i, v : z' = iz$$

Proof. Let a, ℓ, t, u, v be as stated in the theorem. The elements a, ℓ, t, u generate Γ. The subgroup Γ has index two in Γ_1. Since $v \notin \Gamma$ the elements $1, v$ can be taken as coset representatives for the normal subgroup Γ in Γ_1. Hence every element of Γ_1 can be written as $v^\epsilon T$ where $T \in \Gamma$ and $\epsilon = 0$ or $\epsilon = 1$. It follows that a, ℓ, t, u, v are a set of generators for Γ_1.

From computations with the linear fractional transformations we get the relations

$$v^4 = (va)^2 = 1, v^2 = \ell, tv = vu, uv = vt^{-1} \qquad (5)$$

holding in Γ_1.

Using the set of relations (5) we can transform an arbitrary word in a, ℓ, t, u, v and move the v terms all the way to the left. Therefore we can rewrite any word in a, ℓ, t, u, v as a word $v^\epsilon T$ where $\epsilon = 0$ or $\epsilon = 1$ and T is a word in a, ℓ, t, u. Since the set of relations

$$a^2 = \ell^2 = (a\ell)^2 = (t\ell)^2 = (u\ell)^2 = (at)^3 = (ua\ell)^3 = 1, [t, u] = 1 \qquad (6)$$

provides a complete set of relations for the subgroup generated by a, ℓ, t, u, which is Γ, it follows that the set of relations (5) together with (6) provide a complete set of relations for Γ_1. Therefore the presentation given in the statement of the theorem is a presentation for Γ_1.

The following corollary, whose proof is embedded in the proof of the last theorem, describes the group theoretical structure of Γ_1.

Corollary 1. *The extended Picard group $\Gamma_1 = PGL(2, \mathbb{Z}[i])$ is an extension of the Picard group by a cyclic group of order 2. Each element in Γ_1 can be written as $v^\epsilon T$ where $\epsilon = 0$ or $\epsilon = 1$, v is the transformation $z' = iz$ and $T \in \Gamma$.*

Using this structure, the lemmas in the previous section can be reobtained, since an element of trace 0 must have order 2 or 4 and if it has determinant i it must be conjugate to v or v^3.

Finally we would like to thank Anja Moldenhauer for her assistance in preparing and editing the paper.

References

1. B. Fine. A Note on the Two-Square Theorem *Can. Math. Bulletin*, 20: 93–94, 1977.
2. B. Fine. Sums of Squares Rings *Can. J. Math*, 29: 159–160, 1977.
3. B. Fine. Cyclotomic Equations and Square Properties in Rings *Int. J. Nath. and Math. Sci.*, 9: 89–95, 1986.
4. B. Fine. *The Algebraic Theory of the Bianchi Groups* Marcel Dekker 1989
5. B. Fine and G. Rosenberger. *Algebraic Generalizations of Discrete Groups* Marcel Dekker, 2001.
6. B. Fine and G. Rosenberger. *Number Theory: An Introduction via the Distribution of Primes* Birkhauser, 2006.
7. G. Kern-Isberner and G. Rosenberger. A note on numbers of the form $n = x^2 + Ny^2$ *Arch. Math.*, 43, 148–156, 1984.
8. G. Kern-Isberner and G. Rosenberger. Einige Bemerkungen über Untergruppen der PSL(2,$\mathbb{C}$), *Resultate der Mathematik*, 6, 40–47, 1983.

Cryptosystems Using Automorphisms of Finitely Generated Free Groups

Anja Moldenhauer[1], *Gerhard Rosenberger*[2]

Abstract. This paper introduces a newly developed private key cryptosystem and a public key cryptosystem. In the first one, each letter is encrypted with a different key. Therefore, it is a kind of a one-time pad. The second one is inspired by the ElGamal cryptosystem. Both presented cryptosystems are based on automorphisms of free groups. Given a free group F of finite rank, the automorphism group $Aut(F)$ can be generated by Nielsen transformations, which are the basis of a linear technique to study free groups and general infinite groups. Therefore Nielsen transformations are introduced.

1 Introduction

The topic of this paper is established in the area of mathematical cryptology, more precisely in group based cryptology. We refer to the books [1], [7] and [13] for the interested reader. The books [1] and [7] can also be used for a first access to the wide area of cryptology.

We introduce two cryptosystems, the first one is a private key cryptosystem (one-time pad) and the second one is a public key cryptosystem. We require that the reader is familiar with the general concept of these types of protocols. In cryptology it is common to call the two parties who want to communicate privately with each other Alice and Bob.

Throughout the paper let F always be a free group $F = \langle X \mid \ \rangle$ of finite rank. Both cryptosystems are based on free groups F of finite rank and automorphisms of F. It is known that the group of all automorphisms of F, $Aut(F)$, can be generated by Nielsen transformations (see [3]).

We first review some basic definitions concerning regular Nielsen transformations and Nielsen reduced sets and we give additional information which is also important for the understanding of the paper.

Both cryptosystems use automorphisms of a free group F of finite rank. Thus, a random choice of these automorphisms is practical. An approach for this random choice using the Whitehead-Automorphisms is given. Therefore, the Whitehead-Automorphisms are reviewed. Note, that the Whitehead-Automorphisms generate the Nielsen transformations and vice versa, but the use of the Whitehead-Automorphisms is more practical if a random choice of automorphisms is required.

---

[1] Fachbereich Mathematik, Universität Hamburg, Bundesstrasse 55, 20146 Hamburg, Germany, anja.moldenhauer@uni-hamburg.de

[2] Fachbereich Mathematik, Universität Hamburg, Bundesstrasse 55, 20146 Hamburg, Germany, gerhard.rosenberger@math.uni-hamburg.de

After this, a private key cryptosystem using Nielsen transformations and Nielsen reduced sets is introduced. An example and a security analysis for this private key cryptosystem is given. Finally we explain a public key cryptosystem which is inspired by the ElGamal cryptosystem, from which we describe two variations and give an example.

The new cryptographic protocols are in part in the dissertation [12] of A. Moldenhauer under her supervisor G. Rosenberger at the University of Hamburg.

2 Preliminaries for Automorphisms of Free Groups

We now review some basic definitions concerning regular Nielsen transformations and Nielsen reduced sets and we give additional information which will be used later on (see also [3], [9] or [10]).

Let F be a free group on the free generating set $X := \{x_1, x_2, \ldots, x_q\}$ and let $U := \{u_1, u_2, \ldots, u_t\} \subset F$, $q, t \geq 2$. A **freely reduced word** in X is a word in which the symbols x_i^ϵ, $x_i^{-\epsilon}$, for $\epsilon = \pm 1$ and $i = 1, 2, \ldots, q$, do not occur consecutively. We call q the **rank** of F. The free generating set X is also called a **basis** of F. The elements u_i are freely reduced words with letters in $X^{\pm 1} := X \cup X^{-1}$, with $X^{-1} := \{x_1^{-1}, x_2^{-1}, \ldots, x_q^{-1}\}$.

Definition 1.
*An **elementary Nielsen transformation** on*
$U = \{u_1, u_2, \ldots, u_t\} \subset F$ *is one of the following transformations*

(T1) *replace some u_i by u_i^{-1};*
(T2) *replace some u_i by $u_i u_j$ where $j \neq i$;*
(T3) *delete some u_i where $u_i = 1$.*

*In all three cases the u_k for $k \neq i$ are not changed. A (finite) product of elementary Nielsen transformations is called a **Nielsen transformation**. A Nielsen transformation is called **regular** if it is a finite product of the transformations (T1) and (T2), otherwise it is called **singular**. The regular Nielsen transformations generate a group. The set U is called **Nielsen-equivalent** to the set V, if there is a regular Nielsen transformation from U to V. Nielsen-equivalent sets U and V generate the same group, that is, $\langle U \rangle = \langle V \rangle$.*

Now, we agree on some notations. We write $(T1)_i$ if we replace u_i by u_i^{-1} and we write $(T2)_{i.j}$ if we replace u_i by $u_i u_j$. If we want to apply the same Nielsen transformation (T2) consecutively t-times we write $[(T2)_{i.j}]^t$ and hence replace u_i by $u_i u_j^t$. In all cases the u_k for $k \neq i$ are not changed.

Definition 2.
*A finite set U in F is called **Nielsen reduced**, if for any three elements v_1, v_2, v_3 from $U^{\pm 1}$ the following conditions hold:*

(N0) $v_1 \neq 1$;
(N1) $v_1 v_2 \neq 1$ *implies* $|v_1 v_2| \geq |v_1|, |v_2|$;
(N2) $v_1 v_2 \neq 1$ *and* $v_2 v_3 \neq 1$ *implies* $|v_1 v_2 v_3| > |v_1| - |v_2| + |v_3|$.

Here $|v|$ denotes the **free length** of $v \in F$, that is, the number of letters from $X^{\pm 1}$ in the freely reduced word v.

Remark 1. We say that any word w with finitely many letters from $X^{\pm 1}$ has **length** L if the number of letters occurring is L. The length of a word w is greater than or equal to the free length of the word w. For freely reduced words the length and the free length are equal. If a word w is not freely reduced then the length is greater than the free length of w.

Proposition 1. [3, Theorem 2.3] or [9, Proposition 2.2]
If $U = \{u_1, u_2, \ldots, u_m\}$ is finite, then U can be carried by a Nielsen transformation into some V such that V is Nielsen reduced. We have $\operatorname{rank}(\langle V \rangle) \leq m$.

Proposition 2. [10, Corollary 3.1]
Let H be a finitely generated subgroup of the free group F on the free generating set X. Let $U = \{u_1, u_2, \ldots, u_t\}$, u_i words in X, be a Nielsen reduced set. Then, out of all systems of generators for H, the set U has the shortest total x-length, that is $\sum_{i=1}^{t} |u_i|$.

Remark 2. If F_V is a finitely generated subgroup of $F = \langle X \mid \ \rangle$, with free generating set $V = \{v_1, v_2, \ldots, v_N\}$, v_i words in X, then there exist only finitely many Nielsen reduced sets $U_i = \{u_{i_1}, u_{i_2}, \ldots, u_{i_N}\}$, $i = 1, 2, \ldots, \ell$, to V, which are Nielsen-equivalent. With the help of a lexicographical order $<_{lex}$ (see for instance [3, Proof of Satz 2.3]) the smallest set U_s, in the set of all Nielsen reduced sets $U_{Nred}^V := \{U_1, U_2, \ldots, U_\ell\}$ to V, can be uniquely marked. With the use of regular Nielsen transformations it is possible to obtain this marked set U_s starting from any arbitrary set in U_{Nred}^V.

Proposition 3. [3, Korollar 2.10]
Let F be the free group of rank q. Then, the group of all automorphisms of F, $Aut(F)$, is generated by the elementary Nielsen transformations (T1) and (T2).
More precisely: Each automorphism of F is describable as a regular Nielsen transformation between two bases of F, and, each regular Nielsen transformation between two bases of F defines an automorphism of F.

Remark 3. In [16] an algorithm, using elementary Nielsen transformations, is presented which, given a finite set S of m words of a free group, returns a set S' of Nielsen reduced words such that $\langle S \rangle = \langle S' \rangle$; the algorithm runs in $\mathcal{O}(\ell^2 m^2)$ time, where ℓ is the maximum length of a word in S.

Theorem 1. [3, Satz 2.6]
Let U be Nielsen reduced, then $\langle U \rangle$ is free on U.

For the next lemma we need some notations. Let $w \neq 1$ be a freely reduced word in X. The initial segment s of w which is "a little more than half" of w (that is, $\frac{1}{2}|w| < |s| \leq \frac{1}{2}|w| + 1$) is called the **major initial segment** of w. The **minor initial segment** of w is that initial segment s' which is "a little less than half" of w (that is, $\frac{1}{2}|w| - 1 \leq |S'| < \frac{1}{2}|w|$). Similarly, **major** and **minor terminal segments** are defined.

If the free length of the word w is even, we call the initial segment s of w, with $|s| = \frac{1}{2}|w|$ the **left half** of w. Analogously, we call the terminal segment s' of w with $|s'| = \frac{1}{2}|w|$ the **right half** of w.

Let $\{w_1, w_2, \ldots, w_n\}$ be a set of freely reduced words in X, which are not the identity. An initial segment of a w-symbol (that is, of either w_i or w_i^{-1}, which are different w-symbols) is called **isolated** if it does not occur as an initial segment of any other w-symbol. Similarly, a terminal segment is isolated if it is a terminal segment of a unique w-symbol.

Lemma 1. [10, Lemma 3.1]
Let $M := \{w_1, w_2, \ldots, w_m\}$ be a set of freely reduced words in X with $w_j \neq 1$, $1 \leq j \leq m$. Then M is Nielsen reduced if and only if the following conditions are satisfied:

1. Both the major initial and major terminal segments of each $w_i \in M$ are isolated.
2. For each $w_i \in M$ of even free length, either its left half or its right half is isolated.

Definition 3.
Let F be a free group of rank q and let G be a free subgroup of F with rank m. An element $g \in G$ is called a **primitive element** of G, if a basis U of G with $g \in U$ exists.

Proposition 4. [14]
The number of primitive elements of free length k of the free group $F_2 = \langle x_1, x_2 \mid \ \rangle$ (and therefore, in any group $F_q = \langle x_1, x_2, \ldots, x_q \mid \ \rangle$, $q \geq 2$) is:

1. more than $\frac{8}{3\sqrt{3}} \cdot (\sqrt{3})^k$ if k is odd;
2. more than $\frac{4}{3} \cdot (\sqrt{3})^k$ if k is even.

Theorem 2. [2]
If $P(q, k)$ is the number of primitive elements of free length k of the free group $F_q = \langle x_1, x_2, \ldots, x_q \mid \ \rangle$, $q \geq 3$, then for some constants c_1, c_2, we have

$$c_1 \cdot (2q - 3)^k \leq P(q, k) \leq c_2 \cdot (2q - 2)^k.$$

Definition 4. [15]
A subgroup H of F is called **characteristic** in F if $\varphi(H) = H$ for every automorphism φ of F.

For $n \in \mathbb{N}$ let $\mathbb{Z}_n := \mathbb{Z}/n\mathbb{Z}$ be the ring of integers modulo n. The corresponding residue class in $\mathbb{Z}_n$ for an integer β is denoted by $\overline{\beta}$ (see also [1]).

Definition 5. [1]
Let $n \in \mathbb{N}$ and $\overline{\beta}, \overline{\gamma} \in \mathbb{Z}_n$. A bijective mapping $h : \mathbb{Z}_n \to \mathbb{Z}_n$ given by $x \mapsto \overline{\beta} x + \overline{\gamma}$ is called a **linear congruence generator**.

Theorem 3. [1] (Maximal period length for $n = 2^m$, $m \in \mathbb{N}$)
Let $n \in \mathbb{N}$, with $n = 2^m$, $m \geq 1$ and let $\beta, \gamma \in \mathbb{Z}$ such that $h : \mathbb{Z}_n \to \mathbb{Z}_n$, with $x \mapsto \overline{\beta}x + \overline{\gamma}$, is a linear congruence generator. Further let $\alpha \in \{0, 1, \ldots, n-1\}$ be given and $x_1 = \overline{\alpha}$, $x_2 = h(x_1)$, $x_3 = h(x_2)$, ...
Then the sequence $x_1, x_2, x_3, \ldots$ is periodic with maximal periodic length $n = 2^m$ if and only if the following holds:

1. β is odd.
2. If $m \geq 2$ then $\beta \equiv 1 \pmod 4$.
3. γ is odd.

Theorem 4. [8]
Let F be a free group with countable number of generators $x_1, x_2, \ldots$. Corresponding to x_j define

$$M_j = \begin{pmatrix} -r_j & -1 + r_j^2 \\ 1 & -r_j \end{pmatrix}$$

with $r_j \in \mathbb{Q}$ and

$$r_{j+1} - r_j \geq 3$$
$$r_1 \geq 2.$$

Then G^* generated by $\{M_1, M_2, \ldots\}$ is isomorphic to F.

3 The Random Choice of the Automorphisms of $Aut(F)$

Let $F = \langle X \mid \; \rangle$ be the free group on the free generating set X with $|X| = q$. The cryptosystems we develop are based on automorphisms of F. These automorphisms should be chosen randomly. It is known, see Proposition 3, that the Nielsen transformations generate the automorphism group $Aut(F)$. For a realization of a random choice procedure the Whitehead-Automorphisms will be used.

A fixed set of randomly chosen automorphisms is part of the key space for the private key cryptosystem.

Definition 6. *Whitehead-Automorphisms:*

1. Invert the letter a and leave all other letters invariant:

$$i_a(b) = \begin{cases} a^{-1} & \text{for } a = b \\ b & \text{for } b \in X \setminus \{a\}. \end{cases}$$

There are q **Whitehead-Automorphisms** of this type.
2. Let $a \in X$ and L, R, M be three pairwise disjoint subsets of X, with $a \in M$. Then the tuple (a, L, R, M) defines a **Whitehead-Automorphisms** $W_{(a,L,R,M)}$ as follows

$$W_{(a,L,R,M)}(b) = \begin{cases} ab & \text{for } b \in L \\ ba^{-1} & \text{for } b \in R \\ aba^{-1} & \text{for } b \in M \\ b & \text{for } b \in X \setminus (L \cup M \cup R). \end{cases}$$

There are $q \cdot 4^{q-1}$ automorphisms of this type.

Note, that $W^{-1}_{(a,L,R,M)} = i_a \circ W_{(a,L,R,M)} \circ i_a$.

With this definition it is clear how the Whitehead-Automorphisms can be generated as a product of regular Nielsen transformations. Conversely, the Whitehead-Automorphisms generate the group of the Nielsen transformations and therefore also the automorphism group $Aut(F)$ (see also [4]). With the Whitehead-Automorphisms it is simple to realize a random choice of automorphisms. We now give an approach for this choice.

An approach for choosing randomly automorphisms of $Aut(F)$:

Let $X = \{x_1, x_2, \ldots, x_q\}$ be the free generating set for the free group F.

1. First of all it should be decided in which order an automorphism f_i is generated by automorphisms of type i_a and $W_{(a,L,R,M)}$. For this purpose an automorphism of type i_a is identified with a zero and $W_{(a,L,R,M)}$ with a one. A sequence of zeros and ones is randomly generated. This sequence is translated to randomly chosen Whitehead-Automorphisms and hence presents an automorphism $f_i \in Aut(F)$. This translation is as follows:

2.1. For a zero in the sequence we generate i_a randomly: choose a random number z, with $1 \leq z \leq q$; hence an element $a \in X$ must be chosen to declare the automorphism. Then it is $a := x_z$ and hence x_z is replaced by x_z^{-1} and all other letters are invariant.

2.2. For a one in the sequence we generate $W_{(a,L,R,M)}$ randomly: choose a random number z, with $1 \leq z \leq q$. Hence it is $a := x_z$. Moreover it is $a \in M$. After this the disjoint sets $L, R, M \subset X$ are chosen randomly. One possible approach is the following:

   (a) Choose random numbers z_1, z_2 and z_3 with

   $$0 \leq z_1 \leq q - 1,$$
   $$0 \leq z_2 \leq q - 1 - z_1,$$
   $$0 \leq z_3 \leq q - 1 - z_1 - z_2.$$

   If we are in the situation of $z_1 = z_2 = z_3 = 0$ we get the identity id_X. If this case arises a random number $\tilde{z}$ from the set $\{1, 2, \ldots, q\} \setminus \{z\}$ is chosen and hence the element $x_{\tilde{z}}$ is assigned randomly to one of the sets L, R or M; therefore the identity is avoided.
   It is

   $$|L| = z_1, \qquad |R| = z_2, \qquad |M| = z_3 + 1.$$

   (b) Choose z_1 pairwise different random numbers $\{r_1, r_2, \ldots, r_{z_1}\}$ of the set $\{1, 2, \ldots, q\} \setminus \{z\}$. Then L is the set

   $$L = \{x_{r_1}, x_{r_2}, \ldots, x_{r_{z_1}}\}.$$

(c) Choose z_2 pairwise different random numbers $\{p_1, p_2, \ldots, p_{z_2}\}$ of the set $\{1, 2, \ldots, q\} \setminus (\{z\} \cup \{r_1, r_2, \ldots, r_{z_1}\})$. Then R is the set

$$R = \{x_{p_1}, x_{p_2}, \ldots, x_{p_{z_2}}\}.$$

(d) Choose z_3 pairwise different random numbers $\{t_1, t_2, \ldots, t_{z_3}\}$ of the set $\{1, 2, \ldots, q\} \setminus (\{z\} \cup \{r_1, r_2, \ldots, r_{z_1}\} \cup \{p_1, p_2, \ldots, p_{z_2}\})$. Then M is the set

$$M = \{x_{t_1}, x_{t_2}, \ldots, x_{t_{z_3}}\} \cup \{a\}.$$

Remark 4. If Alice and Bob use Whitehead-Automorphisms to generate automorphisms on a free group with free generating set X they should take care, that there are no sequences of the form

1. $i_a \circ i_a = id_X$,
2. $W_{(a,L,R,M)} \circ \underbrace{i_a \circ W_{(a,L,R,M)} \circ i_a}_{=W^{-1}_{(a,L,R,M)}} = id_X$ or

$\underbrace{i_a \circ W_{(a,L,R,M)} \circ i_a}_{=W^{-1}_{(a,L,R,M)}} \circ W_{(a,L,R,M)} = id_X,$

for the automorphism f_j. They also should not use Whitehead-Automorphisms sequences for f_j, which cancel each other and so be vacuous for the encryption.

4 Private Key Cryptosystem Based on Automorphisms of Free Groups F

Before Alice and Bob are able to communicate with each other, they have to make some arrangements.

Public Parameters

They first agree on the public parameters.

1. A free group F with free generating set $X = \{x_1, x_2, \ldots, x_q\}$, with $q \geqslant 2$.
2. A plaintext alphabet $A = \{a_1, a_2, \ldots, a_N\}$, with $N \geqslant 2$.
3. A subset $\mathcal{F}_{aut} := \{f_1, f_2, \ldots, f_{2^{128}}\} \subset Aut(F)$ of automorphisms of F is chosen. It is $f_i : F \to F$ and the f_i, $i = 1, 2, \ldots, 2^{128}$, pairwise different, are generated with the help of 0-1-sequences (of different length) and random numbers as described in Section 3. The set $\mathcal{F}_{aut}$ is part of the key space.
4. They agree on a linear congruence generator $h : \mathbb{Z}_{2^{128}} \to \mathbb{Z}_{2^{128}}$ with a maximal period length (see Definition 5 and Theorem 3).

Remark 5. If the set $\mathcal{F}_{aut}$ and the linear congruence generator h are public Alice and Bob are able to change the automorphisms and the generator publicly without a private meeting. The set $\mathcal{F}_{aut}$ should be large enough to make a brute force search ineffective.

Another variation could be, that Alice and Bob choose the number of elements in the starting set $\mathcal{F}_{aut}$ smaller than 2^{128}, say for example 2^{10}. These starting automorphism set $\mathcal{F}_{aut}$ should be chosen privately by Alice and Bob as their set of seeds and should not be made public. Then Alice and Bob can extend publicly the starting set $\mathcal{F}_{aut}$ to the set $\mathcal{F}_{aut_1}$ of automorphisms such that $\mathcal{F}_{aut_1}$ contains, say for example, 2^{32} automorphisms. The number of all elements in $\mathcal{F}_{aut_1}$ should make a brute force attack inefficient. The linear congruence generator stays analogously, just the domain and codomain must be adapted to, say for example, $\mathbb{Z}_{2^{32}}$. Because of Theorem 3 Alice and Bob get at all times a linear congruence generator with maximal periodic length.

Private Parameters

Now they agree on the private parameters.

1. A free subgroup F_U of F with rank N and the free generating set $U = \{u_1, u_2, \ldots, u_N\}$ is chosen where U is a minimal Nielsen reduced set (with respect to a lexicographical order) and the u_i freely reduced words in X. Such systems U are easily to construct using Theorem 1 and Lemma 1 (see also [3] and [9]). It is $\mathcal{U}_{Nred}$ the set of all minimal Nielsen reduced sets with N elements in F, which is part of the key space.
2. They use a one to one correspondence

$$A \to U$$
$$a_j \mapsto u_j \quad \text{for } j = 1, \ldots, N.$$

3. Alice and Bob agree on an automorphism $f_{\overline{\alpha}} \in \mathcal{F}_{Aut}$, hence α is the common secret starting point $\alpha \in \{0, 1, \ldots, 2^{128} - 1\}$, with $x_1 = \overline{\alpha} \in \mathbb{Z}_{2^{128}}$, for the linear congruence generator. With this α they are able to generate the sequence of automorphisms of the set $\mathcal{F}_{aut}$, which they use for encryption and decryption, respectively.

The key space: The set $\mathcal{U}_{Nred}$ of all minimal (with respect to a lexicographical order) Nielsen reduced subsets of F with N elements. The set $\mathcal{F}_{Aut}$ of 2^{128} randomly chosen automorphisms of F.

Protocol

Now we explain the protocol and look carefully at the steps for Alice and Bob.

Public knowledge: $F = \langle X \mid \ \rangle$, $X = \{x_1, x_2, \ldots, x_q\}$ with $q \geq 2$; plaintext alphabet $A = \{a_1, a_2, \ldots, a_N\}$ with $N \geq 2$; the set $\mathcal{F}_{Aut}$; a linear congruence generator h.

Encryption and Decryption Procedure:

1. Alice and Bob agree privately on a set $U \in \mathcal{U}_{Nred}$ and an automorphism $f_{\overline{\alpha}} \in \mathcal{F}_{Aut}$. They also know the one to one correspondence between U and A.

2. Alice wants to transmit the message

$$S = s_1 s_2 \cdots s_z, \quad z \geqslant 1,$$

with $s_i \in A$ to Bob.

2.1. Alice generates with the linear congruence generator h and the knowledge of $f_{\overline{\alpha}}$ the z automorphisms $f_{x_1}, f_{x_2}, \ldots, f_{x_z}$, which she needs for encryption. It is $x_1 = \overline{\alpha}, x_2 = h(x_1), \ldots, x_z = h(x_{z-1})$.

2.2. The encryption is as follows

$$\text{if } s_i = a_t \quad \text{then } s_i \mapsto c_i := f_{x_i}(u_t), \quad 1 \leqslant i \leqslant z, \quad 1 \leqslant t \leqslant N.$$

Recall that the one to one correspondence $A \to U$ with $a_j \mapsto u_j$, for $j = 1, 2, \ldots, N$, holds. The ciphertext

$$C = f_{x_1}(s_1) f_{x_2}(s_2) \cdots f_{x_z}(s_z)$$
$$= c_1 c_2 \cdots c_z$$

is sent to Bob. We call c_j the ciphertext units. We do no cancellations between c_i and c_{i+1}, for $1 \leqslant i \leqslant z - 1$.

3. Bob gets the ciphertext

$$C = c_1 c_2 \cdots c_z,$$

and the information that he has to use z automorphisms of F from the set $\mathcal{F}_{Aut}$ for decryption. He has now two possibilities for decryption.

3.1.a. With the knowledge of $f_{\overline{\alpha}}$, the linear congruence generator h and the number z, he computes for each automorphism f_{x_i}, $i = 1, 2, \ldots, z$, the inverse automorphism $f_{x_i}^{-1}$.

3.1.b. With the knowledge of $f_{\overline{\alpha}}$, the set $U = \{u_1, u_2, \ldots, u_N\}$, the linear congruence generator h and the number z, he computes for each automorphism f_{x_i}, $i = 1, 2, \ldots, z$, the set

$$U_{f_{x_i}} = \{f_{x_i}(u_1), f_{x_i}(u_2), \ldots, f_{x_i}(u_N)\}.$$

Hence, with the one to one correspondence between U and A, he gets a one to one correspondence between the letters in the alphabet A and the words of the ciphertext depending on the automorphisms f_{x_i}. This is shown in Table 1.

Table 1. Plaintext alphabet $A = \{a_1, a_2, \ldots, a_N\}$ corresponded to ciphertext alphabet $U_{f_{x_i}}$ depending on the automorphisms f_{x_i}

	$U_{f_{x_1}}$	$U_{f_{x_2}}$	$\cdots$	$U_{f_{x_z}}$
a_1	$f_{x_1}(u_1)$	$f_{x_2}(u_1)$	$\cdots$	$f_{x_z}(u_1)$
a_2	$f_{x_1}(u_2)$	$f_{x_2}(u_2)$	$\cdots$	$f_{x_z}(u_2)$
$\vdots$	$\vdots$	$\vdots$	$\cdots$	$\vdots$
a_N	$f_{x_1}(u_N)$	$f_{x_2}(u_N)$	$\cdots$	$f_{x_z}(u_N)$

3.2. With the knowledge of the Table 1 or the inverse automorphisms $f_{x_i}^{-1}$, respectively, the decryption is as follows

if $c_i = f_{x_i}(u_t)$ then $c_i \mapsto s_i := f_{x_i}^{-1}(c_i) = a_t$, $\quad 1 \leqslant i \leqslant z, \quad 1 \leqslant t \leqslant N$.

He generates the plaintext message

$$S = f_{x_1}^{-1}(c_1) f_{x_2}^{-1}(c_2) \cdots f_{x_z}^{-1}(c_z)$$
$$= s_1 s_2 \cdots s_z,$$

with $s_i \in A$, from Alice.

Remark 6. The cryptosystem is a polyalphabetic system. A word $u_i \in U$, and hence a letter $a_i \in A$, is encrypted differently at different places in the plaintext.

Example 1. This example was executed with the help of the computer program GAP and the package "FGA[3]".

First Alice and Bob agree on the **public parameters**.

1. Let F be the free group on the free generating set $X = \{a, b, c, d\}$.
2. Let $\tilde{A} := \{a_1, a_2, \ldots, a_{12}\} = \{A, E, I, O, U, T, M, L, K, Y, B, N\}$ be the plaintext alphabet.
3. A set $\mathcal{F}_{Aut}$ is determined. In this example we give the automorphisms, which Alice and Bob use for encryption and decryption, respectively, just at the moment when they are needed.
4. The linear congruence generator with maximal periodic length is

$$h : \mathbb{Z}_{2^{128}} \to \mathbb{Z}_{2^{128}}$$
$$x \mapsto \bar{5}x + \bar{3}.$$

---

[3] Free Group Algorithms, a GAP4 Package by Christian Sievers, TU Braunschweig.

The **private parameters** for this example are:

1. The free group $F_{\tilde{U}}$ of F with the free generating set

$$\tilde{U} = \{u_1, u_2, \ldots, u_{12}\}$$
$$= \{ba^2, cd, d^2c^{-2}, a^{-1}b, a^4b^{-1}, b^3a^{-2}, bc^3, bc^{-1}bab^{-1},$$
$$c^2ba, c^2dab^{-1}, a^{-1}d^3c^{-1}, a^2db^2d^{-1}\}.$$

It is known, that $a_i \mapsto u_i$, $i = 1, 2, \ldots, 12$, for $u_i \in \tilde{U}$ and $a_i \in \tilde{A}$. The set $\tilde{U}$ is a Nielsen reduced set and the group $F_{\tilde{U}}$ has rank 12. Alice and Bob agree on the starting automorphism $f_{\overline{93}}$, hence it is $x_1 = \overline{\alpha} = \overline{93}$.

We look at the **encryption and decryption procedure** for Alice and Bob.

2. With the above agreements **Alice** is able to encrypt her message

$$S = \text{I LIKE BOB}.$$

Her message is of length 8. She generates the ciphertext as follows:

2.1. She first determines, with the help of the linear congruence generator h, the automorphisms f_{x_i}, $i = 1, 2, \ldots, 8$, which she needs for encryption. It is

$x_1 = \overline{\alpha} \quad = \quad \overline{93}, \quad x_2 = h(x_1) = \quad \overline{468}, \quad x_3 = h(x_2) = \quad \overline{2343},$
$x_4 = h(x_3) = \quad \overline{11718}, \quad x_5 = h(x_4) = \quad \overline{58593}, \quad x_6 = h(x_5) = \overline{292968},$
$x_7 = h(x_6) = \overline{1464843}, \quad x_8 = h(x_7) = \overline{7324218}.$

The automorphisms are described with the help of regular Nielsen transformations, it is

$f_{x_1} = (N1)_3(N2)_{1.4}(N2)_{4.3}(N2)_{2.3}(N1)_3(N2)_{1.4}(N2)_{3.1}$,

$$f_{x_1} : F \to F$$
$$a \mapsto ad^2c^{-1}, \ b \mapsto bc^{-1}, \ c \mapsto cad^2c^{-1}, \ d \mapsto dc^{-1};$$

$f_{x_2} = (N2)_{1.4}(N1)_2(N2)_{2.4}(N2)_{3.1}(N1)_2(N1)_1(N2)_{1.3}[(N2)_{4.3}]^2(N1)_3$,

$$f_{x_2} : F \to F$$
$$a \mapsto d^{-1}a^{-1}cad, \ b \mapsto d^{-1}b, \ c \mapsto d^{-1}a^{-1}c^{-1}, \ d \mapsto d(cad)^2;$$

$f_{x_3} = (N1)_2(N2)_{4.2}(N1)_4(N2)_{2.4}(N1)_2(N2)_{4.2}(N1)_3(N2)_{2.1}(N2)_{3.2}$
$\quad [(N2)_{1.4}]^3(N1)_2(N2)_{4.2}$,

$$f_{x_3} : F \to F$$
$$a \mapsto ab^3, \ b \mapsto a^{-1}d^{-1}, \ c \mapsto c^{-1}da, \ d \mapsto ba^{-1}d^{-1};$$

$f_{x_4} = [(N2)_{3.1}]^2(N1)_2[(N2)_{2.1}]^3(N2)_{2.4}(N2)_{4.2}(N2)_{1.3}$,

$$f_{x_4} : F \to F$$
$$a \mapsto aca^2, \ b \mapsto b^{-1}a^3d, \ c \mapsto ca^2, \ d \mapsto db^{-1}a^3d;$$

$f_{x_5} = (N2)_{1.2}(N1)_3(N1)_1[(N2)_{4.3}]^2(N2)_{1.2}(N1)_2(N1)_3(N2)_{2.4}(N2)_{3.1},$

$f_{x_5} : F \to F$

$a \mapsto b^{-1}a^{-1}b, \ b \mapsto b^{-1}dc^{-2}, \ c \mapsto cb^{-1}a^{-1}b, \ d \mapsto dc^{-2};$

$f_{x_6} = (N1)_1(N2)_{2.3}(N2)_{3.1}(N1)_2(N2)_{1.2}(N2)_{4.2},$

$f_{x_6} : F \to F$

$a \mapsto a^{-1}c^{-1}b^{-1}, \ b \mapsto c^{-1}b^{-1}, \ c \mapsto ca^{-1}, \ d \mapsto dc^{-1}b^{-1};$

$f_{x_7} = [(N2)_{2.1}]^3(N1)_3[(N2)_{4.3}]^3(N1)_1(N2)_{1.2}(N1)_2(N2)_{2.4}(N2)_{3.1},$

$f_{x_7} : F \to F$

$a \mapsto a^{-1}ba^3, \ b \mapsto a^{-3}b^{-1}dc^{-3}, \ c \mapsto c^{-1}a^{-1}ba^3, \ d \mapsto dc^{-3};$

$f_{x_8} = (N2)_{1.4}(N1)_2(N1)_3(N2)_{2.1}[(N2)_{3.4}]^2(N1)_4(N1)_1(N1)_3(N2)_{4.2},$

$f_{x_8} : F \to F$

$a \mapsto d^{-1}a^{-1}, \ b \mapsto b^{-1}ad, \ c \mapsto d^{-2}c, \ d \mapsto d^{-1}b^{-1}ad.$

Note, that the regular Nielsen transformations are applied from the left to the right.

2.2 The ciphertext is now

$C = f_{x_1}(I)f_{x_2}(L)f_{x_3}(I)f_{x_4}(K)f_{x_5}(E)f_{x_6}(B)f_{x_7}(O)f_{x_8}(B)$
$= f_{x_1}(d^2c^{-2})f_{x_2}(bc^{-1}bab^{-1})f_{x_3}(d^2c^{-2})f_{x_4}(c^2ba)f_{x_5}(cd)f_{x_6}(a^{-1}d^3c^{-1})$
$f_{x_7}(a^{-1}b)f_{x_8}(a^{-1}d^3c^{-1})$
$= dc^{-1}d^{-1}a^{-1}d^{-2}a^{-1}c^{-1} \wr d^{-1}bcabd^{-1}a^{-1}cadb^{-1}d \wr$
$(ba^{-1}d^{-1})^2(a^{-1}d^{-1}c)^2 \wr (ca^2)^2b^{-1}a^3daca^2 \wr cb^{-1}a^{-1}bdc^{-2}\wr$
$bca(dc^{-1}b^{-1})^3ac^{-1} \wr a^{-1}(a^{-2}b^{-1})^2dc^{-3} \wr (ab^{-1})^3adc^{-1}d^2$
$= c_1c_2c_3c_4c_5c_6c_7c_8.$

The symbol "$\wr$" marks the end of a ciphertext unit c_i.

3. **Bob** gets the ciphertext

$C = dc^{-1}d^{-1}a^{-1}d^{-2}a^{-1}c^{-1} \wr d^{-1}bcabd^{-1}a^{-1}cadb^{-1}d \wr$
$(ba^{-1}d^{-1})^2(a^{-1}d^{-1}c)^2 \wr (ca^2)^2b^{-1}a^3daca^2 \wr cb^{-1}a^{-1}bdc^{-2}\wr$
$bca(dc^{-1}b^{-1})^3ac^{-1} \wr a^{-1}(a^{-2}b^{-1})^2dc^{-3} \wr (ab^{-1})^3adc^{-1}d^2$

from Alice. Now he knows, that he needs eight automorphisms for decryption.

3.1. Bob knows the set U, the linear congruence generator h and the starting seed automorphism $f_{\overline{93}}$. For decryption he uses tables (analogously to Table 1).

Now, he is able to compute for each automorphism f_{x_i} the set $U_{f_{x_i}}$, $i = 1, 2, \ldots, 8$, and to generate the tables Table 2, Table 3, Table 4 and Table 5.

Table 2. Correspondence: plaintext alphabet to ciphertext alphabet I

	$U_{f_{x_1}}$	$U_{f_{x_2}}$
A	$b(c^{-1}ad^2)^2c^{-1}$	$d^{-1}bd^{-1}a^{-1}c^2ad$
E	$cad(dc^{-1})^2$	$d^{-1}a^{-1}c^{-1}(dca)^2d$
I	$dc^{-1}d^{-1}a^{-1}d^{-2}a^{-1}c^{-1}$	$((dca)^2d)^2cadcad$
O	$cd^{-2}a^{-1}bc^{-1}$	$d^{-1}a^{-1}c^{-1}ab$
U	$(ad^2c^{-1})^3ad^2b^{-1}$	$d^{-1}a^{-1}c^4adb^{-1}d$
T	$(bc^{-1})^2bd^{-2}a^{-1}cd^{-2}a^{-1}$	$(d^{-1}b)^3d^{-1}a^{-1}c^{-2}ad$
M	$b(ad^2)^3c^{-1}$	$d^{-1}b(d^{-1}a^{-1}c^{-1})^3$
L	$bd^{-2}a^{-1}c^{-1}bc^{-1}ad^2b^{-1}$	$d^{-1}bcabd^{-1}a^{-1}cadb^{-1}d$
K	$c(ad^2)^2c^{-1}bc^{-1}ad^2c^{-1}$	$(d^{-1}a^{-1}c^{-1})^2d^{-1}bd^{-1}a^{-1}cad$
Y	$c(ad^2)^2c^{-1}dc^{-1}ad^2b^{-1}$	$(d^{-1}a^{-1}c^{-1})^2dcadc^2adb^{-1}d$
B	$cd^{-2}a^{-1}(dc^{-1})^2d^{-1}a^{-1}c^{-1}$	$d^{-1}a^{-1}c^{-1}(ad^2cadc)^3adcad$
N	$(ad^2c^{-1})^2d(c^{-1}b)^2d^{-1}$	$d^{-1}a^{-1}c^2ad(dca)^2bd^{-1}b(d^{-1}a^{-1}c^{-1})^2d^{-1}$

Table 3. Correspondence: plaintext alphabet to ciphertext alphabet II

	$U_{f_{x_3}}$	$U_{f_{x_4}}$
A	$a^{-1}d^{-1}(ab^3)^2$	$b^{-1}a^3d(aca^2)^2$
E	$c^{-1}daba^{-1}d^{-1}$	$ca^2db^{-1}a^3d$
I	$(ba^{-1}d^{-1})^2(a^{-1}d^{-1}c)^2$	$(db^{-1}a^3d)^2a^{-2}c^{-1}a^{-2}c^{-1}$
O	$b^{-3}a^{-2}d^{-1}$	$a^{-2}c^{-1}a^{-1}b^{-1}a^3d$
U	$(ab^3)^4da$	$(aca^2)^4d^{-1}a^{-3}b$
T	$(a^{-1}d^{-1})^3(b^{-3}a^{-1})^2$	$(b^{-1}a^3d)^3a^{-2}c^{-1}a^{-3}c^{-1}a^{-1}$
M	$a^{-1}d^{-1}(c^{-1}da)^3$	$b^{-1}a^3d(ca^2)^3$
L	$(a^{-1}d^{-1})^2ca^{-1}d^{-1}ab^3da$	$b^{-1}a^3da^{-2}c^{-1}b^{-1}a^3daca^2d^{-1}a^{-3}b$
K	$c^{-1}dac^{-1}ab^3$	$(ca^2)^2b^{-1}a^3daca^2$
Y	$(c^{-1}da)^2ba^{-1}d^{-1}ab^3da$	$(ca^2)^2db^{-1}a^3daca^2d^{-1}a^{-3}b$
B	$b^{-3}a^{-1}(ba^{-1}d^{-1})^3a^{-1}d^{-1}c$	$a^{-2}c^{-1}a^{-1}(db^{-1}a^3d)^3a^{-2}c^{-1}$
N	$(ab^3)^2b(a^{-1}d^{-1})^2b^{-1}$	$aca^3c(a^2db^{-1}a)^2a^2$

Table 4. Correspondence: plaintext alphabet to ciphertext alphabet III

	$U_{f_{x_5}}$	$U_{f_{x_6}}$
A	$b^{-1}dc^{-2}b^{-1}a^{-2}b$	$(c^{-1}b^{-1}a^{-1})^2c^{-1}b^{-1}$
E	$cb^{-1}a^{-1}bdc^{-2}$	$ca^{-1}dc^{-1}b^{-1}$
I	$dc^{-2}dc^{-1}(c^{-1}b^{-1}ab)^2c^{-1}$	$(dc^{-1}b^{-1})^2ac^{-1}ac^{-1}$
O	$b^{-1}adc^{-2}$	$bcac^{-1}b^{-1}$
U	$b^{-1}a^{-4}bc^2d^{-1}b$	$(a^{-1}c^{-1}b^{-1})^3a^{-1}$
T	$(b^{-1}dc^{-2})^3b^{-1}a^2b$	$(c^{-1}b^{-1})^2abca$
M	$b^{-1}dc^{-1}(b^{-1}a^{-1}bc)^2b^{-1}a^{-1}b$	$c^{-1}b^{-1}(ca^{-1})^3$
L	$b^{-1}dc^{-2}b^{-1}abc^{-1}b^{-1}dc^{-2}b^{-1}a^{-1}bc^2d^{-1}b$	$c^{-1}b^{-1}ac^{-2}b^{-1}a^{-1}$
K	$cb^{-1}a^{-1}bcb^{-1}a^{-1}dc^{-2}b^{-1}a^{-1}b$	$ca^{-1}c(a^{-1}c^{-1}b^{-1})^2$
Y	$(cb^{-1}a^{-1}b)^2dc^{-2}b^{-1}a^{-1}bc^2d^{-1}b$	$(ca^{-1})^2dc^{-1}b^{-1}a^{-1}$
B	$b^{-1}ab(dc^{-2})^3b^{-1}abc^{-1}$	$bca(dc^{-1}b^{-1})^3ac^{-1}$
N	$b^{-1}a^{-2}b(dc^{-2}b^{-1})^2$	$(a^{-1}c^{-1}b^{-1})^2d(c^{-1}b^{-1})^2d^{-1}$

Table 5. Correspondence: plaintext alphabet to ciphertext alphabet IV

	$U_{f_{x_7}}$	$U_{f_{x_8}}$
A	$a^{-3}b^{-1}dc^{-3}a^{-1}(ba^2)^2a$	$b^{-1}d^{-1}a^{-1}$
E	$c^{-1}a^{-1}ba^3dc^{-3}$	$d^{-2}cd^{-1}b^{-1}ad$
I	$dc^{-3}dc^{-3}(a^{-3}b^{-1}ac)^2$	$d^{-1}(b^{-1}a)^2(dc^{-1}d)^2d$
O	$a^{-1}(a^{-2}b^{-1})^2dc^{-3}$	$adb^{-1}ad$
U	$a^{-1}(ba^2)^4ac^3d^{-1}ba^3$	$(d^{-1}a^{-1})^5b$
T	$(a^{-3}b^{-1}dc^{-3})^3a^{-1}(a^{-2}b^{-1})^2a$	$(b^{-1}ad)^3adad$
M	$a^{-3}b^{-1}dc^{-3}(c^{-1}a^{-1}ba^3)^3$	$b^{-1}a(d^{-1}cd^{-1})^2d^{-1}c$
L	$a^{-3}b^{-1}dc^{-3}a^{-3}b^{-1}aca^{-3}b^{-1}dc^{-3}a^{-1}ba^3c^3d^{-1}ba^3$	$b^{-1}adc^{-1}d^2b^{-1}d^{-1}a^{-1}b$
K	$c^{-1}a^{-1}ba^3c^{-1}a^{-1}dc^{-3}a^{-1}ba^3$	$(d^{-2}c)^2b^{-1}$
Y	$(c^{-1}a^{-1}ba^3)^2dc^{-3}a^{-1}ba^3c^3d^{-1}ba^3$	$(d^{-2}c)^2d^{-1}b^{-1}d^{-1}a^{-1}b$
B	$a^{-3}b^{-1}a(dc^{-3})^3a^{-3}b^{-1}ac$	$(ab^{-1})^3adc^{-1}d^2$
N	$a^{-1}(ba^2)^2a(dc^{-3}a^{-3}b^{-1})^2$	$(d^{-1}a^{-1})^2d^{-1}(b^{-1}ad)^2d$

3.2. With these tables he is able to generate the plaintext from Alice, it is

$$\begin{aligned}S =& f_{x_1}^{-1}\left(dc^{-1}d^{-1}a^{-1}d^{-2}a^{-1}c^{-1}\right) f_{x_2}^{-1}\left(d^{-1}bcabd^{-1}a^{-1}cadb^{-1}d\right)\\& f_{x_3}^{-1}((ba^{-1}d^{-1})^2(a^{-1}d^{-1}c)^2) f_{x_4}^{-1}\left((ca^2)^2 b^{-1}a^3 daca^2\right)\\& f_{x_5}^{-1}\left(cb^{-1}a^{-1}bdc^{-2}\right) f_{x_6}^{-1}\left(bca(dc^{-1}b^{-1})^3 ac^{-1}\right)\\& f_{x_7}^{-1}\left(a^{-1}(a^{-2}b^{-1})^2 dc^{-3}\right) f_{x_8}^{-1}\left((ab^{-1})^3 adc^{-1}d^2\right)\\=&\text{I LIKE BOB}.\end{aligned}$$

Security

This private key cryptosystem is secure against chosen plaintext attacks and chosen ciphertext attacks. In a chosen plaintext attack, an attacker, Eve, chooses an arbitrary plaintext of her choice and gets the corresponding ciphertext. In a chosen ciphertext attack Eve sees ciphertexts and gets to some of these ciphertexts the corresponding plaintexts (see also [1]).

An eavesdropper, Eve, intercepts the ciphertext

$$C = c_1 c_2 \cdots c_z,$$

with $c_i = f_{x_i}(u_j)$ for some $1 \leqslant j \leqslant N$. If Alice and Bob choose non characteristic subgroups, then it is likely that $c_j \notin F_U$ for some $1 \leqslant j \leqslant z$. Hence the ciphertext units give no hint for the subgroup F_U. Eve knows $L = \sum_{k=1}^{z} |c_k|$, the length of C, because Alice and Bob are doing no cancellations between c_i and c_{i+1}, for $1 \leqslant i \leqslant z-1$.

To break the system Eve needs to know the set U. For this it is likely that she assumes that the ball $B(F, L)$ in the Cayleygraph for F contains a basis for F_U. With this assumption she searches for primitive elements for F_U in the ball $B(F, L)$, $|y| \leqslant L$, $y \in F$. In fact she needs to find N primitive elements for F_U in $B(F, L)$ (these would be primitive elements for F_U in a ball $B(F_U, L)$ for some Nielsen reduced basis for F_U). From Proposition 4 and Theorem 2 it is known that the number of primitive elements grows exponentially with the free length of the elements. Eve chooses sets $M_i := \{m_{i_1}, m_{i_2}, \ldots, m_{i_K}\}$ with $K \geqslant N$ and elements m_{i_j} in $B(F, L)$ and with Nielsen transformations she constructs the corresponding Nielsen reduced sets M_i'. If $|M_i'| = N$ then M_i' is a candidate for U.

The number N is a constant in the cryptosystem, hence it takes $\mathcal{O}(\lambda^2)$ time, with $\lambda := \max\{|m_{j_\ell}| \mid \ell = 1, 2, \ldots, K\} \leqslant L$, to get the set M_j' from M_j with the algorithm [16] (see Remark 3).

The main security certification depends on the fact, that for a single subset of K elements Eve finds a Nielsen reduced set in polynomial running time (more precisely in quadratic time) but she has to test all possible subsets of K elements for which she needs exponential running time.

The security certification can be improved by the next two improvements.

First, Alice and Bob choose in addition an explicit presentation of the ciphertext units c_i as matrices in $SL(2, \mathbb{Q})$. So, they agree on a faithful representation

$$\varphi : F \to SL(2, \mathbb{Q})$$
$$x_i \mapsto M_i,$$

of F into $SL(2, \mathbb{Q})$ (see Theorem 4). The group $G = \varphi(F)$ is isomorphic to F under the mapping $x_i \mapsto M_i$, for $i = 1, \ldots, q$. The ciphertext is now

$$C' = \varphi(c_1)\varphi(c_2) \cdots \varphi(c_z)$$
$$= W_1 W_2 \cdots W_z,$$

a sequence of matrices $W_j \in SL(2, \mathbb{Q})$. The encryption is realizable with a table (as Table 1) if the representation φ is applied to the elements in the table. Therefore Bob gets a table with matrices and hence an assignment from the matrices to the plaintext alphabet depending on the automorphisms f_{x_i}.

Here the additional security certification is, that there is no algorithm known to solve the membership problem (see for instance [10]) for subgroups of $SL(2, \mathbb{Q})$ which are not subgroups in $SL(2, \mathbb{Z})$. B. Eick, M. Kirschner and C. Leedham-Green presented in the paper [5] a practical algorithm to solve the constructive membership problem for discrete free subgroups of rank 2 of $SL(2, \mathbb{R})$. For example, the subgroup $SL(2, \mathbb{Z})$ of $SL(2, \mathbb{R})$ is discrete. But they also mention, that it is an open problem to solve the membership problem for arbitrary subgroups of $SL(2, \mathbb{R})$ with rank $m \geq 2$. Alice and Bob work with subgroups of rank $N \geq 2$. Hence there is in general no algorithm known for Eve to solve the membership problem, in particular there is always no such algorithm known for $N \geq 3$.

Example 2. In this example[4] Alice and Bob agree additionally to Example 1 on a faithful representation. With Theorem 4 they generate the matrices

$$X_1 := \begin{pmatrix} \frac{-7}{2} & \frac{45}{4} \\ 1 & \frac{-7}{2} \end{pmatrix}, \quad X_2 := \begin{pmatrix} \frac{-15}{2} & \frac{221}{4} \\ 1 & \frac{-15}{2} \end{pmatrix} \text{ and } X_3 := \begin{pmatrix} \frac{-23}{2} & \frac{525}{4} \\ 1 & \frac{-23}{2} \end{pmatrix}.$$

These matrices form a basis for a free group G of rank 3. Alice and Bob generate a subgroup G_1 of G with rank 4. The free generating set for G_1 is $\{X_1 X_2, X_3 X_1^2, X_2 X_3 X_2, X_1^{-1} X_2\}$. They choose the faithful representation

$$\varphi : F \to SL(2, \mathbb{Q})$$

$$a \mapsto X_1 X_2 = \begin{pmatrix} \frac{75}{2} & \frac{-1111}{4} \\ -11 & \frac{163}{2} \end{pmatrix}, \qquad b \mapsto X_3 X_1^2 = \begin{pmatrix} -1189 & 3990 \\ 104 & -349 \end{pmatrix},$$

$$c \mapsto X_2 X_3 X_2 = \begin{pmatrix} -2681 & 19966 \\ 360 & -2681 \end{pmatrix}, \qquad d \mapsto X_1^{-1} X_2 = \begin{pmatrix} 15 & -109 \\ 4 & -29 \end{pmatrix}.$$

---

[4] We realized this example with the computer programs Classic Worksheet Maple 16 and GAP. In GAP we used the package "FGA" (Free Group Algorithms, a GAP4 Package by Christian Sievers, TU Braunschweig).

The ciphertext is now

$$C' = \varphi(dc^{-1}d^{-1}a^{-1}d^{-2}a^{-1}c^{-1})\ \varphi(d^{-1}bcabd^{-1}a^{-1}cadb^{-1}d)$$
$$\varphi((ba^{-1}d^{-1})^2(a^{-1}d^{-1}c)^2)\ \varphi((ca^2)^2b^{-1}a^3daca^2)\ \varphi(cb^{-1}a^{-1}bdc^{-2})$$
$$\varphi(bca(dc^{-1}b^{-1})^3ac^{-1})\ \varphi(a^{-1}(a^{-2}b^{-1})^2dc^{-3})\ \varphi((ab^{-1})^3adc^{-1}d^2)$$

$$= \begin{pmatrix} \frac{-429743093559909}{2} & \frac{-6400784021410159}{4} \\ -62588240305379 & \frac{-932216979117085}{2} \end{pmatrix}$$

$$\begin{pmatrix} \frac{-324007033175442303068324391}{2} & \frac{47007695458416827592369656315}{4} \\ -2233263222037105752723291977 & \frac{32400703278301507513861943 61}{2} \end{pmatrix}$$

$$\begin{pmatrix} \frac{-6899014060703475554169965}{2} & \frac{10275697214519152034878560 7}{4} \\ 30172246868510272996 9483 & \frac{-4493988131847945704997109}{2} \end{pmatrix}$$

$$\begin{pmatrix} \frac{-397074726172421275253684843812134445}{2} & \frac{588331876105967022375198589657847337 7}{4} \\ 26659253089426526822952736194350493 & \frac{-395000924306510751052288425218790 757}{2} \end{pmatrix}$$

$$\begin{pmatrix} \frac{4647588840742582 5}{2} & \frac{69223248973640038 9}{4} \\ -3120351373297111 & \frac{-46475896943687759}{2} \end{pmatrix}$$

$$\begin{pmatrix} \frac{-371540858684921774630357681975 99}{2} & \frac{-5533740137946437638980304441045 47}{4} \\ 16249065697537147499109567230 73 & \frac{24201404758781402065719318991873}{2} \end{pmatrix}$$

$$\begin{pmatrix} \frac{-3418963163764785449276501363}{2} & \frac{-5092355335791681521209536364 1}{4} \\ -2307513696294811415403011 25 & \frac{-3436913216344813651054341083}{2} \end{pmatrix}$$

$$\begin{pmatrix} \frac{2739747352948144349387}{2} & \frac{-3962864429658196770961 5}{4} \\ -4020700843122001145 47 & \frac{5815679440792026855107}{2} \end{pmatrix}.$$

Instead of a sequence of words in F Alice sends to Bob a sequence of eight matrices in $\text{SL}(2, \mathbb{Q})$.

For the second improvement Alice and Bob use instead of a presentation of the ciphertext in $\text{SL}(2, \mathbb{Q})$ a presentation of the ciphertext in a free group in $\text{GL}(2, k)$ with $k := \mathbb{Z}[y_1, y_2, \ldots, y_w]$, the ring of polynomials in variables $y_1, y_2, \ldots, y_w$. With the help of a homomorphism $\epsilon^* : \text{GL}(2, k) \to \text{GL}(2, \mathbb{Z})$ and the knowledge of an algorithm to write each element in the modular group $\text{PSL}(2, \mathbb{Z})$, the group of 2×2 projective integral matrices of determinant 1, in terms of s and t they can reconstruct the message. Here,

$$s = \begin{pmatrix} 0 & 1 \\ -1 & 0 \end{pmatrix} \quad \text{and} \quad t = \begin{pmatrix} 1 & 1 \\ 0 & 1 \end{pmatrix}$$

and $\text{PSL}(2, \mathbb{Z}) = \langle s, t \mid s^2 = (st)^3 = 1 \rangle$.

Every finitely generated free group is faithfully represented by a subgroup of the modular group $\text{PSL}(2, \mathbb{Z})$. Especially, the two matrices

$$\begin{pmatrix} 0 & 1 \\ -1 & 2 \end{pmatrix} \quad \text{and} \quad \begin{pmatrix} 2 & 1 \\ -1 & 0 \end{pmatrix}$$

generate a free group of rank two, and this free group certainly contains finitely generated free groups.

This improvement is very similar to the version in [1]. Here, the security certification depends in addition on the unsolvability of Hilbert's Tenth Problem. Y. Matiyasevich proved in [11] that there is no general algorithm which determines whether or not an integral polynomial in any number of variables has a zero.

5 Public Key Cryptosystem Based on Automorphisms of Free Groups F

Now we describe a public key cryptosystem for Alice and Bob which is inspired by the ElGamal cryptosystem (see [6] or [13, Section 1.3]), based on discrete logarithms, that is:

1. Alice and Bob agree on a finite cyclic group G and a generating element $g \in G$.
2. Alice picks a random natural number a and publishes the element $c := g^a$.
3. Bob, who wants to transmit a message $m \in G$ to Alice, picks a random natural number b and sends the two elements $m \cdot c^b$ and g^b, to Alice. Note that $c^b = g^{ab}$.
4. Alice recovers $m = (m \cdot c^b) \cdot \left((g^b)^a \right)^{-1}$.

Let $X = \{x_1, x_2, \ldots, x_N\}$, $N \geqslant 2$, be the free generating set of the free group $F = \langle X \mid \ \rangle$. It is $X^{\pm 1} = X \cup X^{-1}$. The message is an element $m \in S^*$, the set of all freely reduced words with letters in $X^{\pm 1}$. Public are the free group F, its free generating set X and an element $a \in S^*$. The automorphism f should be chosen randomly, for example as it is described in Section 3.

The public key cryptosystem is now as follows:

Public parameters: The group $F = \langle X \mid \ \rangle$, a freely reduced word $a \neq 1$ in the free group F and an automorphism $f : F \to F$ of infinite order.

Encryption and Decryption Procedure:

1. Alice chooses privately a natural number n and publishes the element $f^n(a) =: c \in S^*$.
2. Bob picks privately a random $t \in \mathbb{N}$ and his message $m \in S^*$. He calculates the freely reduced elements

$$m \cdot f^t(c) =: c_1 \in S^* \quad \text{and} \quad f^t(a) =: c_2 \in S^*.$$

He sends the ciphertext $(c_1, c_2) \in S^* \times S^*$ to Alice.
3. Alice calculates

$$\begin{aligned} c_1 \cdot f^n(c_2)^{-1} &= m \cdot f^t(c) \cdot f^n(c_2)^{-1} \\ &= m \cdot f^t(f^n(a)) \cdot (f^n(f^t(a)))^{-1} \\ &= m \cdot f^{t+n}(a) \cdot (f^{n+t}(a))^{-1} \\ &= m, \end{aligned}$$

and gets the message m.

Remark 7. A possible attacker, Eve, can see the elements $c, c_1, c_2 \in S^*$. She does not know the free length of m and the cancellations between m and $f^t(c)$ in c_1. It could be possible that m is completely canceled by the first letters of

$f^t(c)$. Hence she cannot determine m from the given c_1. Eve just sees words, $f^t(a)$ and $f^n(a)$, in the free generating set X from which it is unlikely to realize the exponents n and t, that is, the private keys from Alice and Bob, respectively. The security certification is based on the Diffie-Hellman-Problem.

Remark 8. We give some ideas to enhance the security, they can also be combined:

1. The element $a \in S^*$ could be taken as a common private secret between Alice and Bob. They could use for example the Anshel-Anshel-Goldfeld key exchange protocol (see [13]) to agree on the element a.
2. Alice and Bob agree on a faithful representation from F into the special linear group of all 2×2 matrices with entries in $\mathbb{Q}$, that is, $g : F \to \mathrm{SL}(2, \mathbb{Q})$. Now $m \in S^*$ and Bob sends $g(m) \cdot g(f^t(c)) =: c_1 \in \mathrm{SL}(2, \mathbb{Q})$ instead of $m \cdot f^t(c) =: c_1 \in S^*$; c and c_2 remain the same. Therefore, Alice calculates $c_1 \cdot (g(f^n(c_2)))^{-1} = g(m)$ and hence the message $m = g^{-1}(g(m)) \in S^*$. This variation in addition extends the security certification to the membership problem in the matrix group $\mathrm{SL}(2, \mathbb{Q})$ (see [5]).

Example 3.
This example[5] is a very small one and it is just given for illustration purposes.
Bob wants to send a message to Alice.
The **public parameters** are the free group $F = \langle x_1, x_2, x_3 \mid \ \rangle$ of rank 3, the freely reduced word $a \in F$, with $a := x_1^2 x_2 x_3^{-2} x_2$ and the automorphism $f : F \to F$, which is given, for this example, by regular Nielsen transformations: $f = [(N2)_{1.2}]^2 \ (N2)_{3.2} \ (N1)_3 \ (N2)_{2.3}$, that is,

$$x_1 \mapsto x_1 x_2^2, \quad x_2 \mapsto x_3^{-1}, \quad x_3 \mapsto x_2^{-1} x_3^{-1}.$$

1. Alice's private key is $n = 7$. Thus, she gets the automorphism

$$f^7 : F \to F$$
$$x_1 \mapsto x_1 x_2^2 x_3^{-1} x_2 (x_2 x_3)^2 (x_3 x_2 x_3^2 x_2)^2 x_3 x_2$$
$$x_2 \mapsto x_2^{-1} ((x_3^{-1} x_2^{-1} x_3^{-1})^2 x_2^{-1} x_3^{-1})^2 x_3^{-1} x_2^{-1} x_3^{-2}$$
$$x_3 \mapsto (((x_2^{-1} x_3^{-1})^2 x_3^{-1})^2 x_2^{-1} x_3^{-2})^2 x_2^{-1} (x_3^{-1} x_2^{-1} x_3^{-1})^2 x_3^{-1}.$$

Her public key is

$$c := f^7(a) = (x_1 x_2^2 x_3^{-1} x_2 (x_2 x_3)^2 (x_3 x_2 x_3^2 x_2)^2 x_3 x_2)^2 (x_3^2 x_2)^2$$
$$((x_3 x_2 x_3)^2 x_2 x_3)^2 x_3 x_2 x_3^2 x_2 x_3^{-1}.$$

2. Bob privately picks the ephemeral key $t = 5$ and gets the automorphism

$$f^5 : F \to F$$
$$x_1 \mapsto x_1 x_2^2 x_3^{-1} x_2^2 x_3 (x_3 x_2)^2$$
$$x_2 \mapsto x_2^{-1} (x_3^{-1} x_2^{-1} x_3^{-1})^2 x_3^{-1}$$
$$x_3 \mapsto ((x_2^{-1} x_3^{-1})^2 x_3^{-1})^2 x_2^{-1} x_3^{-2}.$$

---
[5] We used the computer program GAP and the package "FGA" (Free Group Algorithms, a GAP4 Package by Christian Sievers, TU Braunschweig).

His message for Alice is $m = x_3^{-2}x_2^2x_3x_1^2x_2^{-1}x_1^{-1}$. He calculates

$$c_1 = m \cdot f^5(c)$$
$$= x_3^{-2}x_2^2x_3x_1^2(x_2x_3^{-1})^2((x_3^{-1}x_2^{-1}x_3^{-2}x_2^{-1})^2x_3^{-2}x_2^{-1})^2(x_3^{-1}x_2^{-1}x_3^{-1})^2x_3^{-1}x_2^{-1}$$
$$((((x_3^{-1}x_2^{-1}x_3^{-1})^2x_2^{-1}x_3^{-1})^2x_3^{-1}x_2^{-1}x_3^{-1}x_2^{-1}x_3^{-1})^2(x_3^{-1}x_2^{-1}x_3^{-2}x_2^{-1})^2x_3^{-1}$$
$$x_2^{-1}x_3^{-1})^2((x_3^{-1}x_2^{-1}x_3^{-2}x_2^{-1})^2x_3^{-2}x_2^{-1})^2(x_3^{-1}x_2^{-1}x_3^{-1})^2x_3^{-1}x_1x_2^2x_3^{-1}x_2(x_3^{-1}$$
$$(((x_3^{-1}x_2^{-1}x_3^{-2}x_2^{-1})^2x_3^{-2}x_2^{-1})^2(x_3^{-1}x_2^{-1}x_3^{-1})^2x_3^{-1}x_2^{-1})^3(x_3^{-1}x_2^{-1}x_3^{-1})^2$$
$$x_2^{-1}x_3^{-1}((x_3^{-1}x_2^{-1}x_3^{-2}x_2^{-1})^2x_3^{-2}x_2^{-1})^2(x_3^{-1}x_2^{-1}x_3^{-1})^2x_2^{-1})^3x_3^{-1}$$
$$((x_3^{-1}x_2^{-1}x_3^{-2}x_2^{-1})^2x_3^{-2}x_2^{-1})^2(x_3^{-1}x_2^{-1}x_3^{-1})^2x_2^{-1}x_3^{-1}x_2$$

and

$$c_2 := f^5(a) = (x_1x_2^2x_3^{-1}x_2^2x_3(x_3x_2)^2)^2x_3^2x_2(x_3x_2x_3)^2x_3x_2x_3^{-1}.$$

The ciphertext for Alice is the tuple (c_1, c_2).

3. Alice first computes

$$(f^7(c_2))^{-1} = x_2^{-1}(((((x_3x_2)^2x_3)^2x_3x_2x_3)^2x_3x_2(x_3x_2x_3)^2)^2x_3x_2$$
$$((x_3x_2x_3)^2x_2x_3)^2x_3x_2x_3)^2(x_3x_2(((x_3x_2x_3)^2x_2x_3)^2x_3x_2x_3x_2x_3)^2$$
$$(x_3x_2x_3^2x_2)^2x_3)^2x_2(((((x_3^2x_2)^2x_3x_2)^2x_3^2x_2x_3x_2)^2x_3$$
$$(x_3x_2x_3^2x_2)^2x_3x_2)^2x_3(x_3x_2x_3^2x_2)^2x_3^2x_2$$
$$(((x_3x_2x_3)^2x_2x_3)^2x_3x_2x_3x_2x_3)^2(x_3x_2x_3^2x_2)^2x_3$$
$$(x_3x_2^{-1})^2x_2^{-1}x_1^{-1})^2$$

and gets m by

$$m = c_1 \cdot (f^7(c_2))^{-1} = x_3^{-2}x_2^2x_3x_1^2x_2^{-1}x_1^{-1}.$$

6 Conclusion

In comparison to the standard cryptosystems which are mostly based on number theory we explained two cryptosystems which use combinatorial group theory. The first cryptosystem in Section 4 is a one-time pad, which choice of the random sequence for encryption is not number-theoretic. At the moment it is costlier than the standard systems but it is another option for a one-time pad which is based on combinatorial group theory and not on number theory. The second cryptosystem in Section 5 is similar to the ElGamal cryptosystem (see [6]), which is easier to handle. The ElGamal cryptosystem is based on the discrete logarithm problem over a finite field. If this problem should eventually be solved we introduced here an alternative system, which is not based on number theory.

References

1. G. Baumslag, B. Fine, M. Kreuzer, and G. Rosenberger. *A Course in Mathematical Cryptography*. De Gruyter, 2015.
2. A. V. Borovik, A. G. Myasnikov, and V. Shpilrain. Measuring sets in infinite groups. *Contemporary Math. Amer. Math. Soc. 298*, pages 21–42, 2002.
3. T. Camps, V. große Rebel, and G. Rosenberger. *Einführung in die kombinatorische und die geometrische Gruppentheorie*. Berliner Studienreihe zur Mathematik Band 19. Heldermann Verlag, 2008.
4. V. Diekert, M. Kufleitner, and G. Rosenberger. *Diskrete Algebraische Methoden*. De Gruyter, 2013.
5. B. Eick, M. Kirschmer, and C. Leedham-Green. The constructive membership problem for discrete free subgroups of rank 2 of $SL_2(\mathbb{R})$. *LMS Journal of Computation and Mathematics*, 17 (1): pages 345–359, 2014.
6. T. ElGamal. A public key cryptosystem and a signature scheme based on discrete logarithms. *IEEE Transactions on Information Theory*, IT-31: pages 469–473, 1985.
7. M. I. González Vasco and R. Steinwandt. *Group Theoretical Cryptography*. CRC Press, 2015.
8. J. Lehner. *Discontinuous Groups and Automorphic Functions*. Mathematical Surveys Number VIII. American Mathematical Society, Providence, Rhode Island, 1964.
9. R. C. Lyndon and P. E. Schupp. *Combinatorial Group Theory*. Ergebnisse der Mathematik und ihre Grenzgebiete 89. Springer-Verlag, 1977.
10. W. Magnus, A. Karrass, and D. Solitar. *Combinatorial Group Theory*. Pure and Applied Mathematics, A Series of Texts and Monographs Volume XIII. John Wiley & Sons, 1966.
11. Y. Matiyasevich. Solution of the Tenth Problem of Hilbert. *Mat. Lapok*, 21, pages 83–87, 1970.
12. A. I. S. Moldenhauer. *Cryptographic protocols based on inner product spaces and group theory with a special focus on the use of Nielsen transformations*. PhD thesis, University of Hamburg, 2016.
13. A. Myasnikov, V. Shpilrain, and A. Ushakov. *Group-based Cryptography*. Advanced Courses in Mathematics - CRM Barcelona. Birkhäuser Basel, 2008.
14. A. G. Myasnikov and V. Shpilrain. Automorphic orbits in free groups. *J. Algebra*, 269: pages 18–27, 2003.
15. J. J. Rotman. *An Introduction to the Theory of Groups*. Springer, 1995.
16. I. A. Stewart. Obtaining Nielsen Reduced Sets in Free Groups. *Technical Report Series*, 293, 1989.

Evaluation and Capability of Models

Bernhard Thalheim[1], Marina Tropmann-Frick[2]

Abstract. Models are one of the main instruments in Computer Science. The notion of model is however not commonly agreed due to the wide usage of models. It is challenging to find an acceptable and sufficiently general notion of model due to the large variety of known notations. Such notion should incorporate all of the different notations and at the same time should allow to derive the specific notation from the general notion. We introduce a universal parameterised notion of the model. The parameters in this notion support adaptation of the universal notion of the model to the specific notation of interest. We finally apply this notion and this adaptation to development of business process models that are specified in BPMN.

1 The Model - an Artifact and an Instrument

Classical Computer Science research considers models as *artifacts*[3] that are constructed in certain way and prepared for their utilisation according to the purpose under consideration such as construction of systems, verification, optimization, explanation, and documentation.

Creation for a practical purpose means that the main target of model development is its application in utilisation scenarios. Models are considered to be artifacts in a stronger sense. We observe however that models are developed for their utilisation within some scenario. They are functioning in this scenario. That means models are instruments in these scenarios. The notion of an instrument[4] concentrates on this utilisation of models. Models are therefore mainly *instruments* that are effectively functioning within a scenario. The effectiveness is based on an associated set of methods and satisfies requirements of usage of the model.

---

[1] Department of Computer Science, Christian-Albrechts-University Kiel, 24098 Kiel, Germany
[2] Department of Computer Science, Christian-Albrechts-University Kiel, 24098 Kiel, Germany
[3] An artifact is "something that is created by humans usually for a practical purpose" or "something characteristic of or resulting from a particular human institution, period, trend, or individual" or "a product of artificial character due usually to extraneous (as human) agency" [16]. The last meaning of the notion of an artifact is not taken into consideration for models in most sciences and also in Computer Science.
[4] An instrument is among others (1) a means whereby something is achieved, performed, or furthered; (2) one used by another as a means or aid or tool [16].

1.1 Models - The Third Dimension of Science

Models are used as perception models, experimentation models, formal models, conceptual models, mathematical models, computational models, physical models, visualisation models, representation models, diagrammatic models, exploration models, heuristic models, etc. Experimental and observational data are assembled and incorporated into models and are used for further improvement and adaptation of those models. Models are used for theory formation, concept formation, and conceptual analysis. Models are used for a variety of purposes such as perception support for understanding the application domain, for shaping causal relations, for prognosis of future situations and of evolution, for planning, for retrospection of previous situations, for explanation and demonstration, for preparation of management, for optimisation, for construction, for hypothesis verification, and for control of certain environments.

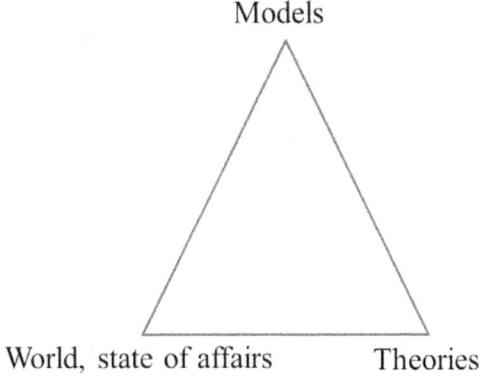

Fig. 1. Models - The third dimension of science

Models are one of the main instruments in scientific research. They are considered to be the *third dimension of science* [26][5] (Figure 1). They provide a tool for research and have an added value in research, e.g. for construction of systems, for education, for the research process itself. Their added value is implicit but can be estimated based on the capability, potential and capacity

---
[5] The title of the book [4] has inspired this observation.

of the model. Models are common culture and practice in sciences. However, each discipline has developed its own modelling expertise and practice.

Models are often language based. Their syntax uses the namespace and the lexicography from the application domain. Semantics is often implicit. The lexicology can be inherited from the application domain and from the discipline. Models do not need the full freedom for interpretation. The interpretation is governed by the purpose of the model within the research scenario, is based on disciplinary concerns (postulates, paradigms, foundations, commonsense, culture, authorities, etc.) and is restricted by disciplinary practices (concepts, conceptions, conventions, thought style and community [6], good practices, methodology, guidelines, etc.). Models combine at least two different kinds of meaning in the namespace: referential meaning establishes an interdependence between elements and the origin ('what'); functional meaning is based on the function of an element in the model ('how'). The pragmatics of a model depends on the community of practice, on the context of the research task and especially on the purpose or function of the model.

A model can be used for different purposes and various usage scenarios. Therefore, a model is typically also extended by views or viewpoints that reflect certain parts of the model and that hide details which are not necessary. This reflection is often only provided in a non-systematic or implicit way. Additionally, we need a refinement notion, methods for combination and for evaluation of models.

1.2 Scenarios of Model Utilisation

Models are used as an instrument in some utilisation scenario. At the same time, the model might be useless and not productive in other scenarios. Their function in these scenarios is a combination of functions such as explanation, optimization-variation, validation-verification-testing, reflection-optimization, exploration, hypothetical investigation, documentation-visualization, and description-prescription as a mediator between a reality and an abstract reality that developers of a system intend to build.

Traditionally, purposes or goals are considered first. The purposes and the goals are used to determine the functions of a model. This approach is centered around the purpose or goal and requires a definite understanding of the purpose and goal. Purposes and goals are often underspecified or blurry at the beginning. They become more clear after the model is being used. Compared to this approach, it is simpler to understand the application cases of a model and thus the utilisation scenarios. In this case we may derive the functions that a model has in these scenarios. Therefore, we use the approach that the functions of the model determine the *purposes* of the deployment of the model.

1.3 The Storyline of the Paper

The large variety of model notations (see, for instance, [13, 23, 30]) does not allow to transfer experience gained with one notation to other notations. Methods for utilisation or development are therefore mainly bound to one notation. Each subdiscipline has therefore its own understanding of modelling. It would however be beneficial to have a general notion of model that can be adapted to the specific notations of interest.

We introduce in Section 2 a universal notion of a model. This notion is based on the understanding of a model as an instrument in some utilisation scenarios. We only consider well-formed instruments since models must be intuitive and easy to understand. The model definition is based on two general parameter sets, adequateness and dependability. Each of the parameters can be instantiated in dependence of the function that the model should have in a given utilisation scenario within the sub-discipline. This instantiation facility is based on a conception frame for the model notion.

The approach is applied to BPMN modelling in Section 3. We describe the business process modelling approach and derive the capability of this modelling technique. We can now also explicitly describe the obstacles of BPMN modelling. Furthermore, we derive the evaluation procedure for the BPMN approach in Section 4.

This approach to modelling in Computer Science can now be used as a starting point of a theory of modelling (Section 5). We start with some, often implicitly given restrictions that a model has, esp. its burden by the background and by the directives. The evaluation of models also supports a statement on not-supported utilisations, called anti-profile. Finally, the conception frame can also be used for development of question forms that support model specification.

2 The Universal Notion of the Model

There are many notions of models. Each of them covers some aspects and concentrates on some properties such as the mapping, analogy, truncation, pragmatism, amplification, distortion, idealisation, carrier, added value, and purpose properties [11, 17, 18, 21]. The main property is however the *function property: The model suffices in its function in the utilisation scenarios that are requested.* This property results in the following notion of the model [25, 27, 29].

2.1 The Model Notion

Models have several *essential properties* that qualify an instrument as a model [22, 24]:

Definition 1. *An instrument is* well-formed *if it satisfies a well-formedness criterion.*

Definition 2. *A well-formed instrument is* adequate *for a collection of origins if (i) it is analogous to the origins to be represented according to some analogy criterion, (ii) it is more focused (e.g. simpler, truncated, more abstract or reduced) than the origins being modelled, and (iii) it is sufficient to satisfy its purpose.*

Definition 3. *Well-formedness enables an instrument to be* justified: *(i) by an empirical corroboration according to its objectives, supported by some argument calculus, (ii) by rational coherence and conformity explicitly stated through formulas, (iii) by falsifiability that can be given by an abductive or inductive logic, and (iv) by stability and plasticity explicitly given through formulas.*

Definition 4. *An instrument is* sufficient *by a* quality *characterisation for internal quality, external quality and quality in use or through quality characteristics [20] such as correctness, generality, usefulness, comprehensibility, parsimony, robustness, novelty etc. Sufficiency is typically combined with some assurance evaluation (tolerance, modality, confidence, and restrictions).*

Definition 5. *A well-formed instrument is called* dependable *if it is sufficient and is justified for some of the justification properties and some of the sufficiency characteristics.*

Definition 6. *An instrument is called* model *if it is* adequate *and* dependable. *The adequacy and dependability of an instrument is based on a* judgement *made by the community of practice.*

Definition 7. *An instrument has a* background *consisting of an undisputable grounding from one side (paradigms, postulates, restrictions, theories, culture, foundations, conventions, authorities) and of a disputable and adjustable basis from other side (assumptions, concepts, practices, language as carrier, thought community and thought style, methodology, pattern, routines, commonsense).*

Definition 8. *A model is used in a* context *such as discipline, a time, an infrastructure, and an application.*

The model notion can be depicted in Figure 2 based on the following conceptions:

a fundament* or *background with
- the grounding, and
- the (meta-)basis,

four governing directives given by
- the artifacts or better origins to be represented by the model,
- the deployment or profile of the model such as goal, purpose or functions,
- the community of practice (CoP) acting in different roles on certain rights through some obligations, and
- the context of time, discipline, application and scientific school,

two pillars which provide
- methods for development of the model, and
- methods for utilisation of the model,

and finally
the model utilisation scenario for the deployment of the model in the given application.

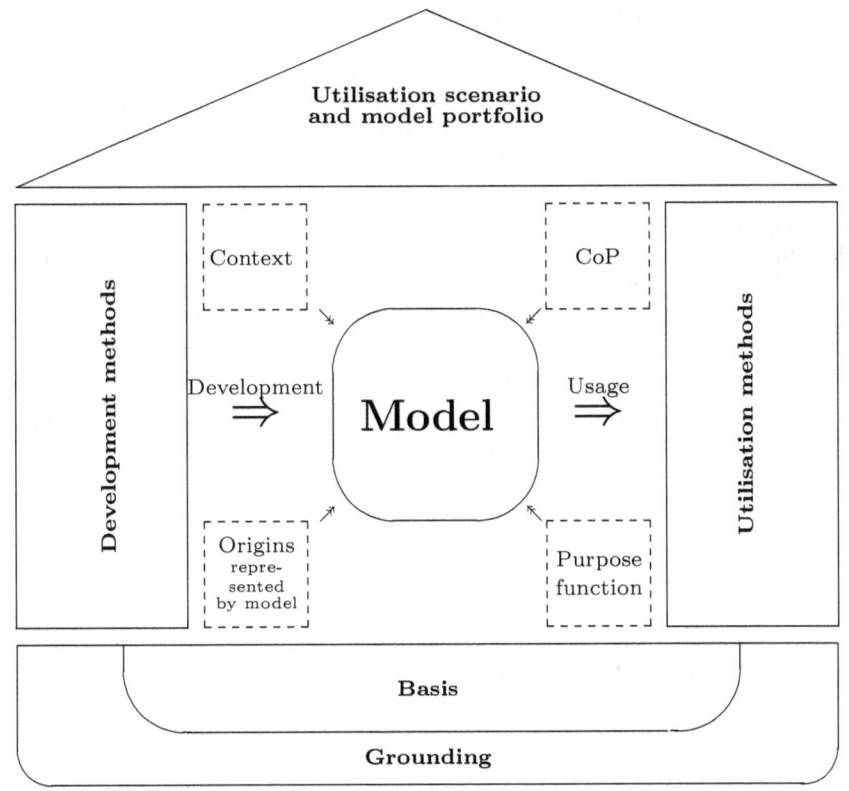

Fig. 2. Facets of the model notion

The *model house* in Figure 2 abstracted from its full version [24, 27] displays these different facets of the model. The house consists of a cellar (basis in Figure 2) and a fundament (grounding in Figure 2), two pillars (development resp. utilisation methods), four driving or governing forces (origins, purpose of function, community of practice, context), and finally the deployment roof (utilisation scenario). The *grounding* is typically implicitly assumed and not disputable. It contains paradigms, the culture in the given application area, the background, foundations and theories in the discipline, postulates, (juristic and other) restrictions, conventions, and the commonsense. The *basis* is the main part of the background and is typically disputable.

Definition 9. *A fully-specified model is* function-purpose-goal invariant *if the model can be used instead of the origins in the given scenario and have the same goal, the same purpose, and the same function. A model is* solution-faithful *if the solution of the problem solved with the model is analogous in the world of the origins based on the analogy criterion that is used for stating adequacy.*

2.2 The Conception Frame for the Model Notion

The model notion covers many different aspects. It might thus be of interest whether there is a guideline for development of models. Models are artifacts that can be specified within a W*H-frame [5] that extends the classical rhetorical frame introduced by Hermagoras of Temnos[6]. Models are primarily characterised by W^4: wherefore (purpose), whereof (origin), wherewith (carrier, e.g. language), and worthiness ((surplus) value). The secondary characterisation dimensions are given by: (1) stakeholder: by whom, to whom, whichever; (2) additional properties of the application domain: wherein, where, for what, wherefrom, whence, what; (3) solution: how, why, whereto, when, for which reason; and (4) context: whereat, whereabout, whither, when.

A practical guideline may just

1. start with fixation of two directives: origins to be represented and community of practice that accepts this model;
2. restrict the model utilisation scenario and the usage model to those that are really necessary and thus derive the purpose and function of the instrument;
3. define adequateness and dependability criteria of the instrument within the decision set made so far;
4. explicitly describe the background of the model, i.e. its undisputable grounding and the selected basis; and
5. explicitly specify the context for utilisation of the model.

The model development and utilisation depends in this case from:

Judgements of some members of the CoP to deploy the instrument as a model for some origin based on an assessment (deployability, rigidity, modality, confidence) within a CoP, utilisation scenario, and within a context.
Utilisation scenarios and use spectra accepted for the instrument with
functions of the instrument in utilisation scenario,
roles and deployment of the instrument in those scenario, and
resulting purposes and goals for the utilisation.
The instrument as such with some appreciation
as a well-shaped instrument on the basis
– of some criteria in dependence on intended utilisation and criteria for:
  • what is accepted in a CoP, and
  • what is syntactically, semantically, pragmatically well-shaped,
that fits to the intended use, and
is appropriate for the utilisation scenarios and the use spectra.

The orientation also reflect our understanding of a model as an instrument.

---

[6] Quis, quid, quando, ubi, cur, quem ad modum, quibus adminiculis (Who, what, when, where, why, in what way, by what means), The Zachman frame uses a simplification of this frame.

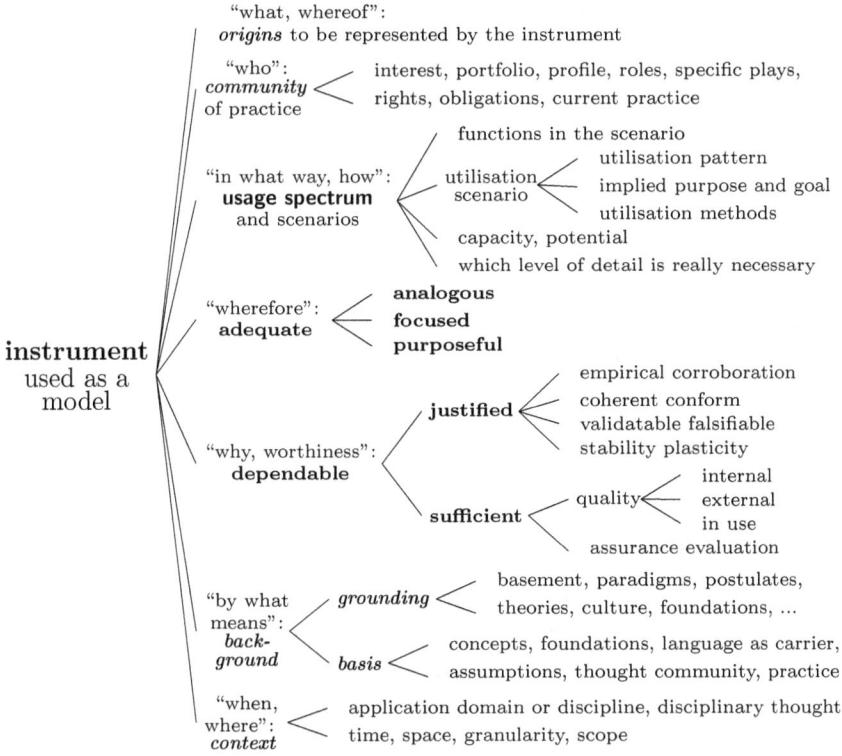

Fig. 3. Conception frame for systematic development of a model

3 BPMN Diagrams as Models

The Business Process Modeling and Notation (BPMN) language [8,14] is a conceptual business process specification language and is standardized by the Object Management Group (OMG). There are many different languages for description of business stories (e.g. SiteLang), of business rules (e.g. business use cases), and of workflows that are essentially specifications of business processes, activities of participants, utilisation with resources, and of communication among the participants. Languages such as S-BPM, BPMN, and EPC concentrate on different aspects of business processes, vary in scope and focus, use different abstraction levels, and are thus restricted in the capacity and potential for modelling. Most of the existing languages evolved over their lifespan and extensionally added features, more features, and other features again. BPMN is not an exception for this kind of overloading.

A business process consists of an ordered set of one or more activities (tasks) which collectively realize a business objective or policy goal. A workflow is the executable specification of a business process. It may describe all or some of the five aspects of business processes [15]:

(1) control flow description for the partial order of the activities, events or steps;
(2) organisation description with participants, theirs roles and plays within the processes, their rights and obligations, their resources, and their assignments;
(3) the data viewpoint description with an association to process elements and access rights for participants;
(4) the functional description that specifies semantics, pragmatics, and behaviour of each element of the workflow, e.g. the operations to be performed, pre- and postconditions, priority, triggers, and time frames for the operations;
(5) the operational assignment of programs that support all elements of the workflow.

The entire modelling process is based on a local-as-design perspective. A holistic or global view on a diagram collection is the task of a designer and becomes problematic in the case of specification evolution.

BPMN 2.0 defined four different kinds of diagrams for workflow specification. We shall briefly review these diagrams in the sequel. The diagram in Figure 4 combines these different aspects. It describes the accomplishment of requirements issued by a customer.

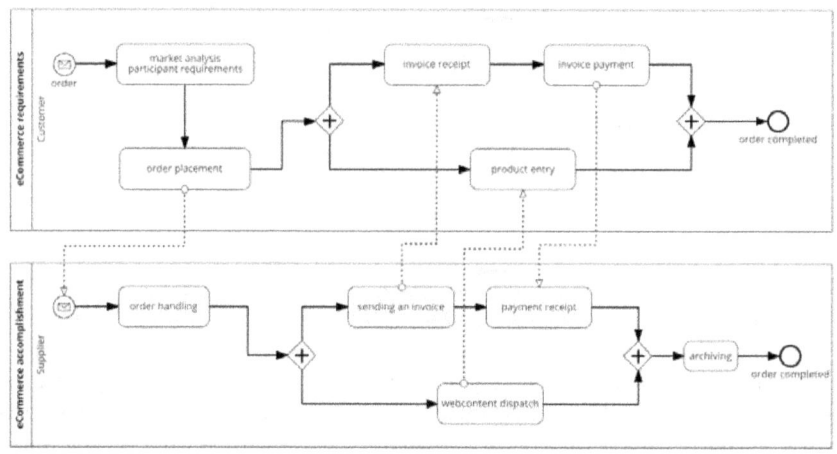

Fig. 4. Fulfillment of customer demands by vendors

3.1 Diagrams in BPMN

Process Diagrams. Process diagrams (also called orchestration diagrams) describe the stepwise task flow for one agent. A task flow might reflect different roles of an agent. These roles are separated by swimlanes. Processes are either public or private. Public processes can also be abstractions of private processes that represent the detailed control and task flow for a singleton agent. Main process elements are (a) atomic or complex activities for direct representation of stepwise actions of an agent, (b) gates for exclusive, non-exclusive, event-based or parallel splitting and joining of the control flow, (c) events for the start of a workflow, for the end of a workflow path, for the complete end of the entire workflow, and an intermediate event for representation of interaction events with agents outside the workflow, and (d) control flow arrows for representation of the order of process elements. Basic activities reflect abstract, service, send, receive, user, manual, business rule, or script tasks. Complex tasks reflect an entire sub-workflow, loops, parallel or sequential multiple executions, ad-hoc workflows, transactions, or specific exception handling workflows such as compensation. Interaction reflects message exchange, timer interaction, escalation enforcement, compensations, conditional interactions, links, and signals. Interaction can be sequential or parallel multiple. Interaction events may also be boundary events for a complex activity. All elements can be explicitly annotated by comments, by consumed data, or by produced data objects or data stores.

Choreography Diagrams. Choreography diagrams describe the message exchange among agents with reference to sending and receiving events, the message issue, and the graph-based representation of the partial order among these messages.

Collaboration Diagrams. Collaboration use choreography diagrams and process diagrams for explicit binding of senders and receivers of messages to black-box abstractions of agent workflows and abstract from message issues.

Conversation Diagrams. Conversation diagrams survey communication flow among agents as a birds view. They allow to derive dependences among process diagrams of agents.

3.2 Capability of BPMN Diagrams

BPMN modelling becomes nowadays a standard for typical business applications. Therefore, the capability of processes must be specified and well understood. It is thus necessary to know what is the ability to achieve a good model through a set of controllable and measurable features.

BPMN diagrams require a work-around for a number of conceptions such as macro-state, history, and system architecture. There are redundancies in the language itself that lead to flavour- or taste-oriented programming due to the overwhelming number of elements, construct excess and overload, e.g. groups, pool and lane, transformations, off-page connectors. The structuring becomes unclear since activities can be itself a workflow or a collection of workflows. This rather specific kind of abstraction should not be mixed with abstraction in general. Exception handling is completely confusing and only partially defined.

BPMN diagrams can represent only 8 out of 43 workflow resource pattern [10]. The data aspect is provided through properties of tasks, processes, and sub-processes. Their interrelationship is left to the developer community. It is the task of the developers to keep in mind the entire picture of the BPMN diagram collection.

BPMN uses an informal approach to semantics description what has been a matter of confusion. A formal approach to BPMN semantics can however be developed [1–3].

Furthermore, there is no conception of well-formed diagrams. Decomposition and composition is left to the developer. BPMN does not properly support the aspects (2), (4), and (5). The data aspect (3) is partially represented.

3.3 Deficiencies of Diagrams and Diagrammatical Reasoning

Diagrams are not universal for modelling. It is often claimed that diagrams are simple to use, are easy to interpret, have an intuitive semantics, are unique within a user community and have thus a unique pragmatics, and are thus powerful instruments. We observe however a number of obstacles that must be resolved before accepting a diagram as a model, e.g. the following ones:

- *Habituation versus unfamiliarity*: Diagram should be familiar to their users, have a unique semantics and pragmatics without any learning effort. Readers of diagrams must be literal with them.
- *Ambiguity of interpretation versus well-formedness*: Diagrams should not confuse by multiple interpretations (e.g. arrows), by instability and by context-dependence of form-content relations.
- *Incremental graphical construction*: Diagrams should follow the same construction pattern as the origin and should concentrate on typicality.
- *Naturalness of local reasoning*: Local-as-design approaches presuppose locality within the world of origins.
- *Unfamiliarity with non-linear behaviour*: Users are mainly linearly reasoning. Non-linear reasoning should be supported in a specific form.
- *Additional and supplementary elements without meaning*: Diagrams of use elements which do not have a unique or any meaning, e.g. colours, shapes, grid forms for lines etc.
- *Hidden dimensions within the diagram*: Diagrams cannot reflect all aspects although there are essential ones, e.g. time.
- *Representation as fine and visual art*: Finding a good representation is a difficult task and should be supported by a culture of modelling.

All these obstacles are observed in the case of BPMN diagrams [10].

Diagrams must be developed on the principles of visual communication, of visual cognition and of visual design [12]. The culture of diagramming is based on a clear and well-defined design, on visual features, on ordering, effect, and delivery, and on familiarity within a user community.

One of the main obstacles of diagrams is the missing abstraction. The simplest way to overcome it is the development of a model suite [19] consisting of a generic model and its refinement models where each of them is adequate and dependable. Generic models [31] reflect the best abstraction of all models within a model suite.

4 Evaluation of the BPMN Approach

BPMN is a powerful diagrammatic languages that uses more than 100 modelling elements. The same situation in the reality or the implemented system can be specified by a variety of diagrams. Since a theory of diagram equivalence is missing, [27] introduced seven evaluation methods for models:

- PURE – SMART – CLEAR evaluation for the goal-purpose-function evaluation of an instrument in a given application context, for given artifacts to be represented, for a given community of practice, and for a given profile (goal, purpose, and function) under consideration of the utilisation scenarios;
- PEST evaluation for assessment of internal, external, and quality of use;
- QUARZSAND evaluation for assessment of the model development, and
- SWOT – SCOPE evaluation for description of the potential of the model, i.e. the general properties of a given instrument or the modelling method.

Since we did not explore the directives in detail nor the adequateness and dependability of an instrument that is a candidate model, we concentrate on the last two methods in this section. The evaluation of adequacy and dependability has been developed in [28]. We concentrate here on the capacity and potential of the BPMN approach.

4.1 The Capacity of the BPMN Approach

Capacity is a strategic measure whereas the potential is a tactical one. The potential can be used to derive the added value of a utilisation of a model within a given scenario. The potential allows to reason on the significance of a model within a given context, within a given community of practice, for a given set of origins, and within the intended profile.

The capacity relates an instrument to utilisation scenarios or the usage spectra. We answer the questions whether the instrument functions well and beneficial in those scenarios, whether it is well-developed for the given goals and purposes, whether it can be properly, more focused, comfortably, simpler and intelligible applied in those scenarios instead of the origins, and whether the instrument can be adapted to changes in the utilisation. The answers to these questions determine the main content or cargo, the comprehensiveness, and the authority or general value of a model. Another important aspect is the solution-faithfulness of the instrument. The capacity is an essential element of the model cargo, especially of the main content of the model.

BPMN diagrams can be used in description, prescription, explanation, documentation, communication, negotiation, inspiration, exploration, definition, prognosis, reporting and other scenarios. We discover that communication, negotiation, and inspiration are supportable. Description, prescription, and definition can be supported if the BPMN diagrams are enhanced and a precise semantics of all BPMN elements is commonly used in all four kinds of diagrams. The adequacy and especially the analogy to the origins (i.e. storyboards or business processes) is assumed to be based on homomorphy what is rarely achieved. This homomorphy is suitable if all processes are completely and in detail specified and all variations and exceptions are consistent.

The general utility of BPMN diagrams becomes rather low if the specific background of the modelling approach is not taken into consideration. BPMN diagrams are process-oriented, based on an orthogonal separation of flow element into activities, gates, and events, differentiate actors within their roles, and support communication among actors based on message exchange. The execution semantics is based on a token interpretation of control flow. Actors are isolated in their execution if binding is not done through message exchange or implicit hidden resource conditions. Data and resource are however local. All processes are potentially executed in parallel. The local-as-design approach might be appropriate if business processes are not intertwined. The concentration on the same abstraction level restricts the applicability of BPMN modelling. Generic workflows [31] provide a solution to this limitation.

4.2 The Potential of the BPMN Approach

The potential describes the (in-)appropriateness of a modelling approach within the directives. The suitability of BPMN diagrams depends on whether the application and the context support the local-as-design approach, on whether the demands of the community of practice can be satisfied, on whether the instrument is adequate (analogous, focused, purposeful), on whether the goals can be achieved with the given instrument, on the fruitfulness of the instrument compared with other instruments, and on the threats and obstacles of utilisation of BPMN diagrams.

4.3 The SWOT Evaluation of the Potential

The SWOT analysis is a high-level method that allows to evaluate the general quality of an instrument and its general assumptions of deployment.

Strengths. The BPMN approach is standardised and uses a large variety of constructs. It thus allows development of detailed models. It has a high expressibility. Both intra- and inter-organisational aspects can be represented. The approach is well supported by tools.

Weaknesses. The large variety of competing elements is also a weakness. The complexity and integration of diagrams may cause solution-unfaithfulness. The language requires high learning efforts. Processes that are dynamic at runtime cannot be modelled. Exchange among tools is an open problem.

Opportunities. Most business processes can be adequately described due to the variety of elements. The standardisation provides at least a base semantics.

Threats and Risks. None-technical users might be unable to cope with diagrams. Work-arounds hinder comprehensibility. Vendors define their own extensions. The BPMN standard does not completely define the execution.

4.4 The SCOPE Evaluation of the Potential

The SCOPE analysis of a model embeds the model into the application context, refines the capacity evaluation of an instrument, and considers the community of practice and their specific needs.

Situation. BPMN diagram suites provide some kind of formalisation of business processes. Communication is specified to a certain degree. Control flow is well-represented.

Competence. BPMN diagrams must be combined with other models since the other four aspects (organisation, data flow, functions, operational assignment) are only partially reflected.

Obstacles. Typical challenges of BPMN modelling are the specification complexity, diagram coherence, exception handling, and the development of an execution semantics. There is no common agreement on well-formedness of diagrams.

Prospects. A separate BPMN diagram is easy to read and to interpret.

Expectations. The BPMN approach can be combined with local-as-design-oriented conceptual data models, storyboards, business rule specification and other modelling approaches as one kind of models within a model suite.

4.5 The Resulting Potential of the BPMN Approach

The BPMN diagram has a high potential for communication and negotiation utilisation scenario. The potential for system construction within a description-prescription scenario is however rather limited due to missing co-design support. A similar inappropriateness can be stated for explanation, prognosis, exploration, definition, and reporting scenario. The potential within a documentation scenario is rather small. The highest potential of the BPMN approach can be however observed for inspiration scenario. The process, choreography, conversation, and collaboration diagrams are an appropriate means for an implementation plan based on inspiring diagrams.

Similar to SPICE assessments [7], we may rate maturation of a model and a modelling approach to: (0) ad-hoc , (1) informal, (2) systematic and managed, (3) standardised and well-understood, and (4) optimising and adaptable, and (5) continuously improvable styles. The evaluation shows that the BPMN approach has not yet reached level (2). This observation leads us to the conclusion that PURE-SMART-CLEAR and PEST evaluations are heavily dependent from the directives for BPMN diagram modelling.

A model must be of high utility, must have a high added value, and should have a high potential. These parameters also depend on the well-formedness of the instrument. The BPMN approach can be enhanced by criteria for well-formedness for syntactical, semantical and pragmatical well-shaped diagrams [28].

5 Towards a Theory of Modelling

5.1 Models Burdened with Directives and Background

The directives and the background (see Figure 2) heavily influence the way how a model is constructed, what is taken into consideration and what not, which rigidity is applied, which basis and grounding is taken for granted, and which community of practice accepts this kind of model.

The model incorporates these influences without marking them in an explicit form. The model is laden or *burdened* by these decisions. Additionally, models are composed of elements that are selected, changed and adapted within a development process. Figure 5 depicts elements of this burden and this development history.

5.2 The Anti-Profile of an Instrument as a Model

We may now directly conclude that an instrument might or might not be adequate and dependable for any utilisation scenario due to its insufficiency to function in this scenario.

Definition 10. *A* utilisation pattern *of an instrument describes the form of usage of an instrument, the discipline of usage, the applications in which the instrument might be used, and the conditions for its utilisation.*

Definition 11. *A* utilisation scenario *consists of a utilisation pattern and a number of functions a specific instrument might play in this utilisation pattern.*

Definition 12. *A* usage spectrum *consists of collection of utilisation scenarios. A* portfolio *of an instrument combines the usage spectra.*

Definition 13. *A* profile *of an instrument as a model consists of the goals, the purposes, and the functions of the instrument within a portfolio.*

We can now roughly describe an anti-profile of a model and resulting *utilisation proscriptions* of a an instrument as a model by answering the following questions:

- For which scenarios is the instrument useless?
- In which of the following scenarios efficiency and effectiveness is not given for the instrument: description & prescription, realisation & coding, theory development, theory refinement, causality consideration, inexplicability, demonstration, prediction, explanation, mastering of complexity, understanding, or ... ?
- Are essential parameters of the origins missing? Are some of the essential parameters only represented via mediating or dependable parameters? Are there dummy or pseudo dependences among the parameters?
- What cannot be adequately represented? Is the dependability really sufficient? In which case users need a special understanding and education? Which tacit knowledge is hidden in the instrument?

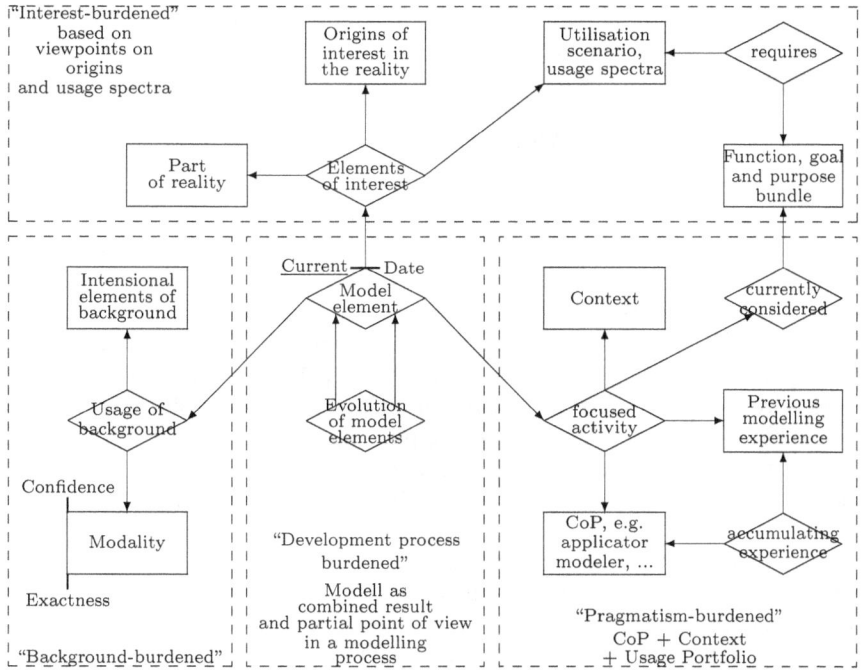

Fig. 5. Models are burdened by their development history, the background and the directives

- In which cases the instrument cannot be effectively used? What must a user respect and obviate before using the instrument?
- Which biases and which background are palmed off? Which assumptions, postulates, paradigms, and schools of thought are hidden and not made explicit? Models might condition conclusions.

Since models are instruments their utilisation conditions conclusions and results. Therefore, it is appropriate to describe the anti-profile of a model as well.

5.3 Questions to Answer Before Using an Instrument as a Model

The rhetoric frame and its extension to the W*H frame [5] can now be used for derivation of questions one must answer before using the model:
- What is the function of the model in which scenario? What are consequential purposes and goals? What are anti-goals and anti-purposes?
- Which origins are going to be represented? Which are not considered? Does the model contain all typical, relevant and important features of the origins under consideration and only those?
- Is the instrument adequate and dependable within the utilisation scenario? What are the parameters for adequacy and dependability? How purpose-invariance and solution-faithfulness is going to be defined?

– What kind of reasoning is supported? What not? What are the limitations? Which pitfalls should be avoided?
 – Do you want to have a universal model that contains all and anything? Would it be better to use a model suite where each of the models represent some specific aspects? What about the nonessential aspects?

6 Conclusion

A general understanding of the notion of a model has been already started with development of Computer Science. Milestones are the papers and books by H. Stachowiak (1980ies and 1990ies), B. Mahr (2000ies until 2015), W. Steinmüller (1993), and R. Kaschek (since 1990ies) [9, 11, 17, 18, 21]. These notions treat models from a phenomenological point of view through properties that a model should have (e.g. as main properties: mapping or analogy, truncation, pragmatic properties). We need however also an explicit definition of the notion of model. Such general notion has been developed in a series of papers, e.g. [22, 24, 25, 27, 29].

The model notion is universal one and based on two parameter sets for adequateness and dependability. The parameter sets seem to be complex and need a methodological support. This paper develops such a support facility based on the notion of a conception frame. The practicality of the approach is demonstrated for the workflow specification language BPMN. BPMN shares the positive treatment with most other formal or informal languages in Computer Science. The capacity and thus the restrictions or obstacles are not explicitly communicated. We see however that the evaluation, capacity, potential, and capability can be explicitly provided based on our approach.

Since the model notion is a mathematical definition, it seems to be achievable to develop a theory of modelling in the sense of a theory. In this paper, we only discuss two components of such a theory: the explicit description of the background of models and the anti-profile. The conception frame for the model definition may also be used for derivation of question forms that a modeller can use before delivering an instrument as a model to a community of practice. The development of a full theory is however a research issue for the next decades.

References

1. E. Börger, O. Sörensen, and B. Thalheim. On defining the behavior of or-joins in business process models. *Journal of Universal Computer Science*, 15(1):3–32, 2009.
2. E. Börger and B. Thalheim. A method for verifiable and validatable business process modeling. In *Software Engineering*, LNCS 5316, pages 59 – 115. Springer, 2008.
3. E. Börger and B. Thalheim. Modeling workflows, interaction patterns, web services and business processes: The ASM-based approach. In *ABZ*, volume 5238 of *Lecture Notes in Computer Science*, pages 24–38. Springer, 2008.
4. S. Chadarevian and N. Hopwood, editors. *Models - The third dimension of science*. Stanford University Press, Stanford, California, 2004.

5. A. Dahanayake and B. Thalheim. *Correct Software in Web Applications and Web Services*, chapter W*H: The conceptual Model for services, pages 145–176. Texts & Monographs in Symbolic Computation. Springer, Wien, 2015.
6. L. Fleck. *Denkstile und Tatsachen, edited by S. Werner and C. Zittel*. Surkamp, 2011.
7. ISO/IEC. Information technology - process assesment. parts 1-5. IS 15504, 2003-2006.
8. ISO/IEC. International organization for standardization: Information technology ? object management group: Business process model and notation. http://www.omg.org/spec/BPMN/ISO/19510/PDF/, 2013.
9. R. Kaschek. *Konzeptionelle Modellierung*. PhD thesis, University Klagenfurt, 2003. Habilitationsschrift.
10. F. Kossak, C. Illibauer, V. Geist, J. Kubovy, C. Natschläger, T. Ziebermayr, T. Kopetzky, B. Freudenthaler, and K.-D. Schewe. *A Rigorous Semantics for BPMN 2.0 Process Diagrams*. Springer, 2014.
11. B. Mahr. Information science and the logic of models. *Software and System Modeling*, 8(3):365–383, 2009.
12. T. Moritz. *Visuelle Gestaltungsraster interaktiver Informationssysteme als integrativer Bestandteil des immersiven Bildraumes*. PhD thesis, HFF Berlin-Babelsberg, 2006.
13. I. Nissen and B. Thalheim. *Wissenschaft und Kunst der Modellierung: Modelle, Modellieren, Modellierung*, chapter Bedeutung, Entwicklung und Einsatz, pages 3–28. De Gryuter, Boston, 2015.
14. OMG. Object management group: Business process model and notation (BPMN) 2.0. http://www.omg.org/spec/BPMN/2.0, 2011.
15. H. Pichler and J. Eder. Business process modeling and workflow design. In *The Handbook of Conceptual Modeling: Its Usage and Its Challenges*, chapter 8, pages 259–286. Springer, Berlin, 2011.
16. J.E. Safra, I. Yeshua, and et. al. *Encyclopædia Britannica*. Merriam-Webster, 2003.
17. H. Stachowiak. Modell. In Helmut Seiffert and Gerard Radnitzky, editors, *Handlexikon zur Wissenschaftstheorie*, pages 219–222. Deutscher Taschenbuch Verlag GmbH & Co. KG, München, 1992.
18. W. Steinmüller. *Informationstechnologie und Gesellschaft: Einführung in die Angewandte Informatik*. Wissenschaftliche Buchgesellschaft, Darmstadt, 1993.
19. B. Thalheim. *The Conceptual Framework to Multi-Layered Database Modelling based on Model Suites*, volume 206 of *Frontiers in Artificial Intelligence and Applications*, pages 116–134. IOS Press, 2010.
20. B. Thalheim. Towards a theory of conceptual modelling. *Journal of Universal Computer Science*, 16(20):3102–3137, 2010. http://www.jucs.org/jucs_16_20/towards_a_theory_of.
21. B. Thalheim. The theory of conceptual models, the theory of conceptual modelling and foundations of conceptual modelling. In *The Handbook of Conceptual Modeling: Its Usage and Its Challenges*, chapter 17, pages 547–580. Springer, Berlin, 2011.
22. B. Thalheim. The science and art of conceptual modelling. In A. Hameurlain et al., editor, *TLDKS VI*, LNCS 7600, pages 76–105. Springer, Heidelberg, 2012.
23. B. Thalheim. The conception of the model. In *BIS*, volume 157 of *Lecture Notes in Business Information Processing*, pages 113–124. Springer, 2013.
24. B. Thalheim. The conceptual model ≡ an adequate and dependable artifact enhanced by concepts. In *Information Modelling and Knowledge Bases*, volume XXV of *Frontiers in Artificial Intelligence and Applications, 260*, pages 241–254. IOS Press, 2014.

25. B. Thalheim and A. Dahanayake. A conceptual model for services. In *Invited Keynote, CMS 2015, ER 2015 workshop*, LNCS 9382, pages 51–61, Berlin, 2015. Springer.
26. B. Thalheim and I. Nissen, editors. *Wissenschaft und Kunst der Modellierung*. De Gruyter, Ontos Verlag, Berlin, 2015.
27. B. Thalheim and I. Nissen. *Wissenschaft und Kunst der Modellierung: Modelle, Modellieren, Modellierung*, chapter Ein neuer Modellbegriff, pages 491–548. De Gryuter, Boston, 2015.
28. B. Thalheim and I. Nissen. *Wissenschaft und Kunst der Modellierung: Modelle, Modellieren, Modellierung*, chapter Fallstudien zum Modellbegriff, pages 549–602. De Gryuter, Boston, 2015.
29. B. Thalheim and M. Tropmann-Frick. The conception of the conceptual database model. In *ER 2015*, LNCS 9381, pages 603–611, Berlin, 2015. Springer.
30. M. Thomas. Modelle in der Fachsprache der Informatik. Untersuchung von Vorlesungsskripten aus der Kerninformatik. In *DDI*, volume 22 of *LNI*, pages 99–108. GI, 2002.
31. M. Tropmann-Frick, B. Thalheim, D. Leber, G. Czech, and C. Liehr. Generic workflows - a utility to govern disastrous situations. In *Information Modelling and Knowledge Bases*, volume XXVI of *Frontiers in Artificial Intelligence and Applications, 272*, pages 417–428. IOS Press, 2014.

Part II

Rationality and Non-monotonic Reasoning

On a Minimal Logic of Default Conditionals

James Delgrande[1], Bryan Renne[2]

Abstract. We study a minimal conditional logic for defaults based on the idea that $\varphi \Rightarrow \psi$ is true in a theory if φ makes ψ no less probable than $\neg \psi$. The model theory (at least in the finite case, which we deal with here) is straightforward: A model is composed of a set of possible worlds, where each possible world is assigned a probability, and a valuation function determines which atomic sentences are true at each world. Of interest is a set of sound principles, and other principles that may be derived from this set. This set of principles is not complete; we discuss the issue of completeness by considering the approach with respect to approaches to qualitative probability. As well, we compare the approach to other conditional approaches, noting that the present framework is (clearly) weaker than the system of preferential entailment.

1 Introduction

In Knowledge Representation and Reasoning, a major focus concerns dealing with *conditionals*, broadly taken. This can be for the purposes of default reasoning, including qualitative and quantitative approaches to uncertainty, belief revision, and others. In this area, Gabriele has been a leading researcher with numerous results covering a wide spectrum of areas and applications. Among other works, [14] provides a summation of work in the 1990s, while more recent work has addressed general semantics (probabilistic and otherwise) of conditionals [2], probabilistic inference [21], first-order issues [16], and qualitative approaches based on ranking functions [15], among other areas. This paper then is intended as a small reflection on such approaches, asking what may consititute a minimal approach to a logic of conditionals.

The general idea in conditional reasoning is that a conditional is an object that one can reason about, either as a connective in some language, or as a nonmonotonic consequence operator. Thus for example "an adult is (typically) employed" might be represented as $a \Rightarrow e$, where $\Rightarrow$ represents some default or normality conditional, often given as a binary modal operator in some object language. Analogously, one might represent such an operator as a nonmonotonic consequence operator $a \mathrel{|\!\sim} e$.[3] The idea in either case is to provide a logic for reasoning about such conditionals and so, in such approaches, one can typically derive other defaults from a given set of defaults. There has been widespread agreement concerning just what principles should constitute a base or minimal logic. Further, it has been suggested that this minimal logic forms

---

[1] Simon Fraser University, Canada
[2] Vancouver, Canada
[3] The formal underpinnings of such a conditional logic or nonmonotonic consequence operator are similar enough that we will treat them as notational variants here.

a "conservative core" of defaults that should be present in *any* approach to nonmonotonic reasoning.

The resulting logic has appeared in a number of guises, but is perhaps best known as System P of *preferential reasoning* [17]; see also [3, 12]. The fundamental intuition of such logics is that a conditional $\varphi \Rightarrow \psi$ has the informal reading that ψ is true in the most normal state of events in which φ is true. Slightly more formally, this means that, assuming that one has some way of ordering worlds according to how "normal" they are, $\varphi \Rightarrow \psi$ is true just if ψ is true at every least (which is to say "most normal") φ world in the ordering (and if φ is necessarily false, then $\varphi \Rightarrow \psi$ is vacuously true). While this approach is appealing, and seems to capture nicely the idea of a weak conditional in some naïve scientific theory [7], it nonetheless is a very strong statement to say that *any* approach to weak conditionals should satisfy the principles of this logic.

Indeed it is not difficult to come up with an interpretation of a default conditional that is substantially weaker than that of System P and related approaches. We propose here a probabilistic interpretation of defaults in which $\varphi \Rightarrow \psi$ has interpretation that φ makes ψ no less probable than $\neg \psi$. We then suggest that the "minimal" condition for accepting a default conditional $\varphi \to \psi$ is that, on the evidence φ, ψ is a likely outcome; that is to say, that ψ is more likely than $\neg \psi$ given φ. If φ has non-zero probability then the conditinoal $\varphi \to \psi$ can be defined in terms of our earlier conditional as $(\varphi \Rightarrow \psi) \wedge \neg(\varphi \Rightarrow \neg \psi)$.

In this paper we explore basic principles underlying such a minimal logic of defaults, which we call CL. The main semantic intuition is that $\varphi \Rightarrow \psi$ is true just if $P(\varphi \wedge \psi) \geq P(\varphi \wedge \neg \psi)$, where $P(\varphi)$ is the probability of φ. A formal semantics is given, and an axiomatisation is provided; as well, various derived principles are obtained. While the axiomatisation is sound, it is not complete. While completeness is not difficult to obtain, we suggest that a *satisfactory* account may not be so straightforward. Nonetheless, it is clear that the logic CL may be trivially embedded in the logic described in [8], which does provide a full account of qualitative probability.

2   Background

Over the years, much attention has been paid to conditional approaches to default reasoning. Such approaches address defeasible conditionals whose meaning is typically based on notions of preference among worlds or interpretations. Hence, the default that a bird normally flies can be represented propositionally as $b \Rightarrow f$ (or, again, as a nonmonotonic consequence operator, $b \mathrel{|\!\sim} f$). These approaches are typically expressed in a modal logic in which the connective $\Rightarrow$ is a binary modal operator, and where the intended meaning of $\varphi \Rightarrow \psi$ is approximately "in the least worlds (or most preferred worlds) in which φ is true, ψ is also true". Possible worlds (or, again, interpretations) are arranged in at least a partial preorder, reflecting this notion of normality or preferredness on the worlds.

There has been a remarkable convergence or agreement on what inferences ought to be common to all such systems, and in the literature a seeming diversity of conditional approaches (based in turn on disparate intuitions) essentially allow the same inferences. These include approaches based on intuitions from probability theory such as ϵ-entailment [19] (or 0-entailment or p-entailment [1]), from qualitative possibilistic logic [9], as well as modal-logic based approaches such as preferential entailment [17], *C4* [18], *CT4* [3], and *S* [4]. Consequently it has been suggested that the resulting set of inferential principles may be taken as specifying a *conservative core* [20] that arguably should be common to *all* default inference systems. One expression of this logic of conditionals is as follows. The logic includes classical propositional logic (that is, includes modus ponens and substitution of provable equivalents, as well as all truth functional tautologies as axioms) along with the following rules and axioms:

LLE   From $\vdash \varphi \equiv \psi$ infer $\vdash (\varphi \Rightarrow \gamma) \equiv (\psi \Rightarrow \gamma)$.
RW   From $\vdash \psi \supset \gamma$ infer $\vdash (\varphi \Rightarrow \psi) \supset (\varphi \Rightarrow \gamma)$

Ref $\vdash \varphi \Rightarrow \varphi$
And $\vdash ((\varphi \Rightarrow \psi) \wedge (\varphi \Rightarrow \gamma)) \supset (\varphi \Rightarrow \psi \wedge \gamma)$
Cut $\vdash ((\varphi \Rightarrow \psi) \wedge (\varphi \wedge \psi \Rightarrow \gamma)) \supset (\varphi \Rightarrow \gamma)$
CM $\vdash ((\varphi \Rightarrow \psi) \wedge (\varphi \Rightarrow \gamma)) \supset (\varphi \wedge \psi \Rightarrow \gamma)$
Or $\vdash ((\varphi \Rightarrow \gamma) \wedge (\psi \Rightarrow \gamma)) \supset (\varphi \vee \psi \Rightarrow \gamma)$

The semantics of these approaches is usually phrased in terms of a possible worlds model, in which possible worlds are ranked by a notion of relative normality or unexceptionalness. The underlying modal logic is generally taken to be *S4* [13] (also called *KT4* [6]), in which accessibility between worlds is given by a reflexive, transitive binary relation. A conditional $\varphi \Rightarrow \psi$ is true at a world w just when, for every world accessible from w, there is an accessible world in which $\varphi \wedge \psi$ is true and $\varphi \supset \psi$ is true at all worlds that are less or equally exceptional, or if there are no accessible φ worlds. Thus, "a bird flies", $b \Rightarrow f$, is true if, in the least b-worlds (if such exist), $b \supset f$ is true. Since a penguin is a bird (either $\Box(p \supset b)$ or $p \Rightarrow b$) but a penguin doesn't fly ($p \Rightarrow \neg f$), this means that the least exceptional penguin-worlds are more exceptional than the least bird-worlds. This in turn means that at the least b-worlds (if there be such) that $\neg p$ is true, which informally justifies the entailment that $b \Rightarrow \neg p$ holds.

The resulting logic is weak. For example, the following relations do not hold:

Strengthening: From $\varphi \Rightarrow \gamma$ infer $\varphi \wedge \psi \Rightarrow \gamma$.
Transitivity: From $\varphi \Rightarrow \psi$ and $\psi \Rightarrow \gamma$ infer $\varphi \Rightarrow \gamma$.
Contraposition: From $\varphi \Rightarrow \gamma$ infer $\neg \gamma \Rightarrow \neg \varphi$.
Modus ponens: From $\varphi \Rightarrow \psi$ and φ infer ψ.

Nor of course would one want these principles to always hold for defaults. Thus, for strengthening, we do not want to conclude, given that a bird flies, that a bird with a broken wing flies. Counterexamples for the other principles can similarly be easily found. As well, the undesirability of unrestricted modus ponens suggests that some extra-logical mechanism is required for deriving "reasonable" consequents of default conditionals.

So, as indicated, research in conditional approaches to defaults has for the most part taken this approach as providing a basic set of desirable properties for defaults (and indeed, as mentioned, this set has been suggested as a minimal set that any such approach should satisfy). And certainly this approach captures compelling and natural intuitions regarding notions of normality. However, there are certainly other interpretations of defaults. Among others, we have already noted the probabilistic interpretation, that a default $\varphi \Rightarrow \psi$ is accepted as being true if, on the evidence that φ, ψ is not less likely than $\neg\psi$. We explore this alternative next, followed by a discussion in the next section.

3 The Approach

3.1 Syntax and Semantics

Let $\mathbf{P} = \{a, b, c, \ldots\}$ be a set of *atomic sentences* or simply *atoms*. L_{PC} is the language of *classical propositional logic*, formed in the usual way from the logical symbols $\neg$ and $\supset$, with symbols $\wedge$, $\vee$, and $\equiv$ introduced by definition. The symbol $\top$ is taken to be some propositional tautology, and $\bot$ is defined as $\neg\top$. The language L_{CL}, of *minimal conditional logic*, is L_{PC} extended with the binary operator $\Rightarrow$ where, for formula $\varphi \Rightarrow \psi$, we have $\varphi, \psi \in L_{\mathsf{PC}}$. More formally, we have the following grammar:

$$\varphi := p \mid (\neg\varphi) \mid (\varphi \supset \varphi) \mid (\psi \Rightarrow \psi) \quad (L_{\mathsf{CL}})$$
$$\psi := p \mid (\neg\psi) \mid (\psi \supset \psi) \quad (L_{\mathsf{PC}})$$
$$p \in \mathbf{P}$$

The formula $(\varphi \supset \psi)$ denotes material implication. The formula $(\varphi \Rightarrow \psi)$ denotes a default conditional, with intended interpretation that φ makes ψ no less probable than $\neg\psi$. Note that the default conditional may not be nested (for example, $((\varphi \Rightarrow \psi) \Rightarrow \chi)$ is not a formula). Parentheses may be freely dropped when no confusion arises. As well, the operator $\Rightarrow$ is assumed to bind least strongly and $\neg$ most strongly. So for example $\neg p \wedge q \Rightarrow p \vee q$ is the same as $((\neg p) \wedge q) \Rightarrow (p \vee q)$.

We next define the notion of a *model*. This is simply a (S5-like) possible worlds model where each world is assigned a probability.

Definition 1 (Models). *A probability model is a structure $M = (W, P, V)$ satisfying the following.*

- *W is a nonempty finite set (of possible worlds).*
- *$P : W \to [0, 1]$ maps each world $w \in W$ to a probability $P(w)$ subject to the condition of* unit sum:

$$\sum_{w \in W} P(w) = 1.$$

- *$V : \mathbf{P} \to 2^W$ is a* valuation *that maps atomic sentences to subsets of W.*

A pointed probability model *is a pair (M, w) consisting of a probability model $M = (W, P, V)$ and a world $w \in W$.*

Intuitively, $P(w)$ specifies the probability that, according to the agent, w may be the actual world. For a valuation, the intuition is that $V(p)$ specifies those worlds at which p is true.

For each $S \subseteq W$, define

$$P(S) = \sum_{w \in S} P(w).$$

For $A, B \subseteq W$ with $P(B) \neq 0$, we let $P(A|B) = P(A \cap B)/P(B)$ be the probability of A conditional on B. In writing conditional probability, worlds may be identified with singleton sets.

Definition 2 (Semantics). *Let (M, w) be a pointed probability model with $M = (W, P, V)$.*
Define $[\cdot]_M : L_{\mathsf{CL}} \to 2^W$ by

$$[\varphi]_M \doteq \{w \in W \mid M, w \models \varphi\} .$$

The relation $\models$ between (M, w) and L_{CL}-formulas is given as follows.

$$\begin{aligned}
M, w &\models p & &\text{iff } w \in V(p) \\
M, w &\models \neg\varphi & &\text{iff } M, w \not\models \varphi \\
M, w &\models \varphi \supset \psi & &\text{iff } M, w \not\models \varphi \text{ or } M, w \models \psi \\
M, w &\models \varphi \Rightarrow \psi & &\text{iff } P([\varphi]_M \cap [\psi]_M) \geq P([\varphi]_M \cap [\neg\psi]_M)
\end{aligned}$$

In the following, we freely drop the subscript M on $[\varphi]_M$ when doing so ought not to cause confusion.

A formula φ is *valid in M*, written $M \models \varphi$, just if $M, w \models \varphi$ for every $w \in W$. A formula φ is *valid*, written $\models \varphi$, just if $M \models \varphi$ for every probability model M. A set of formulas Γ *logically entails* a formula φ, written $\Gamma \models \varphi$, just if for every model M, $[\Gamma]_M \subseteq [\varphi]_M$.

We define the following abbreviation in L_{CL}:

$$\varphi \to \psi \text{ abbreviates } (\varphi \Rightarrow \bot) \lor ((\varphi \Rightarrow \psi) \land \neg(\varphi \Rightarrow \neg\psi)) .$$

This *strict (conditional) default* is read, "in the case that φ is not impossible, φ makes ψ more probable than $\neg\psi$."

We have the following straightforward result that, if (M, w) is a pointed probability model, then

$$M, w \models \varphi \to \psi \quad \text{iff} \quad P([\varphi]) = 0 \text{ or } P([\varphi] \cap [\psi]) > P([\varphi] \cap [\neg\psi]) .$$

We can provide further definitions as follows. First, if necessity is associated with having probability 1, then we can define:

$$\Box\varphi \doteq \neg\varphi \Rightarrow \bot.$$

(Analogously, we could defined the separate notion that φ is necessarily true if φ is true in all possible worlds, though we do not explore this here.) Moreover, we can define that φ has probability not less than that of ψ, by:

$$\varphi \geq \psi \doteq (\varphi \equiv \neg\psi) \Rightarrow (\varphi \wedge \neg\psi) \qquad (1)$$

That is, the right hand side is true just if the probability of $(\varphi \equiv \neg\psi) \wedge (\varphi \wedge \neg\psi)$ is no less than the probability of $(\varphi \equiv \neg\psi) \wedge \neg(\varphi \wedge \neg\psi)$. But this means that $P(\varphi \wedge \neg\psi) \geq P(\neg\varphi \wedge \psi)$ which in turn is the same as $P(\varphi) \geq P(\psi)$.

3.2 Axiomatic Theory

The logic CL is based on propositional logic, in that it contains propositional tautologies and the rule of inference modus ponens. As well, it contains the following rules of inference and axioms.

Rules of Inference:

(**LLE**) From $\varphi \equiv \varphi'$ infer $\varphi \Rightarrow \psi \equiv \varphi' \Rightarrow \psi$
(**RW**) From $\psi \supset \psi'$ infer $\varphi \Rightarrow \psi \supset \varphi \Rightarrow \psi'$
(**LW**) From $\varphi' \supset \psi$ infer $\varphi \Rightarrow \psi \supset (\varphi \vee \varphi' \Rightarrow \psi)$

Axioms:

(**Con**) $\neg(\top \Rightarrow \bot)$
(**Ref**) $\varphi \Rightarrow \varphi$
(**D**) $(\varphi \wedge \varphi' \Rightarrow \psi) \wedge (\varphi \wedge \neg\varphi' \Rightarrow \psi) \supset \varphi \Rightarrow \psi$
(**NR**) $\varphi \Rightarrow \psi \supset (\varphi \wedge \varphi' \Rightarrow \psi) \vee (\varphi \wedge \neg\varphi' \Rightarrow \psi)$
(**RS**) $(\varphi \Rightarrow \psi) \supset (\varphi \Rightarrow \varphi \wedge \psi)$
(**LS**) $(\varphi \Rightarrow \psi \wedge \chi) \supset (\varphi \wedge \psi \Rightarrow \chi)$
(**Trans**) $[(\varphi \equiv \neg\psi \Rightarrow \varphi \wedge \neg\psi) \wedge (\psi \equiv \neg\chi \Rightarrow \psi \wedge \neg\chi)] \supset (\varphi \equiv \neg\chi \Rightarrow \varphi \wedge \neg\chi)$

The first two rules, for *left logical equivalence* and *right weakening* are standard in most conditional logics and, indeed, it would be difficult to find a "reasonable" interpretation of a conditional that violated them. The other rule LW (for *left weakening*) is intriguing, first because it has the appearance of a dual of RW, but for a result regarding the antecedent of a weak conditional and, second, because it can also be seen as a weaker version of the condition Or.

Concerning the axioms, first, Con expresses nontriviality, while Ref expresses semantically that every formula φ has probability ≥ 0. Axiom D is essentially reasoning by cases with respect to the antecedent of a conditional; as well it can also be seen as a weaker version of Or. In fact, D can be rewritten as:

If $\neg(\varphi_1 \wedge \varphi_2)$ then $(\varphi_1 \Rightarrow \psi \wedge \varphi_2 \Rightarrow \psi) \supset \varphi_1 \vee \varphi_2 \Rightarrow \psi$.

So without the implicit mutual exclusion of the antecedents of the premises in D, the consequent is falsifiable.

NR is again something of a dual of D, and gives a weaker version of strengthening the antecedent. The axioms RS and LS are also something of duals, where

RS gives a condition for right strengthening and LS similarly gives a condition for left strengthening. Trans simply reflects transitivity of $\leqslant$ in the semantics, recalling that $(\varphi \equiv \neg\psi) \Rightarrow (\varphi \wedge \neg\psi)$ can be interpreted as asserting that the probability of φ is no less than that of ψ.

As well we obtain the following:

Theorem 1. *The following are derivable in* CL:

(SC)   From $\varphi \supset \psi$ infer $\varphi \Rightarrow \psi$
(LI)   From ψ infer $\varphi \Rightarrow \psi$
(SWC) $\varphi \to \psi \supset \varphi \Rightarrow \psi$
(WCEM) $\neg(\varphi \Rightarrow \bot) \supset ((\varphi \to \psi) \supset \neg(\varphi \to \neg\psi))$
(RF) $\varphi \Rightarrow \bot \supset \varphi \Rightarrow \psi$
(MOD) $\neg\varphi \Rightarrow \bot \supset \psi \Rightarrow \varphi$
(RW') $(\varphi \Rightarrow \psi_1 \wedge \psi_2) \supset \varphi \Rightarrow \psi_1$
(RLW) $(\varphi \Rightarrow \psi) \supset (\varphi \vee \chi \Rightarrow \psi \vee \chi)$
(RImp) $\varphi \Rightarrow \psi \supset \varphi \Rightarrow (\varphi \supset \psi)$

SC is *supraclassicality*; this is another condition that any "reasonable" weak conditional should satisfy; similarly for LI, which is a simple consequence of SC. SWC relates our strong and weaker versions of probabilistic defaults, while WCEM is *weak conditional excluded middle*. RF is immediate from RW, but reflects an intuitive result, that if a formula φ has probability 0, then it implies any other formula. MOD can be written equally well as $\Box\psi \supset \varphi \Rightarrow \psi$. RW' is simply a restating of RW that appears frequently in the conditional logic literature. RLW is perhaps less common and can be seen perhaps as yet another weaker version of Or (this time implicitly making use of the premiss $\varphi \Rightarrow \varphi$). RImp should be compared with RS; given that $\varphi \Rightarrow \psi$ is true, RImp gives the weakest consequence of φ that holds (in terms of φ and ψ) while RS gives the strongest such condition. Last, it can be noted that in the above theorem, if the connective $\Rightarrow$ is uniformly replaced by $\to$ then the resulting formulas are also theorems.

While the various axioms, rules of inference, and derived principles are presumably all intuitively acceptable, the resulting system is nonetheless weak, and results obtained in other conditional clearly may not obtain here. Consider the standard example that birds fly, penguins do not fly, and penguins are birds, which we can represent by:

$$B \to F, \quad P \to \neg F, \quad \Box(P \supset B). \tag{2}$$

We can readily infer for example that birds that are not penguins don't fly: $B \wedge \neg P \to F$. However, in an approach such as that of preferential reasoning, one can also derive that birds are normally not penguins. This inference is absent here, and it is straightforward to set up a counterexample in terms of a model: Assume that we have atoms $P = \{B, P, F\}$ with the expected informal interpretation, and identify the eight possible worlds with the possible truth assignments to the atoms. Assume that we have the following probability assignment to worlds:

$$P(B\neg PF) = .25, \quad P(BPF) = .3, \quad P(B\neg P\neg F) = .1, \quad P(BP\neg F) = .35.$$

All other worlds must have probability 0. It is easily verified that the formulas in (2) are satisfied. Moreover, it can be verified that $B \to P$ is satisfied in this model, i.e. that birds are (probably) penguins.

4 Discussion

Given that the previous section provided an axiomatisation that was sound but not complete, a natural question is, how difficult would it be to obtain a complete axiomatisation? The answer is that while completeness would not be difficult to obtain, a *satisfactory* axiomatisation may be a challenge.

To this end, there has been previous work on approaches to *qualitative probability*; this work involves adding a binary operator $\succeq$ to the language of (usually) propositional logic, such that $\varphi \succeq \psi$ has the interpretation that the probability of φ is not less than that of ψ. Gärdenfors [11] provides an axiomatisation of one such approach. Given that Gärdenfors' definition of a model is the same as our's in the finite case, if we can show that his axiomatisation is derivable from our's via the definition (1) then we would be done: our axiomatisation would imply his, and his is complete with respect to the class of models that we consider.

Gärdenfors' axiomatisation contains, among other things, a countable sequence of axiom schemes that he calls A4(m). A4(m) is difficult to understand (Krister Segerberg, who originally proposed the schema, calls it "formidable") and it grows exponentially with a scheme's position in the sequence. So it is straightforward to obtain a complete axiomatisation in our approach; one just needs to add the schema A4(m) (which refers to necessity, which we in turn have defined in terms of $\Rightarrow$). However, given that A4(m) is (in our opinion at least) not particularly satisfactory, the resulting axiomatisation would similarly be unsatisfactory.

In a recent paper [8], we have reformulated this line of work on qualitative probability that finishes with [11], and developed a logic, called LQP, with a sound and complete axiomatisation that avoids the schema A4(m). However, to do so, the connective $\succeq$ is generalised to a n,m-ary connective. For example, a theorem in our approach is:

$$\varphi \oplus \psi \approx (\varphi \vee \psi) \oplus (\varphi \wedge \psi)$$

which can be read as "the summed probability of φ and ψ is the same as the summed probability of $\varphi \vee \psi$ and $\varphi \wedge \psi$". So, if we could extend the axiomatisation of CL so that it implies the axiomatisation of LQP, we would obtain a completeness result. However, the difficulty now is that the definition (1) yields a binary operator $\succeq$ and it is not clear how to generalise this definition so that the results in LQP that rely on $\succeq$ having arbitrary arity can be obtained. So to sum up, obtaining a completeness result for CL is straightforward; obtaining a satisfactory completeness result is less straightforward.

Turning next to a comparison with the system P of preferential entailment, it is unsurprising that several of the principles in P are unsound in CL:

Theorem 2. *None of the following is sound.*

(**And**) $(\varphi \Rightarrow \psi_1 \wedge \varphi \Rightarrow \psi_2) \supset \varphi \Rightarrow (\psi_1 \wedge \psi_2)$
(**Or**) $(\varphi_1 \Rightarrow \psi \wedge \varphi_2 \Rightarrow \psi) \supset (\varphi_1 \vee \varphi_2) \Rightarrow \psi$
(**Cut**) $((\varphi \Rightarrow \psi) \wedge (\varphi \wedge \psi \Rightarrow \chi)) \supset (\varphi \Rightarrow \chi)$
(**CM**) $(\varphi \Rightarrow \psi_1 \wedge \varphi \Rightarrow \psi_2) \supset (\varphi \wedge \psi_1) \Rightarrow \psi_2$

On the other hand, it can be observed that the rules of CL are rules or derived rules in the system P of preferential reasoning, while the axioms are theorems of P. This gives:

Theorem 3. CL *is weaker than* P.

We obtain the immediate result:

Corollary 1. CL *is strictly weaker than* P.

Given that our connective $\Rightarrow$ of conditional logic arguably captures an intuitive notion of "default", we conclude that the system P of preferential entailment (or any of its equivalent guises) is too strong to provide a "conservative core" that *any* definition of default should adhere to.

Regarding other conditional logics, Chellas [5] investigates basic conditional logics. The focus is on a logic called CK; it consists of the rule LLE and the following rule that generalises RW:

RCK From $\vdash \psi_1 \wedge \cdots \wedge \psi_n \supset \psi$ infer $\vdash (\varphi \Rightarrow \psi_1) \wedge \cdots \wedge (\varphi \Rightarrow \psi_n) \supset (\varphi \Rightarrow \psi)$

Interestingly, CK neither contains nor is contained by CL. For example Ref is an axiom of CL but not a theorem of CK, whereas And is a theorem of CK but not of CL.

With regards to the semantic specification of $\Rightarrow$, a more general formulation is given via *plausibility measures* [10]. A *plausibility space* is a tuple $(W, \mathcal{F}, Pl)$ where W is a set of worlds, $\mathcal{F}$ is an algebra over W, and Pl maps sets in $\mathcal{F}$ to a set D of *plausibility values*. D in turn is a total order with maximum and minimum element, $\top$ and $\bot$ respectively. Pl has the constraints:

- $Pl(\emptyset) = \bot$;
- $Pl(W) = \top$; and
- if $U \subseteq V$ then $Pl(U) \leq Pl(V)$.

Then the fact that an agent believes U conditional on V can be given by $Pl(U|V) > Pl(\overline{U}|V)$. This then amounts to the same semantic condition (though with notation taken from probability theory) as in Definition 2, though in a weaker framework. Friedman and Halpern don't develop an axiomatisation of this base conditional, but rather consider those conditions that can be added so that system P is obtained.

The net result is that it can be suggested that CL provides a minimal "core" of reasonable default principles that ought to be satisfied by any approach to commonsense conditional defaults. That is, while there are systems that are more general than CL, none to date would appear to capture an intuitive notion of "commonsense conditional".

5 Conclusion

We have presented a minimal conditional logic for defaults, based on the idea that $\varphi \Rightarrow \psi$ is true in a theory just if φ makes ψ no less probable than $\neg\psi$. The model theory reflects this intuition directly. As well, a set of basic, sound principles is presented along with other principles that may be derived from this set. While the set of principles is not complete, completeness can be easily obtained by adopting a principle first articulated by Segerberg. How completeness may be obtained in a more perspicuous way is an open question. As well, we briefly compare the approach to other conditional approaches, noting that the framework is (clearly) weaker than the system P of preferential entailment.

References

1. E. Adams. *The Logic of Conditionals*. D. Reidel Publishing Co., Dordrecht, Holland, 1975.
2. Christoph Beierle and Gabriele Kern-Isberner. Semantical investigations into nonmonotonic and probabilistic logics. *Annals of Mathematics and Artificial Intelligence*, 65(2-3):123–158, 2012.
3. C. Boutilier. Conditional logics of normality: A modal approach. *Artificial Intelligence*, 68(1):87–154, 1994.
4. J.P. Burgess. Quick completeness proofs for some logics of conditionals. *Notre Dame Journal of Formal Logic*, 22(1):76–84, 1981.
5. B.F. Chellas. Basic conditional logic. *Journal of Philosophical Logic*, 4:133–153, 1975.
6. B.F. Chellas. *Modal Logic*. Cambridge University Press, 1980.
7. James Delgrande. What's in a default? thoughts on the nature and role of defaults in nonmonotonic reasoning. In Gerhard Brewka, Victor W. Marek, and Miroslaw Truszczynski, editors, *Nonmonotonic Reasoning: Essays Celebrating its 30th Anniversary*. College Publications, 2011.
8. James P. Delgrande and Bryan Renne. The logic of qualitative probability. In Q. Yang and M. Wooldridge, editors, *Proceedings of the 24th International Joint Conference on Artificial Intelligence (IJCAI)*, pages 2904–2910. AAAI Press, 2015.
9. D. Dubois, J. Lang, and H. Prade. Possibilistic logic. In D. M. Gabbay, C. J. Hogger, and J. A. Robinson, editors, *Nonmonotonic Reasoning and Uncertain Reasoning*, volume 3 of *Handbook of Logic in Artifical Intelligence and Logic Programming*, pages 439–513. Oxford, 1994.
10. N. Friedman and J. Halpern. Plausibility measures and default reasoning. *Journal of the ACM*, 48(4):649–685, 2001.
11. P. Gärdenfors. Qualitative probability as an intensional logic. *Journal of Philosophical Logic*, 4(2):171–185, 1975.
12. Hector Geffner and Judea Pearl. A framework for reasoning with defaults. In Henry E. Kyburg Jr., Ronald P. Loui, and Greg N. Carlson, editors, *Knowledge Representation and Defeasible Reasoning*, volume 5 of *Studies in Cognitive Systems*, pages 69–87. Springer, 1990.
13. G.E. Hughes and M.J. Cresswell. *A New Introduction to Modal Logic*. Routledge., London and New York, 1996.
14. G. Kern-Isberner. *Conditionals in Nonmonotonic Reasoning and Belief Revision*, volume 2087 of *Lecture Notes in Artificial Intelligence*. Springer Verlag, 2001.

15. Gabriele Kern-Isberner and Christian Eichhorn. Structural inference from conditional knowledge bases. *Studia Logica*, 102(4):751–769, 2014.
16. Gabriele Kern-Isberner and Matthias Thimm. A ranking semantics for first-order conditionals. In *Proceedings of the European Conference on Artificial Intelligence*, pages 456–461. IOS Press, 2012.
17. S. Kraus, D. Lehmann, and M. Magidor. Nonmonotonic reasoning, preferential models and cumulative logics. *Artificial Intelligence*, 44(1-2):167–207, 1990.
18. P. Lamarre. S4 as the conditional logic of nonmonotonicity. In *Proceedings of the Second International Conference on the Principles of Knowledge Representation and Reasoning*, pages 357–367, Cambridge, MA, April 1991.
19. J. Pearl. *Probabilistic Reasoning in Intelligent Systems: Networks of Plausible Inference*. Morgan Kaufman, San Mateo, CA, 1988.
20. J. Pearl. Probabilistic semantics for nonmonotonic reasoning: A survey. In *Proceedings of the First International Conference on the Principles of Knowledge Representation and Reasoning*, pages 505–516, Toronto, May 1989. Morgan Kaufman.
21. Matthias Thimm and Gabriele Kern-Isberner. On probabilistic inference in relational conditional logics. *Logic Journal of the IGPL*, 20(5):872–908, 2012.

Alternatives to the Ramsey Test

Sven Ove Hansson[1]

Abstract. The Ramsey test is often assumed to provide us with adequate logical principles both for conditionals and for non-monotonic inference. In this contribution several alternatives to the Ramsey test are introduced. It is proposed that modified versions of the Ramsey test should be used for some conditionals and that an alternative test, the co-occurrence test, should be used for non-monotonic inference.

1 Introduction

Conditional statements, non-monotonic inference, and belief revision are commonly seen as closely connected with each other, if not interdefinable. Their interrelations have given rise to a considerable literature.

Conditionals and belief revision. The currently most discussed and well-known of these connections is that between conditionals and belief revision. Conditionals are standardly explicated in terms of belief revision via the Ramsey test. The test was first proposed in a famous footnote by Frank Ramsey:

> "If two people are arguing 'If p will q?' and are both in doubt as to p, they are adding p hypothetically to their stock of knowledge and arguing on that basis about q." (Ramsey 1931, p. 247)

Thus, "if p then q" holds if and only if one would believe in q if one revised one's beliefs to believe in p. Let $\rightarrowtail$ denote "if... then...", K the current belief set, and $*$ belief revision. Then the Ramsey test connects belief revision with conditional sentences as follows:

$p \rightarrowtail q$ holds if and only if $q \in K * p$ (Ramsey test, first version)

The test has an alternative version presupposing that the conditional sentences that are supported in the belief state represented by the belief set K are also elements of K:

$p \rightarrowtail q \in K$ if and only if $q \in K * p$ (Ramsey test, second version)

As was shown by Gärdenfors (1986), the second version of the test is incompatible with certain common requirements on the revision operation $*$.[2] The

---

[1] Division of Philosophy, Royal Institute of Technology, 100 44 Stockholm, Sweden, soh@kth.se

[2] Hans Rott (1989) constructed an elegant variant of Gärdenfors's theorem. The assumptions are, in addition to the second version of the Ramsey test:
(E) there are a belief set K and two logically independent sentences q_0 and q_1 such

reader is referred to Rott (1989) and Hansson (1992) for discussions of these requirements and to Fermé and Hansson (2011, pp. 304-306) for a brief summary of recent developments. Models in which conditional sentences can be included in belief sets have been proposed by Gabriele Kern-Isberner (2004) and Hansson (2015).

Conditionals and non-monotonic inference. The connections between conditional sentences and non-deductive (inductive, non-monotonic) reasoning were pointed out long before formalized treatments of the two notions were developed. For instance, John Stuart Mill (1856/1973), who used "the words hypothetical and conditional ... synonymously" (p. 83) pointed out that in a hypothetical statement, "[w]hat is asserted is not the truth of either of the propositions, but the inferribility of the one from the other" (ibid.).

In the early developments of modern non-monotonic logic, connections with the logic of conditionals were a common topic. Ginsberg (1986) proposed that non-monotonic inference systems should be based on the logic of (counterfactual) conditionals. Similarly, Delgrande (1988) and Boutilier (1990) provided accounts of default reasoning that were based on conditional logic. However, although the two concepts appear to be closely related, it was soon realized that they do not coincide. In their seminal paper on non-monotonic inference, Kraus, Lehmann and Magidor (1990) noted that "conditional logic considers a binary intensional connective that can be embedded inside other connectives and even itself, whereas we [in non-monotonic reasoning] consider a binary relation symbol that is part of the metalanguage." (p. 170) They also pointed out that non-monotonic reasoning can be obtained as the "flat (i.e. nonnested) fragment of a conditional logic" (p. 171). This analysis has been confirmed by other authors, such as Wobcke (1995, p. 73). It can now be described as the standard view that the logic of non-monotonic inference coincides with a logic of non-nested conditional sentences.

Belief revision and non-monotonic inference. Makinson and Gärdenfors (1991) suggested a simple method for deriving a non-monotonic inference relation from an operation of revision. According to their proposal, "q follows non-monotonically from p" means that q holds if an "arbitrary but fixed background theory" (p. 189) is revised by p. Denoting that theory by K and non-monotonic consequence by $\mathrel{|\!\sim}$ we obtain

---

that $q_0 \vee q_1 \notin K$, $q_0 \vee \neg q_1 \notin K$, $\neg q_0 \vee q_1 \notin K$, and $\neg q_0 \vee \neg q_1 \notin K$,
(S) $p \in K * p$,
(C) if $K \not\vdash \bot$ and $p \not\vdash \bot$ then $K * p \not\vdash \bot$ and
(P) If $\neg p \notin K$, then $K \subseteq K * p$.
It follows from (E) that $q_0 \notin \mathrm{Cn}(K \cup \{q_0 \vee q_1\})$ (since otherwise $q_0 \vee \neg q_1 \in K$). Then (P) yields $q_0 \vee q_1 \in (\mathrm{Cn}(K \cup \{q_0 \vee q_1\})) * \neg q_0$. Combining this with $\neg q_0 \in (\mathrm{Cn}(K \cup \{q_0 \vee q_1\})) * \neg q_0$ that follows from (S) we obtain $q_1 \in (\mathrm{Cn}(K \cup \{q_0 \vee q_1\})) * \neg q_0$. The Ramsey test yields $\neg q_0 \rightarrowtail q_1 \in \mathrm{Cn}(K \cup \{q_0 \vee q_1\})$, thus $\neg q_0 \rightarrowtail q_1 \in \mathrm{Cn}(K \cup \{q_0\})$. One more application of the Ramsey test yields $q_1 \in \mathrm{Cn}(K \cup \{q_0\}) * \neg q_0$. In the same way we obtain $\neg q_1 \in \mathrm{Cn}(K \cup \{q_0\}) * \neg q_0$. Thus $\mathrm{Cn}(K \cup \{q_0\}) * \neg q_0$ is inconsistent, contrary to (E) and (C).

$$p \mathrel{\vert\!\sim} q \text{ holds if and only if } q \in K * p$$

which is of course the Ramsey test in a new guise. Importantly, the new guise includes a fixed background theory that precludes the iteration of the relation $\mathrel{\vert\!\sim}$. This connection between non-monotonic inference and belief revision has been further studied by Gärdenfors and Makinson (1994), Wobcke (1995), del Val (1997), Rott (2001, esp. 111-119) and others. Stalnaker (1994, p. 18) maintains that it is "more plausible" to take theories of non-monotonic inference such as that of Lehmann and Magidor to be expressions of belief revision than to treat them as representations of consequence relations.

In a highly clarifying passage, Gabriele Kern-Isberner described the difference in emphasis between belief revision and non-monotonic inference as follows:

"Nonmonotonic reasoning and belief change theory are closely related in that they both deal with reasoning under uncertainty and try to reveal sensible lines of reasoning in response to incoming information. The crucial difference between both areas is the role of the current epistemic state which is only implicit in nonmonotonic reasoning, but explicit and in fact in the focus of interest in belief revision. So the correspondences between axioms of belief change and properties of nonmonotonic inference operations are usually elaborated only in the case that revisions are based on a fixed theory (cf. (Makinson and Gärdenfors 1991)), and very little work has been done to incorporate iterated belief revision in that framework." (Kern-Isberner 2008)

All three notions. Some authors have commented on the interrelations among all three notions. Katsuno and Satoh (1991) pointed out that there is an underlying striving for minimality at play in all three areas. However, as they also noted, the similarity between the three areas is not always obvious due to differences in terminology and notation. Likewise, Wobcke (1995) identified minimal change as the common core of the three concepts. With this he referred to minimal change not only of the belief set but also of the relation of entrenchment.

Are belief revision, conditional statements, and non-monotonic inference close enough to be adequately representable in a unified framework covering them all? And if so, are they also interdefinable and in that case how? These are the issues on which I hope to throw some light on the following pages. The focus will be on the intuitive underpinnings of potential ways to connect conditionals and non-monotonic inferences with belief revision. I will propose some alternative ways in which such connections can be constructed, but the formal investigation of these proposals will be deferred to some other occasion.

Section 2 is devoted to the interpretation of the traditional Ramsey test. In Section 3, descriptor revision is briefly introduced. It is a more versatile framework than traditional sentential revision, and that versatility will be needed in the sections that follow. Section 4 introduces elicited conditionals and Section 5 provides an account of the context-dependence of conditionals. Section 6 introduces the co-occurrence test, an alternative to the Ramsey test that appears to be more truthful to the notion of non-monotonic inference. Section 7 concludes.

2 How the Ramsey test connects with belief revision

There are many varieties of conditional sentences ("if... then..."-sentences), and several ways to classify them in terms of their meanings have been put forward. (Arlo-Costa 2014, Hansson 1995) A simple typology proposed by Lindström and Rabinowicz is particularly useful. They divided conditionals into two groups: ontic and epistemic (doxastic) conditionals. The crucial difference is that "ontic conditionals concern hypothetical modifications of the *world*, but epistemic conditionals have to do with hypothetical modifications of our *beliefs* about the world" (Lindström and Rabinowicz 1992, p. 225). The following examples illustrate the difference:

> If you had been sober, then the accident would not have happened. (*ontic conditional*)

> If she received the message then she will be here in a few minutes. (*epistemic conditional*)

In English the antecedents of epistemic conditionals are commonly expressed with an indicative verb form and those of ontic conditionals with a subjunctive verb form. However, this connection between meaning and mood only holds in some languages. Furthermore, as was pointed out by Hans Rott (1999), the correlation between meaning and mood is far from perfect in English. For philosophical purposes the distinction between subjunctive and indicative conditionals should therefore be replaced by that between ontic and epistemic ones. As the quotation from Ramsey shows, he referred to epistemic conditionals, not ontic ones. In what follows I will focus on epistemic conditionals since they are closer related with belief revision that the ontic ones.

Stalnaker (1968) proposed a possible world semantics for (counterfactual) conditionals: $p \rightarrowtail q$ holds if and only if q holds in the p-world (world in which p is true) that is closest to the actual world. Lewis (1973) modified this analysis by allowing for a tie between several p-worlds that are all maximally close to the actual world. According to Lewis, $p \rightarrowtail q$ holds if and only if q holds in all the p-worlds that are closest to the actual world. It should be noted that Stalnaker's possible world semantics transformed the Ramsey test to a criterion for ontic rather than epistemic conditionals. He was quite explicit about this, and reported that he attained "the transition from belief conditions to truth conditions" by means of possible worlds, "since a possible world is the ontological analogue of a stock of hypothetical beliefs". (Stalnaker 1968, p. 102) Lewis's account can be interpreted as expressing either ontic or epistemic conditionals. For the latter interpretation, note that $p \rightarrowtail q$ holds if and only if q is an element of the intersection of the p-worlds that are closest to the actual world. The intersection of a set of possible worlds is a belief set (logically closed set of sentences).

Long before these possible world models were introduced, another formal approach to conditionals was explored. It can be called the derivability theory of conditionals.[3] The first fully developed statement of this approach seems to be a 1946 paper by Roderick Chisholm, and the standard reference is a paper by Nelson Goodman from 1947. Goodman's proposal was essentially that for $p \rightarrowtail q$ to hold, there must be some set S of true sentences that is compatible with p and such that q follows from S and p. This refers to ontic conditionals, but epistemic variants of the derivability approach have been proposed by Mackie (1965) and Gauker (1987). In an epistemic version, S should be replaced by a set of sentences that the agent believes in and that is compatible with p, in other words a subset of the belief set K that does not imply $\neg p$. The removal of $\neg p$ from K can be interpreted as belief contraction, and it will then follow that $p \rightarrowtail q$ holds if and only if $K \div \neg p$ together with p implies q (where $\div$ denotes belief contraction), in other words:

$p \rightarrowtail q$ holds if and only if $q \in \mathrm{Cn}((K \div \neg p) \cup \{p\})$,

where Cn represents classical consequence. In the belief revision literature it is commonly assumed that revision is derivable from contraction via the Levi identity (Alchourrón et al 1985),

$K * p = \mathrm{Cn}((K \div \neg p) \cup \{p\})$.

It then follows that the contraction-based interpretation of the derivability theory of conditionals is equivalent with the revision-based account that is more commonly referred to in recent literature on the Ramsey test.

3 Descriptor revision and descriptor conditionals

Descriptor revision is a new model of belief change that was introduced in Hansson (2014a). Its construction differs from that of the AGM model in two important respects. First, the inputs are not sentences in the object language but instead metalinguistic sentences constructed with a belief predicate $\mathfrak{B}$. For any sentence p in the object language, $\mathfrak{B}p$ signifies that p is believed. A belief descriptor (henceforth: descriptor) is a set of molecular combinations of such $\mathfrak{B}$-sentences, for instance $\{\neg\mathfrak{B}p, \neg\mathfrak{B}\neg p\}$ means that neither p nor its negation is believed and $\{\mathfrak{B}p \vee \mathfrak{B}q\}$ that either p or q is believed. With this notation, a wide variety of belief change operations can be subsumed under a unified operation $\circ$. Then $K \circ \{\mathfrak{B}p\}$ corresponds to revision by the sentence p, $K \circ \{\neg\mathfrak{B}p\}$ to contraction of the sentence p, $K \circ \{\mathfrak{B}p \vee \mathfrak{B}\neg p\}$ to making up one's mind about p, $K \circ \{\neg\mathfrak{B}p, \mathfrak{B}q\}$ to replacement p by q, etc.

The other major difference is that the selection function (choice function) selects among potential outcomes, i.e. among belief sets, rather than among remainders or possible worlds. Several ways to construct selection functions for descriptor revision are available. (Hansson 2014a, 2014b, 2015) Here the focus will be on a simple construction, *centrolinear revision*. It is based on an ordering

---

[3] It is more commonly known under Lewis's (1973) term "the metalinguistic theory", but as noted by Kit Fine (1975, p. 451), there is "nothing essentially metalinguistic" in it.

$\leq$ of the set $\mathbb{X}$ that consists of the potential outcomes of changes of the original belief set K. It is assumed that K is the highest-ranked ($\leq$-minimal) element of $\mathbb{X}$. For each descriptor Ψ, the outcome $K \circ \Psi$ of revising K by Ψ is equal to the (unique) $\leq$-minimal element of $\mathbb{X}$ that satisfies Ψ.[4] More precisely:

> An operation $\circ$ on a belief set K is a *centrolinear revision* if and only if there is a set $\mathbb{X}$ of belief sets with $K \in \mathbb{X}$ and a relation $\leq$ on $\mathbb{X}$, such that (i) $K \leq X$ for all $X \in \mathbb{X}$, and (ii) for all descriptors Ψ: $K \circ \Psi$ is the unique $\leq$-minimal element of $\mathbb{X}$ that satisfies Ψ, unless Ψ is unsatisfiable within $\mathbb{X}$, in which case $K \circ \Psi = K$.

Descriptor revision has been axiomatically characterized with the following five axioms: ($\Vdash$ is a relation of satisfaction; if K is a belief set and Ψ a descriptor, then $K \Vdash \Psi$ denotes that K satisfies Ψ.)

$K \circ \Psi = \mathrm{Cn}(K \circ \Psi)$ (closure)
$K \circ \Psi \Vdash \Psi$ or $K \circ \Psi = K$ (relative success)
If $K \circ \Xi \Vdash \Psi$ then $K \circ \Psi \Vdash \Psi$ (regularity)
If $K \Vdash \Psi$ then $K \circ \Psi = K$ (confirmation)
If $K \circ \Psi \Vdash \Xi$ then $K \circ \Psi = K \circ (\Psi \cup \Xi)$ (cumulativity)

Closure ensures that the outcome of the operation is a belief set. Both relative closure and regularity are weakened versions of the success postulate ($K \circ \Psi \Vdash \Psi$) that is too strong. According to relative success, the operation leaves the initial belief set unchanged if it does not implement the input descriptor. According to regularity, if there is some revision that results in Ψ being satisfied, then the revision by Ψ does so. Confirmation ensures that revision by a condition that is already satisfied involves no change. Cumulativity says that if revision by Ψ leads to Ξ being satisfied, then it makes no difference if we revise only by Ψ or by both Ψ and Ξ at the same time.

In the language of belief descriptors, an epistemic conditional $p \rightarrowtail q$ can be interpreted as saying that revision by $\mathfrak{B}p$ will lead to a belief state satisfying $\mathfrak{B}q$. In formal terms we can express this by introducing a new conditional relation $\Rightarrow$ that operates on descriptors, and write $\mathfrak{B}p \Rightarrow \mathfrak{B}q$ instead of $p \rightarrowtail q$. It is then but a small step to let $\Rightarrow$ operate also on descriptors of other forms, and we obtain *descriptor conditionals* that we can use to express conditional statements such as:

> $\mathfrak{B}p \Rightarrow \neg \mathfrak{B}q$ (If the agent acquires belief in p then she will not believe in q.)
> $\neg \mathfrak{B}p \Rightarrow \neg \mathfrak{B}q$ (If the agent gives up her belief in p then she will lose her belief in q.)
> $\mathfrak{B}p \vee \mathfrak{B}\neg p \Rightarrow \mathfrak{B}\neg p$ (If the agent makes up her her mind about p then she will believe that p is not the case.)

There is an obvious way to extend the Ramsey test to descriptor conditionals:

> $\Psi \Rightarrow \Xi$ holds if and only if $K \circ \Psi \Vdash \Xi$

---

[4] For this to work, $\leq$ has to be a linear ordering and satisfy a well-foundedness condition, see Hansson 2014a, p. 959.

Standard (sentential) Ramsey test conditionals are of course a special case, obtainable by defining $p \mapsto q$ as $\mathfrak{B}p \Rightarrow \mathfrak{B}q$.[5] The logic of descriptor conditionals that are based on some centrolinear revision has been axiomatized with the following four postulates (Hansson 2015):

> If Ψ and Ψ' are satisfied in exactly the same potential outcomes of belief change, then $\Psi \Rightarrow \Xi$ holds if and only if $\Psi' \Rightarrow \Xi$ holds. (left logical equivalence)
>
> For all Ψ there is some $Y \subseteq \mathcal{L}$ such that for all Ξ: $\Psi \Rightarrow \Xi$ holds if and only if $Y \Vdash \Xi$ (unitarity)
>
> $\Psi \Rightarrow \Psi$ holds (reflexivity), and
>
> If $\Psi \Rightarrow \Xi$ holds, then $\Psi \Rightarrow \Phi$ holds if and only if $\Psi \cup \Xi \Rightarrow \Phi$ does so (cumulativity)

4 Elicited conditionals

Sometimes we are unwilling to express a conditional judgment, but can nevertheless be induced to do so:

> THE NEW COACH: If we replace Susan by Dorothy as a central defender, will the team as a whole play better?
> THE RECENTLY RETIRED COACH: That is very difficult to say, I do not really know.
> THE NEW COACH: Yes, I know this is difficult, but I really need your opinion. Can you think it over?
> THE RECENTLY RETIRED COACH (after thinking for a while): Well, yes. The team as a whole will play better if you replace Susan by Dorothy.

Let p denote that Susan is replaced by Dorothy and q that the team as a whole improves its play. One way to interpret this dialogue is that first when revising her belief by $\mathfrak{B}p$, the retired coach arrived at a belief set $K \circ \mathfrak{B}p$ that satisfied neither $\mathfrak{B}q$ nor $\mathfrak{B}\neg q$. Then she reconsidered the issue, but now with the requirement to arrive at a belief set satisfying either $\mathfrak{B}q$ or $\mathfrak{B}\neg q$. We can express this as an extended success condition. She was no longer searching for the closest belief set satisfying $\mathfrak{B}p$ but the closest belief set satisfying both $\mathfrak{B}p$ and $\mathfrak{B}q \vee \mathfrak{B}\neg q$. We can generalize this pattern by defining the following conditional:

> $p \mapsto\!\!\!\to q$ is an abbreviation of $\{\mathfrak{B}p, \mathfrak{B}q \vee \mathfrak{B}\neg q\} \Rightarrow \mathfrak{B}q$

We can call this an *elicited conditional*. With centrolinear semantics, it will be weaker than the standard sentential conditional, i.e. it holds that

---

[5] Neither of the conditions (S), (C), or (P) of footnote 1 is satisfied in this construction. Therefore, the Gärdenfors impossibility theorem does not apply.

If $p \rightarrowtail q$ then $p \rightarrowtail\!\!\!\!\rightarrow q$,

but the reverse implication does not hold.

The distinction beween $\rightarrowtail$ and $\rightarrowtail\!\!\!\!\rightarrow$ can easily be expressed in the framework of descriptor revision, but it cannot be straightforwardly expressed with possible world semantics. The cases in which $p \rightarrowtail\!\!\!\!\rightarrow q$ holds but $p \rightarrowtail q$ does not hold are the cases when the closest p-containing belief set contains neither q nor $\neg q$, but of course there are no possible worlds in which neither q nor $\neg q$ holds.

There are interesting differences between the logic of the standard sentential Ramsey conditional $\rightarrowtail$ and that of the elicited conditional $\rightarrowtail\!\!\!\!\rightarrow$. Within the framework of centrolinear revision as described above, the former satisfies the following postulate:

If $p \rightarrowtail q_1$ and $p \rightarrowtail q_2$, then $p \rightarrowtail (q_1 \& q_2)$ (*And*)

However, the corresponding principle for the elicited conditional,

If $p \rightarrowtail\!\!\!\!\rightarrow q_1$ and $p \rightarrowtail\!\!\!\!\rightarrow q_2$, then $p \rightarrowtail\!\!\!\!\rightarrow (q_1 \& q_2)$

does not hold in general.

5 Context-dependent conditionals

It is well-known from the literature that many conditionals are context dependent. One of the best examples was provided by Lewis (1973, p. 1):

If kangaroos had no tails, they would topple over.

In a discussion on the principles of mechanics we would have good reasons to assent to this statement. However, in a discussion on evolutionary biology we would have equally good reasons to disapprove of it. In such a context, we would say that if kangaroos had no tails, then their bodies would have had a different weight distribution, so that they would not topple over. The context dependence of conditionals has been referred to as the shiftability problem (Goldstick 1978).[6] Several other examples have been given in the literature.[7]

I will propose a very simple analysis of this problem, based on what seems to be a quite natural reaction to a statement such as:

Kangaroos have no tail. (p)

---

[6] Other early discussions of this problem can be found in Lewis 1979, p. 465 and Nute 1981, p. 134-135.

[7] For instance: "If frogs were mammals, they would have mammae." – "If frogs were mammals, they would be the only ones not to have mammae." (Williamson 1969) "If I had been John Keats, I should not have been able to write the *Ode to a Nightingale*.' – "If I had been John Keats, then I should have been the man who wrote the *Ode to a Nightingale*." (Goldstick 1978, pp 5-6)

Due to its indeterminateness we cannot expect epistemic agents to know how to revise by it. In consequence, (sentential) revision by p will be unsuccessful, in other words $p \notin K * p$. Such epistemic behaviour is in conflict with the AGM theory, according to which revision must satisfy an exceptionless success postulate ($p \in K * p$ for all p). However, as I have argued extensively elsewhere (Hansson 1999) this is not a realistic feature of belief revision. There are input sentences that a rational agent may well reject, either because they are too unrealistic or because they are too vague. Our sentence p belongs to the the latter category.

The Ramsey test for conditionals requires that revision is successful. It does not make sense to evaluate $p \rightarrowtail q$ based on whether q holds in $K * p$ unless the latter set actually contains p.[8] But now consider the two statements:

Kangaroos have suddenly lost their tails. (s)
Kangaroos have lost their tails in an evolutionary process. (e)

Both $p\&s$ and $p\&e$ are much more specified, and arguably they are specified enough to allow for successful revision, i.e. $p\&s \in K*(p\&s)$ and $p\&e \in K*(p\&e)$. Both these revisions provide us with a belief set that has a clear answer to the question whether kangaroos will topple over (q).

We can now see why the conditional is evaluated differently in different contexts. If a stranger at a party suddenly asks me: "Would kangaroos topple over if they had no tail?", then I will not be able to answer the question since I do not know how to revise by the sentence p. However, if I am asked the same question in a physics class I will assume that revision by $p\&s$ is intended, whereas in a biology classroom I will interpret it as referring to revision by $p\&e$. Both these revisions are successful and therefore each of them will provide an answer to the question whether q holds after the revision.

However, this solution requires an adjustment of the underlying operation of revision that was described in Section 3. Due to the postulate of regularity that holds for centrolinear revision, it follows from $p \in K*(p\&s)$ that $p \in K*p$, and this would of course block the solution just described. In order for that solution to work, the model will have to be modified so that regularity does not hold.

Regularity is closely related to the following property that holds for (the strict part $<$ of) the relation $\leqq$ on which centrolinear revision is based:

If $X, Y \in \mathbb{X}$, then exactly one of $X < Y$, $Y < X$ and $X = Y$ holds (trichotomy)

In a formal implementation of the above account of the context-dependence of conditionals, we will have to relax trichotomy. In our case, let X_s be the closest $p\&s$-world and X_e the closest $p\&e$-world. Clearly, $X_s \neq X_e$, and for our solution to work neither $X_s < X_e$ nor $X_e < X_s$ can hold. We can still base descriptor revision on a relation, but that relation will have to be weaker than that of centrolinear revision.

---

[8] Similarly, evaluating $\Psi \Rightarrow \Xi$ based on whether $K \circ \Psi \Vdash \Xi$ only makes sense if $K \circ \Psi \Vdash \Psi$.

6   Non-monotonic inference

Let us return to the Ramsey test, as originally proposed. It employs the following criterion for approval of the conditional sentence "if p then q":

If the agent revises her beliefs by p, then she will believe that q.

Now consider the following alternative criterion:

If the agent comes to believe that p, then she will believe that q.

Obviously, an agent can come to believe in p not only as the result of revising by p but also as the result of revising her beliefs by some other input. The alternative criterion requires that q is an element not only of $K * p$ but also of other belief sets containing p. It refers to the co-occurrence of q with p, not only in the belief set resulting from revision by p but also in other belief sets containing p.

I will call the test based on this criterion the *co-occurrence test*. Furthermore, I propose that it corresponds better to non-monotonic inference (as distinct from conditional sentences) than what the Ramsey test does. The major reason for this is that the co-occurrence test is concerned with whether we will *in general* (given our present epistemic commitments) believe in q if we believe in p, not only whether we will do so in one single specific case. Such a concept of general inferribility seems to connect fairly well with the notion of non-monotonic inference. It should also be noted that the co-occurrence test has standard classical inference as a limiting case, something that should be expected of a notion of non-monotonic inference. (The classical limiting case is that in which all consistent sets of sentences are elements of the outcome set, and every belief set containing p is invoked when the test is performed.) Consequently, in what follows I will take the co-occurrence test to be a test for the statement $p \mathrel{|\!\sim} q$.

The co-occurrence test needs to be specified with respect to which of the belief sets containing p we should include in the analysis. A simple answer would be to include all elements of $\mathbb{X}$ that contain p. However, such an approach would deviate too far away from the notion of non-monotonic inference. It is an essential feature of non-monotonic reasoning that comparatively remote possibilities are not taken into account. When you conclude from "Tweety is a bird" that "Tweety can fly" that is precisely because you do not take remote possibilities into account. The degree of remoteness referred to here is relative to the antecedent. Some of the possibilities that are too remote to be taken into account when considering "Tweety is a bird" would be quite close at hand when considering "Tweety is a bird who was born in Antarctica".

Therefore, when evaluating non-monotonic inferences with p as the antecedent we have to consider not only $K * p$ that is the highest ranked (most plausible, $\leq$-minimal) p-containing belief set, but also a band of other p-containing belief sets that are less plausible than $K * p$ but still reasonably plausible.[9]

---

[9]   From a formal point of view, this proposal is related to the proposals by Nute (1975) and Schlossberger (1978, p. 80) that in possible world semantics, the assessment of

That band has $K*p$ as its inner limit, and since it does not extend indefinitely we can assume that it also has an outer limit. In formal terms, for each potential outcome X there will be another potential outcome $\partial(X)$ that is the outer limit of the "plausibility band" that has X as its inner limit. Intuitively, the plausibility band consists, in addition to X, of all the belief sets that are less plausible than X but only moderately so. The delimiting function ∂ should expectedly satisfy the following two properties:

$X \leq \partial(X)$
If $X \leq Y$ then $\partial(X) \leq \partial(Y)$

We can now express the co-occurrence test in a fully precise manner:

$p \mathrel{\mid\!\sim} q$ holds if and only if it holds for all $Y \in \mathbb{X}$ that if $K*p \leq Y \leq \partial(K*p)$ and $p \in Y$, then $q \in Y$.

This recipe can be straightforwardly extended so that it can take sets as antecedents. Let X_A be the highest ranked ($\leq$-minimal) potential outcome that contains the set A of sentences. Then:

$A \mathrel{\mid\!\sim} q$ if and only if it holds for all $Y \in \mathbb{X}$ that if $X_A \leq Y \leq \partial(X_A)$ and $A \subseteq Y$, then $q \in Y$.

This reformulation is important since we can use it to define a non-monotonic inference operation C such that that $q \in C(A)$ if and only if $A \mathrel{\mid\!\sim} q$. Such a non-monotonic inference operation is an important tool for studying non-monotonic inference and its relationship to classical consequence (as expressed by the consequence operation Cn). (Makinson 1994, 2005) (The need for this extension to sets of sentences is, by the way, another reason why non-monotonic inference should not be assumed to coincide with the non-nested fragment of conditional logic.)

7 Conclusion

At the outset we asked two questions about the relationships between belief revision, conditional statements, and non-monotonic inference: (1) Are these three notions closely enough related to be adequately representable in a unified framework covering them all? (2) If so, are they also interdefinable and in that case how?

Concerning the first question, the analysis proposed here suggests a positive answer. In particular, a framework based on an ordering ($\leq$) on the set of potential outcomes of belief change ($\mathbb{X}$) can be used for an analysis of all three concepts.

However, the above discussion suggests a negative answer to the second question. The interrelationships among the three concepts appear to be more

---

^a a conditional sentence should refer not only to the antecedent-satisfying possible worlds that are most similar to the actual world but to all those that are sufficiently similar.

intricate and open-ended than what has often been assumed. Conditional sentences can be elicited and/or context-dependent. Then the standard Ramsey test does not apply, and no other recipe for interdefinability seems to be available. Non-monotonic inference refers not only to the most plausible way to accommodate the antecedent but to a band of reasonably plausible ways to do this. Therefore, it does not seem to coincide with the flat fragment of conditional logic. The outer limits of what is a reasonably plausible belief set (for the purposes of non-monotonic inference) do not seem to be derivable from belief revision or from conditional logic, and therefore the logic of non-monotonic inference does not seem to be interdefinable with either of these. In summary, although belief revision, conditional statements, and non-monotonic inference are closely interrelated, each of them has important specific features that would have to be disregarded in any attempt to define one of them in terms of another.

References

Alchourrón, Carlos, Peter Gärdenfors, and David Makinson (1985) "On the logic of theory change: partial meet contraction and revision functions", *Journal of Symbolic Logic* 50:510-530.

Arlo-Costa, Horacio (2014) "The logic of conditionals". In Edward N. Zalta (ed.) *Stanford Encyclopedia of Philosophy*. Center for the Study of Language and Information (CSLI), Stanford University, 2014.
http://plato.stanford.edu/archives/sum2014/entries/logic-conditionals.

Boutilier, Craig (1990) "Conditional Logics of Normality as Modal Systems", pp. 594-599 in *AAAI-90: Proceedings. National Conference on Artificiell Intelligence, American Association for Artificial Intelligence*. Menlo Park: AAAI Press.

Chisholm, Roderick M. (1946) "The contrary-to-fact conditional", *Mind* 55:289-307.

Delgrande, James P. (1988) "An approach to default reasoning based on a first-order conditional logic: Revised report", *Artificial Intelligence* 36:63-90.

del Val, Alvaro (1997) "Non monotonic reasoning and belief revision: syntactic, semantic, foundational and coherence approaches", *Journal of Applied Non-Classical Logics* 7: 213-240.

Fermé, Eduardo and Sven Ove Hansson (2011) "AGM 25 years. Twenty-Five Years of Research in Belief Change", *Journal of Philosophical Logic* 40:295-331.

Fine, Kit (1975) *Critical Notice* [Review of Lewis, *Counterfactuals*.] *Mind* 84:451-458.

Gärdenfors, Peter (1986) "Belief revisions and the Ramsey test for conditionals", *Philosophical Review* 95:81-93.

Gärdenfors, Peter, and David Makinson (1994) "Nonmonotonic inference based on expectations", *Artificial Intelligence* 65:197-245.

Gauker, Christopher (1987) "Conditionals in context", *Erkenntnis* 27:293-321.

Ginsberg, Matthew L. (1986) "Counterfactuals", *Artificial Intelligence* 30:35-79.

Goldstick, D. (1978) "The truth-conditions of counterfactual conditional sentences", *Mind* 87:1- 21.

Goodman, Nelson (1947) "The problem of counterfactual conditionals", *Journal of Philosophy* 44:113-128.

Hansson, Sven Ove (1992) "In defense of the Ramsey test", *The Journal of Philosophy* 89:522-540.

Hansson, Sven Ove (1995) "The Emperor's New Clothes. Some recurring problems in the formal analysis of counterfactuals", pp. 13-31 in G. Crocco, L. Fariñas del Cerro and A. Herzig (eds) *Conditionals: from Philosophy to Computer Science*. Oxford: Clarendon Press.

Hansson, Sven Ove (1999) "A Survey of Non-Prioritized Belief Revision", *Erkenntnis* 50:413-427.

Hansson, Sven Ove (2014a) "Descriptor Revision", *Studia Logica* 102:955-980.

Hansson, Sven Ove (2014b) "Relations of Epistemic Proximity for Belief Change", *Artificial Intelligence* 217:76-91.

Hansson, Sven Ove (2015) "Iterated Descriptor Revision and the Logic of Ramsey Test Conditionals", *Journal of Philosophical Logic*, in press.

Katsuno, Hirofumi, and Ken Satoh (1991) "A unified view of consequence relation, belief revision and conditional logic", pp. 406-412 in John Mylopoulos and Ray Reiter (eds) *Proceedings of the Twelfth International Conference on Artificial Intelligence (IJCAI 1991)*, Darling Harbour, Sydney, Australia, 24-30 August 1991. San Mateo, CA.: Kaufmann.

Kern-Isberner, Gabriele (2004) "A thorough axiomatization of a principle of conditional preservation in belief revision", *Annals of Mathematics and Artificial Intelligence* 40:127-164.

Kern-Isberner, Gabriele (2008) "Linking Iterated Belief Change Operations to Nonmonotonic Reasoning", pp. 166-176 in *Proceedings, Eleventh International Conference on Principles of Knowledge Representation and Reasoning*.

Kraus, Sarit, Daniel Lehmann, and Menachem Magidor (1990) "Nonmonotonic reasoning, preferential models and cumulative logics", *Artificial intelligence* 44:167-207.

Lewis, David (1973) *Counterfactuals*. Oxford: Blackwell, Oxford, 1973.

Lewis, David (1979) "Counterfactual dependence and time's arrow", *Nous* 13:455-476.

Lindström, Sten and Wlodek Rabinowicz (1992) "Belief revision, epistemic conditionals and the Ramsey test", *Synthese* 91:195-237.

Mackie, John Leslie (1966) "Counterfactuals and causal laws", pp. 66-80 in R. J. Butler (ed.) *Analytical Philosophy*, First Series. Oxford: Oxford University Press6.

Makinson, David (1994) "General Patterns in Nonmonotonic Reasoning", pp. 35-110 in Dov M. Gabbay, Christopher John Hogger, and John Alan Robinson (eds) *Handbook of Logic in Artificial Intelligence and Logic Programming*, vol. 3. Oxford University Press.

Makinson, David (2005) *Bridges from Classical to Nonmonotonic Logic*. London: College Publications.

Makinson, David and Peter Gäardenfors (1991) "Relations between the logic of theory change and nonmonotonic logic", pp. 185-205 in André Fuhrmann and Michael Morreau (eds) *The Logic of Theory Change*. Springer: Berlin.

Mill, John Stuart (1856) *A system of logic, ratiocinative and inductive, being a connected view of the principles of evidence and the methods of scientific investigation*. London: J.W. Parker & Son. Quoted after Mill, John Stuart (1973) *Collected works of John Stuart Mill*, vol. 7, ed. J.M. Robson. Toronto: University of Toronto Press.

Nute, Donald (1975) "Counterfactuals and the similarity of worlds", *Journal of Philosophy* 12:773-778.

Nute, Donald (1981) "Introduction", *Journal of Philosophical Logic* 10:127-147.

Ramsey, Frank Plumpton (1931) *Foundations of Mathematics and Other Logical Essays*. New York: Routledge.

Rott, Hans (1989) "Conditionals and theory change: Revisions, expansions, and additions", *Synthese* 81:91-113.

Rott, Hans (1999) "Moody conditionals: Hamburgers, switches, and the tragic death of an American president", pp. 98-112 in Jelle Gerbrandy, Maarten Marx, Maarten de Rijke, and Yde Venema (eds) *Essays dedicated to Johan van Benthem on the occasion of his 50th birthday*. Amsterdam: Amsterdam University Press. http://www.illc.uva.nl/j50/contribs/rott.

Rott, Hans (2001) *Change, choice and inference: A study of belief revision and nonmonotonic reasoning*. Oxford: Clarendon Press.

Schlossberger, Eugene (1978) "Similarity and counterfactuals", *Analysis* 38:80-82.

Stalnaker, Robert C. (1968) "A Theory of Conditionals", pp. 98-112 in N. Rescher (ed.) *Studies in Logical Theory, American Philosophical Quarterly Monograph Series* no. 2. Oxford: Blackwell.

Stalnaker, Robert C. (1994) "What is a nonmonotonic consequence relation?", *Fundamenta Informaticae* 21:7-21.

Williamson, Colwyn (1969) "Analysing counterfactuals", *Dialogue* 8:310-314.

Wobcke, Wayne (1995) "Belief revision, conditional logic and nonmonotonic reasoning", *Notre Dame Journal of Formal Logic* 36:55-102.

Can non-monotonic logics model human reasoning?

Marco Ragni[1]

Abstract. For decades classical propositional and first order logic have been the normative framework in the psychology of reasoning, i.e., any deviating answer was considered wrong. Recent experimental research demonstrates that logically naïve reasoners can show the ability to reason non-monotonically. To model such human inferences non-monotonic reasoning systems can, despite their obvious application in the field of artificial intelligence (AI), provide a framework for cognitive modeling as well. The objective of this article is to review some core findings from psychology and advertising cognitive science as a new and important application field.

1 Introduction

A central aspect of intelligence, be it natural or artificial, is the ability to *reason* about given information, i.e., the ability to gain new information from existing knowledge. Human reasoning is among core cognitive abilities. It is a central issue in psychological research and gaining more and more interest in artificial intelligence (AI) [44, 33, 46, 14, 2]. Dating back more than 2500 years to Aristotle the question became pertinent what a correct logical inference is. Aristotle dealt with syllogistic reasoning that is reasoning about quantified assertions of the form "All men are mortal" with quantifiers like "All", "Some", "None", "Some ... not". The work by Aristotle demonstrates, in fact, two aspects that will be relevant later on: It is possible to analyze what inferences can be correctly drawn and human "inferences" can deviate from the "correct" answers making such an analysis otherwise obsolete.

Syllogistic inferences have been regarded for very long as the "only" way of correct thinking and such inferences have been considered normative, i.e., deviations from the system have been considered as erroneous. This normativity has been imported into the psychology of reasoning: Since the beginning of experimental psychological research of human reasoning (e.g., [45]) logic and especially first order logic has been regarded as the gold standard. Inhelder and Piaget [17] even assumed humans to develop into full first order logic capabilities.

Human reasoning is often divided and analyzed in the categories deductive, inductive, and abductive reasoning. By far most work in psychology and cognitive science has focused on "deductive reasoning" [3]. Human deductive reasoning, however, does not simply realize the laws of classical logic, but is

---

[1] Foundations of Artificial Intelligence, Technical Faculty, Georges-Koehler-Allee, Albert-Ludwigs-University Freiburg, `ragni@informatik.uni-freiburg.de`

rather more diverse and sometimes entirely different from it. We will later see that other, non-monotonic logics, such as Łukasiewicz's logic [29] are better suited to explain findings like the Wason Selection Task [47] or the Suppression Task [5].

The way of how humans *ought to reason correctly* is easier to describe and to model than the way of how humans *do* in fact reason, i.e., to identify the underlying mental representations and reasoning processes. While the first – to identify formal reasoning systems – is situated in logic, the second requires a multi-disciplinary approach with experimental methods from psychology, formal methods from AI (and philosophy), and modeling approaches from cognitive science. To understand some specifics of human reasoning to possibly enrich the formal approaches for applying them to human reasoning the rest of the paper will be guided by some core research questions:

1. Is human reasoning non-monotonic?
2. Is there an influence of content?
3. Is e.g., System P cognitively-adequate?

In AI, reasoning systems that satisfy some formal and relevant desires are analyzed and often applied to specific research questions that can even cover a model for commonsense reasoning. Cognitive modeling, the core method of cognitive science, is constrained by assumptions about mental representation and aims to integrate findings from psychological experiments; introspection is not considered a source for modeling as it can lead to illusions. The human reasoning process can be considered as a black-box with inputs (typically the problems participants are presented with) and the output (the given answer). The modeling task consists in uncovering, or more precisely to reverse engineer the human reasoning system. The goal of this reverse engineering process consists in finding an algorithmic cognitive theory that can explain empirical data and even predict new results. The generated data from empirical studies can contain the answers of participants and, hence, shows deviating answers from a reasoning system. But data can include response times, by assuming that a more difficult problem takes longer because additional mental operations or working memory is necessary, it can contain eye-tracking data for measuring attention. And recently even data about brain activations revealing the neural localization of processes, allowing to narrow down the kind of data that is processed (e.g., visual information in the visual cortex etc.). A model that attempts to integrate information and to explain *what* happens *when* on a process level is a cognitive model. Cognitive architectures like ACT-R[2] and the recent Neural Engineering Framework (NEF)[3] provide basic assumptions about how the different systems are represented and how information is distributed – they form the equivalent of a cognitive programming language. Nonetheless, such architectures and cognitive modeling in general requires an implementable theory about cognitive processes. The previously mentioned formal theories can serve as a test-bed for a cognitive theory of human reasoning. The advantage is that AI-theories are formalized and make clear predictions that can be falsified.

---

[2] http://act-r.psy.cmu.edu/
[3] http://nengo.ca/

2 Is human reasoning non-monotonic?

A consequence from the systematic deviation of humans from classical logical inference mechanisms is that propositional logic is not appropriate to model human reasoning. At least we need to have a reasoning formalism that contains the notion of plausability to model human everyday reasoning that is non-monotonic (e.g., [20, 43, 34]). Only recently there has been a shift away from using classical logic as the normative framework in the psychology of reasoning. The classical logic consequence relation $\models$ with $\alpha \models \beta$ represents: if α is true, then β *must* also be true. In contrast the non-monotonic consequence relation $\mid\!\sim$ with $\alpha \mid\!\sim \beta$ means: if α is true, then *typically* β is true as well. And most importantly for modeling human reasoning e.g., in the case of the suppression task $\alpha \mid\!\sim \beta$ does not imply $\alpha \wedge \alpha' \mid\!\sim \beta$. One approach is to find rules characterizing $\mid\!\sim$: for example, if $\alpha \mid\!\sim \beta$ and $\alpha \mid\!\sim \gamma$, then $\alpha \mid\!\sim \beta \wedge \gamma$.

One of the most analyzed experiments that is supporting non-monotonic reasoning is the Suppression Task. [5] showed that graduate students with no training in formal logic suppressed previously drawn conclusions when additional information is available. Interestingly, in some instances the previously drawn conclusions were valid while in others conclusions were invalid with respect to classical two-valued logic. Consider the following example (cp. the descriptions in Table 1). A first experimental group received only the conditional and the categorical statement, i.e., $e \rightarrow l$ and e . They drew the modus ponens inference in about 95% of the cases inferring that she studies late in the library l. A similar pattern appeared in experimental group 2 that received the conditional, the alternative argument, and the categorical statement $e \rightarrow l$ and e and $t \rightarrow l$. In this case participants had two conditionals and the categorical statement and they drew as group 1 the modus ponens *She will study late in the library* in about 98% of the cases.

(1) (α) *If she has an essay to write, then she will study late in the library* and
    (β) *If she has a textbook to read, then she will study late in the library* and
    (δ) *She has an essay to write.*

Group 3, however, received the conditional, the additional, and the categorical statement.

(2) (α) *If she has an essay to write, then she will study late in the library* and
    (γ) *If the library stays open, she will study late in the library* and
    (δ) *She has an essay to write.*

But they drew the modus ponens inference in less than 38% of the cases – $e \rightarrow l$ and e and $o \rightarrow l$, i.e., only 38% responded that she will study late in the library. This shows that although the conclusion is still correct (in a classical logical sense), it is suppressed by an additional conditional. In other words the new information led to a decrease of about 60% to apply the modus ponens. It is an excellent example of the human capability to draw *non-monotonic* inferences. This example explains why this experiment is called the Suppression Task – new information suppresses conclusions.

Table 1. Three modus ponus problems of the Suppression Task [5]. Three groups received three kinds of problems consisting of the categorical information and the conditional alone ($\alpha\delta$), the categorical, conditional and alternative statement ($\alpha\beta\delta$), or the categorical, conditional and additional argument ($\alpha\gamma\delta$).

Conditional (α)	Alternative (β)	Additional(γ)	Categorical (δ)
If she has an (e)ssay to write, then she will study late in the (l)ibrary	If she has a (t)extbook to read, then she will study late in the (l)ibrary	If the library stays (o)pen, then she will study late in the (l)ibrary	She has an (e)ssay to write

Another recent approach investigated if a multi-valued logic is able to explain human reasoning. One of the main criticisms that Bayesian researchers make against the mental model theory is that problems are typically framed such that "a statement or premise holds or does not hold," while in everyday reasoning we typically have degrees of certainty. Despite the great success of two valued logics in artificial intelligence or cognitive science, the truth of a statement or premise cannot always be determined and so it might be cognitively plausible to introduce a third truth value, namely *unknown/undefined*. The first three-valued logic has been introduced by [29]. This logic differs from e.g., the Kleene approach [26] in the interpretation how u implies u is interpreted, in the latter case it is evaluated as u. But using only a third value does not necessarily lead to cognitive adequacy. The way of how knowledge is represented in a knowledge base is important the so-called *conceptual adequacy* [43, 10]. Premise information needs to be adequately represented in a knowledge base. And if the goal is logical reasoning it needs an appropriate logical form. In this sense [43, 10] used logic programs to represent the information. And, based on the logical representation the inferences need to be cognitively-adequate (*inferential adequancy*: [43, 10]), i.e., the operations need to resemble human operations.

For the conceptual adequacy the conditional *if she has an essay to finish, she will study late in the library* or short ($l \leftarrow e$) is often (mentally) represented by $l \leftarrow e \wedge \overline{ab_1}$, where ab_1 is an *abnormality* predicate which expresses that l holds if e holds and nothing abnormal is known. Dietz and colleagues [10] used this for the Lukasiewicz logic and modeled the suppression task of [43]. In Table 2 two logic programs are presented for the two examples of the Suppression Task (cp. [10], and Table 1).

($\alpha'\beta'\delta'$) If she has an essay to write (e) and nothing abnormal (ab_1) is known,
    then she will study late in the library (l) and
    If she has a textbook to read (t) and nothing abnormal (ab_2) is known,
    then she will study late in the library and
    She has an essay to write.

Table 2. The computational logic approach to the two problems $\alpha\beta\delta$ and $\alpha\gamma\delta$; ab are abnormality predicates; WCS is the weak completion semantics; Table adapted from [7].

Problems	$\alpha\beta\delta$	$\alpha\gamma\delta$
Program	$l \leftarrow e \wedge \overline{ab_1}$	$l \leftarrow e \wedge \overline{ab_1}$
	$l \leftarrow t \wedge \overline{ab_2}$	$l \leftarrow o \wedge \overline{ab_3}$
	$ab_1 \leftarrow \bot$	$ab_1 \leftarrow \overline{o}$
	$ab_2 \leftarrow \bot$	$ab_3 \leftarrow \overline{e}$
	$e \leftarrow \top$	$e \leftarrow \top$
WCS	$l \leftrightarrow (e \wedge \overline{ab_1})$	$l \leftrightarrow (e \wedge \overline{ab_1})$
	$\vee (t \wedge \overline{ab_2})$	$\vee (o \wedge \overline{ab_3})$
	$ab_1 \leftrightarrow \bot$	$ab_1 \leftrightarrow \overline{o}$
	$ab_2 \leftrightarrow \bot$	$ab_3 \leftrightarrow \overline{e}$
	$e \leftrightarrow \top$	$e \leftrightarrow \top$
Inferred (True in Least Model)	$\{e, l\}$	$\{e\}$
Percentage participants inferring l	96%	38%

($\alpha'\gamma'\delta$) *If she has an essay to write (e) and nothing abnormal (ab_1) is known, then she will study late in the library (l) and*
If the library stays open (o) and nothing abnormal (ab_3) is known, she will study late in the library (l) and
She has an essay to write.

The abnormality predicates (e.g., ab_1) represent abnormal cases: For instance, ab_1 is true when *the library does not stay open* and ab_3 is true when *she does not have an essay to finish*. The abnormality ab_3 must be understood in the context of the problem description, i.e., the second conditional refers to the task mentioned in the first conditional. The logic programs and the inferences can be found in Table 2. The so-called weak-completion semantics process works as follows [16]:

1. Replace all clauses with the same head by a disjunction of the body elements, i.e., $A \leftarrow B_1, \ldots, A \leftarrow B_n$ by $A \leftarrow B_1 \vee \ldots \vee B_n$.
2. Replace all occurrences of $\leftarrow$ by $\leftrightarrow$.

The resulting set of equivalences is called the *weak completion* and the model intersection property holds for weakly completed programs [16] guaranteeing the existence of a least model.

From a psychological perspective the weak completion process reminds of an interpretation of a conditional as a biconditional. A finding from [21] show that even in those cases where content and context are neutral, a conditional can be sometimes interpreted as a conditional and sometimes as a biconditional. Hence, in the case of the suppression task, at least, the weak completion process can resemble an operation that might have a cognitive counterpart.

While in AI and philosophy a reasoning system is chosen, in humans non-monotonicity could be triggered by the awareness of alternative facts (see above,

Table 3. Results for the modus ponens analysis of the suppression task [39].

	Group A	Group B	Group C	Group D
	If X then Y	If X then *mostly* Y	If X then Y	If X then Y
				If V then Y
				If Z then $\neg Y$
	X	X	Mostly X	X
Conclusion	Y (88%)	Y (83%)	Y (75%)	Y (83%)
Likeliness	Y (87%)	Y (67%)	Y (70%)	Y (76%)

the library can be closed), this can be influenced by additional knowledge from background knowledge (if exceptions are obvious). A second possibility could be that some keywords can trigger non-monotonic reasoning processes more than others. If we express, for instance, that a bird can typically fly then we may trigger the processing of non-monotonicity. But what are relevant non-monotonic keywords?

In another experiment[4] participants were presented with an assertion like *Normally, objects of type A are also of type B*. The task of the participants were to rate: "How many percent of objects of type A are of type B according to your general understanding of the word "normally"? This has been done systematically for the keywords: "All, usually, most, the majority of, few, seldom, normally, typically, some, the minority of, rarely, no" in an interval based (highest and lowest number on a scale from 0 to 100) and a point based interpretation (absolute value). This showed that the best keywords allowing exceptions were "normally, mostly, and usually". Based on this we performed another experimental study investigating the appropriate position of triggering keywords and its influence on the acceptance of the four reasoning types (modus ponens, denial of antecedent, affirmation of the consequent, modus tollens). The research question was: Can we trigger degrees of non-monotonicity, e.g., influence the degree of certainty/uncertainty? To control for background knowledge we used the same problems like [5] and an abstracted alien scenario and compared it.

Above are the results for the simplest reasoning inference (modus ponens). It shows that keywords can modulate the monotonicity, that non-monotonicity triggering keywords attached to a conditional are somehow absorbed by the conditional but not by the categorical statement. If participants have to rate how likely they believe their answer to be correct they drop more for the inconcrete qualitative keyword "most" then for Group 4 where concrete cases are given. In other words the uncertainty is greater by using keywords.

3 Is there an influence of content?

Let us consider in this approach two formally relevant questions: Are the problems closed under isomorphy? In other words does a structurally identical problem show the same performance on human reasoning? If this is the case then

---
[4] Conducted together with Gregory Kuhnmünch

this could support both a semantic and a syntactic modeling approach. In reasoning about conditionals two important findings are relevant: The Wason Selection Task [47] in the social and abstract case and the belief bias finding. Let us consider the first.

> Participants are shown 4 cards and on each card there is a number on one side and a letter on the other side. Since the cards are laying on the table the participant sees only one side. Two of the cards show letters "A" and "D" and two other cards show numbers "2" and "7". The participant's task is to select the cards that need to be turned over in order to test the truth of the statement "If the letter side shows an "A" then the other side shows a "2". How many cards and which have to be turned in order to show that the rule hold or does not hold? (cp. [47])

Answers from the participants about the number of cards to be turned range from 1 to 4 and all combinations of answers were given. But a pattern appears: Most participants would turn the "A" since the number on the back could falsifiy the conditional and then most want to flip the card with number "2". Of course the card "7" is the correct answer as only this can falsify the conditional, but less than 10% chose this together with an "A". Possible explanations vary from heuristics and information gain [34] to a biconditional interpretation of the conditional statement [30] of the logically naïve reasoners (this terms means only that participants in such experiments do not have any training in logic). One explanation was that the domain is too abstract and the possible lack of background knowledge was why participants made this mistake. For this reason Johnson-Laird and colleagues [22] examined the influence of background knowledge and were able to show that a similar problem was solved correctly. Consider the following problem:

> You are a police officer and have to check if guests adhere to the following rule "If someone drinks alcohol then this person mus be over 21." And there are 4 guests sitting in the restaurant at this time (that are represented on cards as above). On the first card there is a person drinking beer, on another card a water, the third card has the age of 22 on it and the last one has 17 years on it. Which cards would you turn to check the rule.

Cosmides and Tooby [6] could show that in this formulation over 70% get it right. The most widely accepted explanation is that humans are better in detecting deviations from social rules. Since the context is crucial, this result shows that we are not just applying rules, regardless of the domain. Human reasoning is context dependent.

Other reasoning difficulties incorporate that the problem space is sometimes too large to be fully represented in working memory [41]. A possible leading to the use of heuristics. Heuristics do not require a complete problem representation and manipulation. This task-dependent benefit is sometimes offset by a disadvantage: in spite of the existence of an optimal solution, none or only a suboptimal solution is found. We will later see that ternary logic can explain the behavior satisfactorily [9].

Taken together the human reasoning process does not consist of some identifiable rules only. There must be more than content-independent rules making semantic interpretations much more likely. But we will see that a pure model-based approach has as well some problems – at least not all models are created equal. Let us briefly consider the notion of *mental models* and how it is used in the literature. According to [18, p. 932] "a mental model is [...] an internal representation of a state of affairs in the external world." This definition allows for any kind of models, even mathematical models. *Mental models*, however, consist of elements that represent premise information by a partially ordered set of tokens. For instance, let us consider the implication in the mental model theory [18]:

Assertion	Mental model	Explicit models
If A then B	A B	A B
	...	¬A ¬B
		¬A B

The initial mental model, is the mental model the average reasoner often constructs first. In this case for the implication it is both the antecedent [A] and the consequence [B] being true at the same time. The construction of the initial mental model can, however, depend on background knowledge, semantics of the premise information, or level of expertise [19]. The ellipsis in the table represent the additional information that other models are possible. They can be "fleshed out" [19], if necessary. In this case the mental model is reminiscent of a variation of a truth table:

Assertion	A B
If A then B	+ +
	– –
	– +

The mental model theory is inspired by a semantic approach and by truth-tables (compare the A and Bs in the Table above with the '+' in the composition table). The alternative models Mental model theory adheres to at least three additional principles: the *principle of truth*, the *principle of iconicity*, and the ordering of the models. The principle of truth claims that humans cannot represent what is false (what does not hold, i.e., all entries in the truth table with '–'), but only what is true. This principle eliminates the third row in the truth table. And, as [13] explicate, the nature of mental models as *iconic*, mental models typically do not use the "+" or "–" notation, instead preferring *abstract tokens* that function as place holders for any kind of events, terms, or objects such as houses or fruits[1]. And the models do not appear in an arbitrary order but follow a sequence of interpretations. In the case above, reasoners prefer the case ¬A ¬B over ¬A B [23]. It could be interesting to relate this in future work to c-representations [24] and to ranking functions [42].

Another point is: If a conclusion sounds likely to be true (even if it is logically false) then participants do more likely accept this conclusion than a conclusion (even if it is logically true) contradicts current beliefs. Consider the following two examples from [11]: "If the Queen dies then Prince Charles will become King." has a high believability, while "If the allied troops pull out of Iraq, then it will become a democracy" has a lower believability. And even for the modus ponens case (e.g., the Queen dies or the allied troops pull out of Iraq) participants drew the inference less likely (92% in the first case or 72% in the lower believability case).

This demonstrates that the human reasoning process must be distinguished for likely and unlikely conclusions – especially if we reason under time pressure [11, 25]. But the way of how humans understand and interpret logical connectives makes a difference. We have seen that in the abstract case of the Wason Selection Task participants may interpret conditionals as biconditionals depending on the kind of problems, e.g., in causal reasoning this is very often interpreted biconditionally.

A main difference between formal approaches and human reasoning is that in logic the semantics is introduced while in every day life we learn the semantics. Additionally, there are communicative principles that make a difference to the logical understanding. For instance, if we hear that "some tax-payers are defrauders" we assume that there are "some tax-payers who are not defrauders". But the existential quantifier in logic does not make such an implication. This is a consequence of Gricean implicature [15] that supports that human communication tends to be as informative as possible (among other principles). But if someone states that there are only "some tax-payers", then there must be "some tax-payers" for whom this property does not hold. Otherwise it follows from the principles of Grice that the person communicating this information would have used the quantifier all. In other words a reasoning system must be sensitive for the interpretation of connectives and quantifiers. So it is possible to discern different levels of reasoning. There is the level of logic, there is level of communication principle and pragmatics, and these have an influence on the given answer.

Findings from the believability to Gricean Implicature demonstrate that a simple reverse-engineering of the human reasoning process from a response is not possible. Furthermore it demonstrates the need for a formal investigation of the reasoning processes. Despite the factor that content and language have on the reasoning process we need to know what could describe the inference part.

4 Is System P cognitively adequate?

In Artificial Intelligence (AI) a wide range of non-monotonic logics have been developed in the last years, e.g., default logics [40], auto-epistemic logics [32], circumscription logic [31], and logic programming approaches [12]. Different logical systems, such as System C for cumulative logics and System P for preferential reasoning, are determined by their set of inference rules or operators on defaults [27]. For example, the Or-rule ("if α normally implies γ and β normally implies γ, then α or β normally imply γ") is part of System P, but not part of System C. Therefore, empirical support for this rule will reduce the plausibility of System C as an adequate human non-monotonic reasoning theory. Default logic [40] becomes too implausible if the Or-rule holds. Since Default Logic does not satisfy the rule "Cautious Monotonicity" a further distinction between AI-systems and cognitively adequate systems is possible. There are several non-monotonic approaches in AI but so far there are only few attempts to test empirically which frameworks could be used to model the human inference process. System P has received special attention since these rules can be considered a minimal set of rationality postulates [4]. Work from Pfeifer and Kleiter [36] argues for a probabilistic interpretation of System P. They found some support for System P testing in the first case reasoning about intervals and demonstrating for instance the Cut-Rule. Recent research [28] investigated which of 23 non-monotonic inference rules humans draw and which are rather neglected. Participants received abstract properties of figures (any combination possible) of *forms* e.g., quadratic, circular, triangular, *colors*, e.g., violet, green, blue, and *filling*, e.g., shaded, checkered, filled out. Hence, there was no closed world by instruction assumed. The case scenario was a fictive one: "Imagine you are a visitor of factory observing the properties of produced artifacts". Participants received two premises as follows:

> Premise 1: The figures that are quadratic and violet normally are shaded.
> Premise 2: The quadratic figures normally are violet.

What follows?

The task of the participants was to identify from a given set of answer possibilities a conclusion. For the problem above a conclusion consistent by the Cut-rule with system C, CL, and P would be *"the quadratic figures normally are shaded"*.

Name	Element in Reasoning System	Formula	Drawn by participants [28]
Reflexivity	C, CL, P	$\dfrac{}{\alpha\mid\!\sim\alpha}$	n/a
Left Logical Equivalence	C, CL, P	$\dfrac{\models \alpha\leftrightarrow\beta,\ \alpha\mid\!\sim\gamma}{\beta\mid\!\sim\gamma}$	46%
Right Weakening	C, CL, P	$\dfrac{\models \alpha\to\beta,\ \gamma\mid\!\sim\alpha}{\gamma\mid\!\sim\beta}$	46%
Cut	C, CL, P	$\dfrac{\alpha\mid\!\sim\beta,\ \alpha\wedge\beta\mid\!\sim\gamma}{\alpha\mid\!\sim\gamma}$	38%
Cautious Monotonicity	C, CL, P	$\dfrac{\alpha\mid\!\sim\beta,\ \alpha\mid\!\sim\gamma}{\alpha\wedge\beta\mid\!\sim\gamma}$	15%
Loop	CL, P	$\dfrac{\alpha_0\mid\!\sim\alpha_1, \alpha_1\mid\!\sim\alpha_2,\dots,\alpha_k\mid\!\sim\alpha_0}{\alpha_0\mid\!\sim\alpha_k}$	26%
Or	P	$\dfrac{\alpha\mid\!\sim\gamma,\ \beta\mid\!\sim\gamma}{\alpha\vee\beta\mid\!\sim\gamma}$	65%

The study [28] demonstrates that there are inference rules in the three systems C, CL, and P that only very view reasoners drew. Another interesting finding is that few reasoners in the experiment applied cautious monotonicity.

But, cautious monotonicity holds in Systems C, Cl, and P and some others. So is System P cognitively inadequate? Not necessarily. The findings demonstrates that there are cases in which humans did reason non-monotonically did not trigger cautious monotonicity. An interesting further empirical investigation is to test, if and under which circumstances hold very cautious, cautious, and rational monotony in the human reasoning process.

5 Discussion & new research directions

In the psychology of reasoning there are numerous theories about the way humans reason. Some of these theories are formalized and others not. To be able to equip and develop cognitive systems we need to have formal and implemented theories that can explain, reproduce, and predict human inferences. A sensible step could be to start with existing non-monotonic reasoning formalisms and test how good they predict the behavioral data. The reported experiments show that it is possible to trigger non-monotonic reasoning in humans (by keywords and by providing exceptions). It seems worth to investigate if there is a degree of non-monotonicity that can be triggered by the number of exceptions. In other words the difference in the reasoning problems with an additional or alternative conditional in the experiment by Byrne [5] is probably not a dichotomous distinction but might be a continuum. Additional experimental research is necessary to show how much this can trigger cautious reasoning, and if, this may then even correspond with the reasoners use of other formal non-monotonic reasoning systems. This could offer an explanation why, for instance, there are differences between [36] and [28]. In future experiments the number of possible exceptions must be strictly controlled.

The reported analysis of System P shows that some of the participants can be in a cautious reasoning mode while others are certainly not. Besides inter-individual differences between reasoners (the analysis in [28] shows some internal consistency for each reasoner) – there is still additional research necessary to identify in which circumstances humans use such cautious reasoning and if content or a formulation as a social rule (like in the Wason Selection Task) can lead to differences. The findings show limitations of a pure rule-based approach to be cognitively-adequate. Some findings indicate that humans build so-called preferred mental models in spatial reasoning [38] and in syllogistic reasoning [37] and it may hold for conditional reasoning as well. Here c-representations and ranks may provide a very fruitful path to go.

Another approach is the use of multi-valued logics. Recent analysis demonstrate that using Lukasiewicz logic together with a weak completion semantics could explain findings for conditional reasoning and as well for spatial reasoning [8] and syllogistic reasoning [35]. Several research questions are still open, e.g., from the formal properties about semantics (when do humans use weak completion?) or do they reason *skeptically* or *credulously* [10]? Is it possible to include ranking functions on the models? How can alternative models be generated? What are the differences to alternative non-monotonic logics?

A non-monotonic logic alone does not capture all cognitive processes: For instance in the belief bias humans do not always reason only in a logical sense from premises to conclusion but they may use heuristics. One heuristic is that in some cases where a conclusion is given a reasoner checks only if a putative conclusion is possible or likely (even based on background information) statement that is consistent with the given knowledge, the premises, is interpreted as the conclusion. This "conclusion" might be read out from the least model of the premises. This could be a possible explanation why so many participants hold erroneously findings for true.

The objective of this article was to give an overview on some relevant aspects from a cognitive perspective that may help to develop "cognitive" theories based on existing formalisms to explain and predict inferences drawn by humans. Non-monotonic logics offer many advantages for cognitive theories: they are already formalized, implementable, and their logical properties are known. But adaptations of the theories are necessary and a thorough benchmark from cognitive science is still missing to evaluate theories.

6 Acknowledgments

This work was supported by DFG-Grant RA1934 2/1 as part of the priority program "New Frameworks of Rationality" (SPP 1516) and a Heisenberg DFG fellowship RA1934 3/1. The author is very grateful to two anonymous reviewers.

References

1. B. Bara, M. Bucciarelli, and V. Lombardo. Model theory of deduction: a unified computational approach. *Cognitive Science*, 25:839–901, 2001.
2. Chitta Baral. *Knowledge representation, reasoning and declarative problem solving*. University Press, Cambridge, 2003.
3. S. Beller and H. Spada. Denken. In G. Strube, editor, *Wörterbuch der Kognitionswissenschaft*. Klett-Cotta-Verlag, 2001.
4. Salem Benferhat, Didier Dubois, and Henri Prade. Possibilistic and standard probabilistic semantics of conditional knowledge bases. *Journal of Logic and Computation*, 9(6):873–895, 1999.
5. R. M. Byrne. Suppressing valid inferences with conditionals. *Cognition*, 31:61–83, 1989.
6. L. Cosmides and J. Tooby. Cognitive adaptations for social exchange. In J. H. Barkow, L. Cosmides, and J. Tooby, editors, *The Adapted Mind: Evolutionary Psychology and the Generation of Culture*, pages 163–228. Oxford, New York, NY, 1993.
7. E.-A. Dietz, S. Hölldobler, and Marco Ragni. A Simple Model for the Wason Selection Task. In T. Barkowsky, Marco Ragni, and F. Stolzenburg, editors, *Report Series of the Transregional Collaborative Research Center SFB/TR 8 Spatial Cognition*, number 032-09/2012 in Transregional Collaborative Research Center SFB/TR 8 Spatial Cognition, 2012.
8. EA Dietz, S Hölldobler, and R Höps. A computational logic approach to human spatial reasoning. Technical report, Technical Report KRR-2015-02, TU Dresden, International Center for Computational Logic, 2015.
9. Emanuelle-Anna Dietz, Steffen Hölldobler, and Marco Ragni. A Computational Logic Approach to the Abstract and the Social Case of the Selection Task. In

Leora Morgenstern, Ernest Davis, and Mary-Anne Williams, editors, *11th International Symposium on Logical Formalizations of Commonsense Reasoning*, 2013.
10. Emmanuelle-Anna Dietz, Steffen Hölldobler, and Marco Ragni. A Computational Approach to the Suppression Task. In N. Miyake, D. Peebles, and R.P. Cooper, editors, *Proceedings of the 34th Annual Conference of the Cognitive Science Society*, pages 1500–1505, Austin, TX, 2012. Cognitive Science Society.
11. Jonathan St. B. T. Evans, Simon J. Handley, and Alison M. Bacon. Reasoning under time pressure. *Experimental Psychology*, 56(2):77–83, 2009.
12. Michael Gelfond and Vladimir Lifschitz. The stable model semantics for logic programming. In R. A. Kowalski and K. A. Bowen, editors, *Proceedings of the International Logic Programming Conference and Symposium, 5th ICLP/SLP*, pages 1070–1080, Cambridge, MA, 1988. MIT Press.
13. G. P. Goodwin and Philip N. Johnson-Laird. Reasoning about relations. *Psychological Review*, 112(2):468–492, 2005.
14. James G Greeno and Herbert A Simon. Problem solving and reasoning. Technical report, DTIC Document, 1988.
15. H.P. Grice. Logic and conversation. *Syntax and Semantics. 3: Speech Acts*, pages 41–58, 1975.
16. Steffen Hölldobler and Carroline Dewi Kencana Ramli. Logic programs under three-valued Łukasiewicz semantics. In Patricia M. Hill and David Scott Warren, editors, *Logic Programming, 25th International Conference, ICLP 2009*, volume 5649 of *Lecture Notes in Computer Science*, pages 464–478, Heidelberg, 2009. Springer.
17. B. Inhelder and J. Piaget. *The Growth of Logical Thinking from Childhood to Adolescence: An Essay on the Construction of Formal Operational Structures*. Developmental psychology. Routledge, 1958.
18. P. N. Johnson-Laird and R. M. J. Byrne. *Deduction*. Erlbaum, Hillsdale, NJ, 1991.
19. Philip N. Johnson-Laird. *How We Reason*. Oxford University Press, New York, 2006.
20. Philip N. Johnson-Laird. Deductive reasoning. *Wiley Interdisciplinary Reviews: Cognitive Science*, 1(1):8–17, 2010.
21. Philip N. Johnson-Laird, R. M. J. Byrne, and W. Schaeken. Propositional reasoning by model. *Psychological Review*, 99(3):418–439, 1992.
22. Philip N. Johnson-Laird, P. Legrenzi, and M. S. Legrenzi. Reasoning and a sense of reality. *British Journal of Psychology*, 1972.
23. PN Johnson-Laird, Sangeet S Khemlani, and Geoffrey P Goodwin. Logic, probability, and human reasoning. *Trends in cognitive sciences*, 19(4):201–214, 2015.
24. Gabriele Kern-Isberner. A thorough axiomatization of a principle of conditional preservation in belief revision. *Annals of Mathematics and Artificial Intelligence*, 40:127–164, 2004.
25. K. C. Klauer, J. Musch, and B. Naumer. On belief bias in syllogistic reasoning. *Psychological Review*, 107(4):852–884, 2000.
26. S. C. Kleene. On notation for ordinal numbers. *The Journal of Symbolic Logic*, 3(4):150–155, 1938.
27. Sarit Kraus, D. Lehmann, and M. Magidor. Nonmonotonic reasoning, preferential models and cumulative logics. *Artificial Intelligence Journal*, 44:167–207, 1990.
28. Gregory Kuhnmünch and Marco Ragni. Can Formal Non-monotonic Systems Properly Describe Human Reasoning? In P. Bello, M. Guarini, M. McShane, and B. Scassellati, editors, *Proceedings of the 36th Annual Conference of the Cognitive Science Society*, pages 1222–1228. Austin, TX: Cognitive Science Society, 2014.

29. Jan Łukasiewicz. O logice trójwartościowej. *Ruch Filozoficzny*, 5:169–171, 1920. English translation: On three-valued logic. In: Łukasiewicz J. and Borkowski L. (ed.). (1990). *Selected Works*, Amsterdam: North Holland, pp. 87–88.
30. K. I. Manktelow. *Reasoning and Thinking*. Psychology Press, Hove, UK, 1999.
31. John McCarthy. Circumscription: A non-monotonic inference rule. *Artificial Intelligence*, 13(1):2, 1980.
32. Drew McDermott and Jon Doyle. Non-monotonic logic i. *Artificial intelligence*, 13(1):41–72, 1980.
33. Allen Newell. Reasoning, problem solving and decision processes: The problem space as a fundamental category. Technical report, Computer Science Department CMU, 1979.
34. M. Oaksford and N. Chater. *Bayesian rationality: The probabilistic approach to human reasoning*. University Press, Oxford, 2007.
35. Luis Moniz Pereira, Emmanuelle-Anna Dietz, and Steffen Hölldobler. Contextual abductive reasoning with side-effects. *arXiv preprint arXiv:1405.3713*, 2014.
36. Niki Pfeifer and GernotD. Kleiter. Coherence and nonmonotonicity in human reasoning. *Synthese*, 146(1-2):93–109, 2005.
37. Marco Ragni, Sangeet Khemlani, and PN Johnson-Laird. The evaluation of the consistency of quantified assertions. *Memory & Cognition*, 42(1):1–14, 2014.
38. Marco Ragni and Markus Knauff. A theory and a computational model of spatial reasoning with preferred mental models. *Psychological Review*, 120(3):561–588, 2013.
39. Marco Ragni, Gregory Kuhnmünch, and Barbara Kuhnert. Suppressed conclusions: An analysis. in prep.
40. Raymond Reiter. A logic for default reasoning. *Artificial Intelligence*, 13(1):81–132, April 1980.
41. L. J. Rips. *The psychology of proof: Deductive reasoning in human thinking*. The MIT Press, Cambridge, MA, 1994.
42. Wolfgang Spohn. *The Laws of Belief: Ranking Theory and Its Philosophical Applications*. Oxford University Press, Oxford, UK, 2012.
43. K. Stenning and M. Lambalgen. *Human reasoning and cognitive science*. Bradford Books. MIT Press, Cambridge, MA, 2008.
44. Robert J. Sternberg. Reasoning, Problem Solving, and Intelligence. Technical report, DTIC Document, 1980.
45. G. Störing. *Experimentelle Untersuchungen über einfache Schlussprozesse*. W. Engelmann, 1908.
46. P. C. Wason. Problem solving and reasoning. *British Medical Bulletin*, 27(3):206–210, 1971.
47. P. C. Wason and Philip N. Johnson-Laird. *Thinking and reasoning*. Penguin Books, 1968.

Enumerative Induction

Wolfgang Spohn[1,2]

Abstract. This is a philosophical paper about enumerative induction, a basic rule of inductive or defeasible reasoning. It shows that it is almost satisfactorily accounted for in probability theory and even more adequately in ranking theory. As such it certainly deals with the foundations of formal rationality and should be of interest not only to philosophers, but also to the Artificial Intelligence community, which is deeply engaged in inductive reasoning as well.

1 Introduction

Enumerative induction is a kind of inference: from "the 1st F is G, the 2nd F is G, ..., the n-th F is G" or from "all observed Fs are G" infer all "Fs are G" or even "it's a law that all Fs are G". This is obviously not a deductively valid inference. Rather, it is a defeasible or nonmonotonic inference: add the premise "the $n+1$st or the next observed F is not G", and you would reject the conclusion. So, is it an inductively valid inference? Hard to say; we have no good criterion of inductive validity. For some it is the most basic inductive inference of all. For some it is much too primitive and not anything we can rely on in our inductive practice. Still others think that it leads into contradictions straightaway.

In any case, it is a most suggestive and most venerable inductive inference rule. Since mankind can think, it uses this rule. We generalize all the time, we wouldn't survive without doing so, and this finds its immediate expression in this rule. As indicated, the rule is contested and has been much discussed. The present situation, though, is strangely silenced. I guess, most philosophers think that this rule has finally found its Bayesian home. And even though this home may not be fully comfortable, they feel we need no longer be concerned.

Well, we shouldn't take this to be the end of the story. There is at least an afterword to it. As this paper will explain, the ranking-theoretic home is even more adequate. We will find that within this framework the above inferences hold good, even with the stronger nomological conclusion and without danger of contradiction. The afterword will not be particularly complicated. In my view, it's just a matter of an adequate conceptualization—which ranking theory is able to provide.

---

[1] Fachbereich Philosophie, Universität Konstanz, 78457 Konstanz, Germany

[2] I dedicate this paper to Gabriele Kern-Isberner. I am not well suited to grasp all the merits of her work. However, from my perspective I am very happy and grateful for the brilliant uses she has made of ranking theory invented by me in 1983, for instance in her Habilitationsschrift (1999), in Kern-Isberner (2004), and many papers thereafter.

The plan of the paper is as follows: First, it should, and does, deliver a bit of background. Section 2 will give a very brief historical sketch of enumerative induction. This history is interwoven with the history of inductive skepticism, as indicated in section 3. Section 4 will explain how enumerative induction finally arrived at its Bayesian home, not without frictions, though. Whenever there is a probabilistic story, there is also an analogous ranking-theoretic story. This holds in the present case, too. So, section 5 will tell that analogous story, in which the Bayesian frictions will simply vanish. Section 6 concludes with two observations concerning Goodman's new riddle of induction and concerning the alleged apriority of the uniformity of nature.

The Bayesians already discovered that their (slightly distorted) version of enumerative induction is entailed by symmetry considerations (and minor additional premises). This entailment will stand out even more clearly in the ranking-theoretic story. Thus, enumerative induction is not a basic inductive inference, as it has seemed through centuries, but is derived from even more basic features of our inductive constitution.

2 A Few Historical Remarks About Enumerative Induction

Human thinking generalizes. And as soon as reflection sets in, one wonders what one is really doing there. So it is not surprising that this generalizing inference was an important topic already in early Indian philosophy. How early is hard to say, because there is often a big time gap between thinking and writing down in Indian philosophy. In any case, there was a debate how many positive instances were required for drawing this inference, and there was a doctrine that one positive instance, one F that is a G, may suffice, at least when F and G stand for the right kind of universals. There was also a clear awareness of the danger of overgeneralization. The generalizations need to be hedged by something like varying the conditions and noting the ensuing differences. Only then one can hope to specify the right kind of F (which may, of course, be logically complex taking account of all the conjectured necessary and sufficient conditions).[3]

The inference was also a concern in Greek philosophy. Aristotle discussed it in his *Topoi* and his *Analytica Priora* under the label *epagoge*. Being aware that it is not cogent, but only more or less persuasive, he tried to find conditions of admissibility. The inference was also an issue between the Epicureans and the Stoics. The more empirically inclined Epicureans tried to spell out conditions under which the inference acquires more certainty, while the more rationalistic Stoics emphasized its ineliminable uncertainty. It's no surprise that already the ancient Skeptics complained that the inference lacks any justification. We owe, by the way, the term *inductio* to Cicero who thus trans-lated the Aristotelian term.[4]

---

[3] Cf., e.g., Smart (1967, pp. 164ff.)
[4] Cf. Ruzicka (1976, pp. 323-325).

The rediscovery of Greek philosophy in the Middle Ages naturally led to a rich discussion of enumerative induction in Scholastic philosophy.[5] Still, all those treatments appear quite academic. The topic acquired real methodological importance only with the rise of empirical science in the 16th century, and, moreover, theoretical wit with the rise of what can be properly called probability theory in the 17th century. And, no doubt, the herald of the new inductive methods was Francis Bacon.

Bacon, particularly in (1620), developed quite sophisticated canons of the scientific method or of inductive reasoning, which served as a sort of blueprint for John Stuart Mill's much more elaborate theory in his (1843, Book III, Chs. 8–10) more than 200 years later. One might say that Bacon's ideas were guided by trust and mistrust in enumerative induction alike. Unguarded generalization would be childish. Hence, he developed what he called the Table of Affirmation, the Table of Negation, and most importantly, the Table of Comparison, which Mill turned into this Method of Concomitant Variations. These refinements are required, since inductive generalizations are counter-acted by eliminations through counter-instances. And this initiates a search for ever more detailed and accurate generalizations.[6]

In modern terms, not available at those times, one might say that the conclusion of enumerative induction is only a ceteris paribus law: ceteris paribus, all Fs are G. And then Bacon's and Mill's sophisticated methodology might be understood as trying to spell out how to substantiate this sweeping reference to ceteris paribus conditions. Not that this remark would clarify much; all the present treatments of ceteris paribus laws are just as tentative as those historical inductive canons.[7] It is noteworthy, though, that the problems ad-dressed now and then are the same.

The advent of probability theory definitely widened the perspective. Probabilistic inference was inductive inference *par excellence*. And the probabilistic generalization of enumerative induction was straightforward. It is called the statistical inference or the straight rule and says: if m of the n observed Fs are G, then infer that the (statistical) probability of an F being G is m/n. Surely, this is beset with all the problems of enumerative induction and more.

Still, the widening of the perspective was most important. The old writings appeared to conceive of induction only in the form of enumerative induction. So-called eliminative induction is also a very old idea, certainly to be found, e.g., already in Bacon's and definitely in Mill's methods, and gained central importance in Popper's (1934) philosophy of science. However, it is a deductive inference, a virtue Popper has continuously emphasized, with the merely negative conclusion that some generalizations are false. Thus, it was only in the 18th century, after the advent of probability theory, that methodologists

---

[5] Cf. Ruzicka (1976, pp. 326f.).

[6] Cf. also Cranston (1967, pp. 238f.).

[7] See Spohn (2014) which contains a very brief critical overview of those treatments as well as my explication of ceteris paribus conditions in ranking-theoretic terms, i.e., in terms to be applied here to enumerative induction as well. The paper also contains a simple learning algorithm for ceteris paribus conditions, which I did not relate to the historical canons. Maybe one should.

started clearly discerning various forms of inductive inference. Bayesian hegemony, though, the idea that all forms of inductive reasoning can be reduced to probabilistic reasoning was a much later thought, taking shape only in the late 20th century.

At the same time, the multitude of inductive inferences casted doubt on the role of enumerative induction. Maybe it is just too primitive, and there are superior inductive methods? For instance, Peirce started conceiving of abduction; and nowadays, many philosophers of science favor the idea that the inference to the best explanation (IBE = abduction) is such a superior inductive inference, which is characteristic of the modern sciences.[8] Clarity about the nature of IBE is quite disproportionate, though, to this emphasis. Thus, calling enumerative induction primitive is at least ambiguous. It may mean: not sufficiently elaborated to be useful. But it may still mean: basic – so that it must be accounted for, before one can hope to do justice to more sophisticated forms of inductive inference.

Another issue came to the fore in the vigorous debate between so-called inductivists and deductivists in the 19th century, apparently not only between philosophers, but also within the scientific disciplines themselves. Inductivists recommended enumerative induction or some of its sophistications as a method of generating scientific hypotheses. Deductivists took this to be absurd; finding hypotheses is a matter of informed imagination, in the first place, and of checking then whether the deductive consequences are the desired ones. Newton could never have arrived at his laws by enumerative induction! Following Popper (1934), it was concluded in the 20th century that this debate was situated mainly within the context of discovery, which is the wrong context, anyway. The appropriate context rather is the context of justification. Induction should not be taken as a heuristic method; it is a matter of justifying or confirming the consequent by the premises. In fact, I perceive here an ambiguity in the notion of inference. Inference as a process might lead us to novel insights, whereas inference as a mere relation or connection between sentences or propositions invites us to assess the quality of that connection. I am entirely on the side of the latter interpretation. One must be aware, however, that it is quite a step to conceive of inference as confirmation.

3 Brief Remarks on Inductive Skepticism

The previous point reminds us of the fact that inductive inference was accompanied by inductive skepticism at all times. I have already mentioned the ancient Skeptics. The undeniable point is, simply, that the conclusion of an inductive inference might be false while its premises are true. This is indeed a defining characteristic of inductive or ampliative inference! So, what good reason is there to believe in the conclusion?

In our modern times it was David Hume who developed inductive skepticism in so masterly a manner, in his *Treatise* (1739) as well as in his *Enquiry* (1748). His basic point again was that inductive inference cannot be deductively cogent.

---

[8] See, e.g., Lipton (1991).

However, he made very clear that there is also no way around this basic point. There is also no inductive justification of inductive inference. Such justification would just use and hence pre-suppose the forms of inference to be justified. Causal inference, another main target of Hume, is no better off than inductive inference in general.

A probabilistic interpretation of inductive inference doesn't help, either. We might say that an inductive conclusion is at least very probable. But in which sense? If "very probable" means "high relative frequency of truth", i.e., if "all observed Fs are G", is to entail at least "most Fs are G", this entailment is as unjustified as before. If, however, "very probable" signifies only our high confidence, nothing is gained; it is just the justification of this high confidence which is at issue.[9]

Hume finally observed that all would be fine if we could presuppose the uniformity of nature as a most general law. Yet, how could we ever assume such a law? We could arrive at it only by a higher-order inductive generalization. This is no solution, no way to ground our inductive inferences. And so Hume acquiesced in his inductive skepticism. Inductive inferences are nothing but our habits of thought. We pursue the habits we have. What else could we do? But don't ask for justification.[10]

Kant tried to do better by trying to establish the uniformity of nature as an a priori principle of thought, without which any kind of experience would be impossible. In his terms this took the form of a general law of causality.[11] Let's not deepen the issue now. Surely, though, one might say that Hume convinced more philosophers than Kant did.

Later on it became clear that even the law of the uniformity of nature wouldn't help. This was the radicalization of Hume's inductive skepticism by Goodman (1946). He invented unnatural generalizations like "all emeralds are grue". His point then was that there are countless diverging generalizations that all agree with the observed facts. Suppose that G and G' disagree only for unobserved Fs and that all observed Fs were G as well as G'. Hence, by enumerative induction, all Fs are G as well as G'. But this can't be, provided there are Fs as yet unobserved. So, which of the two generalizations should we prefer? It won't do to call G natural and G' unnatural. That's precisely the issue. Thus the uniformity of nature may take countless shapes, and we have apparently no criterion for conjecturing rather this than that shape. In other words, not only are all inductive inferences not deductively cogent, they all seem equally bad.[12]

---

[9] Cf. Salmon (1966, pp. 5ff. and pp. 48ff.)

[10] This is the quite common psychologistic interpretation of Hume. I don't think that it is really fair to Hume. I rather see the exercise of reason in his "associations" and "habits of thought", an exercise that is clearly rationally reconstructible.

[11] This is the famous second analogy in Kant (1781/87, B232), which states the a priori principle: "All alterations take place in conformity with the law of connection of cause and effect". This entails the uniformity of nature: "That which follows or happens must follow according to a universal rule from that which was contained in the previous state" (B245).

[12] Quine (1960, §17) and elsewhere made the insightful remark that all language use builds on a two-fold induction concerning nature as well as our own behavior or

Goodman's solution of his 'new riddle of induction' was roughly the same as Hume's. He also referred to the habits of thought or rather to the entrenched social practices. The riddle provoked an intense discussion. However, I haven't seen anything moving essentially beyond the original skeptical solution.[13]

What's the present status of inductive skepticism? My attitude – which seems wide-spread, though I am perhaps overoptimistic – is this: Basically, one has to accept Hume's skeptical solution. Just referring to the habits of thought, to social practice, etc. is, however, too psychologistic, too empirically minded, more defeatist than necessary. There are quite a number of normative principles or rationality postulates guiding our inductive behavior. And those principles have strong normative foundations; at least they allow for reasonable and sophisticated discussion. They need not uniquely deter-mine our inductive behavior. But they provide severe constraints, and how far-reaching and consequential they are is an open, fruitful, and constructive issue. I will return to this below.

4 The Present Status of Enumerative Induction

If this is really the present representative attitude concerning (skepticism about) inductive inference in general, what does this specifically entail for enumerative induction? What is its present status?

The first observation is a bit surprising, I find: The last 40 years have seen an explosion of formal inductive theorizing. For centuries probability theory was the only game in town, and now there are default logic, belief revision theory, Dempster-Shafer belief functions, possibility theory, formal learning theory, ranking theory, indeed all kinds of defeasible or non-monotonic reasoning.[14] However, to the best of my knowledge these theories hardly took a stance towards enumerative induction; this was simply no topic for them. This is a blunt contrast to the fact that in all the past centuries inductive inference was mainly about enumerative induction.

Well, there are a few exceptions. Ranking theory is one; I'll come to this. Pollock (1990) just accepted the statistical inference – as mentioned, the straightforward probabilistic generalization of enumerative induction – among his representative collection of basic defeasible inference rules without deepening the insight in this inference. Baumgartner (2009) promotes a research tradition proposing algorithms for projecting causal structures or deterministic generalizations from singular data; this may indeed be called sophisticated enumerative induction.[15] And most importantly, formal learning theory as initiated by

---

concerning the reference of signs and the signs them-selves. Apply skepticism to this double induction, and you end up with meaning skepticism á la Kripke (1982).

[13] Freitag (2015), though, is one of the cutest recent contributions.

[14] Halpern (2003) gives a beautiful systematization of a broad spectrum of theories, and Huber, Schmidt-Petri (2009) provide a very useful anthology.

[15] Again, this should, but hasn't been compared with the algorithm proposed in Spohn (2014).

Kevin Kelly is fore-most a most elaborate account of enumerative induction; this is still an active research program[16], which I cannot further discuss here.

These exceptions are definitely worth attending, but they still seem minority projects. Therefore it is entirely appropriate, I think, to say that the present status of enumerative induction is represented by its Bayesian account. What is this account? Let me first introduce some terminology so that we can be a little bit more precise:

Let us replace the language of predicate logic informally used so far by the set-theoretic language of variables. Let X be some generic variable taking values in some range V. And Let $X_1, X_2, \ldots$ be an infinite series of possible realizations of X. Thus $X_1, X_2, \ldots$ is an infinite series of variables all taking values in V. For instance, we could consider an infinite series of Fs and let X_n take value 1 or 0 according to whether or not the n-th F has G. Or we could consider an infinite series of objects or experiments and let X_n take values $1, \ldots, 4$ according to whether the n-th object has, or the n-th experiment results in, F & G, F & non-G, non-F & G, or non-F & non-G. The generalization that all Fs are G then translates into the assertion that none of the variables X_n takes value 2. Or X_n could represent which value the n-th particle takes in some state space S or which trajectory in S^T it takes through S during the times in T. So, we observe the behavior of the first n variables $X_1, \ldots, X_n$, and the issue of enumerative induction is what to infer from this concerning the behavior of the further variables $X_{n+1}, X_{n+2}, \ldots$.

Now, a possible world, or a possible course of events, as far as it can be represented by the variables, is just a sequence of values $v_1, v_2, \ldots$ in V which the variables $X_1, X_2, \ldots$ might take. For convenience I shall assume that V is finite (but not a singleton, of course). Let $\mathbb{N}$ be the set of non-negative integers, $\mathbb{N}^+$ the set of positive integers, and $\mathbb{N}^\infty = \mathbb{N} \cup \{\infty\}$. So $W = V^{\mathbb{N}^+}$ may be taken as the set of possible worlds or possible courses of events. Even if V is finite, W is very rich, indeed uncountable. W may also be taken as the domain of the variables $X_1, X_2, \ldots$. Then, for any $\mathbf{v} = (v_1, v_2, \ldots) \in W$, we may define $X_n(\mathbf{v}) = v_n$. Let $\{X_n = v\} = \{\mathbf{v} \in W \mid X_n(\mathbf{v}) = v\}$ and $\{X_n \in U\} = \{\mathbf{v} \in W \mid X_n(\mathbf{v}) \in U\}$, respectively, represent the proposition that X_n takes the value v or some value in $U \subseteq V$. All these *atomic* propositions generate a σ-algebra $\mathfrak{A}$ of propositions over W. Our study of enumerative induction must hence focus on the general propositions $G_U = \bigcap_{n \in \mathbb{N}^+} \{X_n \in U\}$ that all variables take values in U.

This is all the algebraic material we need. So, what is the Bayesian account of enumerative induction? As a first step this means to study the issue entirely in terms of subjective probabilities, i.e., in terms of a (σ-additive) probability measure P on $\mathfrak{A}$ and its rational behavior. I have already mentioned a second step: Within the context of justification, inductive inference should be interpreted as confirmation. A third step then is Carnap's explication of confirmation as probabilistic positive relevance.[17] Thus, enumerative induction turns into the following claim: The proposition $\{X_1 \in U\} \cap \ldots \cap \{X_n \in U\}$ is positively relevant to the proposition G_U.

---

[16] Cf, e.g., Kelly (2008).
[17] See Carnap (1950/62, Chs. VI and VII).

At this point, an obstacle emerged: the problem of the null confirmation of laws. Under the assumptions of Carnap's so-called λ-continuum of inductive methods (which includes symmetry as introduced below) each infinite generalization G_U (for $U \subset V$) provably receives probability 0.[18] Hence, no proposition can be positively relevant to any contingent infinite generalization.

Hintikka (1966) ingeniously circumvented this problem with his so-called two-dimensional continuum of inductive methods. Here, universal generalizations receive a positive a priori probability and can thus be confirmed by positive instances.[19] Somehow, though, Hintikka's ideas have not been well received. I am wondering why; I don't know of any telling refutation. Perhaps my impression was a shared one, namely that Hintikka's two-dimensional continuum was designed ad hoc in order to yield the intended results. Probabilities are anchored in reality in relative frequencies. So, inductive probabilities are made for somehow estimating or approaching relative frequencies. Then, however, there does not seem to be any good reason for favoring extreme relative frequencies, 1 and 0, in such a way that they get a positive a priori weight, whereas otherwise only intervals of relative frequencies get positive a priori weight. Granted, strict laws are peculiar. However, I think the specific characteristics of strict laws are better captured in the picture developed below.

Perhaps, though, it was the strong conception of inductive logic in general which fell out of favor, and with it Hintikka's proposal. Be this as it may, the main probabilistic line was simply Carnap's: If the infinite generalization has probability 0, who cares about the infinite generalization? What counts are the cases within our life span or, simply, the next instance. Thus, in a fourth step, Carnap transformed enumerative induction into his principle of positive instantial relevance: $\{X_1 \in U\} \cap \ldots \cap \{X_n \in U\}$ is positively relevant to $\{X_{n+1} \in U\}$. Or, more generally, however $X_1, \ldots, X_n$ realize, $\{X_{n+1} \in U\}$ is positively relevant to $\{X_{n+2} \in U\}$. Quite some transformation!

Let's be a bit more exact and distinguish various notions of instantial relevance. For any possible world $\mathbf{v} = (v_1, v_2, \ldots)$ let's abbreviate $\{X_1 = v_1\} \cap \ldots \cap \{X_n = v_n\}$ by $E_n(\mathbf{v})$. Then P satisfies PIR_n (the *principle of positive instantial relevance, nonconditional version*) iff for any possible evidence $E_n(\mathbf{v})$ and any non-empty set of values $U \subset V$

$$P(\{X_{n+2} \in U\} \mid \{X_{n+1} \in U\} \cap E_n(\mathbf{v})) > P(\{X_{n+2} \in U\} \mid E_n(\mathbf{v})).$$

And P satisfies PIR_c (*the principle of positive instantial relevance, conditional version*) iff for any such E_n and U

$$P(\{X_{n+2} \in U\} \mid \{X_{n+1} \in U\} \cap E_n(\mathbf{v})) >$$
$$P(\{X_{n+2} \in U\} \mid \{X_{n+1} \notin U\} \cap E_n(\mathbf{v})).$$

Clearly, in the probabilistic case PIR_n and PIR_c are equivalent. Still, I have introduced the distinction, because it will make a difference in the ranking-theoretic case. Moreover, let's say that P satisfies NNIR_n (the *principle of non-negative instantial relevance, nonconditional version*) iff P satisfies the

---

[18] CF. Carnap (1950/62, §110F).

[19] See also Kuipers (1978) for an elaborate study of Hintikka's theory.

inequality for PIR_n when $>$ is replaced by $\geq$. And similarly, P satisfies $NNIR_c$ (the *principle of non-negative instantial relevance, conditional version*) iff P satisfies the inequality for PIR_c when $>$ is replaced by $\geq$. Finally, P satisfies IR_n or, respectively, IR_c (*instantial relevance*) iff $>$ is replaced by $\neq$ in the relevant inequalities.

So, the upshot of the Bayesian transformation so far is that enumerative induction is explicated as the probabilistic PIR ($= PIR_n$ or PIR_c). The crux of the Bayesian account lies now in the fifth and final step: Enumerative induction need not be axiomatically assumed as a basic inductive rule, as it seemed all the centuries before. Rather, it is entailed and hence justified by more basic assumptions. The crucial assumption is *symmetry* or *exchangeability*: P is *symmetric* (with respect to $X_1, X_2, \ldots$) iff for any $n \in \mathbb{N}^+$, any permutation π of $\{1,\ldots,n\}$, and any values $v_1,\ldots,v_n$ in V

$$P(\{X_{\pi(1)} = v_1\} \cap \ldots \cap \{X_{\pi(n)} = v_n\}) = P(\{X_1 = v_1\} \cap \ldots \cap \{X_n = v_n\}).$$

According to symmetry, what counts probabilistically is only how many variables take which value. Which specific variables do so, does not make any probabilistic difference. This is extremely plausible when we miss any special information about specific variables (or the objects or experiments they represent).[20] So I need not repeat here the overwhelming credibility of this assumption.

The first result now is that symmetry entails NNIR ($= NNIR_n$ or $NNIR_c$). So, we may secondly conclude that symmetry plus IR ($= IR_n$ or IR_c) entail PIR, i.e., the Bayesian version of enumerative induction. Why assume IR? This is overwhelmingly plausible as well; without IR we could not learn anything at all from our observations. However, we could also argue that IR is entailed by the so-called Reichenbach axiom, which says this: For any possible world $\mathbf{v} = (v_1, v_2, \ldots)$ and any value $v \in V$ let $rf(\mathbf{v}, v)$ denote the relative frequency with which v occurs among the first n values $v_1, \ldots, v_n$. Then

$$\lim_{n \to \infty} [P(\{X_{n+1} = v\} \mid E_n(\mathbf{v})) - rf(\mathbf{v}, v)] = 0.$$

In other words, the probability that the next variable will take value v converges to the relative frequency of v with increasing evidence. Again such limiting learning behavior seems reasonably required. For all these results see, e.g., Humburg (1971).

As explained there, these results ultimately ground in de Finetti's fundamental representation theorem from (1937), which says that any symmetric probability measure over $\mathfrak{A}$ is a *unique* mixture of *Bernoulli measures* over $\mathfrak{A}$, according to which all variables X_n have the same distribution and each variable X_n is probabilistically independent of all the other variables. We might

---

[20] This kind of symmetry is also assumed in all conceptions of inductive logic. Above, though, when saying that inductive logic fell out of favor I referred to stronger conceptions which assume symmetry not only with respect to objects, but also with respect to predicates, i.e., in our terminology, with respect to the possible values in V. This is indeed very questionable, but not relevant here.

conceive of such a Bernoulli measure as a statistical hypothesis about the objective probabilities governing $X_1, X_2, \ldots$. In this interpretation, which is not de Finetti's, the representation theorem says that our subjective assessment P, if symmetric, uniquely corresponds to a second order distribution over ($=$ mixture of) possible statistical hypotheses. The final twist then is that we can verify from the beginning whether our P has the appropriate limiting behavior as required by the Reichenbach axiom. That is, if the so-called carrier of that second-order distribution ($=$ the smallest topologically closed set having measure 1) corresponding to P is the space of all Bernoulli measures over $\mathfrak{A}$, then P satisfies the Reichenbach axiom.[21]

All in all, this is a beautiful justificatory story for the Bayesian version of enumerative induction and thus indeed huge progress. In view of this story I am no longer impressed by Hume's skepticism. We have now found good reasons why we must assume PIR, and we may do without an argument to the effect that higher subjective probability somehow entails higher relative frequency of truth. And we should follow the good reasons. Indeed, we even know that our probabilities will converge to the relative frequency of truth, although we must grant that at no point can we be sure where the point of convergence comes to lie.

In this way, the topic seems to have come to a rest, and we may be content with the Bayesian account of enumerative induction. Really?

5 A More Adequate Ranking-Theoretic Account

I don't think that we should be fully satisfied. We need not put up with all the Bayesian transformations. In particular, with the step to PIR, positive *instantial* relevance, we have lost all reference to generality so characteristic of enumerative induction. Indeed, enumerative induction is originally foreign to probabilistic epistemology, and not only because it is much older. Its Bayesian naturalization is only due to the fact that confirmation theory took an exclusively probabilistic shape. This was not always so. Hempel (1945), the modern classic of confirmation theory, started a research program of so-called qualitative confirmation theory. However, this program turned out to be infeasible. At least, this was the conclusion beautifully summarized in Niiniluoto (1972). Since, we are left with a probabilistic confirmation theory.

I think that the abandoning of Hempel's program was premature. Confirmation theory can also be developed within ranking theory, and this comes much closer to Hempel's original intentions of a qualitative confirmation theory. There is no place here to develop this claim on a larger scale. Let us see, though, how enumerative induction fares within ranking-theoretic confirmation theory. So, we still conceive of inductive inference as confirmation, and we may still stick to Carnap's explication of confirmation as positive relevance. However, we now replace probabilistic by ranking-theoretic positive relevance. How does this work in detail?

---

[21] For all this, see again Humburg (1971).

We deal with the same atomic propositions as before. However, we now assume $\mathfrak{A}$ to be the complete algebra generated by these atomic propositions. Next, instead of the probability measure P we consider a negative ranking function κ on $\mathfrak{A}$. κ is a *negative ranking function* on $\mathfrak{A}$ iff κ is a function from $\mathfrak{A}$ into $\mathbb{N}^\infty$ such that for all $A, B \in \mathfrak{A}$: (a) $\kappa(W) = 0$ and $\kappa(\varnothing) = \infty$, (b) $\kappa(A \cup B) = \min\{\kappa(A), \kappa(B)\}$ (*minimitivity*). Moreover, let's assume that κ is *completely minimitive*, i.e., that for all $\mathfrak{B} \subseteq \mathfrak{A}$ $\kappa(\bigcup \mathfrak{B}) = \min_{B \in \mathfrak{B}} \kappa(B)$.[22]

If the algebra $\mathfrak{A}$ were finite, there would be no point in considering complete minimitivity. However, we attend to generalizations, which are infinite Boolean combinations of atomic propositions (and we need not restrict ourselves to countable combinations, as we actually do). Hence, we must take a stance on how a ranking function should behave vis á vis such infinite propositions. I think it is reasonable to assume complete minimitivity, as I have more extensively defended in Spohn (2012, pp. 73f.).[23]

The standard interpretation of a negative ranking function κ is as degrees of disbelief (where disbelieving means taking to be false). This is why those functions are called negative. They don't take negative values, but their positive values express negative facts. Hence, I disbelieve A iff $\kappa(A) > 0$, and I believe A iff $\kappa(\overline{A}) > 0$.[24] Still, my belief can be more or less firm, as expressed by the ranks. It is convenient to define two-sided ranks. The two-sided ranking function τ belonging to the negative ranking function κ is defined by $\tau(A) = \kappa(\overline{A}) - \kappa(A)$. Thus, A is taken to be true or false or neither according to whether $\tau(A) > 0$ or < 0 or $= 0$.

Conditional negative ranks are defined as $\kappa(B \mid A) = \kappa(A \cap B) - \kappa(A)$, provided $\kappa(A) < \infty$. Thereby, minimitivity can be expressed as saying $\min\{\kappa(A \mid A \cup B), \kappa(B \mid A \cup B)\} = 0$. This means that given the disjunction you can't take both disjuncts to be false; this is obviously rationally mandated. In my view, this clearly extends to infinite disjunctions, i.e., to complete minimitivity. Finally, conditional two-sided ranks are defined as $\tau(B \mid A) = \kappa(\overline{B} \mid A) - \kappa(B \mid A)$. For all fuller explanations of the basics of ranking theory I must refer to Spohn (2012, Ch. 5).

Now we are prepared to study enumerative induction in ranking-theoretic terms.[25] Again, I assume that we only deal with symmetric ranking functions,

---

[22] This definition goes back to my Habilitationsschrift Spohn (1983, sect. 5.3.) Its first appearance in English is in Spohn (1988), where negative ranking functions were still called ordinal conditional functions. Theory and applications of these functions are comprehensively presented in Spohn (2012).

[23] There would have been no point in considering σ-minimitivity, since it is equivalent to the apparently stronger complete minimitivity. And then it also fits better to build ranking functions on complete instead of σ-algebras. For all these niceties see Spohn (2012, pp. 72ff.).

[24] One may also define a stricter notion of belief by saying that A is believed iff $\kappa(\overline{A}) > z$ for some threshold $z > 0$. This well accounts for the vagueness of the notion of belief (or disbelief). However, whatever the threshold, belief is always consistent and deductively closed. See Spohn (2012, pp. 76f.) for details. Here, we may well neglect this point.

[25] The basic reason why the ranking-theoretic story will be similar to the Bayesian story is quite obvious from the axioms, according to which the minimum, the

for the same overwhelming reasons as in the probabilistic case. κ is *symmetric* iff κ satisfies the above symmetry condition with κ replacing P, i.e., if for any $n \in \mathbb{N}^+$, any permutation π of $\{1, \ldots, n\}$, and any values $v_1, \ldots, v_n \in V$

$$\kappa(\{X_{\pi(1)} = v_1\} \cap \ldots \cap \{X_{\pi(n)} = v_n\}) = \kappa(\{X_1 = v_1\} \cap \ldots \cap \{X_n = v_n\})$$

Then we have a first nice surprise: the credibility of a generalization is the very same as that of its instances. More formally, if κ is symmetric, then $\tau(G_U) = \tau(X_n \in U)$ for all $n \in \mathbb{N}^+$. This is a direct consequence of complete minimitivity. Or more generally: if $G_{>n,U} = \bigcap_{k>n} \{X_k \in U\}$ is the relevant generalization restricted to the future and if $E_n(\mathbf{v})$ is any evidence about the first n variables, then $\tau(G_{>n,U}|E_n(\mathbf{v})) = \tau(X_k \in U | E_n(\mathbf{v}))$ for all $k > n$.[26] Of course, this holds not only for the ranks specified, but also for rank comparisons as required for confirmatory relations. Sloppily stated, this means that generalization is automatically built in into symmetric ranking functions as a consequence of the basic rationality postulates of ranking theory. Hence, we need not despair of the 'null confirmation' of laws, we need not take Carnap's escape route to instantial relevance, and we need not choose ad hoc measures á la Hintikka (1966). I take this to be a first important advantage of the ranking-theoretic account.

The next surprise, however, is less pleasant. We can transfer all the above notions of (positive, non-negative) instantial relevance in the nonconditional or the conditional version to ranking theory simply by substituting the two-sided ranking function τ (not the negative ranking function κ) for P in the defining conditions above. (Relevance is more succinctly expressible in terms of τ.) So, we know what PIR_n, PIR_c, NNIR_n, and NNIR_c mean in ranking theoretic terms. Let us also assume that our ranking function κ is *regular* in the sense that $\kappa(E_n(\mathbf{v})) < \infty$ for all $n \in \mathbf{N}^+$ and $\mathbf{v} \in W$. So, ranks conditional on $E_n(\mathbf{v})$ are always defined. Only then are we guaranteed to be able to learn from any kind of evidence. Then we have first to observe that, in contrast to the probabilistic case, the conditional and nonconditional versions of these notions are not equivalent. Rather, PIR_n entails PIR_c, and NNIR_c entails NNIR_n, but the reverse entailments do not hold. So, we have to make a choice. As I have argued extensively in Spohn (2012, pp. 106f.), confirmation (or the reason relation, as I call it there) is more adequately captured by the conditional versions. So, is PIR_c ranking-theoretically entailed by symmetry, as it is probabilistically? If so, the above extension to future generalizations $G_{>n,U}$ would already conclude our business?

On the contrary; this is the unpleasant surprise. We have the following theorem: There is no regular symmetric ranking function on $\mathfrak{A}$ satisfying PIR_c.[27]

---

sum, and the difference of ranks, respectively, roughly correspond to the sum, the product, and the quotient of probabilities. For the precise formal relation between ranks and probabilities see Theorem 10.1 in Spohn (2012, pp. 203f.) which explains the formal similarities as well as the subtle formal differences between the two theories.

[26] For the simple proof see Spohn (2012, p. 282).

[27] This is Theorem 12.9 of Spohn (2012, p. 283). It is not trivial at all. For a proof see there p. 298.

We might settle for the weaker $NNIR_c$, which is satisfiable by regular symmetric ranking functions. However, this doesn't look attractive. $NNIR_c$ looks too weak; instantial irrelevance most of the time wasn't what we expected to get. Moreover, in contrast to the probabilistic case, $NNIR_c$ is not entailed by ranking-theoretic symmetry; we would have to additionally stipulate it.

Something seems to have gone badly wrong. The solution I have pursued in Spohn (2012, sect. 12.4–5) is to sharply distinguish between generalizations and laws. Generalizations or regularities are simply propositions, members of the propositional algebra, of the form G_U or $G_{>n,U}$. Laws, by contrast, are something entirely different. The issue is embedded in a large issue of philosophy of science. There are mere regularities like "all gold spheres are smaller than one mile in diameter" (which may well be true) and true laws like "all uranium spheres are smaller than one mile in diameter".[28] What distinguishes lawlike sentences from mere generalizations (whether true or false)?[29] This has proved to be a remarkably recalcitrant problem. Early attempts at this distinction all failed. Thus, it has become apparent that lawlikeness or nomicity is connected with explanatory force, with entailing counterfactual conditionals, and with inductive behavior – three very soft topics in philosophy of science. It is a huge task to sort out all these connections. For us, only the last point is relevant. Somehow, it seems that enumerative induction doesn't extend to all generalizations whatsoever, but applies only to laws or potential laws. This is certainly what all philosophers from earlier centuries would have said who were not aware of the intricacies of that distinction.

So, it seems that we have made a mistake so far by trying to ranking-theoretically reconstruct enumerative induction for generalizations. We should restrict it to laws. But what are laws? We can't evade engaging into this distinction. However, here I have to cut a long story short.[30] The fundamental point is that laws are not propositions at all, contrary to what the mainstream concerning the issue has taken for granted! Laws do not make assertions and do not have truth conditions. This seems to be a bizarre claim; no wonder that the mainstream has not taken it seriously. Still, the claim has respectable philosophical precedence starting with Ramsey (1929).[31]

So, if laws are not propositions, what are they? The general, though obscure slogan is: laws rather are inference tickets. (And here you may again read "inference" as "confirmation"!). What is my ranking-theoretic translation of that slogan and the Ramsey quote? Basically, the idea is very simple: We mentioned above that in the probabilistic case statistical laws or hypotheses for $\mathfrak{A}$

---

[28] The example is from van Fraassen (1989, p. 27), who attributes it to the philosophy of science folklore of the 1960s.

[29] For an in-depth discussion of the issue see, e.g., Lange (2000).

[30] The longer story is told in Spohn (2012, sect. 12.4).

[31] My key witness are the following quotes from Ramsey (1929): "Many sentences express cognitive attitudes without being propositions; and the difference between saying yes or no to them is not the difference between saying yes or no to a proposition" (pp. 135f.). "Laws are not either" (namely propositions) (p. 150). Rather "the general belief consists in (a) A general enunciation, (b) A habit of singular belief" (p. 136).

are represented by Bernoulli measures for $\mathfrak{A}$. Each single case is characterized by a certain (objective) probability distribution, and this is turned into a law in the independent, identically distributed repetition of the single case as represented by the corresponding Bernoulli measure. So, correspondingly, what is a deterministic law? Just the corresponding ranking-theoretic notion. That is, I explicate that the (negative) ranking-function λ is a *subjective law* for $\mathfrak{A}$ iff all variables X_n have the same distribution (in terms of ranks) and each variable X_n is ranking-theoretically independent of all the other variables according to λ. (The latter boils down to instantial irrelevance: for all n and $\mathbf{v}$, given $E_n(\mathbf{v})$ X_{n+1} is irrelevant to X_{n+2}) The analogy to statistical laws is charming; for a philosophical defense of this explication I have to refer to Spohn (2012, Sect. 12.4).[32]

Subjective laws are related to generalizations. Let λ be such a law and define $U = \{v \mid \lambda(X_n = v) = 0\}$. Then λ contains the belief in the generalization G_U and in no stronger generalization; that is, U is the largest subset of V such that $\lambda(\overline{G}_U) > 0$. But of course this belief may be realized in ranking functions in many different ways. Hence, laws represent a very specific attitude towards generalizations.

I have used a disturbing term by calling such a λ a *subjective* law. This is owed to the fact that ranking functions still represent only the epistemic state of some epistemic subject. So, such a law λ is just an epistemic attitude, too. There are ways, though, to objectivize this notion and to turn some subjective laws into objective ones. However, this remark can only be clarified by engaging into what I call the objectivization theory for ranking functions, which I develop in Spohn (2012, ch. 15). Here, we better leave this issue aside and go along with the notion of a subjective law.

Now we can proceed as in the probabilistic case by transferring de Finetti's representation theorem to ranking theory and getting all its benefits. Let me mention first the positive results and then the caveats.[33] We can indeed prove that each regular symmetric ranking function is a unique mixture of subjective laws. Well, roughly; for the caveats see below. That is, if Λ is the set of regular subjective laws for $\mathfrak{A}$ and ρ is any ranking function for Λ, then the mixture κ of Λ by ρ is a regular symmetric ranking function for $\mathfrak{A}$, where this mixture is defined by $\kappa(A) = \min\{\lambda(A) + \rho(\lambda) \mid \lambda \in \Lambda\}$ for all $A \in \mathfrak{A}$. I call ρ an *impact function*, since it tells which impact each subjective law has on the mixture. And reversely – that's the difficult part – if κ is a regular symmetric ranking function for $\mathfrak{A}$, then there is a unique impact function ρ such that κ is the mixture of Λ by ρ.

The next point in my dialectics is this: So far, we have notions of relevance and confirmation only for propositions; evidence may or may not confirm generalizations. However, if laws are not propositions, these notions cannot be applied to laws. The impact functions fill the gap: Let our initial symmetric κ be represented by the impact function ρ. We learn by observing the first n vari-

---

[32] Clearly, though, the independent, identically distributed repetition may be taken as a mathematical explication of Ramsey's 'habit of singular belief'.

[33] The mathematically involved full story, together with all the proofs, is presented in Spohn (2012, Sect. 12.5).

ables and conditionalizing our κ on those observations. The conditionalized κ – let's call it κ' – will no longer be symmetric concerning the first n variables, but it will stay symmetric with respect to all future variables $X_{n+1}, X_{n+2}, \ldots$. The point now is that κ' will be represented by a different impact function ρ'. Thus, the impact of a subjective law λ will change or possibly stay the same from ρ to ρ', and accordingly we can say that λ has been confirmed or disconfirmed or neither.

More precisely, we find this: Let κ_n be the ranking function reached after observing the first n variables. Now we observe X_{n+1} to take the value v and thus move to κ_{n+1}. Let the corresponding impact functions be ρ_n and ρ_{n+1}. Then for any subjective law λ we have

$$\rho_{n+1}(\lambda) - \rho_n(\lambda) = \lambda(X_{n+1} = v) - \kappa_n(X_{n+1} = v).^{34}$$

That is, the impact of λ is upgraded or downgraded precisely to the extent to which the credibility it gives to the observed value of X_{n+1} deviates from the credibility this value has according to κ_n.

This is my ranking-theoretic account of enumerative induction. It does not refer to the next single case, as did the Bayesian account. It even does not refer to generalizations as such. It does refer specifically to (subjective) laws, as originally intended, and indeed in an even quantitatively appealing way. In this way it improves upon the Bayesian account. And it preserves the virtues of the Bayesian account, by not just postulating enumerative induction as a basic inductive inference rule, but by essentially deriving it from the more basic epistemic rationality postulates of regularity and symmetry. In this way, enumerative induction seems to have finally found an even more comfortable home.

Now to the caveats: So far, I have not told the full truth about my ranking-theoretic duplication of de Finetti's representation theorem. Things are more complicated. A first qualification is mathematically interesting, but philosophically neutral, as far as I see. It is not true that each regular symmetric κ is the mixture result of exactly one impact function ρ. We must more cautiously define the notion of what I call a *minimal* mixture, and this in fact turns out to be unique. The second qualification is less harmless. Regularity and symmetry alone do not suffice for representation. We also have to assume that κ is *concave* in a mathematically well-specifiable sense. The problem here is that concavity is required for mathematical reasons, although I cannot present good philosophical or normative reasons why our rank-ing function should be concave. I am optimistic; but the case is mathematically intricate. This is presently just an open flank of my argument. In this respect, the Bayesian account does better, since it relies only on well-justifiable features of our subjective probabilities.[35] So, to be precise, the ranking-theoretic representation theorem says: A ranking function κ for $\mathfrak{A}$ is regular, symmetric, and concave if and only if there is a unique impact function ρ for Λ such that κ is the minimal mixture of Λ by ρ.[36] This displays the two qualifications.

---

[35] For a formal explanation of these differences see footnote 25.
[36] Cf. Theorems 12.14 and 12.18 in Spohn (2012, p. 293 and p. 296).

6 Two Concluding Observations

The ranking-theoretic representation theorem and the confirmation of laws entailed by it have various consequences. Let me finally mention two of them.

First, we may observe that Goodman's new riddle of induction partially evaporates. Our above equation for $\rho_{n+1}(\lambda) - \rho_n(\lambda)$ entails that each subjective law λ for which the observed X_{n+1} taking the value v is a positive instance, i.e., for which $\lambda(X_{n+1} = v) = 0$, is thereby confirmed. So, if $\{X_{n+1} = v\}$ represents that the $n+1$st emerald is green, crazy hypotheses like "all emeralds are grue" are thereby just as well confirmed as plausible laws like "all emeralds are green". We may well grant this; there is no reason to be scared by this observation. Indeed, there is no contradiction; in the positive relevance sense incompatible laws may be simultaneously confirmed. It is only that not all laws can be confirmed. There also are many potential laws that get disconfirmed, natural ones like "all emeralds are blue" and crazy ones like "all emeralds are bleen". If all goes well, "all emeralds are grue" will get disconfirmed soon enough. But, of course, we never know.

By no means may we conclude that all the incompatible hypotheses would be equally good. On the contrary, the a priori impact of the crazy hypotheses is so terribly low that it will stay low even after many confirmations. However, it is not infinitely low, and this is only reasonable. Why should we not be able to discover in the end that there is a hidden connection between the color of an emerald and the time of its first observation (or some other apparently entirely unrelated feature)? However, why should the a priori impact of crazy grue hypotheses be very low? That's the snag. I have not given any reason for this. Impacts are just subjectively chosen, and our choice is clear, whereas the defenders of the grue hypothesis choose entirely different initial impacts. This is why I have said that Goodman's new riddle evaporates only partially. We might have expected a rationalization of our a priori impacts, which, however, I cannot provide. Still, my point stands. There is not any paradox in the fact that many incompatible hypotheses are simultaneously confirmed.[37]

The same point could have been made within a Bayesian framework, however less comfortably. In a Bayesian framework we would have to give non-zero initial weights to all the crazy hypotheses (in order to possibly confirm them), and it matters to our probabilities for the factual propositions which weights we choose. This is the so-called problem of the catch-all hypothesis which some take to be a severe objection to Bayesianism.[38] By contrast, in the ranking-theoretic framework very low impacts of crazy hypotheses need not surface in the ranks for factual propositions. This is what I have called the innocence of the worst explanation in Spohn (2012, Sect. 2012). Hence, my partial response to Goodman's new riddle is more easily maintained in ranking-theoretic terms.

The second and final observation concerns the so-called law of the uniformity of nature: Just as in the probabilistic case we may, somewhat sloppily, say that

---

[37] I neglect here the extension of Goodman's new riddle to meaning skepticism according to which it is unclear whether "all emeralds are green" really means this or actually means "all emeralds are grue".

[38] See, e.g., Earman (1992, Sect. 7.2).

the ranking-theoretic representation theorem shows that our symmetric (and concave) epistemic state is always a mixture of (subjective) laws. We are bound to think in laws! Now, this overstates the case a little bit. Formally, λ_0 defined by $\lambda_0(A) = 0$ for all $A \in \mathfrak{A}$ is also a subjective law. However, it rather represents the belief in complete lawlessness instead of belief in a specific law. (Recall my above remark about which generalizations are believed to hold in subjective laws; λ_0 only contains the belief in the empty or tautological generalization.) Still, all the other subjective laws have substantial content.

The more precise statement now is this. Let the regular, symmetric and concave κ be a mixture of Λ by the impact function ρ. Does the mixture contain λ_0? Maybe. Define $\sup \kappa = \sup\{\kappa(A) \mid A \in \mathfrak{A}\}$. $\sup \kappa$ may be 0, positive, but finite, or ∞. Note that regularity requires only that $\kappa(E_n(\mathbf{v})) < \infty$ for all n and $\mathbf{v}$. But this is compatible with the $\kappa(E_n(\mathbf{v}))$ having no upper bound and even compatible with $\kappa(A) = \infty$ for some infinitary proposition A. So $\sup \kappa = \infty$ is a possibility as well. Now the formal observation is that $\rho(\lambda_0) = \sup \kappa$.[39] What does this mean?

$\sup \kappa = 0$ means that $\kappa(A) = 0$ for all $A \in \mathfrak{A}$. Such a κ is incapable of any inductive inference, and we may well exclude it as a totally unreasonable epistemic state. $\sup \kappa = \infty$ means that $\rho(\lambda_0) = \infty$, i.e., a maximal denial of λ_0; and it entails that λ_0 remains maximally excluded after arbitrary amounts of evidence. (Infinite ranks cannot change.) So, whatever the evidence, we stick to the firm belief that some substantial law will hold. This may well be called an unrevisably a priori belief in lawfulness or the uniformity of nature. The third case is that $\sup \kappa = s$ for some finite $s > 0$. This means that $\rho(\lambda_0) = s > 0$, and since ρ is a negative ranking function, this means disbelief in λ_0. Again, this may be called an a priori belief in the uniformity of nature. But it is not unrevisable; it may be defeated. We may (but need not) receive bewildering evidence that ever more upgrades the impact of λ_0, possibly up to 0. Then we would have reached a state of total perplexity in which we do not dare projecting any substantial law into the future. The last two cases may be actually hard to decide; $\sup \kappa$ may be finite, but so large, and hence the impact of λ_0 so low, that we never actually reach the point of despair.

In any case, this observation shows that the belief in the uniformity of nature need not be as unjustified as Hume has thought. It is rationally required (as much as regularity, symmetry and concavity are rationally required). So, Kant may have been right with his a priori claims. At the same time, though, the remarks show that we have to differentiate. Belief in lawfulness or the uniformity of nature may be unrevisably or defeasibly a priori. And Kant was certainly not aware of the latter possibility.

References

Bacon, F. (1620), Novum Organum, London.
Baumgartner, M. (2009), "Uncovering Deterministic Causal Structure: A Boolean Approach", Synthese 170, 71–96.

---

[39] See Spohn (2012, p. 301).

Carnap, R. (1950/62), The Logical Foundations of Probability, Chicago: Chicago University Press, 2nd ed. 1962.

Cranston, M. (1967), "Bacon, Francis", in: P. Edwards (ed.), The Encyclopedia of Philosophy, Vol. 1, New York: Macmillan, pp. 235–240.

de Finetti, B. (1937), "La Prévision: Ses Lois Logiques, Ses Sources Subjectives", Annales de l'Institut Henri Poincaré 7. Engl. translation: "Foresight: Its Logical Laws, Its Subjective Sources", in: H.E. Kyburg jr., H.E. Smokler (eds.), Studies in Subjective Probability, New York: John Wiley & Sons 1964, pp. 93–158.

Earman, J. (1992), Bayes or Bust? A Critical Examination of Bayesian Confirmation Theory, Cambridge, Mass.: MIT Press.

Freitag, W. (2015), "I bet you'll solve Goodman's Riddle", Philosophical Quarterly 65, 254–267.

Goodman, N. (1946), "A Query on Confirmation", Journal of Philosophy 43, 383–385.

Halpern, J. Y. (2003), Reasoning about Uncertainty, Cambridge, Mass.: MIT Press.

Hempel, C. G. (1945), "Studies in the Logic of Confirmation", Mind 54, 1–26 + 97–121.

Hintikka, J. (1966), "A Two-Dimensional Continuum of Inductive Methods", in: J. Hintikka, P. Suppes (eds.), Aspects of Inductive Logic, Amsterdam: North-Holland, pp. 113–132.

Huber, F., C. Schmidt-Petri (eds.) (2009), *Degrees of Belief. An Anthology*, Oxford: Oxford University Press.

Humburg, J. (1971), "The Principle of Instantial Relevance", in: R. Carnap, R. C. Jeffrey (eds.), Studies in Inductive Logic and Probability, Volume I, Berkeley: University of California Press, pp. 225–233.

Hume, D. (1739), A Treatise Concerning Human Nature, Book I, London.

Hume, D. (1748), An Enquiry into Human Understanding, London.

Kant, I. (1781/87), Kritik der reinen Vernunft; engl. translation: Critique of Pure Reason, London: Macmillan 1929.

Kelly, K. T. (2008), "Ockham's Razor, Truth, and Information", in: P. Adriaans, J. van Benthem (eds.), Handbook of the Philosophy of Science, Vol. 8, Philosophy of Information, Amsterdam: Elsevier, pp. 321–359.

Kern-Isberner, G. (1999), *A Unifying Framework for Symbolic and Numerical Approaches to Non-monotonic Reasoning and Belief Revision.* Fachbereich Informatik der FernUniversität Hagen, Habilitationsschrift.

Kern-Isberner, G. (2004), "A Thorough Axiomatization of a Principle of Conditional Preservation in Belief Revision", *Annals of Mathematics and Artificial Intelligence* 40, 127–164.

Kripke, S. A. (1982), Wittgenstein on Rules and Private Language, Oxford, Blackwell.

Kuipers, T. A. F. (1978), Studies in Inductive Logic and Rational Expectation, Dordrecht: Reidel.

Lange, M. (2000), Natural Laws in Scientific Practice, Oxford: Oxford University Press.

Lipton, P. (1991), Inference to the Best Explanation, London: Routledge, 2nd ed. 2004.

Mill, J. St. (1843), A System of Logic, London.

Niiniluoto, I. (1972), "Inductive Systematization: Definition and a Critical Survey", Synthese 25, 25–81.

Pollock, J. L. (1990), Nomic Probability and the Foundations of Induction, Oxford: Oxford University Press.

Popper, K. R. (1934), Logik der Forschung, Wien: Springer; engl. translation: The Logic of Scientific Discovery, London: Hutchinson, 1959.

Quine, W. V. O. (1960), Word and Object, Cambridge, Mass: MIT Press.

Ramsey, F. P. (1929), "General Propositions and Causality", in: F. P. Ramsey, Foundations. Essays in Philosophy, Logic, Mathematics and Economics, ed. by D. H. Mellor, London: Routledge & Kegan Paul 1978, pp. 133–151.

Ruzicka, R. (1976), "Der Induktionsbegriff von Aristoteles bis Galilei", in: K. Gründer, Historisches Wörterbuch der Philosophie, Band 4, Basel: Schwabe, pp. 323–329.

Salmon, W. C. (1966), The Foundations of Scientific Inference, Pittsburgh: University Press.

Smart, N. (1967), "Indian Philosophy", in: P. Edwards (ed.), The Encyclopedia of Philosophy, Vol. 4, New York: Macmillan, pp. 155–169.

Spohn, W. (1983), *Eine Theorie der Kausalität*, unpublished Habilitationsschrift, Universität München, pdf-version at:
uni-konstanz.de/FuF/Philo/Philosophie/philosophie/files/habilitation.pdf

Spohn, W. (1988), "Ordinal Conditional Functions. A Dynamic Theory of Epistemic States", in: W.L. Harper, B. Skyrms (eds.), *Causation in Decision, Belief Change, and Statistics*, vol.II, Dordrecht; Kluwer, pp. 105–134.

Spohn, W. (2012), The Laws of Belief. Ranking Theory and Its Philosophical Applications, Oxford: Oxford University Press.

Spohn, W. (2014), "The Epistemic Account of Ceteris Paribus Conditions", European Journal for the Philosophy of Science 4, 385–408.

van Fraassen, Bas C. (1989), *Laws and Symmetry*, Oxford University Press, Oxford.

Part III

Problem Solving and Query Answering

On the Simulation Assumption for Controlled Interaction Processing

Joachim Biskup[1], Cornelia Tadros[2]

Abstract. Protecting a logic-oriented information system, inference control for enforcing a confidentiality policy evaluates and censors possible reactions on a request issued by some partner agent. The control first checks whether information to be kept secret would be revealed and then determines a reaction that is both harmless and still most informative. For achieving this goal, the control has to suitably simulate the intended receiver of a reaction. However, seeing the receiver not only as a cooperating partner but also as an "attacker", the control has to face the problem that the actual intentions and capabilities of the receiver might be inherently unknown to it. Accordingly, any control method needs to be based on an appropriate simulation assumption. We investigate and discuss pertinent assumptions for the controlled execution of requests for queries, updates, revisions and views.

1 Introduction

A security mechanism of *inference control* exercised by a data *owner* attempts to prevent that a partner agent will be enabled to achieve an *unwanted gain of information* based on *authorized observations* [25,3]. Whether *authorizations* are expressed as access rights granted for reading or represented as cryptographic keys distributed for decryption, the grantor or distributor has to consider the receiver's options and capabilities to gain confidential information from accessible or decrypted data, respectively, and, as far as necessary, employ additional security mechanisms. In all cases, an authorized agent is also seen as a "too curious" *attacker* who might possibly misuse his authorizations to learn information that should be kept secret to him. Accordingly, the data owner will be denoted as the *defender*.

Most generally speaking, a gain of information might happen by "intelligently connecting" any kind of "a priori knowledge" with the totality of data received over the time. However, of course, an attacker will tell neither his malicious *intentions* nor his logical and computational *capabilities* to the defending owner. Thus, inference control is challenged by a situation like the one roughly visualized in Figure 1. Consequently, on the one hand, inference control has to admit that the attacker's actual properties are hidden. But, on the other hand, inference control necessarily has to consider what the attacker already knows before an interaction and which intelligent connections he could employ

---

[1] Fakultät für Informatik, Technische Universität Dortmund, Germany, joachim.biskup@cs.tu-dortmund.de
[2] Fakultät für Informatik, Technische Universität Dortmund, Germany, cornelia.tadros@cs.tu-dortmund.de

Fig. 1. Situation underlying the desired inference control on the defender's side.

afterwards. There appears to be only one way to resolve the crucial discrepancy between lack and need of knowledge: inference control has to be grounded on clearly stated and convincingly justified *assumptions* about the attacker, which the defender's control later on uses to *simulate* the attacker.

Our considerations about such a *simulation assumption* comprise the following issues:

- basic concepts of simulation assumptions (Section 2),
- distinctions for simulation assumptions (Section 3),
- instantiations of simulation assumptions (Subsec. 4.1 and Subsec. 4.2),
- special situations allowing for simplified instantiations of simulation assumptions (Subsection 4.3),
- collection and justification of simulation assumptions (Section 5), and
- robustness of simulation assumptions (Section 6).

2 Simulation Assumptions for the Defender's Control

Regarding an attacker's *intentions*, one might most conservatively and cautiously proceed under the principle that the attacker will aim at gaining *any* information the owner has specified to be kept confidential in his policy. Regarding the *intelligent connections* and the *a priori knowledge* available to an attacker, one is facing an intricate variety, extremely difficult to grasp, out of which one has to take a carefully decided selection for a specific application of inference control.

Based on that principle and on such a selection at design time, one has to construct a specific control mechanism. Afterwards, once in operation, the control should then proceed autonomously on behalf of the defending owner according to the following, still slightly tentative pattern:

- Inspecting a list of candidates of a (re)action possibly sent to the attacker, for each item in the list, the control *simulates* the assumed behavior of the attacker after having observed that item, and thereby the control determines the respective gain of information the attacker would achieve.
- Right from the beginning, the control *blocks* a potential gain of information that is unwanted according to a specified confidentiality policy by making data crucially needed for the gain unobservable.

So far, the pattern is only tentative for the following reason. When delivered as reaction, a result of the protection mechanism by means of hiding or distorting data might offer further options to the attacker to infer information to be kept secret. And thus, one should better conservatively and cautiously postulate that the attacker is aware of the protection mechanism and, accordingly, could try to simulate it in turn. Hence, the control also has to include such so-called *meta-inferences* into its own simulation.

Summarizing the preceding considerations, it should be evident that the success of inference control against an attacker essentially depends on a *simulation assumption* of the following kind:

> The simulation of the attacker, as performed by the defender's control, "closely corresponds" to the possibilities that are actually available to the attacker.

The following, only grossly sketched examples illustrate the importance of such a simulation assumption. We consider a simple, logic-oriented information system which manages data in form of propositional sentences including the atomic sentences s, c_1 and c_2. The atomic sentence s should be kept secret to the attacker in the sense that, from the point of view of the attacker, the truth of $\neg s$ should always appear to be possible.

In an elementary case, the information system interprets sentences in a *classical, model-theoretic* manner. And for the attacker simulation we assume that the attacker interprets sentences classically as well. If the sentence s is actually *true* and the attacker then queries (the truth value of) s, the control must not return the correct answer but should distort it:

- using the "refusal approach", the control could answer by a non-informative "mum";
- employing the "lying approach", it would respond with the lie $\neg s$.

Otherwise, if the sentence s is actually *false* and the attacker now queries (the truth value of) s, one has to make a crucial distinction under the refusal approach: if the attacker is aware that the sentence s is protected, then the control should also answer with the non-informative "mum", rather than with the only seemingly harmless correct response $\neg s$. The reaction by "mum" makes the actual truth and non-truth of s *indistinguishable*. This is necessary to avoid an attacker's meta-inference of the following kind: whenever he observes the reaction "mum" on the query s, the only explanation for the refusal is the actual truth of s.

Let us further assume that the attacker additionally knows a priori that the implicational sentence $s \Leftarrow c_1$ is *true*. Then, evidently, the control must not return the correct answer to the query about (the truth value of) c_1, if c_1 is actually *true*. For the attacker simulation immediately indicates that the attacker could infer the truth of the protected sentence s from the truth of both c_1 and $s \Leftarrow c_1$ by modus ponens.

In a more complex case, the information system forms valid *beliefs* from stored sentences in a *non-monotonic* and, thus, non-classical way, see, e.g., [1,

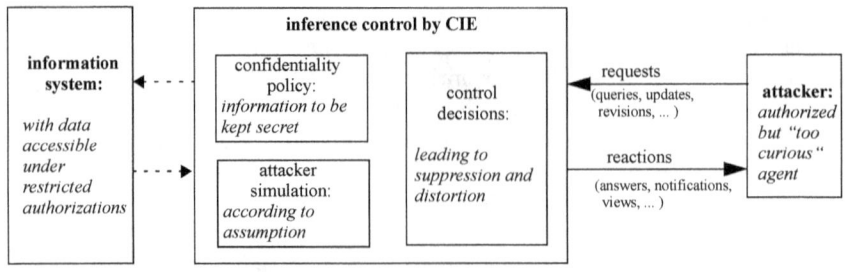

Fig. 2. Inference control by CIE with simulation of a cooperating but attacking agent.

2]. To do so, the system also interprets sentences written as $x|y$, which intuitively express the following: if y is *true*, then x is "mostly" held to be valid. For the attacker simulation, the defending control is then based on the following assumption: in principle, the attacker knows that the information system uses some non-monotonic belief operator but he is not aware of the concretely selected one out of a large range of options. Keeping s confidential can now be given a more sophisticated meaning: form his point of view, the attacker should always imagine that the validity of the belief $\neg s$ is possible.

Moreover, another distinction is important for the simulation, namely whether the attacker is postulated to reason credulously or skeptically about a belief sentence x: a *credulous* attacker argues for the validity of x if *some* belief operator leads to that conclusion; in contrast, a *skeptical* attacker insists that *all* belief operators, out of a specific class under consideration, have to conclude on the validity of x.

The non-monotonicity of a belief operator might lead to consequences that, at first glance, appear somehow unjustified but nevertheless, according to its explicit design, are reasonably wanted. For instance, it might happen that s is believed to be valid, if only c_1 is valid, but $\neg s$ is believed to be valid, if both c_1 and c_2 are valid. Thus, the additional awareness about the validity of c_2 switches the conclusion on the validity of s or $\neg s$. Such a switch might be generated by either all belief operators considered or only some of them. Hence, if an attacker queries the validity of the belief c_1, then the attacker simulation performed by the control, aiming to enforce a conclusion about the validity of $\neg s$, is essentially affected by assumptions on both the attacker's a priori status of c_2 and the flavor of his reasoning, credulous or skeptical.

In this article, we will discuss simulation assumptions for the Controlled Interaction Execution, CIE, a dedicated kind of inference control for logic-oriented information systems which include the (core of) relational databases. Figure 2 visualizes the overall architecture of inference control by CIE, as already partially implemented by a prototype, enforcing a confidentiality policy and being based on an attacker simulation according to a simulation assumption.

For the foundations and outlines of the various forms of CIE investigated so far, the reader is referred to the surveys [5, 6]. The origins of CIE are the seminal proposal of Sicherman/de Jonge/van de Riet [33] to react on harmful *queries* by *refusals* and the work of Bonatti/Kraus/Subrahmanian [21] to prevent leakage of confidentiality alternatively by *lying*. Later on Biskup/Bonatti [8, 7, 9] reconsidered and extended the early contributions, in particular by introducing a reaction scheme which combines refusals and lies and by dealing with open (i.e., SQL-like) queries. Additionally, Biskup et al [19, 12, 20, 4, 14, 18, 13, 15, 17] also dealt with incomplete information systems, updates, revisions of belief, publishing views, mediation, and a kind of database application programming. Recently, Biskup/Bonatti/Galdi/Sauro [10] have suggested an abstract framework for studying problems of expressibility and computational complexity on a most general layer of consideration.

The notion of confidentiality underlying CIE has been formally proved [16] to comply with the concepts of secrecy as summarized by Halpern/O'Neill [27] and Mantel [30]; beyond that, CIE also suggests logic-based enforcement mechanisms. In particular, CIE provides complex declarative means to precisely capture intuitive requirements regarding the secrecy of information specified in a formal confidentiality policy. Basically, and here strongly simplified, the requirements express that over the time, from the point of view of the attacker, the actual, hidden situation (of truth of assertions or validity of beliefs according to the state of the information system) is *indistinguishable* from an alternative, possible situation which is "harmless" with respect to the confidentiality policy. The respective form of a confidentiality requirement is parameterized in the sense that such an indistinguishability property should be guaranteed for all initial values of a series of parameters (true assertions, valid beliefs, protected sentences, authorized interactions, ...).

All control methods developed for CIE and thus based on an attacker simulation, i.e., in the terminology of CIE the parameterized *censors*, are designed to maintain a corresponding *invariant*, provided a suitable *precondition* is satisfied. Moreover, these censors have been mathematically verified by proving the existence of the required alternative harmless situations. In the framework of such a formal treatment, the needed *simulation assumption* could be restated as follows:

> The simulation of the attacker,
> as performed by the defender's chosen censor after its initialization,
> "closely corresponds" to the possibilities
> that are actually available to the attacker.

This understanding of CIE immediately implies the need of the following administration task for any concrete application: on behalf of the owner of the underlying information system, a *security officer* has to select an appropriate censor and to determine the suitable initialization of its parameters, such that afterwards the resulting attacker simulation actually covers what the attacker might attempt to infer about the protected information.

3   Distinctions for Simulation Assumptions

A simulation assumption serves to comprehensively model those aspects of an attacker that are crucially relevant for but *inherently not accessible* to the defending inference control. The central points are the "knowledge" of the attacker and the "intelligent connections" he can employ. In this section we describe important distinctions for these points and their possible instantiations.

While interactions between the underlying information system – protected by inference control – and the cooperating agent – a principally authorized issuer of requests, but also seen as attacker – are taking place, for each incoming *request* the inference control first has to determine whether the functionally anticipated *reaction* could lead to a violation of the confidentiality policy specifically declared for the agent. If a violation is possible, the control further has to modify the anticipated reaction such that the controlled reaction to be actually returned would not be violating. Central for this task is the following check for the functionally anticipated reaction and, if applicable, the candidates for modified reactions: Could the attacking receiver of the reaction *intelligently connect* the *reaction* with already available *knowledge* such that he would succeed to infer a piece of information to be kept confidential – in other words, would the resulting new knowledge contain such a piece of information.

By the very nature of the intuitive concept of "knowledge" there is an overwhelming variety of more concrete understandings of knowledge leading to many and diverse algorithmically processable formalizations. Moreover, it appears to be mandatory to always use the concept of *knowledge* together with a *subject* holding the knowledge. Regarding the simulation assumption, first of all the relevant holder we have to model is the *attacker*. However, the underlying information system and its owner, respectively, and potentially further subjects being involved in the respective interactions, are also holders of knowledge. Moreover, one holder might reason about another holder's understanding of knowledge.

Within the framework of CIE, so far we have made the following distinctions regarding "knowledge" and, of course, in principle we might want to deal with further ones:

1. *"true" knowledge* [8, 9, 20] or (possibly "false") *belief* or suitable combinations and variants thereof [18, 13];
2. *complete* [8, 9, 20] or *incomplete* [19, 14];
3. directly referring to a (real or fictitious) *world* [8, 9, 20] or as view of an *underlying information system* on such a world [19, 14, 18].

Moreover, for CIE we consider that "knowledge" evolves from data by means of "intelligent connections" as follows:

4. On the basis, knowledge is grounded on (primary) *data* that is stored in the underlying *information system* (and thus in the beginning not visible to the attacker) or contained in *messages* for requests and reactions (and thus directly observable by the attacker). Such a message might occur in the course of a variety of interactions: as a (database) *query* and as a direct *answer* to it; as an *update* request or a *revision* request and as the reaction in form of a *notification* about their admissibility and actual internal processing; without any explicit request, as a dedicatedly communicated or even generally published *view*.

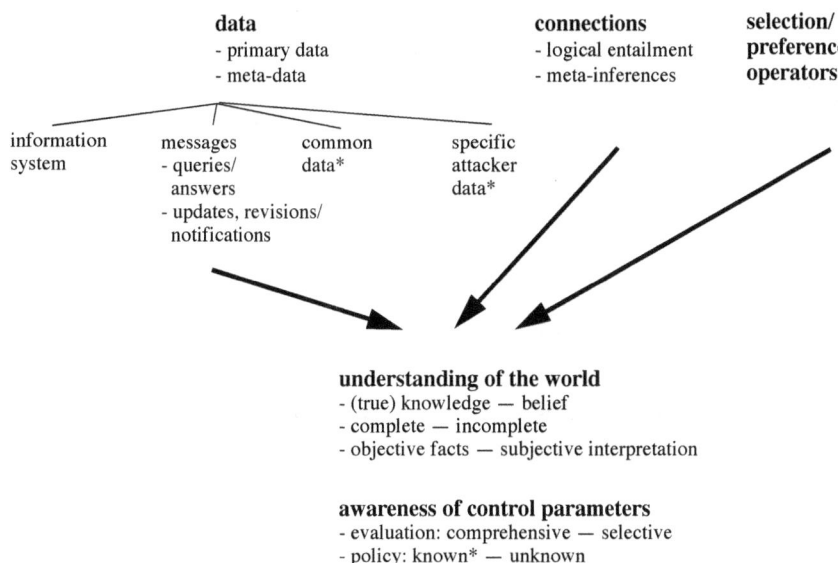

Fig. 3. Simplified overview on the variety of assumptions related to knowledge (* refers to an initialization parameter for a control mechanism, see Section 4).

5. Moreover, such basic knowledge is often complemented by *meta-data* constraining the primary data (and mostly visible to the attacker), in particular by *semantic constraints* declared in the *schema* of the underlying information system or even by an *ontology* agreed on for a cooperation.
6. Further on, specific knowledge might be augmented by data and meta-data on everybody's *common knowledge*.
7. Specific knowledge might also be extended by means of data and meta-data individually available to the attacker, referred to as *attacker knowledge*.
8. Then, inductively, more advanced knowledge is generated from already existing knowledge by means of *connection operators*, in particular including the classical *logical entailment* (or, equivalently as far as possible, symbolic deduction).
9. As a special case relevant for inference control, knowledge might be gained by *meta-inferences* that are based on knowledge about the design of the protection mechanism.
10. Such knowledge generation can be guided by parameterizing the connection operators by *preferences* or related concepts, as seen for non-monotonic logics or when offering the choice of credulous or skeptical reasoning.

Finally, specifically for CIE and related protection mechanisms, we have to consider the "knowledge" about relevant parameters of the inference control itself, including the following ones:

10. The evaluations (truth values, ...) of pieces of information under protection may be required to be kept secret either *comprehensively* or only *selectively*,

where in the latter case a selection might refer to a single value (like either *true* or *false* or *undefined*) or comprise a logical combination (like "*true* or *false*", permitting to know the value *undefined*).

11. The attacker might be *aware* or *not aware* of the *confidentiality policy*, i.e., the collection of pieces of information under protection together with their respective evaluations seen to be sensitive.

If one undertakes to design and operate ambitious inference control, the preceding rough overview might support our general remark about the rich variety of possible assumptions on "knowledge" and "intelligent connections" to be taken seriously. Further simplifying, Figure 3 summarizes the overview.

4 Instantiations of Simulation Assumptions

4.1 Inference Control by Controlled Interaction Execution

The first forms of CIE, as well as most of the more advanced ones, are founded on still rather simple instantiations of a simulation assumption for a respectively parameterized censor:

(1) The attacker unrestrictedly knows on which classical logic with model-theoretic semantics the pertinent underlying information system is based, and he employs that logic as well. In particular, both the information system and the attacker use classical logical entailment to infer conclusions from premises. Accordingly, while simulating the attacker, the censor employs that logic, too. Of course, we have to restrict the considerations to an (at least) decidable fragment of the logic (if not even to a feasible one) in order to profit from an equivalent algorithmic form of symbolic deduction.

(2) The attacker unrestrictedly knows the pertinent declarative form of the confidentiality requirement, expressed as the "existence of an indistinguishable harmless situation". Moreover, to be cautious, the attacker is expected to be an unrestrictedly strong reasoner (not suffering from any undecidability or nonfeasibility) such that he would be able to check any instance of such a confidentiality requirement in an "omnipotent" manner by means of meta-inferences. Regarding the simulating censor, it is based on the conversion of the declarative confidentiality requirement into a provably sufficient invariant that has to be enforced for any single interaction supposing a suitable precondition. Basically, such an invariant asserts that certain sentences of the underlying logic are *not* related by logical entailment. The enforcement is then ensured by determining a suitable reaction.

(3) The preceding points (1) and (2) refer to the "logical connections", in the sense of Section 3 with Figure 3, and are handled equally for all applications and, moreover, selection and preference operators are not part of classical logic. In contrast, for any concrete application and for any specific attacker, meaningful common data and attacker data are individually determined and collected or – more cautiously expressed – are assumed to be correctly guessed by a security officer. The simulating censor gets these data as parameter values during its initialization, as further discussed in Section 5.

(4) In the same way, the simulating censor gets data about those elements of the confidentiality policy that the attacker is supposed to be aware of.

This simple approach to instantiate a simulation assumption has been further and more sophisticatedly elaborated for a more recently developed form of CIE [18], see also [13], and we argue for a strong need of such elaborations for future work on inference control. In contrast to previous forms, the underlying information system is based on a non-monotonic logic of belief. Though the attacker is still assumed to know some basic properties of that logic of belief, it appears to be reasonable to assume that he is not aware of the concrete logic employed. Accordingly, the attacker only knows that some logic out of a specific class is used by the information system to determine the validity of beliefs. Hence, taking his a priori knowledge and his observations into consideration, from his point of view at the best, the attacker can only achieve to approximate the hidden forming of beliefs within the information system. In general, an approximation will tell the attacker that many different belief states of the information system are possible, and the attacker then has to suitably evaluate them according to some criteria.

In turn, the defender's simulating censor is challenged to determine those belief states that are held to be possible by the attacker and the attacker's evaluations of them. Unfortunately, however, the challenge has to be mastered without exactly knowing the attacker's a priori knowledge and evaluation criteria.

4.2 Inference Control in General

Besides CIE, in the rich literature there are reports about research and development of a large variety of forms of inference control for quite different application fields, as summarized, e.g., in [25, 3]. Among others, the following more specific topics have been investigated: information flows caused by constructs of general-purpose programming languages, the (unwanted) exploitation of statistical databases by means of establishing a solvable system of equations, and properties of noninterference of computing systems specified by their possible traces. For these cases, and in many others as well, it has been pointed out that the attacker's a priori knowledge would have a crucial impact on the attacker's success. For example:

- For a program written in some *general-purpose language* with assignment commands and guarded commands, a priori knowledge about possibly occurring values of the input variables might confine the control flow during execution, definitely excluding some of the syntactically possible paths, see, e.g., [22, 23, 31].
- For a *statistical database* specified by the equations expressing the aggregate functions involved, like average over some list of values, a priori knowledge about possibly occurring values in the underlying data set might lead to additional equations that result in the overall solvability of a previously still undetermined system of equations, see, e.g., [24, 34].

- For a *trace-based computing system*, a priori knowledge about possibly occurring traces, definitely excluding those seen to be impossible, might restrict the options to "hide" traces to be kept secret within a larger set of traces, see, e.g., [30, 28, 27].

Though the impact of a priori knowledge on options for coming up with unwanted inferences is well-known, theoretically, unfortunately there are only few studies on practical measurements and mechanisms to actually determine and systematically consider a priori knowledge, as far as we are aware.

4.3 Inference Control for Special Situations

Obviously, the need of relying on a simulation assumption is a weak point of any kind of inference control, whether by CIE or any other form. This insight is somehow disappointing, and it suggests to ask for conditions under which we can totally get rid of a simulation assumption or, at least, are only burdened by a simple and easily manageable instantiation. To begin with, we can think about two extreme cases, both of which are based on a very strong presupposition, namely that, for a specific application, we (i.e., in practical terms, the defenders) are able to be completely aware about the full collection of all possible combinations of a priori knowledge and intelligent connections. The extreme cases can be outlined as follows:

- Control decisions can be taken completely *independently* from the specific capabilities of an attacker. This extreme case would occur in a situation characterized as follows: for any single piece of data evaluated to be harmless per se and thus permitted to be made observable to an attacker, *none* of the collected combinations of a priori knowledge and intelligent connections provides an additional gain of information, beyond just the immediate information of an observation.
- Control decisions can be taken in such a way that for *all* of the collected intelligent combinations the possibly additionally gained information is evaluated to be harmless.

Obviously, these extreme cases are not totally independent, since the event of no additional gain of information should be considered to be harmless. Presumably, in practice there will rarely be such helpful situations. In particular, otherwise, the wordings of the first extreme case suggest that the information contained in any single piece of observable data would be closed under any a priori knowledge and for all intelligent connections, and so would be any set of pieces of observable data. Nevertheless, and somehow surprisingly, there are some special situations known to lead to a helpful situation, and having been thoroughly investigated in the literature, e.g.:

- Information-theoretic *perfect secrecy* of *symmetric encryption* requires that an observer of a single cipher text should not be able to improve his knowledge in form of an a priori probability distribution over the possible plain texts. This requirement can actually be satisfied by bitwise encryption of

a plain text with a one-time pad of the same length employing the XOR-operation [32]. Though normally being too expensive for everyday end-to-end encryption, this technique is being employed in some substep of many cryptographic protocols.
- Complexity-theoretic *semantic secrecy* of *asymmetric encryption* requires, roughly outlined, that a polynomially time-bounded observer of a cypher text, even under a priori knowledge, will only succeed to learn those properties of the pertinent plain text that he could infer without having made the observation at all, at least with high probability. In theory, this requirement has been proven to be satisfiable under some challenging complexity-theoretic assumptions, which are open since quite a long time. In practice, commonly used asymmetric encryption mechanisms appear to provide reasonably good approximation for the time being. See, e.g., [26] for a thorough treatment of this topic.
- *Differential privacy* of an evaluation of a numerical *aggregate function* applied to a database relation requires, again only roughly outlined, that the impact of any possible single tuple (for an identifiable individual) on the returned aggregate values should be negligibly small, with high probability. Accordingly, an observer will not learn anything essential about each of the individuals concerned. Under some assumptions, this requirement can be met by adding some statistical noise, in particular using a Laplace-distribution, see, e.g., [36].

5 Collection and Justification of Simulation Assumptions

As said in the introduction, in the first place an "attacker" is an authorized agent: The data *owner* in control of the "defending" information system has granted some specific access rights to that agent within the framework of some trust relationship or cooperation contract. More specifically, the owner, or a security officer on his behalf, should seriously follow a workflow of *security administration* which at least includes the following tasks:

1. initially, assigning some suitably restricted *trust* to the agent considered for future cooperation and, based on that, granting the *access rights* that are both needed and justifiable;
2. additionally, specifying a dedicated *confidentiality policy*, which expresses a kind of wanted exceptions from the permissions granted before;
3. determining a reasonable *simulation assumption* to be applied for the agent considered and, based on that, selecting and configurating a matching *censor* for a controlled interaction execution that would appropriately confine the information gain achievable by the agent considered;
4. subsequently, while interactions are taking place, observing and recording the *actual behavior* of the agent, who is being both granted access rights and confined by the inference control dedicatedly exercised;
5. repeatedly, *learning patterns* of the agent's *behavior* and evaluating them regarding the goals of the currently granted access rights and the declared confidentiality policy, thereby making a *prognosis*, if applicable including a justifiable *suspicion* on the agent's aim at gaining forbidden information;

6. accordingly, *adapting* the previous trust assignment and *modifying* the current access rights and, consequently, *reconfigurating* the controlled interaction execution in operation.

However, sometimes right from the beginning, the defending owner might prefer to shift his considerations towards *facing misbehavior*, giving the *possibility of attacks* priority. In such cases, for example, first of all the defender would examine attack plans rather than trust suggestions, and instead of starting with granting permissions the defender would investigate the attacking agent's options of non-preventable bad behavior like tapping communications lines. Under those conditions, the tasks of security administration would become closer to those that have been developed for strategic attacker-defender games, in particular for military applications.

As impressively exemplified in the collection [29] about "Adversarial Reasoning", the defender's awareness and learning of the *knowledge* or *belief* of an attacker, and of the attacker's *desires* and *intentions* will play a central role. Crucially important for the defender's success, the defender needs to apply and combine a large variety of methods from many disciplines to handle the resulting challenges, as far as the gap between complex practical requirements on the one hand and simplifying theoretical insights on the other hand can be reconciled at all.

As important as the more general context of adversarial reasoning might be, for the narrower topic of simulation assumptions as studied in this report and focussing on CIE we can exploit some useful particularities:

– As stated before, for the attacker's "desires and intentions" we cautiously assume just the *worst case*.
– Regarding the remaining "knowledge and belief" we restrict to initially establish a *fixed assumption* and to then additionally treat only explicit communication data, rather than periodically reconsidering and revising our assumptions based on an overall evaluation of the attacker's behavior.
– Moreover, as discussed in Section 3, we handle "knowledge and belief" only on a *purely logical level* (selected from an already rich variety of options, though).

Somehow summarizing, so far we thus concentrate on a simulation assumption that refers to (in principle hidden) initial knowledge/belief, (explicitly seen) message data, and (supposedly) invariantly applied intelligent connections. Of course, this concentration should not exclude more ambitious administration procedures including the *cycle* of recording, learning and revising, as described above.

On the basis of a suitably justified simulation assumption for a specific agent considered, an appropriately selected *censor* (in terms of CIE) can be used to control and confine the possible information gain of that agent. In a sense, most of the works about CIE, as surveyed in [5, 6], can be understood as providing policy enforcing censors for exemplarily selected classes of simulation assumptions, aiming to treat a broad scope of instantiations of the distinctions made in this article. In the long range, we hope to achieve a more or less complete covering of all meaningful possibilities to combine such instantiations.

A future adaptable and fully parameterizable CIE-system could then be employed for any application-dependent simulation assumption by means of appropriate initializations and optimizations. Preferably, the actual parameter values would then be at least semi-automatically derived from the specifications. Our achievements so far have already indicated that we could actually approach our long-term goal, for each newly selected instantiation of some distinctions coming up with an innovative censor. For example, we were able to conceptually manage the following transitions and partly also to enhance the prototype under construction accordingly:

- from complete information systems to incomplete ones [19],
- from knowledge management systems to belief management systems [18],
- from structured (relational) data to semi-structured (XML-) data [14],
- from closed "yes/no"-queries to open "give-me-all"-queries [9, 11],
- from queries only to updates and revisions [12, 18],
- from basic operations of information management system to executions of procedural programs [17],
- from dynamic interactions to static view publishing [20, 14, 15].

6 Robustness of Simulation Assumptions

Even if a simulation assumption has suitably been embodied by an effective censor, there will remain an essential uncertainty regarding the overall success: it might just happen that the actual attacker enjoys capabilities that are not covered by the assumption or even completely differ from the postulated ones. Accordingly, ideally one might wish to specify a *class of attackers* as comprehensive as possible, rather than just one instance. For such a class, one then would define one or more partial orders expressing intuitive notions like, e.g., "strength of attacks" or "degree of credulousness", and finally aim at designing a censor that provides effective protection against a large subclass. So far, there is only little insight on this issue available, to be briefly summarized next.

Obviously, a censor will be still effective if the concrete attacker is actually weaker than assumed. For example, a censor designed to deal with an attacker that knows the confidentiality policy in place will also be effective if the concrete attacker does not. Similarly, a censor assuming a credulously inferring attacker will also be successful if the concrete attacker is behaving skeptically.

While one will mostly prefer to postulate a "sophisticated" attacker who would employ his knowledge and most powerful intelligent connections – including those for meta-inferences based on the awareness of the protection mechanism – one could also consider a "plain" attacker, who would literally believe what the interaction data suggests to mean, in particular take lies for truth. Some results of the thesis [35] show the following somehow surprising result under the situation studied there (incomplete propositional information systems with queries only): nearly all censors for that situation designed to deal with a sophisticated attacker are also effective for a plain one (with the only exception of the censor applying uniform lying under a confidentiality policy unknown to the attacker). For these results, the class of attackers has only two elements, which are the highest and lowest, respectively, in the straightforward ordering. Naturally, we would like to have insight about a much richer class.

Recent work [13] has attempted to treat related considerations more systematically, dealing with autonomously interacting and intelligent BDI-like agents. On the one hand, by means of a non-monotonic belief operator, such an agent might generate his own "plausible belief" about the actual state of affairs from his observed but possibly inconsistently appearing "world view". On the other hand, as a defender, that agent might also determine a set of possible "world views" seen to be possible for another agent, based on some assumptions and the observations about the other agent; furthermore, the defending agent can then tentatively inspect a list of message data, each being a candidate to be sent to the other agent, regarding the possibly resulting "plausible belief" on the receiver's side, and finally evaluate the harmfulness of each them with respect to a confidentiality policy.

7   Conclusions

The task of inference control in general, and by CIE in particular, can be considered under two opposed points of view. On the one hand, conceptually, many situations just require to master this very task, namely whenever not only raw "data" but primarily the "information" represented by data should be kept confidential within an environment in which agents might have various kinds of a priori knowledge and are suspected to possibly exploit it actually in an adversarial manner. For example, privacy legislation demands to protect "personal data", i.e., data that is characterized by referring to an identified or *identifiable* human individual. Clearly, the process of identifying an individual from data without explicit identifiers deals with inferences that thus have to be controlled by law.

On the other hand, practically, both the need of simulation assumptions for inference control and the in general inevitably high computational complexity of inference control constitute obstacles that are mostly extremely difficult to overcome. While sometimes complexity issues can be approached by suitable approximations or preprocessing or related techniques, in contrast even then a simulation assumption remains mandatory to appropriately deal with in order to achieve the challenging goal of confidentiality. In fact, the indispensable necessity of a strong simulation assumption leads some researchers to argue that inference control is not practically feasible at all.

Given the strong need of inference control and its conceptual and computational problems, like by our previous and current work, all future efforts should continue to be directed to design and implement socially acceptable and organizationally and algorithmically manageable control procedures. On the same time, such procedures have to sufficiently comply with the great challenges of an important task on the one hand and still satisfiably mind the inevitable obstacles to any implementation of that task on the other hand.

References

1. Christoph Beierle and Gabriele Kern-Isberner. Semantical investigations into nonmonotonic and probabilistic logics. *Ann. Math. Artif. Intell.*, 65(2-3):123–158, 2012.
2. Christoph Beierle and Gabriele Kern-Isberner. *Methoden wissensbasierter Systeme – Grundlagen, Algorithmen, Anwendungen (5. Aufl.)*. Computational intelligence. SpringerVieweg, 2014.
3. Joachim Biskup. Inference control. In Henk C. A. van Tilborg and Sushil Jajodia, editors, *Encyclopedia of Cryptography and Security (2nd Ed.)*, pages 600–605. Springer, Berlin/Heidelberg, 2011.
4. Joachim Biskup. Dynamic policy adaption for inference control of queries to a propositional information system. *Journal of Computer Security*, 20:509–546, 2012.
5. Joachim Biskup. Inference-usability confinement by maintaining inference-proof views of an information system. *International Journal of Computational Science and Engineering*, 7(1):17–37, 2012.
6. Joachim Biskup. Logic-oriented confidentiality policies for controlled interaction execution. In Aastha Madaan, Shinji Kikuchi, and Subhash Bhalla, editors, *Databases in Networked Information Systems, DNIS 2013*, volume 7813 of *Lecture Notes in Computer Science*, pages 1–22. Springer, 2013.
7. Joachim Biskup and Piero A. Bonatti. Controlled query evaluation for enforcing confidentiality in complete information systems. *Int. J. Inf. Sec.*, 3(1):14–27, 2004.
8. Joachim Biskup and Piero A. Bonatti. Controlled query evaluation for known policies by combining lying and refusal. *Ann. Math. Artif. Intell.*, 40(1-2):37–62, 2004.
9. Joachim Biskup and Piero A. Bonatti. Controlled query evaluation with open queries for a decidable relational submodel. *Ann. Math. Artif. Intell.*, 50(1-2):39–77, 2007.
10. Joachim Biskup, Piero A. Bonatti, Clemente Galdi, and Luigi Sauro. Optimality and complexity of inference-proof data filtering and CQE. In Miroslaw Kutylowski and Jaideep Vaidya, editors, *European Symposium on Research in Computer Security, ESORICS 2014, Part II*, volume 8713 of *Lecture Notes in Computer Science*, pages 165–181. Springer, 2014.
11. Joachim Biskup, Martin Bring, and Michael Bulinski. Confidentiality preserving evaluation of open relational queries. In Tadeusz Morzy, Patrick Valduriez, and Ladjel Bellatreche, editors, *Advances in Databases and Information Systems, ADBIS 2015*, volume 9282 of *Lecture Notes in Computer Science*, pages 431–445. Springer, 2015.
12. Joachim Biskup, Christian Gogolin, Jens Seiler, and Torben Weibert. Inference-proof view update transactions with forwarded refreshments. *Journal of Computer Security*, 19:487–529, 2011.
13. Joachim Biskup, Gabriele Kern-Isberner, Patrick Krümpelmann, and Cornelia Tadros. Reasoning on secrecy constraints under uncertainty to classify possible actions. In Christoph Beierle and Carlo Meghini, editors, *Foundations of Information and Knowledge Systems, FoIKS 2014*, volume 8367 of *Lecture Notes in Computer Science*, pages 97–116. Springer, 2014.
14. Joachim Biskup and Lan Li. On inference-proof view processing of XML documents. *IEEE Trans. Dependable Sec. Comput.*, 10(2):99–113, 2013.

15. Joachim Biskup and Marcel Preuß. Inference-proof data publishing by minimally weakening a database instance. In Atul Prakash and Rudrapatna K. Shyamasundar, editors, *Information Systems Security, ICISS 2014*, volume 8880 of *Lecture Notes in Computer Science*, pages 30–49. Springer, 2014.
16. Joachim Biskup and Cornelia Tadros. Policy-based secrecy in the runs & systems framework and controlled query evaluation. In Isao Echizen, Noboru Kunihiro, and Ryôichi Sasaki, editors, *Advances in Information and Computer Security, IWSEC 2010, Short Papers*, pages 60–77. Information Processing Society of Japan (IPSJ), 2010.
17. Joachim Biskup and Cornelia Tadros. Constructing inference-proof belief mediators. In Pierangela Samarati, editor, *Data and Applications Security and Privacy, DBSec 2015*, volume 9149 of *Lecture Notes in Computer Science*, pages 188–203. Springer, 2015.
18. Joachim Biskup and Cornelia Tadros. Preserving confidentiality while reacting on iterated queries and belief revisions. *Ann. Math. Artif. Intell.*, 73(1-2):75–123, 2015.
19. Joachim Biskup and Torben Weibert. Keeping secrets in incomplete databases. *Int. J. Inf. Sec.*, 7(3):199–217, 2008.
20. Joachim Biskup and Lena Wiese. A sound and complete model-generation procedure for consistent and confidentiality-preserving databases. *Theoretical Computer Science*, 412:4044–4072, 2011.
21. Piero A. Bonatti, Sarit Kraus, and V. S. Subrahmanian. Foundations of secure deductive databases. *IEEE Trans. Knowl. Data Eng.*, 7(3):406–422, 1995.
22. Ellis S. Cohen. A formalism for describing information transmission in computational systems. Technical report, Dept. of Computer Science, Carnegie Mellon University, 1976.
23. Dorothy E. Denning and Peter J. Denning. Certification of programs for secure information flow. *Commun. ACM*, 20(7):504–513, 1977.
24. Dorothy E. Denning and Jan Schlörer. Inference controls for statistical databases. *IEEE Computer*, 16(7):69–82, 1983.
25. Csilla Farkas and Sushil Jajodia. The inference problem: a survey. *SIGKDD Explorations*, 4(2):6–11, 2002.
26. Oded Goldreich. *Foundations of Cryptography II – Basic Applications*. Cambridge University Press, 2004.
27. Joseph Y. Halpern and Kevin R. O'Neill. Secrecy in multiagent systems. *ACM Trans. Inf. Syst. Secur.*, 12(1):5.1–5.47, 2008.
28. Dominic Hughes and Vitaly Shmatikov. Information hiding, anonymity and privacy: a modular approach. *Journal of Computer Security*, 12(1):3–36, 2004.
29. Alexander Kott and Willima M. McEneaney, editors. *Adversarial Reasoning: Computational Approaches to Reading the Opponent's Mind*. Chapman & Hall/CRC, Boca Raton, FL, 2007.
30. Heiko Mantel. *A uniform framework for the formal specification and verification of information flow security*. PhD thesis, Universität des Saarlandes, 2003.
31. Andrew C. Myers and Barbara Liskov. Protecting privacy using the decentralized label model. *ACM Trans. Softw. Eng. Methodol.*, 9(4):410–442, 2000.
32. Claude E. Shannon. Communication theory of secrecy systems. *Bell System Technical Journal*, 28(4):656–715, 1949.
33. George L. Sicherman, Wiebren de Jonge, and Reind P. van de Riet. Answering queries without revealing secrets. *ACM Trans. Database Syst.*, 8(1):41–59, 1983.
34. Joseph F. Traub, Yechiam Yemini, and Henryk Wozniakowski. The statistical security of a statistical database. *ACM Trans. Database Syst.*, 9(4):672–679, 1984.

35. Torben Weibert. *A Framework for Inference Control in Incomplete Logic Databases*. PhD thesis, Technische Universität Dortmund, 2008. http://hdl.handle.net/2003/25116.
36. Yin Yang, Zhenjie Zhang, Gerome Miklau, Marianne Winslett, and Xiaokui Xiao. Differential privacy in data publication and analysis. In K. Selçuk Candan, Yi Chen, Richard T. Snodgrass, Luis Gravano, and Ariel Fuxman, editors, *SIGMOD 2012*, pages 601–606. ACM, 2012.

Problem Solving Using the HEX Family[1]

Thomas Eiter[2], Christoph Redl[3], Peter Schüller[4]

Abstract. The HEX formalism has been designed as an extension of answer set programs that offers an abstract interface to access external sources of information and computation, such as the World Wide Web or description logics reasoners. The generic nature makes the extension powerful, which has been exploited in different ways: as an end user problem solving language, as a backend formalism, or as the basis of a richer formalism with possibly increased expressiveness. The increasing spread of HEX is paralleled with the frequently asked questions of what HEX is, and how it can be used for problem solving. In this paper, we aim to answer these questions; we consider different scenarios and provide a methodology for applying HEX from a user perspective. Furthermore, we briefly present a collection of applications based on HEX or derived from it, including sample snippets from associated HEX programs.

1 Introduction

Answer Set Programming (ASP) is a declarative problem solving approach [48, 45, 43], in which a problem is described by the rules of a nonmonotonic logic program, such that the answer sets [33] (i.e., specific models) of the program correspond to the solutions of the problem; the latter can be extracted from the answer sets computed using an ASP solver. With the advent of efficient and expressive such solvers (e.g., smodels [58], dlv [41], ASSAT [44], and GRINGO plus CLASP [31, 30]), this approach has been fruitfully deployed to a growing range of applications in different areas and disciplines, cf. [8].

However, the World Wide Web and trends in distributed systems have created a need for accessing external information sources in a program, ranging from light-weight data access (e.g., XML, RDF, or data bases) to knowledge-intensive formalisms (e.g., description logics reasoners), and even to information sources not based on logical grounds (e.g., dictionaries or route planning services). To cater for this need, *HEX programs* have been introduced in [13] as an extension to nonmonotonic logic programs in which access to external

---

[1] This research has been supported by the Austrian Science Fund (FWF) project P27730 and the Scientific and Technological Research Council of Turkey (TUBITAK) Grant 114E430.
[2] Technische Universität Wien, Institute für Informationssysteme, Knowledge Based Systems Group, Vienna, Austria, eiter@kr.tuwien.ac.at
[3] Technische Universität Wien, Institute für Informationssysteme, Knowledge Based Systems Group, Vienna, Austria, redl@kr.tuwien.ac.at
[4] Marmara University, Faculty of Engineering, Department of Computer Engineering, Istanbul, Turkey, peter.schuller@marmara.edu.tr

sources is possible via designated external atoms, which abstractly define external predicates whose valuation is determined by external computation.

For a simple example, consider the rule

$$pointsTo(X, Y) \leftarrow \&hasHyperlink[X](Y), url(X); \qquad (1)$$

informally, it obtains pairs (X, Y) of URLs, where X actually links Y on the Web. Here, $\&hasHyperlink$ is an *external predicate* associated with an external computation function; X is the input for the latter and Y is a result. Besides single values, also relational information (predicate extensions) can flow from the program to external sources and back; e.g., an extended input $[X, skip]$ in (1) may provide a relation *skip* containing pairs of URLs whose linkage should be omitted. Notably, the output Y may involve values not occurring in the program (known as *value invention*), and Y may influence the input of the atom where in practice, certain safety conditions must be obeyed [15]. This makes the efficient evaluation of HEX programs challenging, for which advanced techniques have been developed [14, 18, 13, 16].

The abstract concept of an external atom has been realized in the open-source software DLVHEX[5] as an API, which allows the user via a plugin mechanism to tailor external atoms for her needs using Python or C++. This makes the system very powerful; depending on the external evaluation cost, HEX programs offer a range of problem solving capacity, from Σ_2^p for polynomial-time external atoms to Turing-completeness in general.

HEX programs and DLVHEX have been used for solving diverse kinds of problems. This frequently raises the questions of prospective users what HEX is after all, and how HEX programs or DLVHEX can be exploited for solving their applications; notably DLVHEX offers a growing suite of library plugins that have been used in different applications.

This paper addresses these questions and focuses on HEX programs as a KR tool for problem solving, which they support at different levels of abstraction, namely as an end user problem solving language, as a backend formalism or as the basis of a richer formalism with possibly increased expressiveness. Besides a methodology for using HEX programs, we further present some examples of HEX applications. More in detail, we proceed on this as follows.

- After recalling in Section 2 HEX programs and briefly addressing some aspects of the DLVHEX system, we show in Section 3 how HEX programs can be used for problem solving. Besides a basic methodology, which is a strict generalization of the ASP methodology, we present typical kinds of external sources and HEX use scenarios.
- In Section 4 we then consider some end user applications which have been realized on top of HEX programs.
- In Section 5 we show how several extensions of the HEX formalism and some of their internals. While some extensions increase the expressiveness and need extensions of the internal evaluation algorithms, many others are in fact only syntactic shortcuts and can be compiled to pure HEX programs.

---

[5] www.kr.tuwien.ac.at/research/systems/dlvhex

- In Section 6 we show applications that use HEX as a backend formalism for the sake of evaluation, sometimes completely hiding HEX from the user.

After a discussion of related work in Section 7, we conclude in Section 8 with an outlook on future topics.

This paper is in honor of Gabriele Kern-Isberner, who has worked extensively on Artificial Intelligence and in particular on knowledge representation and reasoning, and made over many years numerous important contributions to belief change, conditional and nonmonotonic reasoning, reasoning with uncertainty, argumentation, agents etc., based on solid and deep mathematical foundations. Gabriele's excellent book "Methods of Knowledge based Systems – Foundations, Algorithms and Applications" [5], co-authored by Christoph Beierle and now in its fifth edition, demonstrates her comprehension of the field, and provides a coherent and formally guided introduction to the ingredients of knowledge based systems. Unfortunately, the book is only available in German; but in the style of Cato the Elder, Thomas keeps asking her ever since the first edition to have an English translation, so that a much larger audience can read and enjoy it, which would be well-deserved – and that we can use it in our English classes! In turn, we hope that Gabriele may discover the wealth of the HEX family as a tool box for building advanced knowledge based systems, as envisaged in her book and further supported by her many research results.

2   HEX Programs

In this section, we formally introduce the syntax and semantics of HEX programs; for more details and background, see e.g. [23, 24, 13, 55].

2.1   HEX Program Syntax

Let $\mathcal{C}$, $\mathcal{X}$, and $\mathcal{G}$ be mutually disjoint sets of *constants*, *variables*, and *external predicates*, respectively. Usually constants (resp., variables) are denoted with first letter in upper case (resp., lower case), while external predicates start with '&'. Elements from $\mathcal{C} \cup \mathcal{X}$ are called *terms*. An *atom* is a tuple $(Y_0, Y_1, \ldots, Y_n)$, where $Y_0, \ldots, Y_n$ are terms and $n \geq 0$ is the *arity* of the atom. Intuitively, Y_0 is the predicate name, and we often use the more familiar notation $Y_0(Y_1, \ldots, Y_n)$. An atom is *ordinary* (resp., *higher-order*) if Y_0 is a constant (resp., a variable), and it is *ground*, if all its terms are constants.

An *external atom* is of the form

$$\&g[Y_1, \ldots, Y_n](X_1, \ldots, X_m),$$

where $Y_1, \ldots, Y_n$ and $X_1, \ldots, X_m$ are two lists of terms (called *input* and *output* lists, resp.), and $\&g \in \mathcal{G}$ is an external predicate name. We assume that $\&g$ has fixed lengths $in(\&g) = n$ and $out(\&g) = m$ for input and output lists, respectively. In the ground case, the input terms $Y_1, \ldots, Y_n$ intuitively consist of individual constants (e.g. *joe*) and predicate names (e.g. *edge*). An external atom provides a way for deciding the truth value of an output tuple depending on the input tuple and a given interpretation.

A *rule* r is of the form

$$\alpha_1 \vee \cdots \vee \alpha_k \leftarrow \beta_1, \ldots, \beta_n, \mathbf{not}\ \beta_{n+1}, \ldots, \mathbf{not}\ \beta_m, \qquad m, k \geqslant 0,$$

where all α_i are atoms and all β_j are either atoms or external atoms. We let $H(r) = \{\alpha_1, \ldots, \alpha_k\}$ and $B(r) = B^+(r) \cup B^-(r)$, where $B^+(r) = \{\beta_1, \ldots, \beta_n\}$ and $B^-(r) = \{\beta_{n+1}, \ldots, \beta_m\}$. A HEX *program* is a finite set P of rules.

A rule r is a *constraint*, if $H(r) = \emptyset$ and $B(r) \neq \emptyset$; a *fact*, if $B(r) = \emptyset$ and $H(r) \neq \emptyset$; and *nondisjunctive*, if $|H(r)| \leqslant 1$. We call r *ordinary*, if it contains only ordinary atoms. We call a program P *ordinary* (resp., *nondisjunctive*), if all its rules are ordinary (resp., nondisjunctive). Note that facts can be disjunctive.

Example 1. Consider the following program Π_{goto} to decide where to go for a city trip, but exclude cities where the (external) weather report is bad.

$$badweather(rain). \qquad badweather(snow).$$
$$goto(paris) \vee goto(london).$$
$$\leftarrow \&weatherreport[goto](W), badweather(W).$$

We guess where to go and forbid that the externally obtained weather report $\&weatherreport$ indicates bad weather for a city in the extension of $goto$. □

2.2 HEX Program Semantics

The semantics of HEX programs generalizes the well-known answer-set semantics of ordinary programs [33]. Given a HEX program P, its *Herbrand base*, denoted HB_P, is the set of all possible ground versions of atoms and external atoms occurring in P obtained by replacing variables with constants from $\mathcal{C}$. The grounding of a rule r, $grnd(r)$, is defined accordingly, and the grounding of P is given by $grnd(P) = \bigcup_{r \in P} grnd(r)$. Unless specified otherwise, $\mathcal{X}$ and $\mathcal{G}$ are implicitly given by P. Different from the 'usual' ASP setting, the set $\mathcal{C}$ of constants used for grounding a program is only partially given by the program itself; in HEX, external computations may introduce new constants that are relevant for semantics of the program.

An *interpretation relative to* P is any subset $I \subseteq HB_P$ containing no external atoms. We say that I is a *model* of atom $a \in HB_P$, denoted $I \models a$, if $a \in I$.

With every external predicate name $\&g \in \mathcal{G}$, we associate an $(n+m+1)$-ary Boolean function (called *oracle function*) $f_{\&g}$ assigning each tuple $(I, \boldsymbol{y}, \boldsymbol{x})$ where $\boldsymbol{y} = y_1, \ldots, y_n$ and $\boldsymbol{x} = x_1, \ldots, x_m$ either 0 or 1, where $n = in(\&g)$, $m = out(\&g)$, $I \subseteq HB_P$, and $x_i, y_j \in \mathcal{C}$. We say that $I \subseteq HB_P$ is a *model* of a ground external atom $a = \&g[\boldsymbol{y}](\boldsymbol{x})$, denoted $I \models a$, if $f_{\&g}(I, \boldsymbol{y}, \boldsymbol{x}) = 1$. This definition of external atom semantics is very general; indeed an external atom may depend on every part of the interpretation. For practical reasons, external atom semantics is usually restricted such that it depends only on the extension of those predicates in I that are given in the input list.

Let r be a ground rule. Then we say that

1. I satisfies the head of r, denoted $I \models H(r)$, if $I \models a$ for some $a \in H(r)$;

2. I satisfies the body of r $(I \models B(r))$, if $I \models a$ for all $a \in B^+(r)$ and $I \not\models a$ for all $a \in B^-(r)$; and
3. I satisfies r $(I \models r)$, if $I \models H(r)$ whenever $I \models B(r)$.

We say that I is a *model* of a HEX program P, denoted $I \models P$, if $I \models r$ for all $r \in grnd(P)$. We call P *satisfiable*, if it has some model.

Given a HEX program P, the *FLP-reduct* of P with respect to $I \subseteq HB_P$, denoted fP^I, is the set of all $r \in grnd(P)$ such that $I \models B(r)$. Then $I \subseteq HB_P$ is an *answer set of P* if, I is a minimal model of fP^I. We denote by $\mathcal{AS}(P)$ the set of all answer sets of P.

HEX programs are a conservative extension of disjunctive [33] (resp., normal [32]) logic programs under the answer set semantics.

Example 2 (cont'd). Assume that the weather report for *paris* is *sun* and for *london* it is *rain*, then $I \models \&weatherreport[goto](sun)$ if $I \models goto(paris)$, moreover $I \models \&weatherreport[goto](rain)$ if $I \models goto(london)$, and Π_{goto} has one answer set $\{goto(paris)\}$ (we omit atoms in facts of Π_{goto}). If weather reports of both cities are sunny, we additionally obtain the answer set $\{goto(london)\}$. Finally if the weather report for both cities is *snow*, there is no answer set. □

2.3 Usability Issues

When realizing a project with HEX and the DLVHEX reasoner, besides writing the HEX program it is usually also necessary to write a plugin which implements the semantics of external atoms to be used (unless an already existing plugin can be reused). Plugins may be implemented either in Python or C++ using a reasoner API provided by DLVHEX; for details, see [26].

For the sake of performance improvements, external atom semantics implementations can further inject *nogoods* into the solver process, i.e., combinations of truth values of atoms which are inconsistent with the external atom semantics and cannot occur in any answer set. This allows for eliminating inconsistent guesses earlier and might speed up the solving process.

Besides defining external atoms, plugins may also (i) rewrite the input program, and (ii) post-process answer sets. Rewriting the input is useful for creating language extensions (see Section 5) or for changing the behavior of an input program by modifying its code (e.g., for performance reasons or for debugging). Post-processing answer sets allows for translating the answer sets into a more application-specific presentation. This is useful when HEX is used as a backend for other KR formalisms, cf. Section 6.

The DLVHEX user manual [26] presents examples for HEX programs and the corresponding implementations of external atom semantics. Furthermore, it describes different ways of obtaining, building, and installing the DLVHEX solver for Linux (in particular Ubuntu), Mac OS X, and soon Windows.

3 KR Problem Solving using HEX

In this section, we show how the HEX family can be used for declarative problem solving. To this end, we first present the basic methodology in Section 3.1, and show how modeling techniques from ordinary ASP can be generalized to HEX. We provide methodology for using external atoms in Section 3.2 and further distinguish typical kinds of external sources. Roughly, one can classify them as outsourcing of either computation or information, or as a combination thereof. Afterwards, we present three kinds of use case scenarios in Section 3.3. HEX programs can either directly be used as a formalism for modeling end user applications, as a basis for language extensions (i.e., extensions of HEX which are compiled into plain HEX), or as backend for the realization of other KR formalisms. The three use cases are orthogonal to the types of external sources as in any scenario one may use all types of external sources.

3.1 Basic Methodology

Because the HEX family is an extension of ASP, all modeling techniques from ASP may also be used in HEX programs. One of the most important examples is the *guess and check paradigm*, where default negation or disjunctive rules are used to generate a superset of the intended solutions (*guessing part*), and constraints are used to eliminate spurious candidates (*checking part*). For instance, if we assume that facts over predicates *node* and *edge* define a graph, then the well-known graph 3-colorability problem can be solved by guessing all possible colorings of the nodes of a graph using the disjunctive rule

$$g: \quad color(red, X) \lor color(green, X) \lor color(blue, X) \leftarrow node(X), \qquad (2)$$

and eliminating all colorings which assign the same color to adjacent nodes using the constraint

$$c: \quad \leftarrow color(C, X), color(C, Y), edge(X, Y). \qquad (3)$$

However, unlike in ASP, HEX programs allow for using external atoms in addition. They can occur both in the guessing and in the checking part. In the former case, they may be used to import individuals over which guessing is performed. For instance, one may replace the atom $node(X)$ in the body of rule (2) by $\&node[](X)$ to import the nodes of the graph. In the latter case, external atoms can be used in the body of constraints to check given conditions. For instance, rule c may be replaced by

$$c': \quad \leftarrow \mathbf{not}\ \&check[color, edge](), \qquad (4)$$

where $\&check[color, edge]()$ is true if *color* is a valid 3-coloring wrt. *edge* and false otherwise.

The *saturation technique* is an advanced modeling technique for solving problems up to Σ_2^P-completeness, by exploiting the subset-minimality of answer sets for checking whether a property holds *for all* guesses in a search space [20]. A typical example is the check if a graph is *not* 3-colorable, i.e., all possible colorings are invalid. Also here, the checking part may employ external atoms.

For more details about ASP modeling techniques we refer to [20, 29].

3.2 Methodology for Using External Atoms

In general, one can roughly distinguish between two main usages of external sources that we call *computation outsourcing* and *information outsourcing*, respectively, and combinations thereof. We stress that this distinction concerns the usage in applications, as both usages are based on the same language constructs. For each of them we will describe some typical use cases that serve as usage patterns for external atoms when writing HEX programs.

Computation Outsourcing means to send the definition of a subproblem to an external source and retrieve its result. The input to the external source uses predicate extensions and constants to define the problem at hand and the output terms are used to retrieve the result, which can in simple cases also be a Boolean decision.

On-demand constraints are of the form $\leftarrow \&forbidden[p_1,\ldots,p_n]()$ eliminate certain extensions of predicates $p_1,\ldots,p_n$ and are a special case of computational outsourcing, see also the 3-colorability example above. The external evaluation of such a constraint can return reasons for conflicts to the reasoner in order to restrict the search space and avoid reconstruction of the same conflict [14].

This technique avoids explicitly grounding the forbidden combinations of atoms as constraints and reduces the size of the ground program. On-demand constraints have been used for efficient planning in robotics where external atoms verify the feasibility of a 3D motion [56, 35].

Computations that cannot (easily) be expressed by rules. Outsourcing computations also allows for including algorithms which cannot (easily or efficiently) be expressed by rules. As a concrete example, an artificial intelligence agent for the skills and tactics game *AngryBirds* needs to perform physics simulations [10]. This requires floating point computations which can not be done by rules in a practical way (this would either come at the costs of very limited precision or a blow-up of the grounding) therefore the physics simulations are integrated with game playing rules as external atoms in a HEX program.

Complexity lifting. This is another kind of computational outsourcing that allows for realizing computations with a complexity higher than the complexity of ordinary ASP programs. The external atom serves than as an 'oracle' for deciding subprograms. While for the purpose of complexity analysis of the formalism, it is often assumed that external atoms can be evaluated in polynomial time [27][6], as long as external sources are decidable there is no practical reason for limiting their complexity. External sources can also be other ASP or HEX programs, which allows for encoding other formalisms of higher complexity in HEX programs, e.g., *abstract argumentation frameworks* [12].

Information Outsourcing refers, in contrast to computational outsourcing, to external sources which import information, while reasoning itself is done in the logic program.

---

[6] Under this assumption, deciding the existence of an answer set of a propositional HEX program is Σ_2^P-complete.

A typical example can be found in Web resources which provide information for import, e.g., *RDF triple stores* [40] or *geographic data* [47]. More advanced use cases are *multi-context systems*, which are systems of knowledge-bases (*contexts*) that are abstracted to acceptable belief sets (roughly speaking, sets of atoms) and interlinked by *bridge rules* that range across knowledge bases [7]; access to individual contexts has been provided through external atoms [6]. Also sensor data, as often used when planning and executing actions in an environment, is a form of information outsourcing (cf. ACTHEX [4]).

Combinations. It is also possible to outsource computation and information at the same time. A typical example are logic programs with access to Description Logic knowledge bases (DL KB), called *DL-programs* [22]. A DL KB not only stores information, but also provides reasoning services. This allows for interleaving reasoning within the DL KB and the logic program with information that flows across the external atom API in both directions.

3.3 Use Scenarios

One can distinguish between three main types of usages of the HEX formalism. Note that the following classification is orthogonal to the types of external sources above, i.e., each of the following scenarios may make use of various types of external atoms.

End user applications based on HEX. The first scenario is the modeling of *end-user applications*. The HEX language is directly used for modeling a problem at hand and computing its solutions. Note that the problem instance formally consists both of the HEX program and the external sources, but external sources may be reused for different applications if suitable.

The typical procedure when modeling an end user application starts with identifying and realizing the required external sources, followed by writing a HEX program which makes use of these external sources. The two steps may be repeated in order to refine the encoding, i.e., while writing the HEX program, the need for further or modified external sources may arise. In some cases, external atoms of other applications can be reused. Some existing plugins are generic and useful for different applications, e.g., string manipulation functions and an interface to RDF triple stores. We present such applications in Section 4.

HEX language extensions. It turns out that some advanced applications call for additional language features as they can not or not easily by realized in pure HEX programs. A possible relief are language extensions, of which some may be compiled to pure HEX syntax, while others actually increase expressiveness. However, even in cases where language extensions are only syntactic shortcuts, they still not only increase the user comfort but also give the reasoner more specific information about the user's intents (compare this with a constraint $\leftarrow Body$ vs. an equivalent rule $p \leftarrow Body, \textbf{not}\ p$ in ordinary ASP). This can be exploited to improve efficiency. We present such extensions in Section 5.

HEX as backend formalism. Finally, other formalisms may be implemented on top of HEX programs (or extensions thereof) using an appropriate translation. Instead of encoding the end user application directly, its encoding as a

HEX program is automatically generated from a different representation. This step can either be hidden from the user, or can be transparent such that modifications (e.g., extensions or improvements) can be made prior to evaluation. We present existing applications using HEX as a backend in Section 6.

4 End User Applications Based on HEX

In this section, we consider some end applications of HEX programs, which have been conceived in different domains.

In the context of the Semantic Web, HEX was applied to connect SPARQL and RDF querying with logic programming rules [51]. Moreover, HEX was used for archaeological research in order to combine geographical and cultural knowledge from various ontologies [47], and for adapting user interfaces targeted at elderly and disabled people by combining ontologies about user profiles with rules about potential user interface styles [59].

As described above, an important use case of HEX is planning: the DLV^C planning language can use external atoms for determining effects of actions [49]; moreover in robotic planning, external atoms have been used to perform checks on feasibility of actions or action costs [56, 35]. In the following we describe a specific planning application realized with HEX in more detail.

Route planning. While many commercial and free route planning applications exist (Google Maps is currently perhaps the most popular), the supported query types are usually limited. In contrast, an implementation in HEX programs allows for an easy addition of side constraints and thus tailoring to very specific use cases. As a concrete use-case, [16] considered tours with multiple stops (e.g. at shops, a pharmacy, kindergarden, etc) using an external source that supports only point-to-point queries. Side constraints may include restrictions on the order of stops, the tour length, or opening hours at the stops.

Related to route planning is a trip planning scenario. When planning a holiday trip with multiple stops, the order of the stops is often irrelevant, but one wants to spend a certain number of days at each location. However, due to shifts of the dates, the overall price often differs significantly with different sequences. In addition to the sequence of the locations, also other considerations affect the price. E.g. instead of a multi-stop flight through all locations, one may book a return flight to one of them plus local flights from there to the others; sometimes special offers for two-way-tickets make this more attractive. A logic program can automatically generate flight plans according to the constraints and enquire their ticket prices by an external atom that internally uses an online flight booking service. An additional weak constraint can select the cheapest.

AngryHEX. The annual *AIBirds Competition*[7] is a competition for AI agents based on the popular *Angry Birds*[8] game, which is about using a slingshot to shoot birds of different types at pigs placed on a scene in order to destroy them. The pigs are usually protected by obstacles of different types. The game uses a realistic physics simulation, including gravity and statics. In the competition,

---

[7] https://aibirds.org
[8] https://www.angrybirds.com

agents are given the positions and dimensions of the objects in the scene and need to return the angle and velocity for shooting the next bird.

The *AngryHEX* agent [38] is implemented on top of HEX programs. The basic strategy is to maximize the estimated damage to obstacles and pigs for all possible targets. Plain ASP is ill-suited for this application as the computation involves physics simulation and floating point numbers. Therefore, a HEX program was used to realize the basic strategy including the optimal selection of the target, while low-level numeric computations have been outsourced. The agent participated in the competition since 2012 and ranked second in 2015.

HEX programs with nested program calls.

Notably, DLVHEX can be used to 'call' HEX programs from other HEX programs, which we refer to as the *called program* and the *host program*, respectively. Specifically, one can process the collection of answer sets of a different program, and e.g. reason about it. To this end, dedicated external atoms for evaluating subprograms and inspecting their answer sets are available, cf. [53, 25].

When a subprogram call (corresponding to the evaluation of a special external atom) is encountered in the host program, the external atom internally creates another instance of DLVHEX to evaluate the subprogram. The result is then stored in an *answer cache* and gets a unique *handle* which can be later used to reference the result and access its components (e.g., predicate names, literals, arguments) via other external atoms. The subprogram can either be *directly embedded* in the host program, or *stored in a separate file*. In the latter case, code reuse is easy and libraries for solving re-occurring subproblems in ASP applications, e.g., graph problems or combinatorial optimization problems, can be built, where updates are automatically reflected in the call program.

To this end, we use external atoms $\&callhex_n$, $\&callhexfile_n$, $\&answersets$, $\&predicates$, and $\&arguments$, where

$$\&callhex_n[\mathtt{P}, p_1, \ldots, p_n](H) \quad \text{and} \quad \&callhexfile_n[\mathtt{FN}, p_1, \ldots, p_n](H)$$

$n \geqslant 0$, allow to execute a subprogram given by a string $\mathtt{P}$ or in a file $\mathtt{FN}$, respectively; here n specifies the number of predicate names p_i, $1 \leqslant i \leqslant n$, used to define the input facts. When evaluating such an external atom on an interpretation I, the system adds all atoms $p_i(\boldsymbol{t})$ in I as facts to the specified program, creates another DLVHEX instance to evaluate it, and returns a symbolic handle H as result. A *handle* is a unique integer that represents a certain program answer cache entry. For convenience, we omit the subscript n in $\&callhex_n$ and $\&callhexfile_n$ as it is clear from the context.

Example 3. We use two predicates p_1 and p_2 to define the input to the subprogram $\mathtt{sub.hex}$ ($n = 2$), i.e., all atoms over these predicates are added to the subprogram prior to evaluation. The call derives a handle H as result.

$$p_1(x, y); \quad p_2(a); \quad p_2(b);$$
$$handle(H) \leftarrow \&callhexfile[\mathtt{sub.hex}, p_1, p_2](H)$$

In the implementation, handles are consecutive numbers starting at 0. The unique answer set of the program is $\{handle(0), p_1(x,y), p_2(a), p_2(b)\}$. □

Formally, given an interpretation I, $f_{\&callhexfile_n}(I, file, p_1, \ldots, p_n, h) = v$ with $v = 1$ if h is the handle to the result of the program in file *file* augmented with the facts over predicates $p_1, \ldots, p_n$ that are true in I, and $v = 0$ otherwise. The formal notion and use of $\&callhex_n$ to call embedded subprograms is analogous to $\&callhexfile_n$.

Example 4. Consider the following program:

$$h_1(H) \leftarrow \&callhexfile[\text{sub.hex}](H)$$
$$h_2(H) \leftarrow \&callhexfile[\text{sub.hex}](H)$$
$$h_3(H) \leftarrow \&callhex[\text{a; b} \leftarrow \text{not c}](H)$$
□

The rules execute the program **sub.hex** and the embedded program $P_e = \{a; b \leftarrow \text{not } c\}$, with no facts being added. The single answer set is $\{h_1(0), h_2(0), h_3(1)\}$ or $\{h_1(1), h_2(1), h_3(0)\}$ depending on the order in which the subprograms are executed (which is irrelevant). Note that the program in **sub.hex** is called in two places but executed only once; P_e is (possibly) different from **sub.hex** and thus evaluated separately.

Now we want to determine how many answer sets a program has. For this purpose, we design an external atom $\&answersets[PH](AH)$ that associates subprograms with their answer set handles. Formally, for an interpretation I, we have $f_{\&answersets}(I, h_{Prog}, h_{AS}) = v$ with $v = 1$, if h_{AS} is a handle to an answer set of the program with program handle h_{Prog}, and $v = 0$ otherwise.

Example 5. The single rule

$$ash(PH, AH) \leftarrow \&callhex[\text{"a} \vee \text{b} \leftarrow \text{"}](PH), \&answersets[PH](AH)$$

calls the embedded subprogram $P_e = \{a \vee b \leftarrow\}$ and retrieves pairs (PH, PA) of handles to its answer sets; here $\&callhex[\text{"a} \vee \text{b} \leftarrow \text{"}](PH)$ returns a handle $PH = 0$ to the result of P_e, which is passed to the atom $\&answersets[PH](AH)$. The latter returns the handles 0 and 1, as P_e has two answer sets ($\{a\}$ and $\{b\}$). The overall program has thus the single answer set $\{ash(0, 0), ash(0, 1)\}$. As for each program the answer set handles start at 0, only a pair of a program and an answer set handle uniquely identifies an answer set. □

Using the external atoms from above, it is now easy e.g. to count the answer sets of a subprogram by determining the largest valid handle to an answer set. Similarly, external atoms $\&predicates$ and $\&arguments$ can be used to inspect answer sets; we refer to [25] for details.

5 HEX Language Extensions

We now turn to some extensions of HEX programs that are motivated by application needs.

HEX programs with function symbols. Uninterpreted function symbols, as for instance $do(a, s)$ to represent the follow up of a situation s after executing an action a, can be easily realized in HEX using external atoms; we thus can extend the language by such function symbols as syntactic sugar.

Formally, the set $\mathcal{X} \cup \mathcal{C}$ of terms is enriched, given a set $\mathcal{F}$ of function symbols, to the smallest superset $\mathcal{T}$ of $\mathcal{X} \cup \mathcal{C}$ such that for each $f \in \mathcal{F}$ of arity n, it holds that $\{f(t_1,\ldots,t_n) \mid t_1,\ldots,t_n \in \mathcal{T}\} \subseteq \mathcal{T}$; technically, it is possible to let $\mathcal{F} \subseteq \mathcal{C}$.

Using external atoms, it is possible to simulate composition and decomposition of function terms, as described in [9]. For every $k \geqslant 0$, two external predicates $\&comp_k$ and $\&decomp_k$ are defined that have $k+1$ (resp., 1) input arguments and 1 (resp., $k+1$) output arguments; the oracle functions are

$$f_{\&comp_k}(I, f, X_1, \ldots, X_k, T) = f_{\&decomp_k}(I, T, f, X_1, \ldots, X_k) = v,$$

with $v = 1$ if $T = f(X_1,\ldots,X_k)$ and $v = 0$ otherwise. Intuitively, $\&comp_n$ constructs a nested term from a function symbol and its term arguments (possibly nested themselves), and $\&decomp_n$ extracts the function symbol and the term arguments from a nested term.

Concrete occurrences of function terms in rules can now be eliminated by using auxiliary variables and adding appropriate $\&comp_n$ and $\&decomp_n$ atoms to the rule bodies. This will be clear from an example.

Example 6. Consider the following HEX program P with function symbols and its rewriting $T_f(P)$ to a plain HEX program:

P: $q(z); q(y)$ $\qquad T_f(P)$: $q(z); q(y)$
$\quad p(f(f(X))) \leftarrow q(X)$ $\qquad\qquad p(V) \leftarrow q(X), \&comp_1[f,X](U),$
$\quad r(X) \leftarrow p(X)$ $\qquad\qquad\qquad\qquad \&comp_1[f,U](V)$
$\quad r(X) \leftarrow r(f(X))$ $\qquad\qquad r(X) \leftarrow p(X)$
$\qquad\qquad\qquad\qquad\qquad r(X) \leftarrow r(V), \&decomp_1[V](f,X)$

Intuitively, $T_f(P)$ first builds $f(f(X))$ for all X on which q holds using two atoms over $\&comp_1$, and then extracts X from derived $r(f(X))$ facts using a $\&decomp_1$-atom. □

Realizing function symbols on top of external atoms allows for a better control of their processing. For example, the construction of new nested terms may be subject to additional conditions which are integrated into the semantics of the external predicates $\&comp_k$ and $\&decomp_k$. A concrete example is *data type checking*, i.e., checking whether the arguments of a function term are in a given domain. Another example is automatic computation of some argument from others; e.g., in building $roman(8, viii)$ from 8, the first argument is converted to Roman number representation.

HEX programs with action atoms. ACTHEX [4] is an extension of HEX programs which allows for the execution of declaratively scheduled actions. To this end, *action atoms* are introduced to rule heads, which operate on an *environment* and may modify it. The environment can be seen as an abstraction of realms outside the logic program. Thus, in contrast to ASP and HEX programs, which are stateless, ACTHEX allows for modifications of the external environment without wrapping the solver in a procedural language. We here review ACTHEX programs at a glance and refer to [4,28] for details.

Intuitively, the evaluation of an ACTHEX program starts with evaluating it as an ordinary HEX program. Answer sets that optimize an associated objective function are *best models*. The evaluation algorithm selects a single best model which determines a sequence of executable action atoms called the *execution schedule*. These actions are then executed and possibly modify the environment.

The ACTHEX language provides a set $\mathcal{A}$ of *action predicate names*, which start with #. An action atom is of the form $\#g[\boldsymbol{Y}]\{o,p\}[w:l]$, where $\#g \in \mathcal{A}$ is an action predicate name, $\boldsymbol{Y} = Y_1, \ldots, Y_n$ is the input list, $o \in \{b, c, c_p\}$ is the *action option* which declares actions as one of *brave*, *cautious* or *preferred cautious*, and the optional integer attributes p, w, and l are called *precedence*, *weight*, and *level*. Rules and programs are then defined as in ordinary HEX programs but may contain action atoms in rule heads.

The semantics of external atoms is generalized such that the environment may influence its truth value. To this end, a ground external atom $\&g[\boldsymbol{y}](\boldsymbol{x})$ with k-ary input and l-ary output has an associated a $2+k+l$-ary Boolean *oracle function* $f_{\&g}$; the atom $\&g[\boldsymbol{y}](\boldsymbol{x})$ is true wrt. assignment I and environment E, if $f_{\&g}(I, E, \boldsymbol{y}, \boldsymbol{x}) = 1$. Best models are defined based on level and weight of actions, and actions are *executable* wrt. a best model depending on their action option o. An *execution schedule* S_I for a best model I is a sequence of all actions executable in I that respects action precedence. The effect of executing a ground action $\#b[y_1, \ldots, y_n]\{o, r\}[w:i]$ on an environment E is modeled by a $(2+n)$-ary function $f_{\#b}$ that determines a follow-up environment $E' = f_{\#b}(I, E, \boldsymbol{y})$.

Example 7 (from [28]). The following ACTHEX-program controls a robot capable of executing a parameterized action $\#robot$, where an external $\&sensor$ predicate enables to access sensor data.

$$\#robot[clean, kitchen]\{c, 2\}[1:1] \leftarrow night$$
$$\#robot[clean, bedroom]\{c, 2\}[1:1] \leftarrow day$$
$$\#robot[goto, charger]\{b, 1\}[1:1] \leftarrow \&sensor[bat](low)$$
$$night \vee day \leftarrow$$

Informally, in the night the kitchen should be cleaned, and during daytime the bedroom; if the battery is low, the robot needs to go to the charger. The option b makes this action mandatory, while the other actions are by option c only taken if they occur in every answer set; by the disjunctive fact, this is not the case. Note that precedence 1 of $\#robot[goto, charger]\{b, 1\}$ makes the robot recharge its battery (if needed) before any cleaning. □

Use-cases of ACTHEX programs. ACTHEX has been used in several applications; for a more elaborative discussion, we refer to [4] and [28].

Action languages, such as the one by [34], are used to describe the relations between *actions* that modify the state of the world which is described by *fluents*, i.e., predicate that can change over time. Such languages can be captured by ACTHEX, exploiting the precedence attribute of action atoms to model time.

Related to this is *knowledge base update*, as adding and removing statements in knowledge bases maintenance can be modeled by action atoms. The ACTHEX-programmer can in this way reason over knowledge bases and modify them declaratively depending on the current content. Since ACTHEX supports iterative solving, it can be exploited for various use cases such as belief revision, belief merging or implementing the observe-think-act cycle of agents [39].

Agents with *iterative strategies* do not compute solutions in a single shot using an appropriate encoding, but in multiple steps using intermediate solutions. This can be advantageous, in particular if the grounding of a monolithic problem encoding is very large, as holds e.g. for the logic puzzles Sudoku and Reversi. An ACTHEX Sudoku agent that iteratively adds numbers to a cell or excludes them from the set of possible values has the potential to solve larger instances than pure ASP can handle [28].

Constraint HEX programs. *Constraint Answer Set Programming (CASP)* (see e.g. [46, 42]) combines ASP with constraint programming [1]. A well-known implementation is the clingcon system [50], which integrates GRINGO, CLASP and the constraint solver GECODE. Constraints can be encoded in plain ASP using builtin predicates, but this quickly produces groundings of unmanageable size; hence, a genuine support of constraints in ASP is reasonable, which can hide instances of constraint variables in the constraint solver.

Dedicated CASP solvers, however, do not allow to integrate background theories other than constraints. This motivated an integration of CASP with HEX programs to *constraint HEX programs*. Such programs are strictly more general than CASP programs, as besides constraints arbitrary background theories can be accessed via external atoms. Technically, the integration uses a translation of constraint HEX programs into native HEX programs, where constraints are handled using an *SMT-like* [3] approach (also used by clingcon).

Informally, a constraint HEX program may contain besides ordinary and external atoms also *constraint atoms*. The latter are comparisons of arithmetic expressions that consist of (constraint) variables and constants, such as $x + y < 10$. Here, x and y are constraint variables which range over a certain domain. Different from ASP variables, constraint variables are global, i.e., each occurrence in a program is bound to the same value; thus, the atoms $x < 10$ and $x > 20$ can never be jointly true, even if they occur in different rules. Notably, (upper-case) ASP variables can occur in constraint atoms (they are eliminated by the grounder); e.g., in $x + Y > 5$, the ASP variable Y is substituted by ground terms yielding ground constraint atoms.

We omit here a formal definition of constraint HEX programs, but illustrate them by an example.

Example 8. Suppose Alice's restaurant offers daily menus. The menus can be selected based on the price of food and drink, where drink should be cheaper than food and each menu should cost at most 20 Euros; menus of the limit cost are called *exclusive*. This knowledge is encoded by the following program:

$r_1: food(P) \leftarrow \&sql[\text{"Select price from Food"}](P)$

$r_2: drink(P) \leftarrow \&sql[\text{"Select price from Drink"}](P)$

$r_3: max\_price(20)$

$r_4: inMenu(F, D) \vee outMenu(F, D) \leftarrow drink(D), food(F)$

$r_5: \leftarrow D > F, inMenu(F, D)$

$r_6: F + D \leqslant P \leftarrow inMenu(F, D), max\_price(P)$

$r_7: exclusive\_menu \leftarrow inMenu(F, D), max\_price(P), F + D \equiv P$

Here, food and drink prices are represented by atoms $food(\cdot)$ and $drink(\cdot)$, respectively; via the external atoms $\&sql[\cdot](\cdot)$, all prices from the database of the restaurant are loaded. Rule r_4 generates all price combinations of menus, while rule r_5 checks that food is more expensive than drink and rule r_6 that the maximum price is not exceeded. The rule r_7 checks for the existence of an exclusive menu. Note that the constraint atom in the head of rules r_6 is not *derived* to be true if the body is true, but it must *evaluate* to true in this case.

If the database contains prices 18 and 9 for food and 5 for drink, the single answer set contains $inMenu(9, 5)$, and $outMenu(18, 5)$, encoding a single menu of 9 Euros for food and 5 Euros for drink; there is no exclusive menu. □

Constraint HEX programs can be translated into plain HEX programs using a dedicated external atom for constraint checking. The idea is to guess the truth values of all constraint atoms, which are represented using a special predicate $con(\cdot)$, in the program and pass the guess to external constraint checking; the answer set candidate is eliminated if the guess is not compatible, i.e., the corresponding constraints are not satisfiable. This is best illustrated on the previous example.

Example 9 (cont'd). The constraint atoms $D > F$, $F + D \leqslant P$, and $F + D \equiv P$ are represented by $con(D, >, F)$, $con(F, +, D, \leqslant, P)$, and $con(F, +, D, \equiv, P)$, and their negations by $con(D, \leqslant, F)$, $con(F, +, D, >, P)$, and $con(F, +, D, \neq, P)$. The rules r_5–r_7 are now replaced by the following rules:

$r'_5: \leftarrow con(D, >, F), inMenu(F, D)$

$r'_6: con(F, +, D, \leqslant, P) \leftarrow inMenu(F, D), max\_price(P)$

$r'_7: exclusive\_menu \leftarrow inMenu(F, D), max\_price(P), con(F, +, D, \equiv, P)$

$g_1: con(D, >, F) \vee con(D, \leqslant, F) \leftarrow inMenu(F, D)$

$g_2: con(F, +, D, \leqslant, P) \vee con(F, +, D, >, P) \leftarrow inMenu(F, D), max\_price(P)$

$g_3: con(F, +, D, \equiv, P) \vee con(F, +, D, \neq, P) \leftarrow inMenu(F, D), max\_price(P)$

$c: \leftarrow \text{not } \&check[con, sum]()$

The rule r'_i results from r_i by replacing the constraint atom with its guess atom. The rules g_1, g_2 and g_3 guess the truth values of all (ground) constraint atoms, and c checks compatibility of the guess via the constraint solver. Here *sum* is like *con* a special predicate for sums in constraint expressions that is void. □

For more details and discussion, we refer to [54].

HEX$^\exists$ programs. An important feature of HEX programs is that they are capable of value invention, i.e., that new constants are introduced into a program. This relates to existential quantification, as, given an external atom $\&p[\boldsymbol{y}](\boldsymbol{x})$ that evaluates to true, the output values $\boldsymbol{x}$ witness that the formula $\exists \boldsymbol{X} \, \&p(\boldsymbol{y}, \boldsymbol{X})$ is true. If we are just interested in some (arbitrary) such witness $\boldsymbol{x}$, we might write rules that choose one of them; alternatively, one may delegate this choice to the external source, i.e., use a variant $\&p'$ of $\&p$ such that $\&p'[\boldsymbol{y}](\boldsymbol{x})$ holds for a unique $\boldsymbol{x}$. As the choice of $\boldsymbol{x}$ depends on the external source, we obtain in this way *domain-specific existential quantification*. If, as in pure logic, we want to leave concrete witnesses open, we can use a tuple $\boldsymbol{x}' = x_1 \ldots, x_m$ of fresh constants x_i (or *null values*) as generic witness; this amounts to Skolemization for the elimination of function symbols.

Such logical existential quantification is supported in the language of *HEX$^\exists$ programs*, which are finite sets of rules of the form

$$\exists \boldsymbol{X} : p(\boldsymbol{Y}', \boldsymbol{X}) \leftarrow \mathbf{conj}[\boldsymbol{Y}], \qquad (5)$$

where $\boldsymbol{X}$ and $\boldsymbol{Y}$ are disjoint sets of variables, $\boldsymbol{Y}' \subseteq \boldsymbol{Y}$, $p(\boldsymbol{Y}', \boldsymbol{X})$ is an ordinary atom, and $\mathbf{conj}[\boldsymbol{Y}]$ is a conjunction of (possibly default-negated) atoms and external atoms containing all and only the variables $\boldsymbol{Y}$. Semantically, this rule assigns for each ground instance $\mathbf{conj}[\boldsymbol{y}]$ that evaluates to true a tuple $\boldsymbol{x} = x_1, \ldots, x_m$ of *new null values* as above.[9]

A HEX$^\exists$ program Π can be transformed to an equivalent HEX program $T(\Pi)$ by rewriting each rule r of form (5) to the rule

$$p(\boldsymbol{Y}', \boldsymbol{X}) \leftarrow \mathbf{conj}[\boldsymbol{Y}], \&exists^{|\boldsymbol{Y}'|, |\boldsymbol{X}|}[r, \boldsymbol{Y}'](\boldsymbol{X}),$$

where the existential quantifier is replaced by a new external atom $\&exists$ of appropriate input and output arity which uses value invention.

Example 10. Consider the following HEX$^\exists$ program Π, which expresses that each employee X has some office Y:

$$\begin{aligned}
& employee(john). \quad employee(joe). \\
r_1 : & \exists X : \textit{office}(Y, X) \leftarrow employee(Y). \\
r_2 : & \qquad \quad room(X) \leftarrow \textit{office}(Y, X)
\end{aligned}$$

In the translated program $T_\exists(\Pi)$, r_1 is replaced by

$$r_1' : \textit{office}(Y, X) \leftarrow employee(Y), \&exists^{1,1}[r_1, Y](X).$$

One can use HEX$^\exists$ programs to model query answering from existential rules (i.e., $\mathbf{conj}[\boldsymbol{Y}]$ in (5) consists of ordinary atoms). Even if the answer set may be infinite, a finite fragment may suffice for this purpose. For more details see [17].

---

[9] By the underlying unique name assumption, these values do not match with any other values; to model this, equality reasoning would need to be imposed on top.

6 HEX as Backend Formalism

For the third use scenario, we consider some data and knowledge-based formalisms that use (extended) HEX programs as backend.

Multi-context systems. Multi-context systems (MCSs) [7] are a formalism for interlinking multiple knowledge based systems called *contexts*. The formalism abstracts from the knowledge representation language and models context semantics in terms of accepted *belief sets*. The latter are abstractly modeled as naked sets whose elements (i.e., the beliefs) need not bear logical structure. The contexts are interlinked by so called *bridge rules* which add formulas to the knowledge base of a context depending on the presence and/or absence of beliefs from the belief sets of other contexts. The semantics of an MCS is given in terms of *equilibria*, which are global states that consist of acceptable belief sets for each context, such that all bridge rules are satisfied.

Besides computing equilibria, an important reasoning task for MCSs is *inconsistency analysis*; e.g., given a MCS M that lacks equilibria, compute a reason for this inconsistency [19]. Inconsistency explanations can be computed using a HEX program encoding [6] in which external atoms *outsource contextual reasoning* and check whether a context accepts a certain belief set. The use of external atoms in the program is highly cyclic as the saturation technique is employed; the latter is required as the problem is beyond NP and co-NP.

Description Logics plus rules. *Description logics (DLs)* provide a logical formalism for ontologies that are well-suited for the Semantic Web [36] or in medical applications [37]. Ontologies represent classes of objects, referred to as *concepts*, and the relations between objects, called *roles*. Concepts and roles correspond to unary and binary predicates in first-order logic, respectively. A *description logic knowledge base* consists of a *Tbox* (*the terminology*) that defines concepts and roles and represents relations between them, and an *Abox* (*assertions*), that contains specific information on membership of individuals in concepts resp. of pairs of individuals in roles.

Example 11. Suppose *PhDStudent*, *Student* and *Professor* are concepts and *isAssistantOf* is a role. The Tbox may contain the *concept inclusion axiom* $PhDStudent \sqsubseteq Student$, which states that the class of PhD students is a subclass of all students. The Abox contains concept membership assertions like *Professor(smith)* and *PhDStudent(johnson)*, representing that *smith* is a professor and *johnson* a PhD student. An assertion *isAssistantOf(johnson, smith)* states that *johnson* is an assistant of professor *smith*. □

Typical reasoning tasks over description logic knowledge bases include concept and role retrieval, i.e., listing all individuals or pairs of individuals which are members of a given concept or role, respectively. In the example above one may ask for all members of *Student* and expects as answer *johnson* as he is a *PhDStudent* and thus, by the terminological knowledge, also a *Student*.

Combining ontologies and answer set programming is especially valuable as existing domain knowledge can be accessed from logic programs. To this end, *DL-programs* have been developed by [21] which have been implemented on top of HEX programs with dedicated external atoms; where the external source features external atoms for concept and role queries. Prior to query evaluation, concepts and/or roles are enriched by individuals from the ASP program. This allows for advanced reasoning tasks such as terminological default reasoning or closed world reasoning on description logic knowledge bases [11].

As description logics are monotonic, default reasoning can only be realized by the (cyclic) interaction of rules and the DL knowledge base. To this end, appropriate encodings and an implementation were developed [11]. DL-programs have, e.g., been applied in complaint management for e-government [60].

The MELD belief merging system deals with merging *collections of belief sets* [52,53], which are roughly sets of classical ground literals. A merging strategy is defined by tree-shaped *merging plans*, whose leaves are the collections of belief sets to be merged, and whose inner nodes are *merging operators* (provided by the user). The structure is akin to syntax trees of terms. The automatic evaluation of tree-shaped merging plans is based on nested HEX programs; it proceeds bottom-up, where every step requires inspection of the subresults, i.e., accessing the answer sets of subprograms. In fact, the need for such processing has led to develop nested HEX program.

Interactive ASP. The Answer Set Application Programming (ASAP) framework [57] allows for creating interactive applications based on ASP. In this framework, incoming events (e.g., keyboard) are processed by ASP and the application state is managed using fluents (as in planning). An ASAP program is rewritten to a HEX program where each evaluation obtains fluent values and event information via HEX external atoms. Answer sets determine future fluent values by atoms $fl=val@t$, which intuitively means that fluent fl has value val at time t. Programs can contain actions (atoms starting with '@') as in ACTHEX, for example to display the user interface or to quit the program.

Example 12. The following ASAP-program displays a help text and the state (on or off) of a switch. The user can change the state using cursor keys and quit with the Q key.

$$\#initial\ switch=off \leftarrow .$$
$$switch=off@next \leftarrow \&event[\texttt{"key.special"}](\texttt{"Down"}).$$
$$switch=on@next \leftarrow \&event[\texttt{"key.special"}](\texttt{"Up"}).$$
$$@exit(0) \leftarrow \&event[\texttt{"key.normal"}](\texttt{"q"}).$$
$$@drawText(2,2,\texttt{"Up/Down: switch, Q: quit"}) \leftarrow .$$
$$@drawText(5,4,State) \leftarrow switch=State@next.$$

Here the time *next* refers to the state after the currently processed event. □

ASAP is a *hybrid* HEX use scenario: an ASAP-program is rewritten into a HEX program, transforming fluent atoms into regular atoms and adding rules containing external atoms. At the same time an ASAP program can use arbitrary external atoms, e.g., for string processing. ASAP combines computation outsourcing (string processing) with information outsourcing (events and fluents) and uses HEX as a clean interface to the real world.

7 Discussion

In the previous sections, we have considered different types of use case scenarios for formalisms from the HEX family. We now briefly discuss some advantages and drawbacks of these types. However, we remain at the surface and the considerations only serve as a general guideline to select the way for realizing a concrete application using HEX. After that, we consider related work.

7.1 Comparison of the use case types

Unsurprisingly, using the HEX formalism directly provides maximal flexibility to the user, as this allows one to formulate arbitrary rules and to access arbitrary external sources from the rules, provided that the external predicates are defined and implementations are provided as plugins. Depending on the conceptual complexity of the application, this might come at the price of a low user convenience due to the need for heavy-weight and repetitive syntax in the problem encodings. Furthermore, the manual implementation of external source access through a plugin requires some effort (as in general, if one aims at coupling systems via interfaces), and depending on experience and technical skills may be time-consuming; besides development also proper testing and validation of the plugin have to be considered.

In contrast to this, HEX program extensions as well as automated translations of frontends in other KR formalisms into HEX programs provide one with the possibility to use specific language features for expressing particular aspects of a problem. This comes with the advantage of eliminating some repetitive work, where the same rule patterns need not be written over and over again, and in this way also helps to reduce errors that are made by spoiled copy and paste. Furthermore, using designated language constructs gives the solver as in ordinary ASP more insight into the user's intention in writing a certain part of the program, which may be exploited for performance enhancements; guessing such intention respectively discovering program parts or rule patterns that may amount to an intention is expensive in general. For a simple example, the constraint c' in (4) is equivalent to the rule

$$a \leftarrow \mathbf{not}\ \&check[color, edge](), \mathbf{not}\ a; \tag{6}$$

where a is a fresh atom. The explicit form of c' allows us to immediately conclude that $\&check[color, edge]()$ must be true in every answer set; given (6), further analysis of the program is needed.

On the other hand, the use of front end translations provides limited flexibility and one is in general committed to a specific encoding. This can be disadvantageous, if for some targeted application that encoding is not working well (e.g., for performance reasons), or some additional features or aspects should be respected (e.g., some simple preference of alternatives). As a compromise, one can then resort to encodings that are automatically generated, but then manually customized by the user; this has been advocated and used e.g. in [21]. However, this approach requires a proper grasp and understanding of the encoding produced by the translation, which may require considerable effort, the more if—as happens often in practice—optimizations are applied to the code (sometimes without proper documentation, and based on implicit assumptions).

7.2 Related Work

Despite the terminological similarity, customizable functions as supported by GRINGO and so called *frozen atoms* supported by CLASP (sometimes called external atoms) are different from external atoms as supported by HEX. GRINGO supports custom functions (implemented in the scripting languages Lua or Python) which are evaluated during the program grounding and thus compiled away prior to the solving step. They are intended to be used as customizable built-in atoms, but no cyclic dependencies are possible. The frozen atoms of CLASP are sometimes also called *external atoms*; they are protected from optimization, and their truth values can be determined from code (e.g., in Python) that controls the grounding and solving process, for example in advanced techniques such as incremental solving. Using frozen atoms requires that the solver and the truth values of atoms are controlled "from outside" using imperative code. HEX inverts the roles of ASP and imperative code: the DLVHEX solver engine controls the ASP evaluation and evaluates external atom semantics "inside" the ASP evaluation on demand and whenever needed. That is, with HEX external semantics is evaluated within evaluation of ASP semantics, while with CLASP, ASP semantics is evaluated within imperative code that configures the truth values of frozen atoms.

Besides GRINGO and CLASP, there are extensions of ASP towards the integration of specific external sources. Examples are constraint ASP as an integration of ASP with constraint programming as realized e.g. in clingcon [50] and EZCSP [2], or DL-programs as a native combination of ASP with ontologies [22]. In contrast, HEX allows for the integration of arbitrary external sources through a general interface and their flexible combination; the other use cases correspond to special cases thereof.

8 Conclusion

Arriving at the end, we give a brief summary and an outlook on future work.

8.1 Summary

HEX programs extend answer set programs with access to external sources through an API-style interface, which has been fruitfully deployed to various applications. In this paper, we briefly discussed how the formalism can be used for problem solving in KR.

To this end, we first presented the general methodology as a strict generalization of ASP. In particular, the prominent guess and check paradigm can be seamlessly combined with external sources, both in the guessing and in the checking part. Also other ASP techniques, such as saturation, can be used with external sources. We then presented two typical types of external sources for computation outsourcing and for information outsourcing, respectively, and for combinations thereof. We further distinguished three typical use case scenarios of HEX programs, namely for encoding end user applications, as a basis for language extensions, and as a backend for other KR formalisms. For each of the scenarios, we have briefly presented existing applications.

8.2 Open Issues and Future Work

Ongoing work includes the extension of the interface for external sources. While the current interface is convenient, it turns out that the integration of external sources as black boxes inhibits efficient evaluation in many cases. Low-level interfaces might be less convenient, but give the reasoner more insights into the semantics and properties of external atoms, which might be exploited to increase efficiency. This includes the possibility for evaluating external sources under partial interpretations and retrieving partial answers. Such interfaces are currently under development and will be an alternative to the existing ones without replacing them.

Another issue concerns robustness of performance. Currently, in some cases small changes in the encoding and in the run options of the solver influence the efficiency considerably. While such effects are understandable to users who know the algorithms underlying the systems well, inexperienced users may face difficulties in crafting efficient encodings for their applications. The same issue applies to ordinary ASP as well, and thus is not a genuine issue for HEX programs; indeed, [29] states that "crafting an [plain] ASP encoding that also leads to the best possible system performance is yet not as obvious as it might seem." In general, the highly declarative nature of ASP and its intrinsic intractability comes at a computational price, and handling different inputs smoothly requires complex and sophisticated algorithms, where also the use of heuristics is indispensable; the presence of external atoms adds to this difficulty.

This naturally leads to two directions for potential future work. On the solver side, the problem might be mitigated, but may not be expected to be eliminated, by more advanced optimization methods and heuristics, which automatically optimize the input program and determine optimal solver options to process the current input program. On the modeling side, an interesting issue are methodologies for deciding when it makes sense to outsource computation for efficiency reasons, and methodologies for encoding problems with external atoms to ensure the best possible efficiency.

References

1. Krzysztof Apt. *Principles of Constraint Programming*. Cambridge University Press, New York, NY, USA, 2003.
2. Marcello Balduccini. Representing constraint satisfaction problems in answer set programming. In *Workshop on Answer Set Programming and Other Computing Paradigms (ASPOCP) at ICLP*, 2009.
3. Clark Barrett, R. Sebastiani, S. A. Seshia, and C. Tinelli. Satisfiability modulo theories. In *Handbook of Satisfiability*, volume 185, chapter 26, pages 825–885. IOS Press, 2009.

4. Selen Basol, O. Erdem, M. Fink, and G. Ianni. HEX programs with action atoms. In *Technical Communications of the International Conference on Logic Programming (ICLP)*, pages 24–33, 2010.
5. Christoph Beierle and G. Kern-Isberner. *Methoden wissensbasierter Systeme - Grundlagen, Algorithmen, Anwendungen (5. Aufl.)*. Computational Intelligence. Springer/Vieweg, 2014.
6. Markus Bögl, T. Eiter, M. Fink, and P. Schüller. The MCS-IE system for explaining inconsistency in multi-context systems. In *European Conference on Logics in Artificial Intelligence (JELIA)*, pages 356–359, 2010.
7. Gerd Brewka and T. Eiter. Equilibria in heterogeneous nonmonotonic multi-context systems. In *AAAI Conference on Artificial Intelligence*, pages 385–390. AAAI Press, 2007.
8. Gerhard Brewka, T. Eiter, and M. Truszczyński. Answer set programming at a glance. *Commun. ACM*, 54(12):92–103, 2011.
9. Francesco Calimeri, S. Cozza, and G. Ianni. External sources of knowledge and value invention in logic programming. *Annals Math. Artif. Intell.*, 50(3–4):333–361, 2007.
10. Francesco Calimeri, M. Fink, S. Germano, G. Ianni, C. Redl, and A. Wimmer. AngryHEX: an artificial player for angry birds based on declarative knowledge bases. In *National Workshop and Prize on Popularize Artificial Intelligence*, pages 29–35, 2013.
11. Minh Dao-Tran, T. Eiter, and T. Krennwallner. Realizing default logic over description logic knowledge bases. In *European Conference on Symbolic and Quantitative Approaches to Reasoning with Uncertainty*, pages 602–613, 2009.
12. Phan Minh Dung. On the acceptability of arguments and its fundamental role in nonmonotonic reasoning, logic programming and n-person games. *Artificial Intelligence*, 77(2):321–357, 1995.
13. Thomas Eiter, M. Fink, G. Ianni, T. Krennwallner, C. Redl, and P. Schüller. A model building framework for answer set programming with external computations. *Theory and Practice of Logic Programming*, 2015. doi:10.1017/S1471068415000113, http://arxiv.org/abs/1507.01451.
14. Thomas Eiter, M. Fink, T. Krennwallner, and C. Redl. Conflict-driven ASP solving with external sources. *Theory and Practice of Logic Programming*, 12(4-5):659–679, 2012.
15. Thomas Eiter, M. Fink, T. Krennwallner, and C. Redl. Liberal Safety Criteria for HEX-Programs. In Marie des Jardins and Michael Littman, editors, *AAAI Conference on Artificial Intelligence (AAAI)*. AAAI Press, 2013.
16. Thomas Eiter, M. Fink, T. Krennwallner, and C. Redl. Domain expansion for ASP-programs with external sources. Tech. Rep. INFSYS RR-1843-14-02, Inst. f. Informationssysteme, TU Wien, Austria, Sept. 2014.
17. Thomas Eiter, M. Fink, T. Krennwallner, and C. Redl. HEX-programs with existential quantification. In *International Conference on Applications of Declarative Programming and Knowledge Management (INAP)*, 2014.
18. Thomas Eiter, M. Fink, T. Krennwallner, C. Redl, and P. Schüller. Efficient HEX-program evaluation based on unfounded sets. *Journal of Artificial Intelligence Research*, 49:269–321, 2014.
19. Thomas Eiter, M. Fink, P. Schüller, and A. Weinzierl. Finding Explanations of Inconsistency in Multi-Context Systems. *Artificial Intelligence*, 216:233–274, November 2014.
20. Thomas Eiter, G. Ianni, and T. Krennwallner. Answer set programming: A primer. In *Reasoning Web Summer School*, pages 40–110, 2009.

21. Thomas Eiter, G. Ianni, T. Krennwallner, and R. Schindlauer. Exploiting conjunctive queries in description logic programs. *Annals Math. Artif. Intell.*, 53(1-4):115–152, 2008.
22. Thomas Eiter, G. Ianni, T. Lukasiewicz, R. Schindlauer, and H. Tompits. Combining answer set programming with description logics for the semantic web. *Artificial Intelligence*, 172(12-13):1495–1539, 2008.
23. Thomas Eiter, G. Ianni, R. Schindlauer, and H. Tompits. A Uniform Integration of Higher-Order Reasoning and External Evaluations in Answer-Set Programming. In *Proc. IJCAI 2005*, pages 90–96. Professional Book Center, 2005.
24. Thomas Eiter, G. Ianni, R. Schindlauer, and H. Tompits. Effective integration of declarative rules with external evaluations for semantic-web reasoning. In *European Semantic Web Conference*, pages 273–287, 2006.
25. Thomas Eiter, T. Krennwallner, and C. Redl. HEX-Programs with Nested Program Calls. In *Applications of Declarative Programming and Knowledge Management (INAP 2011)*, pages 1–10. Springer, 2013.
26. Thomas Eiter, M. Mehuljic, C. Redl, and P. Schüller. User guide: dlvhex 2.x. Tech. Rep. INFSYS RR-1843-15-05, Inst. f. Informationssysteme, TU Wien, Austria, Sept. 2015.
27. Wolfgang Faber, N. Leone, and G. Pfeifer. Recursive aggregates in disjunctive logic programs: Semantics and complexity. In *European Conference on Logics in Artificial Intelligence (JELIA)*, pages 200–212. Springer, 2004.
28. Michael Fink, S. Germano, G. Ianni, C. Redl, and P. Schüller. ActHEX: implementing HEX programs with action atoms. In Pedro Cabalar and TranCao Son, editors, *International Conference on Logic Programming and Nonmonotonic Reasoning (LPNMR)*, pages 317–322. Springer, 2013.
29. M. Gebser, R. Kaminski, B. Kaufmann, and T. Schaub. *Answer Set Solving in Practice*. Synthesis Lectures on Artificial Intelligence and Machine Learning. Morgan and Claypool Publishers, 2012.
30. Martin Gebser, B. Kaufmann, R. Kaminski, M. Ostrowski, T. Schaub, and M.T. Schneider. Potassco: The Potsdam Answer Set Solving Collection. *AI Commun.*, 24(2):107–124, 2011.
31. Martin Gebser, B. Kaufmann, and T. Schaub. Conflict-driven answer set solving: From theory to practice. *Artificial Intelligence*, 187–188:52–89, 2012.
32. M. Gelfond and V. Lifschitz. The Stable Model Semantics for Logic Programming. In R. Kowalski and K. Bowen, editors, *Logic Programming: Proceedings of the 5th International Conference and Symposium*, pages 1070–1080. MIT Press, 1988.
33. M. Gelfond and V. Lifschitz. Classical Negation in Logic Programs and Disjunctive Databases. *Next Generation Computing*, 9(3–4):365–386, 1991.
34. Enrico Giunchiglia, J. Lee, V. Lifschitz, N. McCain, and H. Turner. Nonmonotonic causal theories. *Artificial Intelligence*, 153:2004, 2004.
35. Giray Havur, G. Ozbilgin, E. Erdem, and V. Patoglu. Geometric Rearrangement of Multiple Movable Objects on Cluttered Surfaces: A Hybrid Reasoning Approach. In *International Conference on Robotics and Automation (ICRA)*, pages 445–452, 2014.
36. J. Heflin and H. Munoz-Avila. Lcw-based agent planning for the semantic web. In A. Pease, editor, *Ontologies and the Semantic Web*, number WS-02-11 in AAAI Technical Report, pages 63–70, Menlo Park, CA, 2002. AAAI Press.
37. Robert Hoehndorf, F. Loebe, J. Kelso, and H. Herre. Representing default knowledge in biomedical ontologies: Application to the integration of anatomy and phenotype ontologies. *BMC Bioinformatics*, 8(1):377, 2007.
38. Giovambattista Ianni, F. Calimeri, S. Germano, A. Humenberger, C. Redl, D. Stepanova, A. Tucci, and A. Wimmer. Angry-HEX: an artificial player for

angry birds based on declarative knowledge bases. *IEEE Trans. Computational Intelligence and AI in Games*, 2015. Accepted for publication.
39. R. Kowalski and F. Sadri. From logic programming towards multi-agent systems. *Annals Math. Artif. Intell.*, 25(3-4):391–419, 1999.
40. Ora Lassila and R.R. Swick. Resource description framework (RDF) model and syntax specification, 1999. www.w3.org/TR/1999/REC-rdf-syntax-19990222.
41. Nicola Leone, G. Pfeifer, W. Faber, T. Eiter, G. Gottlob, S. Perri, and F. Scarcello. The DLV System for Knowledge Representation and Reasoning. *ACM Transactions on Computational Logic (TOCL)*, 7(3):499–562, July 2006.
42. Yuliya Lierler. Relating constraint answer set programming languages and algorithms. *Artificial Intelligence*, 207:1–22, February 2014.
43. Vladimir Lifschitz. Answer Set Programming and Plan Generation. *Artificial Intelligence*, 138:39–54, 2002.
44. Fangzhen Lin and Y. Zhao. ASSAT: computing answer sets of a logic program by SAT solvers. *Artificial Intelligence*, 157(1–2):115–137, 2004.
45. Victor W. Marek and M. Truszczyński. Stable Models and an Alternative Logic Programming Paradigm. In *The Logic Programming Paradigm – A 25-Year Perspective*, pages 375–398. Springer, 1999.
46. Veena S Mellarkod, M. Gelfond, and Y. Zhang. Integrating Answer Set Programming and Constraint Logic Programming. *Annals Math. Artif. Intell.*, 53(1-4):251–287, 2008.
47. Alessandro Mosca and D. Bernini. Ontology-driven geographic information system and dlvhex reasoning for material culture analysis. In *Italian Workshop RiCeRcA at ICLP*, 2008.
48. Ilkka Niemelä. Logic programming with stable model semantics as constraint programming paradigm. *Annals Math. Artif. Intell.*, 25(3–4):241–273, 1999.
49. Davy Van Nieuwenborgh, T. Eiter, and D. Vermeir. Conditional planning with external functions. In *International Conference on Logic Programming and Nonmonotonic Reasoning (LPNMR)*, pages 214–227. Springer, 2007.
50. Max Ostrowski and T. Schaub. ASP modulo CSP: the clingcon system. *Theory and Practice of Logic Programming (TPLP)*, 12(4-5):485–503, 2012.
51. Axel Polleres. From SPARQL to rules (and back). In *International Conference on World Wide Web (WWW)*, pages 787–796. ACM, 2007.
52. Christoph Redl. Development of a belief merging framewerk for dlvhex. Master's thesis, Vienna University of Technology, A-1040 Vienna, Karlsplatz 13, 2010.
53. Christoph Redl, T. Eiter, and T. Krennwallner. Declarative belief set merging using merging plans. In *International Symposium on Practical Aspects of Declarative Languages (PADL)*, pages 99–114. Springer, 2011.
54. Alessandro De Rosis, T. Eiter, C. Redl, and F. Ricca. Constraint answer set programming based on HEX-programs. In *Eighth Workshop on Answer Set Programming and Other Computing Paradigms (ASPOCP 2015)*. https://sites.google.com/site/aspocp15/accepted.
55. Roman Schindlauer. *Answer Set Programming for the Semantic Web*. PhD thesis, Vienna University of Technology, Vienna, Austria, 2006.
56. Peter Schüller, V. Patoglu, and E. Erdem. A Systematic Analysis of Levels of Integration between Low-Level Reasoning and Task Planning. In *Workshop on Combining Task and Motion Planning at ICRA*, 2013.
57. Peter Schüller and A. Weinzierl. Answer Set Application Programming: a Case Study on Tetris. In Marina De Vos, T. Eiter, Yuliya Lierler, and Francesca Toni, editors, *International Conference on Logic Programming (ICLP), Technical Communications*, volume 1433. CEUR-WS.org, 2015.

58. Patrik Simons, I. Niemelä, and T. Soininen. Extending and implementing the stable model semantics. *Artif. Intell.*, 138(1-2):181–234, 2002.
59. Jesia Zakraoui and W.L. Zagler. A method for generating CSS to improve web accessibility for old users. In *International Conference on Computers Helping People with Special Needs (ICCHP)*, pages 329–336, 2012.
60. Hande Zirtiloğlu and P. Yolum. Ranking semantic information for e-government: complaints management. In *International Workshop on Ontology-supported business intelligence (OBI)*. ACM, 2008.

Ranking Answers to Datalog+/− Ontologies based on Trust and Reliability of Subjective Reports

Thomas Lukasiewicz[1], Maria Vanina Martinez[2], Cristian Molinaro[3], Livia Predoiu[4], Gerardo I. Simari[5]

Abstract. The use of preferences in query answering, both in traditional databases and in ontology-based data access, has recently received much attention due to its many real-world applications. In this paper, we tackle the problem of top-k query answering in Datalog+/− ontologies subject to a user's preferences and a collection of (subjective) reports provided by other users. Here, each report consists of scores for a list of features, its author's preferences among the features, as well as other information. These pieces of information of every report are then combined, along with the querying user's preferences and their trust in each report, to rank the query results. We present two alternative such rankings, along with algorithms for top-k (atomic) query answering under these rankings. We also show that, under suitable assumptions, these algorithms run in polynomial time in the data complexity. We finally present more general reports, which are associated with sets of atoms rather than single atoms.

1 Introduction

User preferences have been incorporated both in traditional databases and in ontology-based query answering mechanisms for some time now; the recent change in the way data is created and consumed in the Social Semantic Web has caused this aspect of query answering to receive more attention, since users play a central role in both the knowledge engineering effort and knowledge consumption. In this work, we tackle the problem of preference-based query answering in Datalog+/− ontologies assuming that querying users must rely on subjective reports of observing users to get a complete picture and make a decision. Real-world examples of this kind of situation arise when searching for products, such as smartphones or other mobile devices: users provide some basic

---
[1] Department of Computer Science, University of Oxford, UK, thomas.lukasiewicz@cs.ox.ac.uk
[2] Inst. for Computer Science and Engineering, Univ. Nacional del Sur–CONICET, Argentina, mvm@cs.uns.edu.ar
[3] DIMES, Università della Calabria, Italy, cmolinaro@dimes.unical.it
[4] Department of Computer Science, University of Oxford, UK; Department of Computer Science, Otto-von-Guericke Universität Magdeburg, Germany, livia.predoiu@cs.ox.ac.uk
[5] Inst. for Computer Science and Engineering, Univ. Nacional del Sur–CONICET, Argentina, gis@cs.uns.edu.ar

information in the search interface and receive a list of answers to choose from, each associated with a set of subjective reports (often called reviews) written by other users to tell everyone about their experience. The main problem with this setup, however, is that users are often overwhelmed and frustrated, because they cannot decide which reviews to focus on and which ones to ignore, since it is likely that, e.g., a very negative (or positive) review may have been produced on the basis of a feature that is completely irrelevant to the user who issued the query.

Hence, we study a formalization of ranking answers with the help of the associated sets of subjective reports and of incorporating this ranking into preference-based query answering in Datalog+/− ontologies. Our approach is based on trust and relevance measures to select the best reports to focus on, given the user's initial preferences. We propose ranking algorithms to obtain a user-tailored ranking over the set of query answers. We discuss an approach to preference-based top-k query answering in Datalog+/− ontologies, given a collection of subjective reports. Here, each report contains scores for a list of features, its author's preferences among the features, as well as additional information (e.g., information on the reporter, such as age, nationality, etc.). These pieces of information of every report are then aggregated, along with the querying user's trust in each report, to a ranking of the query results relative to the preferences of the querying user. The querying user's preferences are expressed over the same list of features that reports refer to. We then first present a basic approach to a ranking of the query results, where each atom is associated with the average of the scores of all its reports—the aim is to present a ranking of the (top-k) atoms in the query result to the querying user. Every report's score is the average of the scores of each feature, weighted by (i) the trust value for the feature score by the report and (ii) the relevance of the feature according the querying user's preferences (over the features). Also, to determine a report's score, the relevance of the report to the querying user's preferences is taken into account. Next, we present an alternative approach to a ranking of the query results, where we first select the most relevant reports for the querying user, adjust the scores by the trust measure, and compute a single score for each atom by combining the scores computed in the previous step, weighted by the relevance of the features. We describe algorithms for preference-based top-k (atomic) query answering in Datalog+/− ontologies under both rankings, proving that, under suitable assumptions, the two algorithms run in polynomial time in the data complexity. Finally, we propose and discuss a more general form of reports, which are associated with sets of atoms rather than single atoms.

2 Preliminaries

We now recall the basics on Datalog+/− [6], namely, on relational databases and (Boolean) conjunctive queries, along with tuple- and equality-generating dependencies and negative constraints, the chase (procedure and data structure used for query answering), and ontologies in Datalog+/−. We also define the preference models used.

Databases and Queries. We assume (i) an infinite universe of *(data) constants* Δ (which constitute the "normal" domain of a database), (ii) an infinite set of *(labeled) nulls* Δ_N (used as "fresh" Skolem terms, which are placeholders for unknown values, and can thus be seen as variables), and (iii) an infinite set of *variables* $\mathcal{V}$ (used in queries, dependencies, and constraints). Different constants represent different values (*unique name assumption*), while different nulls may represent the same value. We denote by $\mathbf{X}$ sequences of variables $X_1, \ldots, X_k$ with $k \geq 0$. We assume a *relational schema* $\mathcal{R}$, which is a finite set of *predicate symbols* (or *predicates*). A *term* t is a constant, null, or variable. An *atomic formula* (or *atom*) A has the form $p(t_1, \ldots, t_n)$, where p is an n-ary predicate, and $t_1, \ldots, t_n$ are terms. It is *ground* (resp., *existentially closed*) iff every t_i belongs to Δ (resp., $\Delta \cup \Delta_N$). Every ground atom A is uniquely identified with an id, denoted $id(A)$; sometimes, the id is used to denote the atom itself. We use $\mathcal{H}$ to denote the set of all possible ground and existentially closed atoms.

A *database (instance)* D for a relational schema $\mathcal{R}$ is a set of ground atoms with predicates from $\mathcal{R}$. A *conjunctive query* (CQ, for short) Q over $\mathcal{R}$ is of the form $\exists \mathbf{Y}\, \Phi(\mathbf{X}, \mathbf{Y})$, where $\Phi(\mathbf{X}, \mathbf{Y})$ is a nonempty conjunction of atoms (possibly equalities, but not inequalities) with the variables $\mathbf{X}$ and $\mathbf{Y}$, and possibly constants, but no nulls. A *Boolean CQ* (BCQ) over $\mathcal{R}$ is a CQ where all variables are existentially quantified. Answers to CQs and BCQs are defined via *homomorphisms*, which are mappings $\mu \colon \Delta \cup \Delta_N \cup \mathcal{V} \to \Delta \cup \Delta_N \cup \mathcal{V}$ such that (i) $c \in \Delta$ implies $\mu(c) = c$, and (ii) $c \in \Delta_N$ implies $\mu(c) \in \Delta \cup \Delta_N$. Moreover, μ is naturally extended to atoms, sets of atoms, and conjunctions of atoms. The set of all *answers* to a CQ $Q = \exists \mathbf{Y}\, \Phi(\mathbf{X}, \mathbf{Y})$ over a database D, denoted $Q(D)$, is the set of all tuples t over Δ for which a homomorphism $\mu \colon \mathbf{X} \cup \mathbf{Y} \to \Delta \cup \Delta_N$ exists such that $\mu(\Phi(\mathbf{X}, \mathbf{Y})) \subseteq D$ and $\mu(\mathbf{X}) = t$. The *answer* to a BCQ Q over a database D is *Yes*, denoted $D \models Q$, iff $Q(D) \neq \emptyset$.

Dependencies and Constraints. Given a relational schema $\mathcal{R}$, a *tuple-generating dependency* (TGD) σ is a first-order formula $\forall \mathbf{X} \forall \mathbf{Y}\, \Phi(\mathbf{X}, \mathbf{Y}) \to \exists \mathbf{Z}\, \Psi(\mathbf{X}, \mathbf{Z})$, where $\Phi(\mathbf{X}, \mathbf{Y})$ and $\Psi(\mathbf{X}, \mathbf{Z})$ are conjunctions of atoms over $\mathcal{R}$ (without nulls), called the *body* and the *head* of σ, denoted $body(\sigma)$ and $head(\sigma)$, respectively. Such σ is satisfied in a database D for $\mathcal{R}$ iff, whenever there exists a homomorphism h that maps the atoms of $\Phi(\mathbf{X}, \mathbf{Y})$ to atoms of D, there exists an extension h' of h that maps the atoms of $\Psi(\mathbf{X}, \mathbf{Z})$ to atoms of D. All sets of TGDs are finite here. As TGDs can be reduced to TGDs with only single atoms in their heads, in the sequel, every TGD has w.l.o.g. a single atom in its head. A TGD σ is *guarded* iff it has a body atom that contains all universally quantified variables of σ. The leftmost such atom is the *guard atom* (or *guard*) of σ. A set of TGDs is guarded iff all its TGDs are guarded. *Query answering under TGDs*, i.e., the evaluation of CQs and BCQs on databases under a set of TGDs is defined as follows. For a database D for $\mathcal{R}$, and a set of TGDs Σ on $\mathcal{R}$, the set of *models* of D and Σ, denoted $mods(D, \Sigma)$, is the set of all (possibly infinite) databases B such that (i) $D \subseteq B$ and (ii) every $\sigma \in \Sigma$ is satisfied in B. The set of *answers* for a CQ $Q = q(\mathbf{X}) = \exists \mathbf{Y}\, \Phi(\mathbf{X}, \mathbf{Y})$ to D and Σ, denoted $ans(Q, D, \Sigma)$ (or, for $KB = (D, \Sigma)$, $ans(Q, KB)$), is the set of all ground atoms $q(t)$ such that $q(t) \in Q(B)$ for all $B \in mods(D, \Sigma)$. The *answer* for a BCQ Q to D and Σ is *Yes*, denoted $D \cup \Sigma \models Q$, iff $ans(Q, D, \Sigma) \neq \emptyset$.

Query answering under general TGDs is undecidable [2], even when $\mathcal{R}$ and Σ are fixed [5]. Decidability and data tractability of query answering for the guarded case follows from a bounded tree-width property.

A *negative constraint* (or simply *constraint*) γ is a first-order formula of the form $\forall \mathbf{X}\, \Phi(\mathbf{X}) \to \bot$, where $\Phi(\mathbf{X})$ (called the *body*) is a conjunction of atoms over $\mathcal{R}$ (without nulls). Under the standard semantics of query answering of BCQs in Datalog+/− with TGDs, adding negative constraints is computationally easy, as for each constraint $\forall \mathbf{X}\, \Phi(\mathbf{X}) \to \bot$, we only have to check that the BCQ $\exists \mathbf{X}\, \Phi(\mathbf{X})$ evaluates to false in D under Σ; if one of these checks fails, then the answer to the original BCQ Q is true; otherwise, the constraints can simply be ignored when answering the BCQ Q.

As another component, the Datalog+/− language allows for special types of *equality-generating dependencies* (EGDs). As they can be modeled via negative constraints, we do not discuss them here. We usually omit the universal quantifiers in TGDs, constraints, and EGDs, and assume that sets of dependencies and/or constraints are finite.

The Chase. The *chase* was first introduced for checking implications of dependencies, and later also for checking query containment. By "chase", we refer to both the chase procedure and its output. The TGD chase works on a database via so-called TGD *chase rules* (see [6] for further details and for an extended chase with also EGD chase rules). The (possibly infinite) chase of a database D relative to a set of TGDs Σ, denoted $chase(D, \Sigma)$, is a *universal model*, i.e., there is a homomorphism from $chase(D, \Sigma)$ onto every $B \in mods(D, \Sigma)$ [6]. Thus, BCQs Q over D and Σ can be evaluated on the chase for D and Σ, i.e., $D \cup \Sigma \models Q$ is equivalent to $chase(D, \Sigma) \models Q$. For guarded TGDs Σ, such BCQs Q can be evaluated in polynomial time on an initial fragment of $chase(D, \Sigma)$ of constant depth $k \cdot |Q|$ in the data complexity.

Datalog+/− Ontologies. A *Datalog+/− ontology* $KB = (D, \Sigma)$, where $\Sigma = \Sigma_T \cup \Sigma_E \cup \Sigma_{NC}$, consists of a database D, a set of TGDs Σ_T, a set of EGDs Σ_E, and a set of negative constraints Σ_{NC}. To ensure decidability and data tractability of query answering, we assume that EGDs are *separable*, which means that the interaction between TGDs and EGDs is controlled—this condition can be ensured by the syntactic criterion of *non-conflicting keys*; for details on these conditions, we refer the reader to [6]. Finally, we say that KB is *guarded* iff Σ_T is guarded. The following example illustrates a simple Datalog+/− ontology, used in the sequel as a running example.

Example 1. Consider the following ontology $KB = (D, \Sigma)$:

$\Sigma = \{\sigma_1\colon smartphone(P) \to mobDev(P),\quad \sigma_2\colon tablet(T) \to mobDev(T),$
$\sigma_3\colon eReader(R) \to mobDev(R),\quad \sigma_4\colon smartphone(P) \to \exists O\ os(P, O),$
$\sigma_5\colon eReader(R) \to \exists F\ supports(R, F),\ \sigma_6\colon tablet(T) \to \exists O\ os(T, O)\},$

$D = \{id_1\colon smartphone(p_1),\ id_2\colon mobDev(p_1),\ id_3\colon tablet(t_1),\ id_4\colon mobDev(p_2),$
$id_5\colon os(p_1, and\_6),\ id_6\colon os(p_2, and\_6),\ id_7\colon os(t_1, and\_6)\}.$

This ontology models a very simple knowledge base for electronic devices, which can be used as the underlying model in an online shopping system. Mobile devices can be either smartphones, tablets, or e-book readers; the first two have operating systems, and the latter has supported formats. The database D gives some example instances. ∎

Preference Models. A *preference relation* $\succ$ over a set S is a strict partial order (SPO) over S, i.e., an irreflexive and transitive binary relation over S—we consider these to be the minimal requirements for a useful preference relation. If $a \succ b$, we say that a is *preferred to* b. The *indifference relation* $\sim$ induced by $\succ$ is defined as follows: for any $a, b \in S$, $a \sim b$ iff $a \not\succ b$ and $b \not\succ a$.

A *stratification* of S relative to $\succ$ is an ordered sequence $S_1, \ldots, S_k$, where each S_i is a maximal subset of S such that for every $a \in S_i$ there is no $b \in \bigcup_{j=i}^{k} S_j$ with $b \succ a$. Intuitively, S_1 contains the most preferred elements in S relative to $\succ$; then, S_2 contains the most preferred elements of $S - S_1$, and so on. Stratifications always exist, are unique, and are a partition of S. Elements in stratum S_i have *rank* i. The rank of an element $a \in S$ relative to $\succ$ is denoted as $rank(a, \succ)$.

3 Subjective Reports

Let KB be a Datalog+/− ontology, $a = p(c_1, \ldots, c_m)$ be a ground atom such that $KB \models a$, and $\mathcal{F} = (f_1, \ldots, f_n)$ be a tuple of *features* associated with the predicate p, each of which has a domain $dom(f_i) = [0, 1] \cup \{-\}$. We sometimes slightly abuse notation and use $\mathcal{F}$ to also denote the set of features $\{f_1, \ldots, f_n\}$.

A *report* for a is a triple $(E, \succ_P, I)$, where $E \in dom(f_1) \times \cdots \times dom(f_n)$, $\succ_P$ is an SPO over the elements of $\mathcal{F}$, and I is a set of pairs (*key, value*). Intuitively, reports are evaluations of an entity of interest (atom a) provided by observers. In a report $(E, \succ_P, I)$, E specifies a "score" for each feature (with "$-$" meaning that no score has been provided), $\succ_P$ indicates the relative importance of the features to the report's observer, and I (called *information register*) contains general information about the report itself, and who provided it. Reports will be analyzed by a user, who has his own strict partial order, denoted $\succ_{P_U}$, over the set of features.

Example 2. Consider the mobile device domain from Example 1, and let the features for predicate $mobDev$ be $\mathcal{F} = (Design, Hardware\ Quality, Battery\ Life, Price)$; in the following, we abbreviate these features as *des*, *hwq*, *batt*, and *pri*, respectively.

An example of a report for $mobDev(h_1)$ is $r_1 = (\langle 1, 0, 0.4, 0.1 \rangle, \succ_{P_1}, I_1)$, where $\succ_{P_1}$ is given by the graph in Fig. 1 (left side), and I_1 is a register with fields *age*, *nationality*, and *type of user*, with data $I_1.age = 22$, $I_1.nationality = British$, and $I_1.type = Gamer$.

An example of $\succ_{P_U}$ (i.e., the querying user's preferences over the features) is shown in Fig. 1 (right side). ∎

The set of all reports available is denoted with *Reports*. In the following, we use $Reports(a)$ to denote the set of all reports that are associated with a ground atom a. Given a tuple of features $\mathcal{F}$, we use $SPOs(\mathcal{F})$ to denote the set of all SPOs over $\mathcal{F}$.

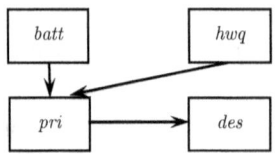 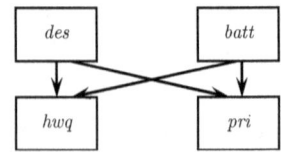

Fig. 1. Preference relation $>_{P_1}$ from Example 2 (left) and $>_{P_U}$, the user's preferences (right).

3.1 Trust Measures over Reports

A user analyzing a set of reports may decide that certain opinions within a given report may be more trustworthy than others. For instance, returning to our running example, the score given for the *battery life* feature of $mobDev(p_1)$ might be considered more trustworthy than the ones given for *design*, e.g., because the reporter declared the former to be among the most preferred features, while the latter is the least preferred, cf. Fig. 1 (left). Another example could be a user that is generally untrustworthy of reports on feature *hardware quality*, because he has learned that many people are more critical than he is when evaluating that aspect of a mobile device, or of reports on feature *price* by business users, because they do not use their own money to pay for their devices. Formally, a *trust measure* is any function $\tau\colon Reports \to [0,1]^n$, where higher values correspond to more trust.

Note that trust measures do not depend on the querying user's own preferences over $\mathcal{F}$ (in $>_{P_U}$); rather, for each report $(E, >_P, I)$, they give a measure of trust to each of the n scores in E depending on P and I. The following is a simple example.

Example 3. Consider again our running example and suppose that the user defines a trust measure τ, which assigns trust values to a report $r = (E, >_P, I)$ as follows:

$$\tau(r) = \begin{cases} 0.25 \cdot \left(2^{-(rank(f_1, >_P)-1)}, \ldots, 2^{-(rank(f_n, >_P)-1)}\right) & \text{if } I.type = \textit{Gamer}; \\ \left(2^{-(rank(f_1, >_P)-1)}, \ldots, 2^{-(rank(f_n, >_P)-1)}\right) & \text{otherwise.} \end{cases}$$

For r_1 from Example 2 and the SPO in Fig. 1 (left), we get

$$0.25 \cdot 2^{-(rank(des, >_{P_1})-1)} = 1/16,$$
$$0.25 \cdot 2^{-(rank(hwq, >_{P_1})-1)} = 0.25 \cdot 2^{-(rank(batt, >_{P_1})-1)} = 0.25,$$

and finally for price, we get $0.25 \cdot 2^{-(rank(pri, >_{P_1})-1)} = 1/8$. ∎

3.2 Relevance of Reports

The other aspect of importance that a user must consider when analyzing reports is how *relevant* they are to their own preferences. For instance, a report given by someone who has preferences that are completely opposite to those of the user should be considered less relevant than one given by someone whose preferences only differ in a trivial aspect. This is inherently different from the trust measure described above, as trust is computed without taking into account the preference relation given by the user issuing the query. Formally, a *relevance measure* is any function ρ: $Reports \times SPOs(\mathcal{F}) \to [0,1]$, where higher output values reflect higher relevance. Thus, a relevance measure takes as input a report $(E, >_P, I)$ and an SPO $>_{P'}$ and gives a measure of how relevant the report is relative to $>_{P'}$; this is determined on the basis of $>_P$ and $>_{P'}$, and can also take I into account.

Example 4. Consider again the running example, and suppose that the user assigns relevance to a report $r = (E, >_P, I)$ according to the function

$$\rho(r, >_{P_U}) = 2^{-\sum_{f_i \in \mathcal{F}} |rank(f_i, >_P) - rank(f_i, >_{P_U})|}.$$

From Fig. 1, e.g., we have that $\rho(r_1, >_{P_U}) = 2^{-1 \cdot (2+1+0+0)} = 0.125$.

Alternatively, a relevance measure comparing the SPO $>_P$ of a report $(E, >_P, I)$ with the user's SPO $>_{P_U}$ may be defined via measuring the similarity between two preference relations as follows.

Example 5. The relevance measure checks to what extent the two SPOs agree on the relative importance of the features in $\mathcal{F}$. Formally, let P_1 and P_2 be SPOs over $\mathcal{F}$. We define a measure of similarity of P_1 and P_2 as follows:

$$sim(P_1, P_2) = \frac{\sum_{1 \leq i < j \leq n} sim(f_i, f_j, P_1, P_2)}{n(n-1)/2},$$

where

$$sim(f_i, f_j, P_1, P_2) = \begin{cases} 1 & \text{if } (f_i, f_j) \in P_1 \cap P_2 \text{ or } (f_j, f_i) \in P_1 \cap P_2 \\ 1 & \text{if } (f_i, f_j) \notin P_1 \cup P_2 \text{ and } (f_j, f_i) \notin P_1 \cup P_2 \\ 0.5 & \text{if } ((f_i, f_j) \in P_1 \Delta P_2 \text{ and } (f_j, f_i) \notin P_1 \cup P_2) \text{ or} \\ & ((f_j, f_i) \in P_1 \Delta P_2 \text{ and } (f_i, f_j) \notin P_1 \cup P_2) \\ 0 & \text{if } (f_i, f_j) \in P_1 \cup P_2 \text{ and } (f_j, f_i) \in P_1 \cup P_2. \end{cases}$$

Note that here, Δ is used to denote the symmetric difference (i.e., $A \Delta B = A \cup B - A \cap B$). More specifically, in the definition of $sim(f_i, f_j, P_1, P_2)$,

- the first condition refers to the case where P_1 and P_2 are expressing the same order between f_i and f_j,
- the second condition refers to the case where both P_1 and P_2 are not expressing any order between f_i and f_j,
- the third condition refers to the case where one of P_1 and P_2 is expressing an order between f_i and f_j and the other is not expressing any order,
- the last condition refers to the case where P_1 and P_2 are expressing opposite orders between f_i and f_j.

Clearly, $sim(P_1, P_2)$ is 1, when P_1 and P_2 agree on everything, and 0, when P_1 and P_2 agree on nothing. Finally, we define a relevance measure by $\rho((E, >_P, I), >_{P'}) = sim(>_P, >_{P'})$ for every report $(E, >_P, I) \in Reports$ and SPO $>_{P'} \in SPOs(\mathcal{F})$. ∎

```
Algorithm RepRank-Basic(KB, Q(X), F, >_{P_U}, τ, ρ, Reports, k)
Input:  Datalog+/− ontology KB, atomic query Q(X), set of features F = {f_1, ..., f_n},
        preference relation of the querying user >_{P_U}, trust measure τ,
        relevance measure ρ, set of reports Reports, k ≥ 1.
Output: Top-k answers to Q.
 1. RankedAns := ∅
 2. for each atom a in ans(Q(X), KB) do begin
 3.    score := 0;
 4.    for each report r = (E, >_P, I) in Reports(a) do begin
 5.       trustMeasures := τ(r);
 6.       score := score + ρ(r, >_{P_U}) · (1/n) · Σ_{i=1}^n E[i] · trustMeasures[i] · (1/rank(f_i, >_{P_U}));
 7.    end;
 8.    score := score/|Reports(a)|;
 9.    RankedAns := RankedAns ∪ {⟨a, score⟩};
10. end;
11. return top-k atoms in RankedAns.
```

Fig. 2. A first algorithm for computing the top-k answers to an atomic query Q according to a given set of user preferences and reports on answers to Q.

4 Query Answering based on Subjective Reports

To produce a ranking based on the basic components presented in Section 3, we must first develop a way to combine them in a principled manner. We first consider queries of the form $Q(\mathbf{X}) = p(\mathbf{X})$, called *atomic queries*, that is, queries with a single atom and no existential variables. The answers to an atomic query $Q(\mathbf{X}) = p(\mathbf{X})$ over KB in *atom form* are defined as $\{p(t) \mid t \in ans(Q(\mathbf{X}), KB)\}$; we still use $ans(Q(\mathbf{X}), KB)$ to denote the set of answers in atom form. At the end of this section, we will discuss how our approach can indeed be applied also to a more general class of queries called *simple*. The problem that we address is the following. The user is given a Datalog+/− ontology KB and has an atomic query $Q(\mathbf{X})$ of interest. The user also supplies an SPO $>_{P_U}$ over the set of features $\mathcal{F}$. Recall that in our setting, each ground atom b such that $KB \models b$ is associated with a (possibly empty) set of reports. As we consider atomic queries, each ground atom $a \in ans(Q(\mathbf{X}), KB)$ is an atom entailed by KB and thus is associated with a set of reports $Reports(a)$. Furthermore, recall that each report $r \in Reports(a)$ is associated with a trust score $\tau(r)$. We want to rank the ground atoms in $Ans(Q(\mathbf{X}), KB)$, that is, we want to obtain a set $\{\langle a_i, score_i \rangle \mid a_i \in ans(Q(\mathbf{X}), KB)\}$ where $score_i$ for ground atom a_i takes into account the set of reports $Reports(a_i)$ associated with a_i, the trust score $\tau(r)$ associated with each report $r \in Reports(a_i)$, and the SPO $>_{P_U}$ over $\mathcal{F}$ provided by the user issuing the query.

4.1 A Basic Approach

A first approach to solving this problem is Algorithm RepRank-Basic (Fig. 2). A score for each atom is computed as the average of the scores of the reports associated with the atom, where the score of a report $r = (E, >_P, I)$ is computed as follows: (i) we first compute the average of the scores $E[i]$ weighted by the trust value for $E[i]$ and a value measuring how important feature f_i is for the user issuing the query (this value is given by $rank(f_i, >_{P_U})$); (ii) then, we multiply the value computed in the previous step by $\rho(r, >_{P_U})$, which gives a measure of how relevant r is relative to $>_{P_U}$. The following is an example of how Algorithm RepRank-Basic works.

Example 6. Consider again the setup from the running example, where we have the Datalog+/– ontology from Example 1, the set *Reports* of the reports depicted in Fig. 3, the SPO $>_{P_U}$ from Fig. 1 (right), the trust measure τ defined in Example 3, and the relevance measure ρ introduced in Example 4. Finally, let $Q(X) = mobDev(X)$.

Algorithm RepRank-Basic iterates through the set of answers (in atom form) to the query, which in this case consists of $\{mobDev(p_1), mobDev(p_2)\}$. For the atom $mobDev(p_1)$, the algorithm iterates through the set of corresponding reports, which is $Reports(mobDev(p_1)) = \{r_1, r_2, r_3\}$, and maintains the accumulated score after processing each report. For r_1, the score is computed as (cf. line 6 in Fig. 2):

$$0.125 \cdot \frac{1}{4} \cdot \left(\frac{1 \cdot 0.0625}{1} + \frac{0 \cdot 0.25}{2} + \frac{0.4 \cdot 0.25}{1} + \frac{0.1 \cdot 0.125}{2}\right) = 0.00527.$$

The score for $mobDev(p_1)$ after processing the three reports is around 0.065205. Analogously, assuming $Reports(mobDev(p_2)) = \{r_4, r_5, r_6\}$, the score for $mobDev(p_2)$ is 0.17388. Thus, the top-2 answer to Q is $\langle mobDev(p_2), mobDev(p_1)\rangle$. ∎

The following result states the time complexity of Algorithm RepRank-Basic. As long as both query answering and computation of the trust and relevance measures can be done in polynomial time, RepRank-Basic also runs in polynomial time.

Proposition 1. *The worst-case time complexity of Algorithm* RepRank-Basic *is $O(m * \log m + (n + |>_{P_U}|) + m * Reports_{max} * (f_\tau + f_\rho + n) + f_{ans(Q(\mathbf{X}),KB)})$, where $m = |ans(Q(\mathbf{X}), KB)|$, $Reports_{max} = \max\{|Reports(a)| : a \in ans(Q(\mathbf{X}), KB)\}$, f_τ (resp. f_ρ) is the worst-case time complexity of τ (resp. ρ), and $f_{ans(Q(\mathbf{X}),KB)}$ is the data complexity of computing $ans(Q(\mathbf{X}), KB)$.*

In the next section, we explore an alternative approach for applying the trust and relevance measures to top-k query answering.

4.2 Leveraging Trust and Relevance to a Greater Extent

A more complex approach consists of using the trust and relevance scores provided by the respective measures in a more fine-grained manner. One way of doing this is via the following steps (more details on each of them are given shortly):

1. Keep only those reports that are most relevant to the user issuing the query, i.e., those reports that are relevant enough to $>_{P_U}$ according to a relevance measure ρ;

Reports $r_1 = (E_1, >_{P_1}, I_1)$ and $r_4 = (E_4, >_{P_1}, I_1)$					
Relevance scores: $\rho(r_1, >_{P_U}) = \rho(r_4, >_{P_U}) = 0.125$					
Features	E_1	E_4	$\tau(r_1) = \tau(r_4)$	$>_{P_1}$	I_1
des	1	0.7	1/16	batt → hwq	$Age = 22$
hwq	0	1	1/4	↓ pri	$Nationality = British$
batt	0.4	0.9	1/4	↓	$Type = Gamer$
pri	0.1	0.8	1/8	des	

Reports $r_2 = (E_2, >_{P_2}, I_2)$ and $r_5 = (E_5, >_{P_2}, I_2)$					
Relevance scores: $\rho(r_2, >_{P_U}) = \rho(r_5, >_{P_U}) = 0.125$					
Features	E_2	E_5	$\tau(r_2) = \tau(r_5)$	$>_{P_2}$	I_2
des	0.3	0.6	1/2	pri	$Age = 59$
hwq	0.7	0.7	1/2	↓	$Nationality = Belgian$
batt	1	1	1/2	↓	$Type = Novice$
pri	0.9	0.1	1	des hwq batt	

Reports $r_3 = (E_3, >_{P_3}, I_3)$ and $r_6 = (E_6, >_{P_3}, I_3)$					
Relevance scores: $\rho(r_3, >_{P_U}) = \rho(r_6, >_{P_U}) = 0.25$					
Features	E_3	E_6	$\tau(r_3) = \tau(r_6)$	$>_{P_3}$	I_3
des	0.25	0.5	1/2	batt hwq	$Age = 18$
hwq	0.2	0.9	1	✕	$Nationality = US$
batt	0.7	1	1	✕	$Type = Intensive$
pri	0.8	0.6	1/2	pri des	

Fig. 3. Reports used in Examples 6 and 7. We assume that each pair of reports (r_1–r_4, r_2–r_5, and r_3–r_6) was generated by the same reviewer—they thus share the SPO and information. In each pair of reports, the first one is for $mobDev(p_1)$ and the second one is for $mobDev(p_2)$.

2. consider the most relevant reports obtained in the previous step and use the trust measure given by the user to produce scores adjusted by the trust measure; and
3. for each atom, compute a single score by combining the scores computed in the previous step with $>_{P_U}$.

The first step can simply be carried out by checking, for each report r, if $\rho(r, >_{P_U})$ is above a certain given threshold. One way of doing the second step is described in Algorithm SummarizeReports (Fig. 4), which takes a trust measure τ, a set of reports Reports (for a certain atom), and a function collFunc. The algorithm processes each report in the input sets by building a histogram of the average score over all reports per range of trust values (as collected in a bucket); for each report, the algorithm applies the trust measure to update each feature's histogram. Once all of the reports are processed, the last step is to collapse the histograms into a single value—this is done by applying the collFunc function, which could simply be defined as the computation of a weighted average for each feature. This single value is finally used to produce the output, which is a tuple of n scores.

Example 7. Consider again the setup from Example 6. Suppose that we want to keep only those reports for which the relevance score is above 0.1 (as per the first step of our more complex approach). Recall that the answers to Q are $\{mobDev(p_1), mobDev(p_2)\}$ and there are six associated reports; in this case we trivially keep all records. Algorithm SummarizeReports has $Reports = \{r_1, r_2, r_3\}$ when called for $mobDev(p_1)$. The histograms built during this call are as follows:

- *des*: value 1 in bucket $[0, 0.1)$ and value 0.275 in bucket $[0.5, 0.6)$;
- *hwq*: value 0.7 in bucket $[0.5, 0.6)$ and value 0.2 in bucket $[0.9, 1]$;
- *batt*: value 0.4 in bucket $[0.4, 0.5)$, value 1 in bucket $[0.5, 0.6)$, and value 0.7 in bucket $[0.9, 1]$;
- *pri*: value 0.1 in bucket $[0.1, 0.2)$, value 0.8 in bucket $[0.5, 0.6)$, and value 0.9 in bucket $[0.9, 1]$.

Assuming that function *collFunc* disregards the values in the bucket corresponding to the lowest trust value (if more than one bucket is non-empty), and takes the average of the rest, we have the following result tuple as the output of Summarize-Reports: $(0.275, 0.45, 0.85, 0.85)$. Analogously, we have tuple $(0.55, 0.8, 1, 0.35)$ for tuple $mobDev(p_2)$ after calling SummarizeReports with $Reports = \{r_4, r_5, r_6\}$. ∎

The following proposition states the time complexity of Algorithm SummarizeReports. As long as the trust measure and the *collFunc* function can be computed in polynomial time, Algorithm SummarizeReports is polynomial time as well.

Proposition 2. *The worst-case time complexity of Algorithm* Summarize-Reports *is* $O(|Reports| * (f_\tau + n) + n * f_{collFunc})$, *where* f_τ *(resp., $f_{collFunc}$) is the worst-case time complexity of τ (resp., collFunc)*.

The following example explores a few different ways in which function *collFunc* used in Algorithm SummarizeReports might be defined.

Example 8. One way of computing *collFunc* is shown in Example 7. There can be other reasonable ways of collapsing the histogram for a feature into a single value. E.g., *collFunc* might compute the average across all buckets ignoring the trust measure so that no distinction is made among buckets, i.e., $collFunc(hists[i]) = \frac{\sum_{b=1}^{10} hists[i](b)}{10}$. Alternatively, the trust measure might be taken into account by giving a weight w_b to each bucket b (e.g., the weights might be set in such a way that buckets corresponding to higher trust scores have a higher weight, that is, $weight_i < weight_j$ for $i < j$). In this case, the histogram might be collapsed as follows $collFunc(hists[i]) = \frac{\sum_{b=1}^{10} w_b * hists[i](b)}{10}$. We may also want to apply the above strategies but ignoring the first k buckets (for which the trust score is lower). Function *collFunc* can also be extended so that the number of elements associated with a bucket is taken into account. ∎

Thus, the second step discussed above gives n scores (adjusted by the trust measure) for each ground atom. Recall that the third (and last) step of the approach adopted in this section is to compute a score for each atom by combining the scores computed in the previous step with $>_{P_U}$. One simple way of

Algorithm SummarizeReports(τ, *Reports*, *collFunc*)
Input: Trust measure τ, set of reports *Reports*, and function *collFunc* that collapses histograms to values in $[0, 1]$.
Output: Scores representing the trust-adjusted average from *Reports*.

1. Init. *hists* as an n-array of empty mappings with keys: $\{[0, 0.1), [0.1, 0.2), \ldots, [0.9, 1]\}$ and values of type $[0, 1]$, where $n = |\mathcal{F}|$ (we use values $1, \ldots, 10$ to denote the keys).
2. Initialize array *bucketCounts* of size $n \times 10$ with value 0 in all positions;
3. for each report $r = (E, >_P, I) \in$ *Reports* do begin;
4.   *trustMeasures* := $\tau(r)$;
5.   for $i = 1$ to n do begin
6.   let b be the key for *hists*$[i]$ under which *trustMeasures*$[i]$ falls;
7.   *hists*$[i](b) := \frac{hist[i](b) * bucketCounts[i][b] + E[i]}{bucketCounts[i][b] + 1}$;
8.   *bucketCounts*$[i][b]$++;
9.   end;
10. end;
11. Initialize *Res* as an n-array;
12. for $i = 1$ to n do
13.   *Res*$[i] :=$ *collFunc*(*hists*$[i]$);
14. return *Res*.

Fig. 4. This algorithm takes a set of reports for a single entity and computes an n-array of scores obtained by combining the reports with their trust measure.

doing this is to compute the weighted average of such scores where the weight of the i-th score is the inverse of the rank of feature f_i in $>_{P_U}$.

Algorithm RepRank-Hist (Fig. 5) is the complete algorithm that combines the three steps discussed thus far. The following continues the running example to show the result of applying this algorithm.

Example 9. Let us adopt once again the setup from Example 6, but this time applying Algorithm RepRank-Hist. Suppose *collFunc* is the one discussed in Example 7 and thus Algorithm SummarizeReports returns the scores $(0.275, 0.45, 0.85, 0.85)$ for $mobDev(p_1)$ and the scores $(0.55, 0.8, 1, 0.35)$ for $mobDev(p_2)$. Algorithm RepRank-Hist computes a score for each atom by performing a weighted sum of the scores in these tuples, which results in: $\langle mobDev(p_1), 1.775 \rangle$ and $\langle mobDev(p_2), 2.125 \rangle$. Therefore, the top-2 answer to query Q is $\langle mobDev(p_2), mobDev(p_1) \rangle$. ∎

Note that the results from the two algorithms are not necessarily the same, since they each use the relevance and trust score differently—the more fine-grained approach adopted by Algorithm RepRank-Hist allows it to selectively use both kinds of values to generate a more informed result.

Proposition 3. *The worst-case time complexity of Algorithm* RepRank-Hist *is:* $O(m * \log m + (n + | >_{P_U} |) + m * (Reports_{max} * f_\rho + f_{sum} + n) + f_{ans(Q(\mathbf{X}), KB)})$, *where* $m = |ans(Q(\mathbf{X}), KB)|$, $Reports_{max} = \max\{|Reports(a)| : a \in ans(Q(\mathbf{X}), KB)\}$, f_ρ *is the worst-case time complexity of* ρ, f_{sum} *is the worst-case time complexity of Algorithm* SummarizeReports *as per Proposition 2, and* $f_{ans(Q(\mathbf{X}), KB)}$ *is the data complexity of computing* $ans(Q(\mathbf{X}), KB)$.

As a corollary to Propositions 1 and 3, we have the following result.

Theorem 1. *If the input ontology belongs to the guarded fragment of Datalog+/−, then Algorithms* RepRank-Basic *and* RepRank-Hist *run in polynomial time in the data complexity.*

```
Algorithm RepRank-Hist(KB, Q(X), F, ><sub>P_U</sub>, τ, ρ, relThresh, collFunc, Reports, k)
Input:  Datalog+/- ontology KB, atomic query Q(X), set of features $\mathcal{F} = \{f_1, \ldots, f_n\}$,
        user preferences $>_{P_U}$, trust measure τ, relevance measure ρ, relThresh ∈ [0, 1],
        function collFunc that collapses histograms to values in [0, 1], set of reports Reports,
        $k \geq 1$.
Output: Top-k answers to Q.
 1. RankedAns := ∅;
 2. for each atom a in ans(Q(X), KB) do begin
 3.    relReps:= select all r ∈ Reports(a) with $\rho(r, >_{P_U}) \geq relThresh$;
 4.    scores:= SummarizeReports(τ, relReps, collFunc);
 5.    finalScore:= $\sum_{i=1}^{n} \frac{scores[i]}{rank(f_i, >_{P_U})}$;
 6.    RankedAns:= RankedAns ∪ {⟨a, finalScore⟩};
 7. end;
 8. return top-k atoms in RankedAns.
```

Fig. 5. Algorithm for computing the top-k answers to an atomic query Q according to a given set of user preferences and reports on answers to Q.

Thus far, we have considered atomic queries. As each ground atom a such that $KB \models a$ is associated with a set of reports, and every ground atom b in $ans(Q(\mathbf{X}), KB)$ is such that $KB \models b$, then reports can be associated with query answers in a natural way. We now introduce a class of queries more general than the class of atomic queries for which the same property holds. A *simple query* is a conjunctive query $Q(\mathbf{X}) = \exists \mathbf{Y}\, \Phi(\mathbf{X}, \mathbf{Y})$ where $\Phi(\mathbf{X}, \mathbf{Y})$ contains exactly one atom of the form $p(\mathbf{X})$, called *distinguished* atom (i.e., an atom whose variables are the query's free variables). For instance, $Q(X) = mobDev(X) \wedge os(X, and\_6)$ is a simple query where $mobDev(X)$ is the distinguished atom. The answers to a simple query $Q(\mathbf{X})$ over KB in *atom form* are defined as $\{p(t) \mid t \in ans(Q(\mathbf{X}), KB)\}$ where the distinguished atom is of the form $p(\mathbf{X})$; we still use $ans(Q(\mathbf{X}), KB)$ to denote the set of answers in atom form. Clearly, for each atom a in $ans(Q(\mathbf{X}), KB)$, it is the case that $KB \models a$.

5 Towards more General Reports

In the previous section we considered the setting where reports are associated with ground atoms a such that $KB \models a$. This setup is limited, since it does not allow us to express the fact that certain reports may apply to whole *sets* of atoms—this is necessary to model certain kinds of opinions often found in reviews, such as "smartphones running the latest version of Android are expensive". We now generalize the framework presented in Sections 3 and 4 to contemplate this kind of reports. A *generalized report* (*g-report*, for short) is a pair $gr = (r, Q(\mathbf{X}))$, where r is a report and $Q(\mathbf{X})$ is a simple query, called the *descriptor* of gr. We denote with *g-Reports* the universe of g-reports. Intuitively, given an ontology KB, a g-report $(r, Q(\mathbf{X}))$ is used to associate report r with every atom a in $ans(Q(\mathbf{X}), KB)$—recall that $KB \models a$ and thus general reports allow us to assign a report to a set of atoms entailed by KB. Clearly, a report for a ground atom a as defined in Section 3 is a special case of a g-report in which the only answer to the descriptor is a.

Example 10. Consider our running example from the mobile device domain and suppose that we want to associate a certain report r with all the mobile devices that run Android 6.0. This can be expressed with a g-report $(r, Q(X))$ where $Q(X) = mobDev(X) \land os(X, and\_6)$, with $mobDev(X)$ being the distinguished atom. ∎

Intuitively, a g-report $gr = (r, Q(\mathbf{X}))$ is a report associated with a set of atoms, i.e., the set of atoms in $ans(Q(\mathbf{X}), KB))$. A simple way of handling this generalization would be to associate report r with every atom in this set. Note that, as in the non-generalized case, it may be the case that two or more g-reports assign two distinct reports to the same ground atom. E.g., we may have a g-report $(r, Q(X))$, where $Q(X) = mobDev(X) \land os(X, and\_6)$, expressing that r applies to all mobile devices that run Android 6.0, and another g-report $(r', Q'(X))$, where $Q'(X) = mobDev(X) \land smartphone(X)$, expressing that r' applies to all mobile devices that are smartphones. In our running example, we would associate r with $mobDev(p_1)$, $mobDev(p_2)$, and $mobDev(t_1)$ and r' to $mobDev(p_1)$.

In the approach just described, the reports coming from different g-reports are treated in the same way—they all have the same impact on the common atoms. Another possibility is to determine when a g-report is in some sense *more specific* than another and to take such a relationship into account (e.g., more specific g-reports should have a greater impact when computing the ranking over atoms). We consider this kind of scenario in the following: we study two kinds of structure that can be leveraged from knowledge contained in the ontology. The first is based on the notion of *hierarchies*, which are useful in capturing the influence of reports in "is-a" type relationships. As an example, given a query requesting a ranking over smartphones running Android, a report for all Samsung smartphones running Android 6.0 should have a higher impact on the calculation of the ranking than a report for all Samsung mobile devices—in particular, the latter might be ignored altogether since it is too general. The second kind of structure is based on identifying subset relationships between the atoms associated with the descriptors in g-reports. For instance, a report for all smartphones running Android is more general than a report for all smartphones running Android 6.0. We now define a partial order among reports based on these notions; we first define hierarchical TGDs as follows. A set of linear TGDs Σ_T is *hierarchical* iff for every $p(\mathbf{X}) \to \exists \mathbf{Y} q(\mathbf{X}, \mathbf{Y}) \in \Sigma_T$ we have that $features(p) \subseteq features(q)$ and there does not exist a database D over $\mathcal{R}$ and TGD in Σ_T of the form $p'(\mathbf{X}) \to \exists \mathbf{Y} r(\mathbf{X}, \mathbf{Y})$ such that $p(\mathbf{X})$ and $p'(\mathbf{X})$ share ground instances relative to D.

In the rest of this section, we assume that all ontologies contain a (possibly empty) subset of hierarchical TGDs. Furthermore, given an ontology $KB = (D, \Sigma)$, where $\Sigma_H \subseteq \Sigma$ is a set of hierarchical TGDs, and two ground atoms a and b, we say that a *is-a* b iff $chase(\{a\}, \Sigma_H) \models b$. For instance, in Example 1, the set $\{\sigma_1, \sigma_2, \sigma_3\} \subseteq \Sigma$ is a hierarchical set of TGDs (assuming that the conditions over the features hold).

Given tuples of features $\mathcal{F}$ and $\mathcal{F}'$ such that $\mathcal{F} \subseteq \mathcal{F}'$ and vectors E and E' over the domains of $\mathcal{F}$ and $\mathcal{F}'$, respectively, we say that E' is a particularization of E, denoted $E' = part(E)$ iff $E'[f] = E[f]$, if $f \in \mathcal{F} \cap \mathcal{F}'$, and $E'[f] = -$,

$$gr_1 = ((r_1, >_{P_1}, I_1), \underline{mobDev}(X) \wedge os(X, and))$$
$$gr_2 = ((r_3, >_{P_3}, I_3), \underline{smartphone}(X) \wedge os(X, and\_5))$$
$$gr_3 = ((r_4, >_{P_1}, I_1), \underline{mobDev}(X) \wedge os(X, and\_6))$$

Fig. 6. A set of general reports (distinguished atoms in the descriptors are underlined).

otherwise. Let $KB = (D, \Sigma)$ be a Datalog+/− ontology, a be a ground atom such that $KB \models a$, and $gr = (r, Q(\mathbf{X}))$ be a g-report with $r = (E, >_P, I)$. If there exists a ground atom $b \in Ans(Q(\mathbf{X}), KB)$ such that a is-a b, then we say that g-report $gr' = ((E', >_P, I), a)$, with $E' = part(E)$, is a *specialization* of gr for a. Clearly, a g-report is always a specialization of itself for every atom in the answers to its descriptor.

Example 11. Let $\mathcal{F}_1$ be the set of features for predicate *mobDev* presented in Example 2, and let $\mathcal{F}_2 = \langle des, hwq, batt, pri, net\_supp \rangle$ be the set of features for predicate *smartphone*, where *net_supp* denotes "supported network types".

Let $gr = (r_1, Q(X))$ be a g-report, where r_1 is the report from Fig. 3 and $Q(X) = mobDev(X) \wedge os(X, and\_6)$. If we consider $a = smartphone(p_1)$ and $b = mobDev(p_1)$, clearly, we have that $b \in Ans(Q(X), KB)$ and a is-a b. Therefore, a specialization of gr for a is $gr' = ((E', >_{P_1}, I_1), a)$, where $E' = \langle 1, 0, 0.4, 0.1, - \rangle$. ∎

Given g-reports $gr_1 = (r_1, Q_1(\mathbf{X}_1))$ and $gr_2 = (r_2, Q_2(\mathbf{X}_2))$, we say that gr_1 is *more general than* gr_2, denoted $gr_2 \sqsubseteq gr_1$, iff either (i) $Ans(Q_2(\mathbf{X}_2), KB) \subseteq Ans(Q_1(\mathbf{X}_1), KB)$; or (ii) for each $a \in Ans(Q_2(\mathbf{X}_2), KB)$, there exists some $b \in Ans(Q_1(\mathbf{X}_1), KB)$ such that a is-a b. If $gr_1 \sqsubseteq gr_2$ and $gr_2 \sqsubseteq gr_1$, we say that gr_1 and gr_2 are *equivalent*, denoted $gr_1 \equiv gr_2$.

Example 12. Consider the g-reports in Fig. 6 and the database in the running example with the addition of atoms $smartphone(p_3)$ and $os(h_3, and\_5)$. We then have:

- $gr_3 \sqsubseteq gr_1$, since $\{mobDev(p_1), mobDev(p_2), mobDev(t_1)\} \subseteq \{mobDev(p_1), mobDev(p_2), mobDev(p_3), mobDev(t_1)\}$;
- $gr_2 \sqsubseteq gr_1$, since for atom $smartphone(p_3)$ (the only answer for the descriptor in gr_2), there exists atom $mobDev(p_3)$ in the answer to the descriptor in gr_1 and $smartphone(p_3)$ is-a $mobDev(p_3)$. ∎

The "more general than" relationship between g-reports is useful for defining a partial order for the set of reports associated with a given ground atom. This partial order can be defined as follows: $gr_1 \sim gr_2$ iff $gr_1 \equiv gr_2$, and $gr_1 > gr_2$ iff $gr_1 \sqsubseteq gr_2$. Here, $a \sim b$ denotes the equivalence between a and b. This relationship can then be used to define weighting functions that allow us to assign importance to reports based on there generality. Such a weighting function could be defined as any function $\omega : g\text{-}Reports \to [0, 1]$ such that: (i) if $gr_1 > gr_2$, then $\omega(gr_1) > \omega(gr_2)$; and (ii) if $gr_1 \sim gr_2$, then $\omega(gr_1) = \omega(gr_2)$. For example, one possible weighting function is defined as $\omega(gr) = 2^{-rank(gr, >)+1}$.

6 Related Work

The study of preferences has been carried out in many disciplines; in computer science, the developments that are most relevant to our work is in the incorporation of preferences into query answering mechanisms. To date (and to our knowledge), the state of the art in this respect is centered around relational databases and, recently, in ontological languages for the Semantic Web [12, 11]. The seminal work in preference-based query answering was that of [10], in which the authors extend the SQL language to incorporate user preferences. The preference formula formalism was introduced in [7] as a way to embed a user's preferences into SQL. An important development in this line of research is the well-known *skyline* operator, which was first introduced in [3]. A recent survey of preference-based query answering formalisms is provided in [14]. Studies of preferences related to our approach have also been done in classical logic programming [8, 9] as well as answer set programming frameworks [4].

The present work can be considered as a further development of the PrefDatalog+/− framework presented in [12], where we develop algorithms to answer skyline queries, and their generalization to k-rank queries, over classical Datalog+/− ontologies. The main difference between PrefDatalog+/− and the work presented here is that PrefDatalog+/− assumes that a model of the user's preferences are given at the time the query is issued. Here, we make no such assumption; instead, we assume that the user only provides some very basic information regarding their preferences over certain features, and that they have access to a set of reports provided by other users in the past. In a sense, this approach is akin to building an ad hoc model *on the fly* at query time and using it to provide a ranked list of results. Compared to [11], in which we consider a related model where several observers provide reports and use the lineage of queries to propagate reports to derived atoms, here we focus on different methods for computing relevance and also additionally incorporate trust into the ranking.

Finally, this work is closely connected to the study and use of provenance in information systems and, in particular, the Semantic Web and social media [13, 1]. Provenance information describes the history of data in its life cycle. Research in provenance distinguishes between *data* and *workflow* provenance; the former explores the data flow within (typically, database) applications in a fine-grained way, while the latter is coarse-grained and does not consider the flow of data within the involved applications. In this work, we consider information provenance via the registers, which however does not fit into the *why*, *how*, and *where* provenance framework typically considered in data provenance research. We take into account (in a fine-grained way) where evaluations and reports within a social media system are coming from (i.e., information about who has issued the report, and what their preferences were) and use this information to allow users to make informed provenance-based decisions.

7 Summary and Outlook

In this paper, we have studied the problem of preference-based query answering in Datalog+/– ontologies under the assumption that the user's preferences are informed by a set of subjective reports representing opinions of others—such reports model the kind of information found, e.g., in online reviews of products, places, and services. We have first introduced a basic approach, in which reports are assigned to ground atoms. We have proposed two ranking algorithms using trust and relevance functions in order to model the different impact that reports should have on the user-specific ranking by taking into account the differences and similarities between the user's preferences over basic features and those of the person writing the report, as well as the person's self-reported characteristics (such as age, gender, etc.). As a generalization, we have then extended reports to apply to entire sets of atoms, so that they can model more general opinions. Apart from the naive approach of simply replicating the general report for each individual atom that it pertains to, we have proposed a way to use the information in the knowledge base to assign greater weights to more specific reports.

Much work remains to be done in this line of research, for instance, exploring conditions over the trust and relevance functions to allow pruning of reports, applying more sophisticated techniques to judging the impact of generalized reports, and the application of existing techniques to allow the obtainment of reports from the actual information available in reviews on the Web. We also plan to implement our algorithms and evaluate them over synthetic and real-world data. Finally, another topic for future research is to formally investigate the relationship between well-known data provenance frameworks and the preference-based provenance framework presented here.

Acknowledgments. This work was supported by the UK EPSRC grants EP/J008346/1, EP/L012138/1, and EP/M025268/1, the EU (FP7/2007-2013) Marie-Curie Intra-European Fellowship PRODIMA, the ERC grant 246858, a Yahoo! Research Fellowship, and funds from Universidad Nacional del Sur and CONICET, Argentina.

References

1. G. Barbier, Z. Feng, P. Gundecha, and H. Liu. *Provenance Data in Social Media*. Morgan and Claypool, 2013.
2. C. Beeri and M. Y. Vardi. The implication problem for data dependencies. In *Proc. of ICALP*, pages 73–85, 1981.
3. S. Börzsönyi, D. Kossmann, and K. Stocker. The skyline operator. In *Proc. of ICDE*, pages 421–430, 2001.
4. G. Brewka. Preferences, contexts and answer sets. In *Proc. of ICLP*, page 22, 2007.
5. A. Calì, G. Gottlob, and M. Kifer. Taming the infinite chase: Query answering under expressive relational constraints. *J. Artif. Intell. Res.*, 48:115–174, 2013.
6. A. Calì, G. Gottlob, and T. Lukasiewicz. A general Datalog-based framework for tractable query answering over ontologies. *J. Web Sem.*, 14:57–83, 2012.

7. J. Chomicki. Preference formulas in relational queries. *ACM Trans. Database Syst.*, 28(4):427–466, 2003.
8. K. Govindarajan, B. Jayaraman, and S. Mantha. Preference logic programming. In *Proc. of ICLP*, pages 731–745, 1995.
9. K. Govindarajan, B. Jayaraman, and S. Mantha. Preference queries in deductive databases. *New Generat. Comput.*, 19(1):57–86, 2001.
10. M. Lacroix and P. Lavency. Preferences: Putting more knowledge into queries. In *Proc. of VLDB*, pages 217–225, 1987.
11. T. Lukasiewicz, M. V. Martinez, C. Molinaro, L. Predoiu, and G. I. Simari. Answering ontological ranking queries based on subjective reports. In *Proc. of SUM*, 2014.
12. T. Lukasiewicz, M. V. Martinez, and G. I. Simari. Preference-based query answering in Datalog+/− ontologies. In *Proc. of IJCAI*, pages 1017–1023, 2013.
13. L. Moreau. The foundations for provenance on the Web. *Found. Trends Web Sci.*, 2(2/3):99–241, 2010.
14. K. Stefanidis, G. Koutrika, and E. Pitoura. A survey on representation, composition and application of preferences in database systems. *ACM Trans. Database Syst.*, 36(3):19:1–19:45, 2011.

Part IV
Belief Revision

Identity Merging and Identity Revision in Talmudic Logic: An Outline Paper

Michael Abraham[1], Israel Belfer[2], Uri Schild[3], Dov Gabbay[4]

Abstract. Suppose we are given a monadic theory **T** about the constants x and y. So **T** is built up in classical logic from monadic predicates $\{P_1, P_2, ...\}$ and the classical connectives and the quantifiers and possibly the equality symbol $=$. For example the theory **T** may have in it $P(x)$ and $\neg P(y)$. We now add to the theory the revision input $x = y$. The new theory may be inconsistent. We need a belief revision mechanism to revise **T** so that it is consistent with the input. This is a very specific form of input and belief revision, of the form which we are calling "identity merging". We present in outline how the Talmud deals with this type of revision.

1 Background and orientation

This paper makes two statements about AGM revision Theory:

1. AGM deals with abstract revision, where the theories **T** we deal with are purely logical and symbolic and without any accompanying interpretation in an application area. So the revision postulates and machinery we use are mathematical and have no guidance from an application area. In practice, theories **T** are formulated with an area in mind and when they need to be revised, we can fall back on the application area to guide us on how to revise them
2. We illustrate our point by looking at Identity Merging theories in Talmudic Logic and examining how Talmudic logic does the revision in this case. Talmudic logic is a typical case of a hierarchy of theories where higher nodes in the hierarchy guide us on how to revise the lower nodes.

Let us start. Suppose we are given a monadic theory **T** about the free variables x and y. If x and y are never quantified upon in **T** they can also be viewed as constants x and y. So **T** is built up in classical logic from monadic predicates $\{P_1, P_2, \ldots\}$ and the classical connectives $\{\neg, \wedge, \vee, \rightarrow\}$ and the quantifiers $\forall, \exists$ and possibly the equality symbol $=$.

For example the theory **T** may have in it $P(x)$ and $\neg P(y)$.

We now add to the theory the revision input $x = y$. The new theory is $\mathbf{T}_{x=y}$, $\mathbf{T} \cup \{x = y\}$ is inconsistent. We need a belief revision mechanism to revise **T** so that it is consistent with the input. We can of course use one of the many

---

[1] Bar-Ilan University, Israel
[2] Bar-Ilan University, Israel
[3] Bar-Ilan University, Israel
[4] Bar-Ilan University, Israel, King's College London, University of Luxembourg

AGM approaches and algorithms for this case, but these are too general and we need a more specialised tailored approach for our case. This is a very specific form of input and belief revision, of the form which we are calling "identity merging". We have knowledge about two distinct individuals (so we believe) and we discover that they are the same individuals. Now we have to reconcile what we know. Another very common case is where two conflicting bodies of laws apply to the same individual case and we need to decide how to proceed. Such cases require specialised procedures possibly tailored for each application area (case study). So formally we have a theory $\mathbf{T}$ and two distinct constants or variables x and y and we add the input $x = y$. We need not necessarily deal with a language with identity. If we do not have "=" in the language, we can still take $\mathbf{T}$ and take a new variable z and substitute in $\mathbf{T}$ the variable z for every free occurrence of x or of y. We will get a new theory which we can denote by $\mathbf{T}(x = y = z)$ which is inconsistent (containing $P(z)$ and $\neg P(z)$) and in need for revision. Note on the notation side, we can regard x and y either as constants or as free variables, this does not matter as long as we do not apply any quantifiers to them. So in the sequel we might talk about constants a or b or variables x and y. The choice depends on stylistic reasons.

We have three options here for revision.

1. Use the well known AGM machinery, [1, 2], which is this case means we choose one of each atomic contradicting pair $\{P(z), \neg P(z)\}$;
2. Use some new approach taking advantage of the specific form of the revision problem.
3. Use the Talmudic Logic approach;

Let us give some examples before we continue.

Example 1. Consider a university sysem with a Rector x and Head of Department of Informatics y. The university has regulations which say among others that:

1. The Rector can offer a position to a candidate and this is legally binding.
2. A Head of Department can offer a position to a candidate (in his department) but it is not legally binding, but is subject to approval by the Rector. The Head of Department should use a standard letter form which makes this clear.

Suppose now that there is a big fight between the Head of Department and his professors and the Head resigns and there is no agreement about a successor. Someone has to run the day-to-day matters of the Department, and so the Rector becomes acting Head of this Department for the time being. The Rector in his capacity as Head, offers a position in the Department to a candidate c. The standard letter which one sends in such a case says that this offer still requires the approval of the Rector but that the Department and its Head are confident that the Rector will approve the offer.

In this case the Head, who writes the letter as Head, is also the Rector, who needs to approve the appointment. The question is:

Is this letter binding or not?[5]

We have:

Rector writes → binding
Heads writes → ¬ binding

If we revise by the input Rector = Head, do we take binding or ¬ binding? Commonsense says that this is a binding offer.

Example 2. This is a real example recently discussed in the American press. It relates to the Boston terrorist bombing.[6] The terrorists were American citizens and so there were two options:

1. Viewed as terrorists, send them to military trial or even to Guantanamo Bay detention camp.
2. Treat them as American citizens and send them through the American legal system.

In principle what is happening here is that we have two bodies of laws and regulations:

$\mathbf{T}_1(a)$ = Rules for a, a typical terrorist
$\mathbf{T}_2(b)$ = Rules for b, a typical US citizen.

The theory is $\mathbf{T} = \mathbf{T}_1(a) \cup \mathbf{T}_2(b)$. The input, forced upon us by the real world, is $a = b$.

The aim of this paper is to formalise and introduce the Talmudic approach. The approach is general and can be used in AI for this case, as an alternative methodology to AGM.

The AGM approach would simply take out from $\mathbf{T}$ one of $\{P(z), \neg P(z)\}$ and restore consistency (assuming P is atomic).

The Talmudic approach will do something different. To introduce it, however, we begin with describing an intermediate non-monotonic approach which is not the Talmudic one, but has an independent interest and would lead into the Talmudic approach.

The non-monotonic approach (ANM vs. AGM) says that the language of $\mathbf{T}$ (i.e. $P_1, P_2, P_3, \ldots$) is only a surface language M and the fact that $\mathbf{T}$ contains $P(x)$ and $\neg P(y)$ stems from deeper level non-monotonic considerations in a deeper language $\mathbb{L}$. When we revise with $x = y = z$, we have to go to the deep level non-monotonic theory governing P, x and y and see what happens

---

[5] This actually happened to D. Gabbay in 1972. The perceptive reader might wonder why the Rector was ambiguous in his letter? Well, it may have been deliberate or he may have wanted to follow clear procedural lines and first write as Head and follow it up as Rector. There is more to it than that. Can the recipient of the letter assume that for all practical purposes he/she actually got the job or is there still a practical possibility that the Rector would say "I approve the appointment as Head of Department but I do not approve it as Rector"?

[6] https://en.wikipedia.org/wiki/Boston_Marathon_bombing, http://www.britannica.com/event/Boston-Marathon-bombing-of-2013.

there and then decide whether to contract $P(x)$ or to contract $\neg P(y)$. Thus the non-monotonic approach is a refinement of the AGM approach, where the choice of which of the contradicting pair $\{P(z), \neg P(z)\}$ to take out is made using the extra non-monotonic machinery available in the system. This is best understood when actually defined. Let us propose an ANM model.

Definition 1.

1. Let $\mathbb{M}$ be the monadic classical predicate language with unary predicates $\{P_1, P_2, \ldots\}$ and variables and constants $\{x, y, z, c_1, c_2, \ldots\}$. We say that $\{P_i, c_j\}$ are the predicates and constants of $\mathbb{M}$. Let $\mathbb{L}$ be an expansion of $\mathbb{M}$ with additional predicates and constants $\{A, B, \ldots\}$ and and $\{d_1, d_2, \ldots\}$.
2. With each constant c and predicate P of $\mathbb{M}$ we associate a family of predicates and constants from $\mathbb{L}$ (which include P and c). Let us use the notation $\mathbb{F}(P, c)$ for this family. For example, let

$$\mathbb{F}(P, a) = \{A\} \cup \{P, a\}$$
$$\mathbb{F}(P, b) = \{B\} \cup \{P, b\}.$$

(We will not repeat "$\{P, c\}$" any more.)

3. Assume that we have a non-monotonic consequence $\Vdash$ governing the language $\mathbb{L}$ and a theory Δ of facts for the new predicates $\{A, B \ldots\}$ of $\mathbb{L}$ which contains $\mathbb{M}$.
4. Assume that our surface theory $\mathbf{T}$ is the result of Δ. Namely

$$P(x) \in \mathbf{T}_{\Delta, \mathbb{F}} \text{ iff } \Delta \restriction \mathbb{F}(P, x) \Vdash P(x).$$

5. We say that $\mathbf{T} = \mathbf{T}_{\Delta, \mathbb{F}}$ is derived from Δ using $\Vdash$.

Example 3. Consider the surface predicates and constants of $\mathbf{T}$ to be P, a, b.
Let
$$\mathbb{F}(P, a) = \{A\}$$
$$\mathbb{F}(P, b) = \{B\}$$

Assume our non-monotonic logic for $\mathbb{L}$ relies on more specificity and that we have a theory Δ with the following rules:

1. $A(a) \to P(a)$
2. $B(b) \to \neg P(b)$
3. $A(b) \land B(b) \to P(b)$
4. $A(a)$
5. $B(b)$

Our theory $\mathbf{T}$ will contain therefore $P(a)$ and $\neg P(b)$. This is because showing $P(a)$ we can use only clauses 1. and 4. and showing $P(b)$ we can use only clauses 2. and 5.

Now let us see what happens if we add the input $a = b$. This changes the language we consider from the separate $\mathbb{F}(P, a)$ and $\mathbb{F}(P, b)$ into the joint $\mathbb{F}' = \mathbb{F}(P, a) \cup \mathbb{F}(P, b)$. Now the clauses to consider from Δ are 1. to 5.

But now, because of more specificity

$$\Delta \restriction \mathbb{F}' \Vdash P(b)$$

and so we revise **T** in view of the input $a = b$ by contracting $\neg P(b)$. Thus we see that whereas ordinary AGM theory allowed for arbitrary choice in either contracting $P(b)$ or contracting $\neg P(b)$, the non-monotonic background theory Δ, recommended contracting $\neg P(b)$. Intuitively we asked ourselves (and asked Δ) where do $\mathbf{T} \vdash P(a)$ and $\mathbf{T} \vdash \neg P(b)$ come from and we made a decision based on the answer.

We realise that perhaps the non-monotonic system may not resolve the issue. We can rely on another level (i.e. another Δ' related to Δ in a similar way to the relation of Δ to **T**) and language to resolve the issue. The details are not so important as the overall approach.

Remark 1. The perceptive reader might ask: "Where does the richer language theory Δ come from? Isn't it a bit artificial, to introduce it just to solve the problem, like a special distance/choice function for AGM revision?

I see how it works and where it comes from in the examples, but in the general case?".

Our answer to this is that indeed laws and regulations come from practical situations where in the background there are undesirable cases to be avoided. So for each specific **T** tailored for a specific application area there will be a corresponding Δ. For a general theory we must stipulate and study a general recursive hierarchy of $\mathbf{T}_n$, where $\mathbf{T}_{n+1}$ acts as the "Δ" of $\mathbf{T}_n$.

Let us now work towards giving a complete formal presentation of the above ideas.

Definition 2. *Let $\mathbb{L}$ be a language containing $\neg$ and let $(\Delta, \Vdash)$ be a non-monotonic theory in $\mathbb{L}$. This means that Δ is a set of formulas of $\mathbb{L}$ and $\Vdash$ is a consequence relation of the form*

$$A_1, \ldots, A_n \Vdash B$$

where A_i, B are formulas of $\mathbb{L}$ and $\Vdash$ satisfies the following conditions:

1. *Reflexivity*
$$A_1, \ldots, A_n \Vdash B, \text{ if } B \in \{A_i\}$$

2. *Cut*
$$A_1, \ldots, A_n X \Vdash B$$

and
$$A_1, \ldots, A_n \Vdash X$$

imply
$$A_1, \ldots, A_n \Vdash B$$

3. *Restricted monotonicity*
$$A_1, \ldots, A_n \Vdash B$$

and
$$A_1, \ldots, A_n \Vdash C$$

imply
$$A_1, \ldots, A_n, B \Vdash C.$$

4. Let $\theta', \theta \subseteq \Delta$ be two finite subsets of Δ.
   We may have $\theta \Vdash B$ but $\theta \cup \theta' \Vdash \neg B$ or alternatively $\theta \Vdash \neg B$ but $\theta \cup \theta' \Vdash B$.
5. $(\Delta, \Vdash)$ is said to be consistent iff for no $B, \theta \subseteq \Delta, \theta$ finite, we have $\theta \Vdash B$ and $\theta \Vdash \neg B$.
6. θ is said to decide a wff B if we have either $\theta \Vdash B$ or $\theta \Vdash \neg B$ (as opposed to neither).
7. θ is said to be minimal theory deciding B, if θ decides B and no $\theta' \subsetneq \theta$ decides B.
8. We say that $\Delta \Vdash B$ iff there is a $\theta \subseteq \Delta$, θ finite, which decides B and for every minimal θ which decides B we have $\theta \Vdash B$.

Definition 3. *1. Let* **T** *be a classical consistent set of wff, (considered as a theory) in a language* M. *Let* Δ *be a consistent non-monotonic set of wffs (considered a theory) in a richer langauge* L $\supseteq$ M. *Let* $\Vdash$ *be its consequence relation. Let* P *be a unary atomic predicate of* M *and let* $a_i, i = 1, \ldots, k$ *be distinct constants of* M. *Let* $\mathbb{F}$ *be a function giving for each* $\alpha = \{P, a_1, \ldots, a_k\}$ *a sublanguge* $\mathbb{F}(\alpha)$ *of* L. *We assume that* $\alpha \subseteq \mathbb{F}(\alpha)$. *Note that we may have*

$$\Delta \upharpoonright \mathbb{F}(\alpha) \Vdash \pm P(a)$$

but

$$\Delta \upharpoonright \mathbb{F}(\beta) \Vdash \mp P(a),$$

for $\beta \supsetneq \alpha$. *This can happen because* $\Vdash$ *is non-monotonic.*
2. Let **T** *and* Δ *be as in (1) above. We say that* Δ *supports* **T** *if the follwing (*) holds for each unary P and constant a:*
(*) **T** $\vdash P(a)$ *iff* $\Delta \upharpoonright \mathbb{F}(P, a) \Vdash P(a)$.
3. Let a, b be two distinct constants of **T**. *Denote by* $\mathbf{T}_{a/b}$ *the theory obtained from* **T** *by replacing every occurrence of a by the constant b.*
4. Similarly, let P, Q be two unary predicates, we let $\mathbf{T}_{P/Q}$ *be the theory obtained by replacing every occurence of P by Q.*

Remark 2. Let **T** be a monadic theory with monadic predicates $\{P_1, \ldots, P_k\}$ and constants $\{c_1, \ldots, c_m\}$. Let **m** be a classical model of the language with $\{P_j, c_i, \neg, \wedge, \vee, \rightarrow, \forall, \exists, =\}$. For this language, any model **m** has to say the following

1. For each c_i and P_j **m** has to say whether $P_j(c_i)$ or $\neg P_j(c_i)$ holds.
2. For each vector ε of $\{0, 1\}$ values of length k, let

$$\alpha_\varepsilon(x) = \bigwedge_{j=1}^{k} P_j^{\varepsilon_j}(x)$$

where $P_j^1(x) = P_j(x)$ and $P_j^0(x) = \neg P_j(x)$, and x is a variable. Then the model **m** has to say whether $\exists x \alpha_\varepsilon(x)$ holds for each ε and, if equality is in the language, **m** has to say how many different such elements exist, $0, 1, 2 \ldots$ or infinity, i.e. for each n, **m** must say whether $\exists x_1, \ldots, x_n \bigwedge_{i \neq j} x_i \neq x_j$.

If there is no equality in the language then a model **m** for the language can be characterised by a wff $\varphi_{\mathbf{m}}$ of the form

$$\varphi_{\mathbf{m}} = \bigwedge_{\varepsilon} \pm \exists x \alpha_\varepsilon(x) \wedge \bigwedge_{i,\varepsilon} \pm \alpha_\varepsilon(c_i).$$

Let us assume we do not have equality.

Then we can assume that any **T** has only a finite number of finite models.

Remark 3. Let **T** be a monadic theory without equality as in Remark 2. Let **m** be a model for the language of the form $\varphi_{\mathbf{m}}$ as in Remark 2.

We now investigate the effect of the additional revision/input information that c_1 and c_2 are the same (i.e. $c_1 = c_2$). We ask whether $\varphi_{\mathbf{m}}$ is still consistent under the substituation of c_2 for c_1 (or $c_1 = c_2$)?

This depends on what conjuncts appear in $\varphi_{\mathbf{m}}$. The critical ones to be watched are triples $\varepsilon, \varepsilon', P$ such that

$$\varphi_{\mathbf{m}} \vdash \alpha_\varepsilon(c_1) \wedge \alpha_{\varepsilon'}(c_2)$$

and such that

$$\alpha_\varepsilon(c_1) \vdash P(c_1) \text{ and } \alpha_{\varepsilon'}(c_2) \vdash \neg P(c_2).$$

In other words, the problem is that the model **m** says for some set of unary predicates $\{P_{i_1}, \ldots, P_{i_n}\}$ the opposing pair $\{\pm P_{i_r}(c_1) \text{ and } \mp P_{i_r}(c_2)\}$.

Since we are claiming $c_1 = c_2$, we need to choose only one of them, if we want to maintain consistency.

We have a similar problem if we input equality of two predicates, say $P_1 = P_2$. There may be some $c_{j_1}, \ldots, c_{j_n}$, which the model says

$$\pm P_1(c_{j_r}) \text{ and } \mp P_2(c_{j_r})$$

again, we have opposing pairs and again we need to choose one of them if we want to maintain consistency.

Our theory of identity merging will tell us how to choose one from each opposing pair and thus maintain consistency. Our identity merging theory is a refinement of AGM for this particular case. AGM does not care how we choose.

Definition 4. *Let* **T** *be a complete and consistent monadic theory with constants* $\{c_i\}$ *and unary predicates* $\{P_j\}$, *and let* $(\Delta, \Vdash)$ *be a supporting theory for* **T** *as in Definition 3. Let* a, b *be two distinct constants and consider* $\mathbf{T}_{a/b}$ *and assume that it is inconsistent. The merge revision of* $\mathbf{T}_{a/b}$ *is performed as follows.*

Since **T** *is complete,* $\mathbf{T}_{a/b}$ *being inconsistent means that either* $\mathbf{T} \vdash P(a) \wedge \neg P(b)$ *or that* $\mathbf{T} \vdash \neg P(a) \wedge P(b)$ *for some predicates* P.

Assume without loss of generality that the former holds. Then we have that

$$\Delta \upharpoonright \mathbb{F}(\{P, a\}) \Vdash P(a)$$
$$\Delta \upharpoonright \mathbb{F}(\{P, b\}) \Vdash \neg P(b)$$

Consider $\theta = \Delta \upharpoonright \mathbb{F}(\{P, a, b\})$ *we may have* $\theta \Vdash P(a)$ *or* $\theta \Vdash \neg P(a)$ *or neither (but not both!). Similarly we have for the case of* $P(b)$. *We may now have that*

θ proves $P(a)$ and θ does not prove $\neg P(b)$ or θ proves $\neg P(a)$ and does not prove $P(b)$ or θ proves $P(b)$ and does not prove $\neg P(a)$ or θ proves $\neg P(b)$ and does not prove $P(a)$. In each of these cases we know how to revise.

If we still have that θ proves $P(a)$ and $\neg P(b)$ or θ proves $\neg P(a)$ and $P(b)$ or that θ proves nothing, then we revise arbitrarily.

Following the discussion of Remark 3, we can make a choice of whether to take $+P(a)$ or $\neg P(a)$ for our revised model. If $\Vdash$ does not tell us which one to take we can make an arbitrary choice. The algorithm is as follows:

1. If $\theta \Vdash P(a)$, then delete all occurrences of $\pm P(b)$ from $\varphi_\mathbf{m}$ to get $\varphi'_\mathbf{m}$.
    If $\theta \Vdash \neg P(b)$ then delete all occurences of $\pm P(a)$ from $\varphi_\mathbf{m}$ to get $\varphi''_\mathbf{m}$.
    Otherwise delete $\pm P(b)$.
    Since we assume that $a = b$, we have that $\varphi'_\mathbf{m} = \varphi''_\mathbf{m}$ and this is our revised model. The revised theory $\mathbf{T}_{a/b}$ is the theory of this model.

Similar considerations will apply to $\mathbf{T}_{P/Q}$.

This completes our discussion of the non-monotonic identity merging method. This method is, however, not how the Talmud handles the case.

The above discussion of the obvious solution now has prepared us for the introduction of the Talmudic approach, as well as providing us with the means of comparison.

A theory can be revised by introducing new items of data which affect what it can prove. A theory can be revised also by cancelling or restricting the proof rules it can use. The latter method is used in resolving logical paradoxes. The data is fixed and leads to a paradox (inconsistency or unintuitive results). So one blocks some of the proofs and thus resolves the paradox. The Talmud revises by using a hierarchy of rules cancellations as we explain in the next section.

2 The Talmudic approach

Let us look at a well known example (x is universally quantified):

1. $\text{Bird}(x) \to \text{Fly}(x)$
2. $\text{Penguin}(x) \to \text{Bird}(x)$
3. $\text{Penguin}(x) \to \neg \text{Fly}(x)$
4. $\text{Penguin}(a)$

We say that in viewing the above data, since Penguin is a more specific bird, then it wins and so we deduce $\neg \text{Fly}(a)$.

Let us look now at the following data:

5. Aeroplane 747 Flight BA101 $\to$ Land at Heathrow
    $A \to L$.
6. Aeroplane 747 Flight BA101 and Bad Weather $\to \neg$ Land at Heathrow
    $A \wedge W \to \neg L$
7. Aeroplane 747 Flight BA101 and Bad Weather and Short on Fuel $\to$ Land at Heathrow
    $A \wedge W \wedge F \to L$

We may look at this again using the principle that the more specific assumptions (i.e. the antecedent of the rule contains more conjuncts than the other rule) win. So if we have only the information that an Aeroplane 747 Flight BA101 wants to land, we conclude that it can land. If we also add the conjunct that the weather is bad then it cannot land and if we even further add the conjunct that it is also short of fuel then it can land.

The Talmud looks at this differently as in Figure 1. W and F are meta-level principles. In the Figure ordinary arrow $\to$ means support and double arrow $\twoheadrightarrow$ means attack.

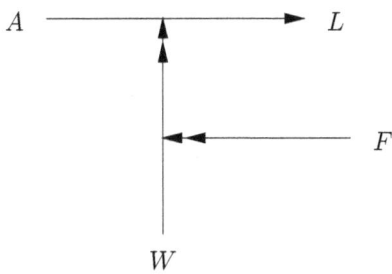

Fig. 1.

The basic principle is $A \to L$. The weather conditions involve a meta-level principle which cancel the arrow leading from A to L.[7]

The fuel shortage is involved in another meta-level principle which cancels the cancellation. So we are not dealing here with more specific knowledge but we are dealing with levels of meta-knowledge and a calculus of cancellations. The appropriate modelling of this is higher level attack and support (argumentation) networks.

So the Talmud uses a calculus of cancellations to resolve identity merging, as opposed to our previous proposal of non-monotonic support.

Let us give some examples.

Example 4. This example is really from Talmudic logic, recast in everyday modern situation.

1. The story runs as follows:
   We have a duty to maintain our homes. We also have the instinct to save money. We believe in professional people doing jobs for us, but if we can do it properly ourselves, then we do it ourselves, and not call the expert and thus save money.
   So, if the kitchen sink is blocked, we do not call a plumber to do the job but do it ourselves and save money (a plumber home visit costs about $50 just to come, independent of the job he does).

---

[7] Think of it as a rule of wisdom based on experience. "Just do not land in bad weather". Another such rule is "If you are short of fuel land as soon as you can".

If the problem is more serious, say a blocked toilet, then better call a plumber and not take the risk of doing the job yourself. This case does need an expert!

We can write these rules in non-monotonic logic as follows
(a) x is blocked $\to$ repair x yourself
(b) x is blocked $\wedge x$ is a serious job $\to$ get John the plumber to repair x
(c) sink is blocked
(d) toilet is blocked
(e) repairing the toilet is a serious job, but not the sink.

The problem with the above is that it implies that we call John the plumber and he repairs the toilet while we repair the sink. Common sense dictates that since the plumber is available he should repair the sink as well! We could add a new clause (f) to help:

f. x is blocked $\wedge$ John the plumber repairs y $\to$ get John the plumber to repair x.

Clause (f) says that if x is blocked and there is any y which John the plumber repairs y[8] then John the plumber to repair x. The format of clause (f) is not the usual monadic one, and does not make the information on x more specific. We can artificially fix this by adding a dummy universal predicate $U(x, y)$ which relates any two elements (something like $((x = y) \vee \neg(x = y)))$ and write (f*)

f*. x is blocked $\bigwedge$ (John the plumber repairs $y \bigwedge U(x, y)) \to$ get John the plumber to repair x.

Now (f*) is more specific than (a) on account of the additional predicate $V(x) =$ (John the plumber repairs $y \bigwedge U(x, y))$ This is clearly a fiddle and it departs from the intuitive understanding of what is going on, which is clearly two meta-principles, namely save money but not at the expense of needed expertise!

Let us see how the Talmudic calculus of cancellations overcomes this problem.

Figures 2 and 3 describe these rules. The description is intuitive and not formal. The meaning of the nodes and arrows can be read intuitively from the figures.

The question arises what to do if both the sink and the toilet are blocked? If we just take the union of the two figures, (i.e. union of Figure 2 and Figure 3) i.e. update that the two plumbers a and b are equal, we will get that we call a plumber, the plumber does the toilet and at the same time we do the sink ourselves. It is more reasonable, however, since the plumber is already coming (and the $50 call fee is to be paid anyway) to have the plumber do the sink as well.

Thus the "merging" of the two cases, i.e. merging of the two figures for the case that both the sink and the toilet are blocked is just a union of the graphs of the two figures. We will get Figure 4.

2. We now explain our notation.
   (a) $x, y, \ldots$ denote objects like $x =$ kitchen sink, $y =$ toilet.

---

[8] We tacitly assume here that they are all in the same, say, apartment building to be considered the same "call" by the plumber.

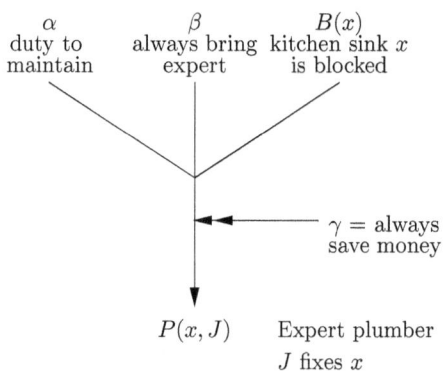

Fig. 2.

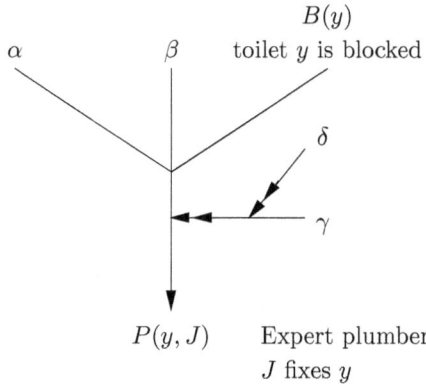

Fig. 3.

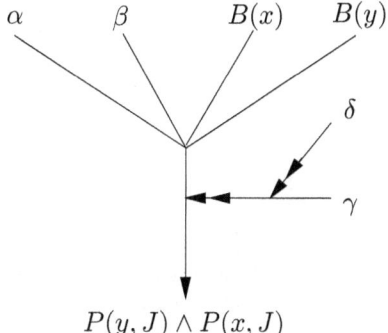

Fig. 4.

$$B(x) \xrightarrow{\pi} P(x,a)$$

Fig. 5.

π and $B(x) \to P(x,a)$

Fig. 6. Alternative notation to Figure 5

(b) B, P denote predicates which when applied to objects give states:
$B(x) =$ kitchen sink is blocked, $B(y) =$ toilet is blocked.
$P(x, z) =$ kitchen sink is repaired by plumber z, $P(y, z) =$ toilet is repaired by plumber z.

(c) α, β, γ are policies. For example:
$\alpha =$ policy to maintain your house
$\beta =$ policy to always use experts
$\gamma =$ policy to always save money
$\delta =$ policy to not take any risk for heavy maintenance jobs, if possible.

(d) A word about our notation: We denote the transition from one state to another by an arrow.
Figure 5 shows such notation. The π annotates the arrow. This means that because of policy π we take action and move from $B(x)$ to $P(x,a)$. It may be that several policies come together and are involved in motivating some action, or it may be the case that some policies may cancel or overrule other policies. So we allow for alternative notation which we can use as well, when there are lots of policies to denote.
Figure 5 can be equivalently presented as Figure 6 or as Figure 7.

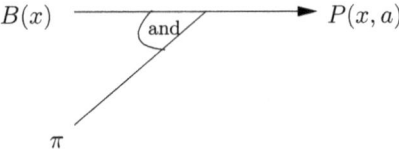

Fig. 7. Alternative notation to Figure 5

(e) Cancellation is done by double arrow.
Figure 8 shows some cancellations from some policies. It has no meaning, just a sample technical figure illustrating the notation.
  i. π_1 and π_2 support together the move from $B(x)$ to $P(x,a)$.
  ii. π_3 cancels the support of π_1 but allows the action to go forward on the basis of π_2.

iii. π_5 cancels the move to $P(x,a)$ no matter what, but also does not think that the support of π_4 to $Q(y)$ is a reason to cancel $\pi_2 \to P(x,a)$.

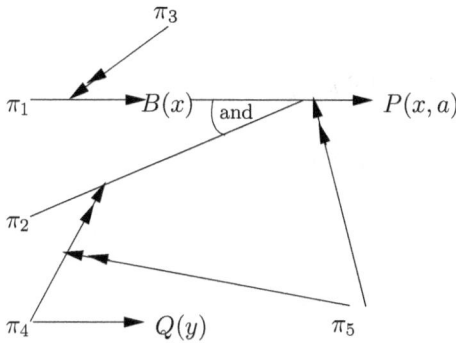

Fig. 8.

Remark 4. The perceptive reader might think that the model of arrow cancellations as presented in Figures 2, 3 and 4 is nothing special and is just a notational variant of defeasible logic with specificity. Thus using the notation of Example 4 we can write a defeasible database Δ with the following universal formulas clauses, with w, z universal variables.

1. $B(w) \wedge \alpha \wedge \beta \wedge \gamma \to \neg P(w,z)$
2. $B(w) \wedge \alpha \wedge \beta \wedge \gamma \wedge \delta(w) \to P(w,z)$

If we instantiate (1) with $w = x, z = a$ and (2) with $w = y, z = b$ we get

1*. $B(x) \wedge \alpha \wedge \beta \wedge \gamma \to \neg P(x,a)$.
   This is Figure 2 with $x =$ sink and $z =$ plumber a.
2*. $B(y) \wedge \alpha \wedge \beta \wedge \gamma \wedge \delta(y) \to P(y,b)$
   This is Figure 3 with $y =$ toilet and $z =$ plumber b.

If we put (1*) and (2*) together in the same database and add the input $a = b = e$, then the database is consistent and the same plumber e will repair the toilet but not the sink. Defeasible logic based on specificity cannot tell us that because we have (2*) with $P(y,e)$ plumber e, we reverse and defeat (1*) and conclude $P(x,e)$ as well.

However, if we use the figures with the cancellation arrows, it is easier to model this feature. Figure 9 sums it all up. This is a predicate argumentation network involving joint attacks and higher order attacks, see [3].

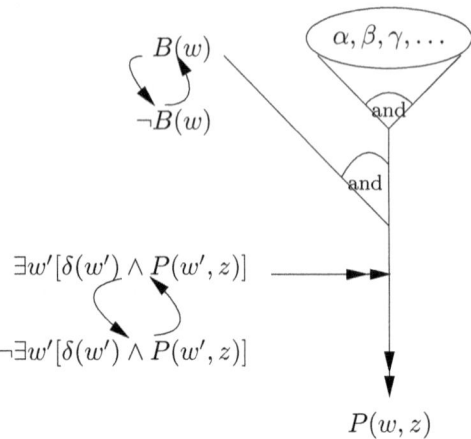

Fig. 9.

3 Conclusion

We presented an outline paper showing how Talmudic logic uses a calculus of cancellation to execute identity merging. In this conclusion section we want to impress upon the reader the schematic advantage of the calculus of cancellation.

Suppose we have two clauses

1. $\alpha \wedge A \rightarrow \exists x C(x)$
2. $\beta \wedge B \rightarrow \exists x \neg D(x)$.

We want to put (1) and (2) together in the same database and

(a) maintain consistency
(b) have the existential quantifiers pick up the same element.

The mechanisms we use are

(i) to take the specificity formulas out of the clauses and consider them as meta-principles, which are subject to being prioritised and apply to them the calculus of cancellations.

So we have

3. $\{\alpha, \beta\} : A \wedge B \rightarrow \exists x C(x) \wedge \exists x \neg D(x)$.

(ii) Convert
$$A \rightarrow \exists x C(x)$$
$$B \rightarrow \exists x \neg D(x)$$
into respective figures and turn (3) into

4. $\{\alpha, \beta\}$: union of Figures.

In the process of taking union of Figures we get that $\exists x$ chooses the same x.

Acknowledgement

The authors are grateful to the referees for their most valuable comments.

References

1. C. E. Alchourròn, P. Gärdenfors, and D. Makinson. On the logic of theory change: Partial meet contraction and revision functions. *Journal of Symbolic Logic*, 50:510–530, 1985.
2. http://en.wikipedia.org/wiki/Belief_revision
3. D. Gabbay. Theory of Semi-instantiation in Abstract Argumentation. To appear *Logica Universalis*. http://arxiv.org/abs/1504.07020

On the Syntactic Representation of Multiple Iterated Belief c-Revision

Salem Benferhat[1], Amen Ajroud[2]

Abstract. Belief revision is the process of revising epistemic states in the light of new information. In this paper epistemic states are represented in the framework of Spohn's Ordinal Conditional Functions (OCF). The input is a consistent set of propositional formulas issued from different and independent sources.

We focus on the so-called multiple iterated belief c-revision recently proposed by Kern-Isberner. We propose a syntactic representation of c-revision when epistemic states are compactly represented by weighted propositional knowledge bases, called OCF knowledge bases.

1 Introduction

Intelligent agents need to update their knowledge, or epistemic states, on the basis of new observations on their environment. This problem is known as the one of belief revision and is axiomatically characterized by the well known AGM postulates [2, 25, 24, 30].

There are at least three issues that need to be addressed when modeling a belief revision problem.

The first issue concerns the representation of initial epistemic states. An epistemic state should at least contain a set of accepted beliefs. It also contains some meta-information which is very useful for defining a meaningful belief revision operation. A simple representational format of an epistemic state is a closed set of propositional formulas, called a belief set. However, epistemic states have in general a complex structure. They can be represented by a total pre-order over a set of propositional interpretations [32], a probability distribution, a possibility distribution [20], an ordinal conditional function or simply an OCF function [47, 49, 3, 9], a set of conditionals [33–35, 5, 26], a partial pre-order over a set of interpretations [11, 48, 43].

The second issue concerns the representation of the input or the new information. The input can be a simple observation over a static world, a result of an action or even a result of an intervention (external action) that forces some variables to take specific values [45, 28]. The input can be simply a propositional formula, a set of propositional formulas, a partial or total pre-order on a set of interpretations, a set of uncertain and mutually exclusive formulas inducing a partition of a set of interpretations [31, 19], etc.

---

[1] CRIL - CNRS UMR 8188, Université d'Artois, Faculté des Sciences Jean Perrin. Rue Jean Souvraz, 62307 Lens, France, benferhat@cril.fr

[2] PRINCE - ISITCOM, Université de Sousse, Route principale n°1, 4011 Hammam-Sousse, Tunisia, amen.ajroud@isetso.rnu.tn|

The third issue concerns the definition of a revision operation where from an initial epistemic state and the input produces a new epistemic state. This new epistemic state should at least satisfy two requirements. Firstly, it should accept the input. Secondly, it should be as close as possible to the initial epistemic state. Depending on the representation of epistemic states and on a the nature of the input, a large number of revision operations (e.g. [10, 23, 29, 40, 44]) or contraction operations (e.g. [1, 13, 17, 18]) has been proposed in the literature. Other approaches that deal with inconsistency have also been proposed in (e.g. [39, 46, 8, 6]).

The framework considered in this paper for representing epistemic states is the one of ordinal conditional functions OCF [47]. We will analyze the so-called multiple iterated propositional c-revision recently proposed in [37]. An important feature of c-revision is that it can be iterated [37, 16, 41] since initial and revised epistemic states are both represented by OCF distributions. The iterated belief c-revision has as input a consistent set of propositional sentences S assumed to be provided by different and independent sources. The revised ordinal conditional distribution should accept all sentences of the input. Independence relations between information sources are represented by the fact that counter-models of S are ranked with respect to the number of falsified propositions in S. Namely, the more an interpretation falsifies formulas in S, the less it is a preferred interpretation.

Multiple iterated belief c-revision has been shown to satisfy all natural and rational properties given in [37]. However, c-revision is only defined at the semantic level (over the set of interpretations) which is impossible to be provided in practice if the set of variables is important. This paper addresses this issue by providing an equivalent characterization of c-revision defined on a compact representation of ordinal conditional functions over the set of interpretations. An OCF over a set of interpretations will be compactly represented by a set of weighted formulas called OCF knowledge bases.

The rest of this paper is organized as follows. Section 2 gives a refresher on OCF distributions and on their compact representations OCF knowledge bases. Section 3 presents the multiple iterated belief c-revision defined on OCF distributions. Section 4 shows how multiple iterated belief c-revision can be directly defined on OCF knowledge bases. Section 5 concludes the paper.

2   OCF-based representations of epistemic states

Let $\mathcal{L}$ denote a finite propositional language and Ω be the set of propositional interpretations. We will denote by ω an element of Ω. Greek letters φ, ψ, ... represent propositional formulas. $\models$ denotes a propositional logic satisfaction relation.

In belief revision, epistemic states represent a set of all available beliefs. There are different representations of epistemic states: an uncertainty (probability, possibility, etc) distributions, a total pre-order over Ω, a partial pre-order over Ω, etc. In this paper, we use ordinal conditional functions (OCF) to represent epistemic states [47–50].

An OCF distribution can be simply viewed as a function that assigns to each interpretation ω of Ω an integer denoted by $k(\omega)$. $k(\omega)$ represents the degree of surprise of having ω as being the real world. $k(\omega) = 0$ means that nothing prevents ω for being the real world. $k(\omega) = 1$ means that ω is somewhat surprising to be the real world. $k(\omega) = +\infty$ simply means that it is impossible for ω to be the real world.

Example 1: Let a and b be two propositional symbols. Table 1 gives an example of an epistemic state represented by an OCF distribution k:

ω	$k(\omega)$
ab	4
$\neg ab$	1
$a\neg b$	1
$\neg a\neg b$	0

Table 1. An example of an OCF distribution

From Table 1, the most normal state of world is the one where both a and b are false. A surprising world (with a degree of surprise 1) is the one where either a or b is true. A more surprising world (with a degree of surprise 4) is the one where both a and b are true.

From an OCF distribution k, one can induce a degree of surprise over formulas φ of $\mathcal{L}$, simply denoted by $k(\varphi)$ and defined by:

$$k(\varphi) = min\{k(\omega) : \omega \in \Omega, \omega \models \varphi\}. \qquad (1)$$

For example, from Table 1 we have $k(\neg a \vee \neg b) = min(k(\neg a\neg b), k(\neg ab), k(a\neg b)) = 0$ while $k(a \vee b) = 1$.

Given an OCF distribution, a set of accepted beliefs is a propositional formulas such that its models are those having minimal surprise degrees in k. In Example 1, the set of accepted beliefs is represented by the propositional formulas $\neg a \wedge \neg b$.

In practice, an OCF distribution k cannot be provided over a set of interpretations Ω (except if the number of propositional variables is small). A compact representation may be provided using for instance the concept of OCF networks [36, 12, 27, 14, 22] or the concept of weighted propositional knowledge bases.

In this paper, we only consider weighted propositional knowledge bases, simply called OCF knowledge bases and denoted by $\mathcal{K}$. An OCF knowledge base is a set of weighted formulas of the form $\mathcal{K} = \{(\varphi_i, \alpha_i) : i = 1,..,n\}$ where φ_i's are propositional formulas and α_i's are positive integers. The higher is the certainty degree α_i, the more certain is the formula φ_i. When $\alpha_i = +\infty$ this means that φ_i represents an integrity constraint that should absolutely be satisfied. Formulas with a certainty degree equal to '0' are not explicitly stated in $\mathcal{K}$. Weighted or prioritized knowledge bases have been intensively used in the literature for handling uncertainty (such as in a possibilistic logic framework [42, 4]) or for handling inconsistency.

Given an OCF knowledge base $\mathcal{K}$, one can induce a unique OCF distribution, denoted by $k_{\mathcal{K}}$ and defined by:

$\forall \omega \in \Omega,$

$$k_{\mathcal{K}}(\omega) = \begin{cases} 0 & if\ \forall(\varphi_i, \alpha_i) \in \mathcal{K},\ \omega \models \varphi_i \\ max\ \{\alpha_i : (\varphi_i, \alpha_i) \in \mathcal{K}, \omega \not\models \varphi_i\} & otherwise. \end{cases} \quad (2)$$

Namely, $k_{\mathcal{K}}(\omega)$ is associated with the highest certain formulas in $\mathcal{K}$ falsified by ω. Models of formulas in $\mathcal{K}$ are considered as the most normal interpretations (hence they have a surprise degree equal to 0). Clearly, the concepts of OCF distributions and OCF knowledge bases are very close to the concepts of possibility distributions and possibilistic knowledge bases used in a possibility theory framework [20], where instead of using a set of integers, the unit interval [0,1] is used.

Example 2: let $\mathcal{K} = \{(\neg a \vee \neg b, 4), (\neg a, 1), (\neg b, 1)\}$. This knowledge base means that we are somewhat certain that a and b are both false and we are even more confident if only one of them is false. One can easily check that applying Equation (2) to the knowledge base $\mathcal{K}$ will simply lead to the OCF distribution given in Table 1. For instance, $k_{\mathcal{K}}(a \wedge b) = max\ \{\alpha_i : (\varphi_i, \alpha_i) \in \mathcal{K}, a \wedge b \not\models \varphi_i\} = max\{4, 1, 1\} = 4$.

3 C-revision of OCF distributions

Several works have been proposed for revising OCF distributions. For instance, in [49, 50] a general form of changing OCF distributions, called transmutations [51], has been proposed. In [34, 35] a revision of OCF distributions with a set of conditionals has also been proposed.

In this section, we focus on a so-called multiple iterated belief c-revision proposed in [37] for revising an OCF distribution with a consistent set of propositional formulas $S = \{u_1, .., u_n\}$. In order to have a faithfull revision operation, each propositional formula u_i is associated with an integer β_i. These integers β_i's are not explicitly stated by the user, but they are implicitly constrained as it will be shown below. More precisely:

Definition 1: Let k be an OCF distribution. Let $S = \{u_1, .., u_n\}$ be a consistent finite set of propositional formulas. Then the propositional c-revision of k with S, denoted by $k * S$, is defined by:

$$\forall \omega \in \Omega, \qquad k * S(\omega) = k(\omega) - k(u_1 \wedge .. \wedge u_n) + \sum_{i=1,\ \omega \models \neg u_i}^{n} \beta_i, \quad (3)$$

where $(\beta_1, .., \beta_n)$ are positive integers satisfying:

$$\forall i, \qquad \beta_i > k(u_1 \wedge .. \wedge u_n) - \min_{\omega \models \neg u_i} \{k(\omega) + \sum_{j \neq i,\ \omega \models \neg u_j}^{n} \beta_j\} \quad (4)$$

The revision process given in Equation (3) first consists in shifting up each interpretation ω with the sum of weights β_i's of propositional formulas u_i that it falsifies. The expression "$-k(u_1 \wedge .. \wedge u_n)$" is a normalization term that guarantees that $min\{k * S(\omega) : \omega \in \Omega\}$ is equal to zero. Propositional formulas from S are assumed to be issued from independent sources. Hence, interpretations will be compared with respect to the number of falsified formulas. This is reflected by the expression "$\sum_{i=1,\ \omega \models \neg u_i}^{n} \beta_i$" in the definition of the resulted revised OCF $k * S$.

Example 3: Let us continue our example and consider the OCF distribution given in Table 1 (which is the same distribution as the one given in [37]). Assume that $S = \{a, b\}$. Let β_1 and β_2 the weights associated with a and b respectively. We have $k(a \wedge b) = 4$ and using Equation (3) we get:

ω	$k * S(\omega)$
ab	0
$\neg ab$	$\beta_2 - 3$
$a \neg b$	$\beta_1 - 3$
$\neg a \neg b$	$\beta_1 + \beta_2 - 4$

Table 2. The result of revising k, given in Table 1, by $S = \{a, b\}$

Using Table 1, Equation (4) gives: $\beta_1 > 4 - min(1 - \beta_2)$ and $\beta_2 > 4 - min(1 - \beta_1)$ which are equivalent to $\beta_1 > 3$ and $\beta_2 > 3$.

Clearly, the c-revision is characterized by a set of parameters (weight). Each set of parameters induces an OCF distribution. In [37] a so-called minimal c-revision has also been proposed. This is obtained by considering only vectors of weights $(\beta_1, .., \beta_n)$ that satisfy Equation (4) and which are pareto-optimal. In the above example, a minimal c-revision is obtained when β_1 and β_2 are both assigned the degree of 4.

Note that the input considered in multiple iterated belief c-revision is different from the notion of uncertain input proposed in [31] for conditioning probability distributions. Indeed, in Jeffrey's rule of conditioning the input represents a partition of the set interpretations Ω, while in c-revision the input is a consistent set of propositional formulas.

4 Syntactic representations of c-revision

The aim of this section is to describe the syntactic representations of multiple iterated belief c-revision when OCF distributions are encoded by means of OCF knowledge bases.

More precisely, let $\mathcal{K}$ be an OCF knowledge base and $k_{\mathcal{K}}$ be its associated OCF distribution obtained using Equation (2). Let S be an input. The aim of this section is to compute, from $\mathcal{K}$ and $S = \{u_1, .., u_n\}$, a new OCF knowledge base

$\mathcal{K}'$ such that:
$$\forall \omega, \quad k_{\mathcal{K}'}(\omega) = k_{\mathcal{K}} * \mathcal{S}(\omega),$$
where $k_{\mathcal{K}'}$ and $k_{\mathcal{K}}$ are the OCF distributions associated with $\mathcal{K}'$ and $\mathcal{K}$ using Equation (2).

To achieve this aim, we proceed in four steps:
- Compute $k(u_1 \wedge .. \wedge u_n)$.
- Compute the syntactic counterpart of adding an OCF distribution k with a binary possibility distribution.
- Compute the syntactic counterpart of the c-revision of $\mathcal{K}$ with $\mathcal{S}$ for a fixed vector of integers $(\beta_1, .., \beta_n)$ associated with formulas of $\mathcal{S}$.
- Provide the syntactic counterpart of the set of inequalities that the weights β_i's should satisfy (see Equation (4)).

The following subsections provide details of each of the above steps.

4.1 Computing $k(u_1 \wedge .. \wedge u_n)$

The aim of this subsection is to compute $k(u_1 \wedge .. \wedge u_n)$ directly from an OCF knowledge base $\mathcal{K}$. As is it shown in the following proposition, computing $k(u_1 \wedge .. \wedge u_n)$ comes down to compute the highest rank α such that formulas of $\mathcal{K}$ having a weight higher than or equal to α are inconsistent with $u_1 \wedge .. \wedge u_n$. More precisely:

Proposition 1: Let $\mathcal{K}$ be an OCF knowledge base and $k_{\mathcal{K}}$ be its associated OCF distribution using Equation (2). Let $\mathcal{S} = \{u_1, .., u_n\}$ be a consistent set of propositional formulas. Let $\mathcal{K}_{\geq \alpha}$ be the α-cut of $\mathcal{K}$ defined by $\mathcal{K}_{\geq \alpha} = \{\varphi_j : (\varphi_j, \gamma_j) \in \mathcal{K}, \gamma_j \geq \alpha\}$. Then:
$$k_{\mathcal{K}}(u_1 \wedge .. \wedge u_n) = max\{\alpha_i : \mathcal{K}_{\geq \alpha_i} \wedge (u_1 \wedge .. \wedge u_n) \text{ is inconsistent}\}.$$

Proof. By definition, we have:
$k_{\mathcal{K}}(u_1 \wedge .. \wedge u_n) = min_{\omega \models u_1 \wedge .. \wedge u_n} k_{\mathcal{K}}(\omega)$

$= min_{\omega \models u_1 \wedge .. \wedge u_n} max\{\alpha_i : (\varphi_i, \alpha_i) \in \mathcal{K}, \omega \not\models \varphi_i\}$

$= min_{\omega \models u_1 \wedge .. \wedge u_n} max\{\alpha_i : (\varphi_i, \alpha_i) \in \mathcal{K}, \omega \models \neg \varphi_i \wedge u_1 \wedge .. \wedge u_n\}$

$= max\{\alpha_i : \mathcal{K}_{\geq \alpha_i} \wedge (u_1 \wedge .. \wedge u_n) \text{ is inconsistent}\}.$

∎

From computational point of view, computing $k(u_1 \wedge .. \wedge u_n)$ needs $O(log_2 m)$ calls to a satisfiability test of a set of clauses, where m is the number of different degrees used in $\mathcal{K}$.

Example 4: Let us continue our example. Recall that $\mathcal{K} = \{(\neg a \vee \neg b, 4), (\neg a, 1), (\neg b, 1)\}$ and that its associated OCF distribution is given in Table 1. Let $\mathcal{S} = \{a, b\}$. From Table 1, we have $k(a \wedge b) = 4$. One can easily check that:
$$max\{\alpha_i : \mathcal{K}_{\geq \alpha_i} \wedge (a \wedge b) \text{ is inconsistent}\} = k(a \wedge b) = 4.$$

Next subsection is devoted to a syntactic computation of the result of adding an OCF distribution with a binary OCF distribution. A binary OCF distribution k' is an OCF distribution where the degree of surprise of each interpretation ω is either equal to 0 (namely, $k'(\omega) = 0$) or is equal to some constant β ($k'(\omega) = \beta$). Intuitively, a binary distribution will represent a weighted formula (u_i, β_i) of the input (models of u_i will have 0 degree, while counter-models of u_i will have a surprise degree equal to β_i).

4.2 Syntactic computations of adding an OCF distribution with a binary OCF distribution

The aim of this section is to provide a syntactic counterpart of:

$$\forall \omega \in \Omega, \qquad k'(\omega) = k(\omega) + \sum_{i=1, \omega \not\models \neg u_i}^{n} \beta_i, \qquad (5)$$

where β_i's are weights associated with each propositional formula u_i of S.

More precisely, our aim is to compute a new knowledge base $\mathcal{K}'$, from $\mathcal{K}$ and $S = \{u_1, .., u_n\}$, such that

$$\forall \omega \in \Omega, \qquad k_{\mathcal{K}'}(\omega) = k'(\omega) = k(\omega) + \sum_{i=1, \omega \not\models \neg u_i}^{n} \beta_i.$$

Equation (5) is clearly a part of the definition of c-revision given by Equation (3).

Let us first denote k_{u_i} the binary OCF distribution associated with (u_i, β_i) and defined by:

$$\forall \omega \in \Omega, \; k_{u_i}(\omega) = \begin{cases} 0 & \text{if } \omega \models u_i \\ \beta_i & \text{otherwise.} \end{cases}$$

Clearly, Equation (5) can be rewritten as:

$$\forall \omega \in \Omega, \qquad k'(\omega) = k(\omega) + k_{u_1}(\omega) + ... + k_{u_n}(\omega). \qquad (6)$$

The following proposition gives the counterpart of combining $k(\omega)$ with some individual and binary distribution k_{u_i}:

Proposition 2: Let $\mathcal{K}$ be an OCF knowledge base. Let (u_i, β_i) be a weighted propositional formula. Let $\mathcal{K}' = \{(u_i, \beta_i)\} \cup \mathcal{K} \cup \{(u_i \vee \varphi_j, \alpha_j + \beta_i) : (\varphi_j, \alpha_j) \in \mathcal{K}\}$.

Then: $\forall \omega \in \Omega, \; k_{\mathcal{K}'}(\omega) = k(\omega) + k_{u_i}(\omega)$, where k and $k_{\mathcal{K}'}$ are the OCF distributions associated with $\mathcal{K}$ and $\mathcal{K}'$ using Equation (2).

Proof. Let $\omega \in \Omega$. We distinguish four cases depending whether ω is a model or not of u_i and formulas in $\mathcal{K}$:

a. $\omega \models u_i$ and $\forall (\varphi_j, \alpha_j) \in \mathcal{K}$ $\omega \models \varphi_j$. Namely, ω is a model of u_i and satisfies all formulas in $\mathcal{K}$. In this case $k_{\mathcal{K}'}(\omega) = 0$ since $k(\omega) = 0$ and $k_{u_i}(\omega) = 0$.

b. $\omega \models u_i$ (hence, $\omega \models u_i \vee \varphi_j$ for each $(\varphi_j, \alpha_j) \in \mathcal{K}$) and $\exists (\varphi_j, \alpha_j) \in \mathcal{K}$ such that $\omega \not\models \varphi_i$. In this case, $k_{\mathcal{K}'}(\omega) = k_{\mathcal{K}}(\omega)$ since $k_{u_i}(\omega) = 0$.

c. $\omega \not\models u_i$ and $\forall (\varphi_j, \alpha_i) \in \mathcal{K}$ we have $\omega \models \varphi_j$ (hence $\omega \models u_i \vee \varphi_j$ for each $(\varphi_j, \alpha_j) \in \mathcal{K}$). Hence, $k_{\mathcal{K}'}(\omega) = k_u(\omega)$ since $k(\omega) = 0$.

d. $\omega \not\models u_i$ and $\exists (\varphi_j, \alpha_j) \in \mathcal{K}$ such that $\omega \not\models \varphi_j$. Namely, ω is neither a model of u_i nor a model of all propositional formulas in $\mathcal{K}$. Then by definition:

$$k_{\mathcal{K}'}(\omega) = max\{\beta_i, max\{\alpha_j : (\varphi_j, \alpha_j) \in \mathcal{K}\ \omega \not\models \varphi_j\}, max\{\alpha_j + \beta_i : (\varphi_j, \alpha_j) \in \mathcal{K}\ \omega \not\models \varphi_j\}\}$$
$$= max\{\alpha_j : (\varphi_j, \alpha_j) \in \mathcal{K}, \omega \not\models \varphi_j\} + \beta_i$$
$$= k_{\mathcal{K}}(\omega) + k_u(\omega).$$

∎

Note that Proposition 2 is similar to the syntactic fusion mode proposed in [7] in a possibility theory framework. From Proposition 2, trivially in the worst case the size of $\mathcal{K}'$ is $2 * |\mathcal{K}| + 1$, and the computation of the OCF knowledge base $\mathcal{K}'$ is done in a linear time with respect to the size of $\mathcal{K}$.

Clearly, the repetitive application of Proposition 2 on each propositional formula u_i of $\mathcal{S}$ allows us to provide the syntactic counterpart of Equation (6).

Example 5: Let us continue our example. We recall that $\mathcal{K} = \{(\neg a \vee \neg b, 4), (\neg a, 1), (\neg b, 1)\}$ and $\mathcal{S} = \{a, b\}$.

Applying Proposition 2 on $\mathcal{K}$ and (a, β_1) gives:

$\mathcal{K}' = \{(\neg a \vee \neg b, 4), (\neg a, 1), (\neg b, 1)\} \cup \{(a, \beta_1)\} \cup \{(a \vee \neg b, 1 + \beta_1)\}$

Again, applying Proposition 2 on $\mathcal{K}'$ and (b, β_2) gives:

$\mathcal{K}'' = \{(\neg a \vee \neg b, 4), (\neg a, 1), (\neg b, 1), (a, \beta_1), (a \vee \neg b, 1 + \beta_1)\} \cup \{(b, \beta_2)\} \cup \{(\neg a \vee b, 1 + \beta_2), (a \vee b, \beta_1 + \beta_2)\}$

Namely

$\mathcal{K}'' \equiv \{(\neg a \vee \neg b, 4), (a, \beta_1), (b, \beta_2), (a \vee \neg b, 1+\beta_1), (\neg a \vee b, 1+\beta_2), (a \vee b, \beta_1+\beta_2)\}$

(since (a, β_1) and $(\neg a \vee b, 1 + \beta_2)$ leads to $(b, min(\beta_1, \beta_2 + 1))$ and (b, β_2) and $(a \vee \neg b, 1 + \beta_1)$ leads to $(a, min(\beta_2, \beta_1 + 1))$)

$\equiv \{(\neg a \vee \neg b, 4), (a \vee \neg b, 1 + \beta_1), (\neg a \vee b, 1 + \beta_2), (a \vee b, \beta_1 + \beta_2)\}.$

Clearly,

From Table 3, we have: $\forall \omega \in \Omega \quad k_{\mathcal{K}''}(\omega) = k_{\mathcal{K}}(\omega) + k_a(\omega) + k_b(\omega).$

The last step is to compute the syntactic computation of c-revision. Namely, to compute the counterpart of Equation (3). This is the aim of next subsection.

Ω	k_a	k_b	$k_{\mathcal{K}}$	$k_{\mathcal{K}''}$
ab	0	0	4	4
$a\neg b$	0	β_2	1	$1+\beta_2$
$\neg ab$	β_1	0	1	$1+\beta_1$
$\neg a\neg b$	β_1	β_2	0	$\beta_1+\beta_2$

Table 3. The resulting of adding OCF $k_{\mathcal{K}}$ with two binary distributions k_a and k_b

4.3 Computing c-revision

The following Lemma will help us in providing the syntactic computation of multiple iterated belief c-revision operation.

Lemma 1: Let $\mathcal{K}$ be an OCF knowledge base and $\mathcal{S} = \{u_1, .., u_n\}$ be a consistent set of propositional formulas. Let $\mathcal{K}' = \{(\varphi_i, \alpha_i - k(u_1 \wedge .. \wedge u_n)) : (\varphi_i, \alpha_i) \in \mathcal{K}\}$. Then:

$$\forall \omega \in \Omega, \quad k_{\mathcal{K}'}(\omega) = k(\omega) - k(u_1 \wedge .. \wedge u_n),$$

where k and $k_{\mathcal{K}'}$ are the OCF distributions associated with $\mathcal{K}$ and $\mathcal{K}'$ respectively.

The proof is immediate since by definition:

$$\forall \omega \in \Omega, \ k_{\mathcal{K}'}(\omega) = max\{\alpha_i - k(u_1 \wedge .. \wedge u_n) : (\varphi_i, \alpha_i) \in \mathcal{K}, \omega \not\models \varphi_i\}$$
$$= max\{\alpha_i : (\varphi_i, \alpha_i) \in \mathcal{K}, \omega \not\models \varphi_i\} - k(u_1 \wedge .. \wedge u_n)$$
$$= k(\omega) - k(u_1 \wedge .. \wedge u_n).$$

Now, to get the syntactic computation of $k * \mathcal{S}$, (the result of applying multiple iterated belief c-revision on k and $\mathcal{S}$ using Equation (3)) it is enough to apply Proposition 2 on each element of $\mathcal{S}$, and then apply Lemma 1.

Clearly, the computation of $\mathcal{K}'$ in Lemma 1 is done in linear time with respect to the size of k (once $k(u_1 \wedge .. \wedge u_n)$ is already computed).

Example 6: Let us continue our example. Recall that from Example 5 we have:

$\mathcal{K}'' = \{(\neg a \vee \neg b, 4), (a \vee \neg b, 1 + \beta_1), (\neg a \vee b, 1 + \beta_2), (a \vee b, \beta_1 + \beta_2)\}$.

Recall that $\mathcal{S} = \{a, b\}$ and $k(a \wedge b) = 4$.
Applying Lemma 1 on $\mathcal{K}''$ we get:

$\mathcal{K}^* = \{(\neg a \vee \neg b, 0), (a \vee \neg b, \beta_1 - 3), (\neg a \vee b, \beta_2 - 3), (a \vee b, \beta_1 + \beta_2 - 4)\}$
$\equiv \{(\neg a \vee b, \beta_2 - 3), (a \vee \neg b, \beta_1 - 3), (a \vee b, \beta_1 + \beta_2 - 4)\}$

One can easily check that:

$$\forall \omega, \quad k_{\mathcal{K}^*}(\omega) = k * \mathcal{S}(\omega),$$

where $k * \mathcal{S}(.)$ is given in Table 2 and $k_{\mathcal{K}^*}$ is the OCF distribution associated with $\mathcal{K}^*$ using Equation (2).

4.4 On the characterization of inequality constraints

It remains now to characterize the constraints bearing on β_i's. Namely, our aim is to directly characterize from $\mathcal{K}$ and $\mathcal{S} = \{u_1, .., u_n\}$ the set of inequalities:

$$\forall i = 1, .., n, \quad \beta_i > k(u_1 \wedge .. \wedge u_n) - \min_{\omega \not\models u_i}\{k(\omega) + \sum_{j \neq i, \omega \not\models \beta_j} \beta_j\}$$

The computation of such inequalities is possible thanks to Propositions (1,2) and to Lemma 1. Indeed, Proposition 1 allows us to compute $k(u_1 \wedge .. \wedge u_n)$. Proposition 2 allows us, for each i, to compute the syntactic counterpart of:

$$\forall \omega, k'(\omega) = k(\omega) + \sum_{i=1, j \neq i}^{n} k_{u_j}(\omega).$$

Using similar steps as in Proposition 1, we get:

$$\min_{\omega \not\models u_i} k'(\omega) = max\{\alpha_i : \mathcal{K}'_{\geqslant \alpha_i} \wedge \neg u_i \text{ is inconsistent}\},$$

where $\mathcal{K}'$ is an OCF knowledge base associated with

$$\forall \omega \in \Omega, \quad k'(\omega) = k(\omega) + \sum_{j=1, j \neq i}^{n} k_{u_j}(\omega).$$

Example 7: Let us finish our example where we have $\mathcal{K} = \{(\neg a \vee \neg b, 4), (\neg a, 1), (\neg b, 1)\}$ and $\mathcal{S} = \{a, b\}$. Recall that we already computed $k(a \wedge b)$ which is equal to 4. Let us now give the inequality relations associated with β_1 and β_2.

For β_1, our aim is to characterize:

$$\beta_1 > k(a \wedge b) + \min_{\omega \models \neg a}\{k(\omega) + k_b(\omega)\}$$

The knowledge base associated with $k(.) + k_b(.)$ is obtained using Proposition 2:

$$\mathcal{K}' = \{(\neg a \vee \neg b, 4), (\neg a, 1), (\neg b, 1), (b, \beta_2), (\neg a \vee b, 1 + \beta_2)\}$$
$$= \{(\neg a \vee \neg b, 4), (\neg b, 1), (b, \beta_2), (\neg a \vee b, 1 + \beta_2)\}$$

Now,

$$\min_{\omega \models \neg a}\{k(\omega) + k_b(\omega)\} = max\{\alpha_i : \mathcal{K}'_{\geqslant \alpha_i} \wedge \neg a \text{ is inconsistent}\}$$
$$= min(1, \beta_2).$$

Similar result for β_2. Hence, β_1 and β_2 should satisfy:

$$\beta_1 > 4 - min(1, \beta_2),$$

and

$$\beta_2 > 4 - min(1, \beta_1).$$

Which are equivalent to
$$\beta_1 > 3 \text{ and } \beta_2 > 3.$$

These inequalities are the same as the ones given in Example 2 but obtained here using OCF knowledge bases.

5 Conclusion

The multiple iterated propositional c-revision, defined in [37], is a revision operator that takes into account the independence relations that may exist between propositional formulas of the input.

This paper shows that c-revision can be equivalently defined using OCF knowledge bases. In particular, we provide an explicit computation of the inequalities associated with certainly degrees attached with formulas of the input.

In this paper, OCF distributions are obtained from OCF knowledge bases using a translation function similar to the one used in possibility theory. A future work is to redefine c-revision when OCF distributions are obtained from OCF knowledge bases using penalty logic [15, 21].

References

1. Adaricheva, K., Sloan, R.H., Szorenyi, B., Turan, G.: Horn belief contraction: remainders, envelopes and complexity. In 13th international Conf. on Principles of Knowledge Representation and Reasoning, (2012)
2. Alchourrón, C., Gärdenfors, P., Makinson, D.: On the logic of theory change: Partial meet contraction and revision functions. J. Symb. Log. 50(2), 510–530 (1985)
3. Beierle, C., Hermsen, R., Kern-Isberner, G.: Observations on the Minimality of Ranking Functions for Qualitative Conditional Knowledge Bases and Their Computation. In: 28th International FLAIRS Conference. (2014)
4. Benferhat, S.: Graphical and Logical-Based Representations of Uncertain Information in a Possibility Theory Framework. In: 4th International Conference on Scalable Uncertainty Management (SUM 2010), pp.3-6 (2010)
5. Benferhat, S., Dubois, D., Prade, H.: Possibilistic and Standard Probabilistic Semantics of Conditional Knowledge Bases. Journal of Logic and Computation. 9(6), 873–895 (1999)
6. Benferhat, S., Dubois, D., Prade, H.: Some syntactic approaches to the handling of inconsistent knowledge bases : A comparative study. Part 2 : the prioritized case, (24), 473–511 (1998)
7. Benferhat, S., Dubois, D., Prade, H.: From semantic to syntactic approaches to information combination in possibilistic logic. In: Aggregation and Fusion of Imperfect Information, Studies in Fuzziness and Soft Computing, pp. 141–151. (B. Bouchon-Meunier, ed.), Physica Verlag (1997)
8. Benferhat, S., Dubois, D., Prade, H.: Argumentative inference in uncertain and inconsistent knowledge bases. In: the 9th Conference on Uncertainty in Artificial Intelligence (UAI 1993), pp. 411–419. David Heckerman and E. H. Mamdani editors, Morgan Kaufmann, (1993)
9. Benferhat, S., Dubois, D., Prade, H., Williams, M.A.: A practical approach to revising prioritized knowledge bases. Studia Logica Journal 70, 105–130 (2002)

10. Benferhat, S., Konieczny, S., Papini, O., Pino Pérez, R.: Iterated revision by epistemic states: axioms, semantics and syntax. In: the 14th European Conf. on Artificial Intelligence (ECAI-00), pp. 13–17. IOS Press, Berlin (2000)
11. Benferhat, S., Largue, S., Papini, O.: Revision of partially ordered information: Axiomatization, semantics and iteration. In Proceedings of the Nineteenth International Joint Conference on Artificial Intelligence (IJCAI'05), pp. 376–381 (2005)
12. Benferhat, S., Tabia, K.: Belief Change in OCF-Based Networks in Presence of Sequences of Observations and Interventions: Application to Alert Correlation. In PRICAI 2010, Trends in Artificial Intelligence, Lecture Notes in Computer Science vol. 6230, pp. 14–26. Springer (2010)
13. Booth, R., Meyer, T., Varzinczak I.J.: Steps in propositional horn contraction. In: Proceedings of the Twenty first International Joint Conference on Artificial Intelligence (IJCAI'09), pp. 702–707. (2009)
14. Darwiche, A., Goldszmidt, M.: On the Relation between Kappa Calculus and Probabilistic Reasoning. In: the 10th Conference on Uncertainty in Artificial Intelligence, (UAI-1994) (1994)
15. Darwiche, A., Marquis, P.: Compiling propositional weighted bases. J. Artif. Intell. 157(1-2), 81–113 (2004)
16. Darwiche, A., Pearl, J.: On the logic of iterated belief revision. J. Artif. Intell. 89, 1–29 (1997)
17. Delgrande, J.P.: Horn clause belief change: Contraction functions. In: Proceedings of the Eleventh International Conference on the Principles of Knowledge Representation and Reasoning (KR'08), pp. 156–165. (2008)
18. Delgrande, J.P., Wassermann R.: Horn clause contraction functions: Belief set and belief base approaches. In: Proceedings of the Twelfth International Conference on the Principles of Knowledge Representation and Reasoning (KR' 10), 2010.
19. Dubois, D., Prade H.: A synthetic view of belief revision with uncertain inputs in the framework of possibility theory. International Journal of Approximate Reasoning 17(2–3), 295–324 (1997)
20. Dubois, D., Prade H.: Possibility theory. Plenum Press, New-York (1988)
21. Dupin De Saint Cyr - Bannay, F., Lang, J., Schiex, T.: Penalty logic and its link with Dempster-Shafer theory. In: Conference on Uncertainty in Artificial Intelligence (UAI 1994), pp. 204–211. Ramon Lopez de Mantaras, David Poole (Eds.), Morgan Kaufmann Publishers (1994)
22. Eichhorn, C., Kern-Isberner, G.: LEG Networks for Ranking Functions. In: 14th edition of the European Conference on Logics in Artificial Intelligence (JELIA 2014), pp. 210–223 (2014)
23. Falappa, M.A., Kern-Isberner, G., Reis, M., Simari, G.R.: Prioritized and Nonprioritized Multiple Change on Belief Bases. J. Philosophical Logic. 41(1), 77–113 (2012)
24. Ferm'e, E., Hansson, S. O.: editors. Journal of Philosophical Logic. Special Issue on 25 Years of AGM Theory, 40(2). Springer Netherlands, (2011)
25. Gärdenfors, P.: Belief Revision. Cambridge University Press (1992)
26. Goldszmidt, M.,Pearl, J.: Qualitative probabilities for default reasoning, belief revision, and causal modeling. J. Artificial Intelligence. 84(1-2), 57–112 (1997)
27. Goldszmidt, M.,Pearl, J.: Reasoning with Qualitative Probabilities Can Be Tractable. In: the 8th Conference on Uncertainty in Artificial Intelligence, (UAI-1992) (1992)
28. Halpern, J., Pearl, J.: Causes and explanations: A structurel model approach. In: the 17th Conf. on Uncertainty in Articcial Intelligence. pp. 194–202. J. Breese, D.K. (ed.) (2001)

29. Hansson, S.O.: Revision of belief sets and belief bases. Handbook of Defeasible Reasoning and Uncertainty Management systems 3, 17–75 (1998)
30. Hansson, S.O.: A Textbook of Belief Dynamics. Kluwer Academic Publishers (1997)
31. Jeffrey, R.C.: The logic of decision. Mc. Graw Hill, New York (1965)
32. Katsuno, H., Mendelzon, A.: Propositional knowledge base revision and minimal change. J. Artif. Intell. 52, 263–294 (1991).
33. Kern-Isberner, G.: A Thorough Axiomatization of a Principle of Conditional Preservation in Belief Revision. Ann. Math. Artif. Intell. 40(1-2), 127–164 (2004)
34. Kern-Isberner, G.: Conditionals in nonmonotonic reasoning and belief revision. Lecture Notes in Artificial Intelligence LNAI, vol. 2087 Springer (2001)
35. Kern-Isberner, G., Eichhorn, C.: Structural Inference from Conditional Knowledge Bases. Studia Logica Journal 102(4), 751–769 (2014)
36. Kern-Isberner, G., Eichhorn, C.: Intensional Combination of Rankings for OCF-Networks. In: 26th International FLAIRS Conference (2013)
37. Kern-Isberner, G., Huvermann, D.: Multiple Iterated Belief Revision Without Independence. In: 29th International FLAIRS Conference, pp. 570–575 (2015)
38. Kern-Isberner, G., Thimm, M.: A Ranking Semantics for First-Order Conditionals. In: the 20th European Conference on Artificial Intelligence (ECAI-2012), pp. 456–461 (2012)
39. Konieczny, S., Lang, J., Marquis, P.: Reasoning under inconsistency: the forgotten connective. In: Proceedings of the Nineteenth International Joint Conference on Artificial Intelligence (IJCAI-05), pp. 484–489. Leslie Pack Kaelbling and Alessandro Saffiotti editors, Edinburgh, Scotland, UK (2005)
40. Konieczny, S., Medina Grespan, M., Pino Pérez R.: Taxonomy of improvement operators and the problem of minimal change. In: Proceedings of the Twelfth International Conference on Principles of Knowledge Representation and Reasoning (KR'10), (2010)
41. Konieczny, S., Pino Pérez R.: A framework for iterated revision. Journ. of Applied Non-classical Logics 10(3-4), (2000)
42. Lang, J.: Possibilistic logic: complexity and algorithms. 5, 179–220. Kluwer Academic Publishers, (2001)
43. Ma, J., Benferhat, S., Liu, W.: Revising partial pre-orders with partial pre-orders: A unit-based revision framework. In: the 13th International Conference on Principles of Knowledge Representation and Reasoning (KR'12). Rome, Italy (2012)
44. Papini, O.: A complete revision function in propositional calculus. In: the 10th European Conference on Artificial Intelligence (ECAI-92), pp. 339–343. (1992)
45. Pearl, J.: Causality: Models, Reasonning and Inference. Cambridge University Press (2000)
46. Qi, G., Liu, W., Bell, D. A.: A revision-based approach to handling inconsistency in description logics. Artif. Intell. Rev. 26(1-2), 115–128 (2006)
47. Spohn, W.: The Laws of Belief: Ranking Theory and Its Philosophical Applications. Oxford University Press (2012)
48. Touazi, F., Cayrol, C., Dubois. D.: Possibilistic reasoning with partially ordered beliefs. Journal of Applied Logic, Elsevier (2015)
49. Williams, M.A.: Iterated theory base change: A computational model. In: the 14th Inter. Joint Conf. On Artificial Intelligence (IJCAI'95) pp. 1541–1547 (1995)
50. Williams, M.A.: On the logic of theory base change. In: the Europ. Workshop on Logics in Artificial Intelligence (JELIA'95), LNCS, vol. 838, pp. 86–105. Springer Verlag (1994)
51. Williams, M.A.: Transmutation of knowledge systems. In: the 4th Inter. Conf. on the Principles of Knowledge Representation and Reasoning (KR'94), pp. 619–629. Morgan Kaufmann, Bonn (1994)

A Set of Operations for Stratified Belief Bases

Marcelo A. Falappa[1], Alejandro Garcia[2], Guillermo Simari[3]

Abstract. This work presents a set of change operators that can be defined in a stratified belief base, which is a belief base where the beliefs stored in it are assigned a value and each stratum stores all the beliefs that have the same value attached. The central idea is to provide a complete set of operations and a way of reasoning over stratified belief bases. We propose pertinent constructions and suggest practical applications.

1 Introduction and Related Work

A stratified belief base (SBB) is a belief base in which all beliefs are assigned a certain value and each stratum contains all the beliefs that share the same appraisal. The concept of value will remain as an abstract notion in this work providing a natural way to label each stratum; the set of labels will be assumed to have a total order. Furthermore, and characterizing our approach that is based in belief revision theory, each stratum is required to be internally consistent.[4] Nevertheless, direct inconsistency between information stored in different strata could remain. Inconsistency among potential beliefs will be handled by the reasoning mechanism that we will introduce, leading to a consistent set of actual beliefs to which the reasoning agent is committed.

Such a stratified belief base provides a suitable foundation for the epistemic state of a rational agent. This epistemic state will contain explicit beliefs, *i.e.*, the ones stored in the stratified belief base, and implicit beliefs, *i.e.*, the ones that can be obtained by inference from the explicit beliefs. For the derivation of the implicit beliefs from such repositories, we will resort to an argumentation system approach under one of Dung-style semantics [13]. Since each stratum is assumed to be consistent, it is necessary to devise a mechanism for revising stratified belief bases for handling the case when conflicting information attempts to be added to the same stratum; the constraint of keeping each stratum consistent will force a rational agent to resolve these conflicts before obtaining inferences from its stratified epistemic base. The current work extends the research presented in [14], where the authors generalize the one-level

---
[1] Institute for Computer Science and Engineering (UNS–CONICET), Department of Computer Science and Engineering, Universidad Nacional del Sur, Argentina, mfalappa@cs.uns.edu.ar
[2] Institute for Computer Science and Engineering (UNS–CONICET), Department of Computer Science and Engineering, Universidad Nacional del Sur, Argentina, ajg@cs.uns.edu.ar
[3] Institute for Computer Science and Engineering (UNS–CONICET), Department of Computer Science and Engineering, Universidad Nacional del Sur, Argentina, grs@cs.uns.edu.ar
[4] For a different, paraconsistent approach see [5–7], see Section 7 for some details.

reuse of beliefs shown in [16] by extending it to a multi-level formalism such as the one presented in [15].

2 Stratified Belief Bases: Preliminaries and Notation

Given an epistemic input, classical belief revision models discard sentences that are inconsistent with that input. This behavior appears to contradict the idea that is necessary to preserve as many of the existing beliefs as possible, a criterion that is known as the *principle of minimal change* which was introduced by Gärdenfors in [19].

The use of stratified beliefs bases as a representational device provides an adequate setup for maintaining more of the existing beliefs albeit perhaps it could be necessary to move them to a different stratum. This multi-level reuse of beliefs presents the chance of taking sentences that are eliminated by the revision process in one stratum and revise the next less-valued stratum with them as new epistemic input to that stratum. This schema for keeping the discarded beliefs from one stratum and reusing them prevents the complete loss of information, creating a *dynamic reclassification of beliefs*, that is, beliefs are dynamically classified using their value upon an epistemic input.

In this paper we will adopt a propositional language $\mathcal{L}$ with a complete set of boolean connectives: $\neg, \wedge, \vee, \rightarrow, \leftrightarrow$. Formulæ in $\mathcal{L}$ will be denoted by lowercase Greek letters: $\alpha, \beta, \delta, \ldots, \omega$. Sets of sentences in $\mathcal{L}$ will be denoted by uppercase Latin letters: $A, B, C, \ldots, Z$. The symbol $\top$ represents a tautology or *truth*, and the symbol $\bot$ represents a contradiction or *falsum*. The characters γ and σ will be reserved to represent selection and incision functions for change operators, respectively. We use a consequence operator $Cn : 2^{\mathcal{L}} \longrightarrow 2^{\mathcal{L}}$ that takes sets of sentences in $\mathcal{L}$ and produces new sets of sentences in $\mathcal{L}$. This operator satisfies *inclusion* ($A \subseteq Cn(A)$), *idempotence* ($Cn(A) = Cn(Cn(A))$), and *monotony* (if $A \subseteq B$ then $Cn(A) \subseteq Cn(B)$). It is assumed that the consequence operator includes classical consequences and verifies the standard properties of *supraclassicality* (if α can be derived from A by deduction in classical logic, then $\alpha \in Cn(A)$), *deduction* ($\beta \in Cn(A \cup \{\alpha\})$ if and only if $(\alpha \rightarrow \beta) \in Cn(A)$) and *compactness* (if $\alpha \in Cn(A)$ then $\alpha \in Cn(A')$ for some finite subset A' of A). In general, we will write $\alpha \in Cn(A)$ as $A \vdash \alpha$.

In our approach, an agent's epistemic state will be represented as a stratified belief base. In such a base, each stratum will contain a set of sentences with the same value and each stratum will be assumed consistent. This decision contrasts with other approaches, such as [5, 6] where a form of argumentation is used as a general inference mechanism from a (potentially) inconsistent stratified belief base rendering a paraconsistent approach.

Definition 1. $\Sigma = (\Sigma_0, \ldots, \Sigma_n)$ *is a stratified belief base* (SBB) *if and only if $\Sigma_i \subseteq \mathcal{L}$ and Σ_i is finite and consistent for all $0 \leqslant i \leqslant n$, and all the sentences in Σ_i have the same value. The strata are considered to be ordered in the following way: the beliefs of Σ_j have a higher assigned value than the ones in Σ_i when $j < i$. Sometimes, slightly abusing the notation, $x \in \Sigma$ will mean that $x \in \Sigma_i$ for some $\Sigma_i \in \Sigma$, $0 \leqslant i \leqslant n$.*

It seems a good design decision for a rational agent, faced with contradictory beliefs of different levels, to believe only the stronger of them disregarding the weaker. Such agent will reserve the first stratum Σ_0 for proper beliefs, while the rest of the layers refer to information that:

1. has assigned a lower level than the information in Σ_0;
2. is believed only when it does not contradict beliefs that are in Σ_0 or that are contained in a stratum of higher level; and:
3. even when it is believed, it will remain more provisional than any belief in Σ_0.

Therefore, in our approach, stratified belief bases are largely composed of information that is not fully believed, and the stratification is produced by the perceived value of that information. This characterization differs from other approaches that consider that beliefs have a certain strength or entrenchment [20, 22, 28].

It is important to note that our approach differs from the research presented in [5, 6].[5] In these works, our abstract notion of value is grounded as reliability and each stratum can be inconsistent. Moreover, the reliability assigned to the elements in the belief base cannot be modified. Our approach considers consistent strata and will allow to change the level of beliefs as a consequence of a change operator.

3 Multiple Revision Subsystem

A multiple revision operator is defined as a belief revision operator capable of accepting a set of sentences as epistemic input [14]. The intention is to use this type of operator as part of the global framework we are introducing. To construct a multi-layer revision operator, first we need to discuss certain general ideas concerning revision operators on each layer of a stratified belief base seen in isolation. Although some related literature on the topic of multiple revision exists, such as the proposed in [23] where multiple revision is reduced to classical AGM sentence revision, or the work of [12] where revision and expansion operators of logic programs by answer set semantics were presented, our approach differs from these in that we consider multiple revision in a more general, logically classic environment. Further discussion on related work on this particular subject can be found in [15]. We will proceed by recalling the main ideas originally presented in that paper and introducing a generalization of its capabilities.

---

[5] Also, [11] presented an approach, that is a precursor of this line of research, generalizing [26, 25] by introducing the idea of preferred subtheories in default reasoning which uses a stratification criterion based on set inclusion to rank the content of knowledge bases.

3.1 The General Setting

The layers in a stratified belief base Σ can be regarded as separated, consistent belief bases. This will allow us to conceptualize the problem as multiple belief base revision.

Let S be a consistent belief base, A and B be consistent sets of sentences, and $*$ be a binary multiple revision operator that takes a belief base and a set of sentences as inputs. The following postulates were proposed for multiple revision operations in [15]:

Inclusion: $S*A \subseteq S \cup A$.

Weak Success: If A is consistent then $A \subseteq S*A$.

Relative Success: $A \subseteq S*A$ or $S*A = S$.

Consistency: If A is consistent then $S*A$ is consistent.

Vacuity 1: If A is inconsistent then $S*A = S$.

Vacuity 2: If $S \cup A \not\vdash \bot$ then $S*A = S \cup A$.

Uniformity 1: Given A and B two consistent sets, for all subset X of S, if $(X \cup A) \vdash \bot$ if and only if $(X \cup B) \vdash \bot$ then $S \backslash (S*A) = S \backslash (S*B)$.

Uniformity 2: Given A and B two consistent sets, for all subset X of S, if $(X \cup A) \vdash \bot$ if and only if $(X \cup B) \vdash \bot$ then $S \cap (S*A) = S \cap (S*B)$.

Relevance: If $\alpha \in S \backslash (S*A)$ then there is a set C such that $S*A \subseteq C \subseteq (S \cup A)$, C is consistent with A but $C \cup \{\alpha\}$ is inconsistent with A.

Core-Retainment: If $\alpha \in S \backslash (S*A)$ then there is a set C such that $C \subseteq (S \cup A)$, C is consistent with A but $C \cup \{\alpha\}$ is inconsistent with A.

The following lemma shows some relations among these postulates [15].

Lemma 1 ([15]).

(a) *If an operator satisfies* relevance *then it satisfies* core-retainment.

(b) *If an operator satisfies* vacuity 1 *and* weak success *then it satisfies* relative success.

(c) *An operator satisfies* uniformity 1 *if and only if it satisfies* uniformity 2.

(d) *If an operator satisfies* inclusion, weak success *and* core-retainment *then it satisfies* vacuity 2.

In order to construct the multiple revision operators, we must define the concept of (A-inconsistent) kernels and (A-consistent) remainder sets [15].

Definition 2. *Let S, A be sets of sentences, where A is consistent. The set of A-inconsistent-kernels of S, noted by $S \perp\!\!\!\perp_\bot A$, is the set of sets X such that:*

1. $X \subseteq S$.
2. $X \cup A$ *is inconsistent.*
3. *For any X', if $X' \subset X \subseteq S$ then $X' \cup A$ is consistent.*

That is, given a consistent set A, $S \perp\!\!\!\perp_\bot A$ is the set of minimal S-subsets inconsistent with A.

Definition 3. *Let S, A be sets of sentences, where A is consistent. The set of A-consistent-remainders of S, noted by $S \perp_T A$, is the set of sets X such that:*

1. $X \subseteq S$.
2. $X \cup A$ is consistent.
3. For any X', if $X \subset X' \subseteq S$ then $X' \cup A$ is inconsistent.

That is, $S \perp_T A$ is the set of maximal S-subsets consistent with A.

Example 1. Suppose that $S = \{p, p \to q, q, \neg r\}$ and $A = \{\neg q, r\}$. Then we have that:

- $S \perp\!\!\!\perp_\perp A = \{\{p, p \to q\}, \{q\}, \{\neg r\}\}$.
- $S \perp_T A = \{\{p\}, \{p \to q\}\}$.

3.2 Prioritized Multiple Kernel Revision

The first construction of multiple revision by a set of sentences is based on the concept of A-inconsistent-kernels. In order to complete the construction, we must define an incision function that cuts in every inconsistent-kernel [15].

Definition 4. *Let S be a set of sentences. Then σ is a consolidated incision function for S ($\sigma : 2^{2^{\mathcal{L}}} \longrightarrow 2^{\mathcal{L}}$) if and only if, for all consistent sets A:*

1. $\sigma(S \perp\!\!\!\perp_\perp A) \subseteq \bigcup S \perp\!\!\!\perp_\perp A$.
2. If $X \in S \perp\!\!\!\perp_\perp A$ then $X \cap (\sigma(S \perp\!\!\!\perp_\perp A)) \neq \emptyset$.

Definition 5. *Let S, A be sets of sentences, A consistent, and σ a consolidated incision function for S. The prioritized multiple kernel revision of S by A that is generated by σ is the operator $*_\sigma$ such that ($*_\sigma : 2^{\mathcal{L}} \times 2^{\mathcal{L}} \longrightarrow 2^{\mathcal{L}}$):*

$$S *_\sigma A = \begin{cases} (S \setminus \sigma(S \perp\!\!\!\perp_\perp A)) \cup A & \text{if } A \text{ is consistent} \\ S & \text{otherwise} \end{cases}$$

Observation 1 *Let S, A be sets of sentences, A be consistent. Suppose that $\alpha \in S$ and $\alpha \in A$. Then $\alpha \notin \bigcup (S \perp\!\!\!\perp_\perp A)$ and, therefore, $A \cap \bigcup (S \perp\!\!\!\perp_\perp A) = \emptyset$.*

From Observation 1 and Definition 4, it follows that all the sentences of A are *protected*, meaning they cannot be considered for removing by the consolidated incision function. That is, a consolidated incision function selects among the sentences of $K \setminus A$ that make $K \cup A$ inconsistent.

Theorem 1. *An operator $*_\sigma$ is a prioritized multiple kernel revision operator for S if and only if it satisfies* inclusion, consistency, weak success, vacuity (1 and 2), uniformity (1 and 2), *and* core-retainment.

Weak Success ensures the input set is accepted when it is consistent. Vacuity 1 ensures that the belief base remains unchanged when the input set is inconsistent. These two postulates resolve a controversial point of AGM revisions such as success which forces α to be accepted in the revision of S by α even though α might be inconsistent.

3.3 Prioritized Multiple Partial Meet Revision

The second construction of multiple revision by a set of sentences is based on the concept of A-consistent-remainders. In order to complete the construction, we must define a selection function that selects the 'best' consistent remainders [15].

Definition 6. *Let S be a set of sentences. Then γ is a consolidated selection function for S ($\gamma : 2^{2^{\mathcal{L}}} \longrightarrow 2^{2^{\mathcal{L}}}$) if and only if, for all sets A:*

1. *If $S \perp_T A \neq \emptyset$ then $\gamma(S \perp_T A) \subseteq S \perp_T A$.*
2. *If $S \perp_T A = \emptyset$ then $\gamma(S \perp_T A) = \{S\}$.*

Observation 2 *Let S, A be sets of sentences, A be consistent. Suppose that $\alpha \in S$ and $\alpha \in A$. Then $\alpha \in X$ for all $X \in S \perp_T A$ and, therefore, $\alpha \in \bigcap (S \perp_T A)$.*

From Observation 2 and Definition 6, it follows that all the sentences of $K \cap A$ are *protected*, meaning that they are included in the intersection of any collection of remainders. That is, a consolidated selection function selects a subset of the set $K \perp_T A$ whose elements all contain the set $K \cap A$.

Definition 7. *Let S, A be sets of sentences, A consistent, and γ a consolidated selection function for S. The prioritized multiple partial meet revision of S by A generated by γ is the operator $*_\gamma : 2^{\mathcal{L}} \times 2^{\mathcal{L}} \longrightarrow 2^{\mathcal{L}}$ such that:*

$$S *_\gamma A = \begin{cases} \bigcap \gamma(S \perp_T A) \cup A & \text{if } A \text{ is consistent} \\ S & \text{otherwise} \end{cases}$$

Theorem 2. *An operator $*_\gamma$ is a prioritized multiple partial meet revision operator for S if and only if it satisfies* inclusion, consistency, weak success, vacuity (1 and 2), uniformity (1 and 2) *and* relevance.

Corollary 1. *Let S be a belief base. If $*$ is a multiple partial meet revision for S then $*$ is a multiple kernel revision for S.*

From the definitions, Lemma 1, and Theorems 1 and 2, it is easily seen that multiple partial meet revision operators are multiple kernel revision operators [15].

The following example, adapted from a similar one presented in [21, page 91], clarifies that the converse of the above corollary does not hold, *i.e.*, that some multiple kernel revision operators are not multiple partial meet revision operators.

Example 2. Let p, q and r be logically independent sentences and $S = \{p, q, r\}$. Suppose that $A = \{\neg p \vee \neg q, \neg p \vee \neg r\}$. Suppose that we are applying a prioritized multiple kernel revision. Then we have that:

$$S \perp\!\!\!\perp_\perp A = \{\{p, q\}, \{p, r\}\}$$

Suppose that $\sigma(S\perp\!\!\!\perp_\perp A) = \{p,r\}$. Then, the prioritized multiple kernel revision of S with respect to A is:

$$S *_\sigma A = (S \backslash \sigma(S\perp\!\!\!\perp_\perp A)) \cup A = \{q, \neg p \vee \neg q, \neg p \vee \neg r\}$$

Now, suppose that we are applying a prioritized multiple partial meet revision. Then we have that:

$$S \perp_\top A = \{\{p\}, \{q,r\}\}$$

If γ is a selection function for S, then $S *_\gamma A = \bigcap(S \perp_\top A) \cup A$. We have three cases:

$\gamma_1(S \perp_\top A) = \{\{p\}\}$ and $S*_{\gamma_1} A = \{p, \neg p \vee \neg q, \neg p \vee \neg r\}$.
$\gamma_2(S \perp_\top A) = \{\{q,r\}\}$ and $S*_{\gamma_2} A = \{q, r, \neg p \vee \neg q, \neg p \vee \neg r\}$.
$\gamma_3(S \perp_\top A) = \{\{p\},\{q,r\}\}$ and $S*_{\gamma_3} A = \{\neg p \vee \neg q, \neg p \vee \neg r\}$.

Thus, we can immediately conclude that $*_\sigma$ is not a multiple partial meet revision.

4 Argumentative Inference from Stratified Data Bases

As was described before, a stratified belief base (SBB) is a collection of strata where each stratum contains a set of sentences which have values assigned. In that scenario, it is possible that the same sentence will be part of different strata and, although it is required that each stratum be consistent, it might be the case that the aggregation of all strata results in an inconsistent set, as it will happen when sentences from different strata are contradictory to each other. When considering a stratified belief base as a belief repository, the problem of obtaining the implicit beliefs becomes important, particulary because the conflict occurs "across" strata. These conflicts must be faced by an inference mechanism capable of handling this type of inconsistent state; argumentative reasoning represents a good alternative to address the inconsistency without falling in the trap of deriving any sentence in the language, *i.e.*, *"ex falso quodlibet"*. Having that aim in mind, we will make use of a formalism previously introduced in [14]; this formalism defines an inference operator that acts over a stratified belief base making use of an argumentation framework to obtain conclusions from a stratified belief base which is potentially inconsistent across strata.

Conceptually, an argument supports a conclusion or claim from a set of premises through some form of reasoning that links these premises with the argument's claim. An argumentation-based inference mechanism considers arguments in favor and against a given conclusion in a confrontational process, extending the analysis to the components of the arguments. Our proposal will introduce a particular argumentation system where all the components will be specified and the argumentative process (akin to a debate) will be precisely described. With the argumentation framework defined, a revision operator for a stratified belief base will be introduced in Section 5 which will permit to consider the addition of information to the SBB, keeping each SBB's stratum locally consistent and dealing with the possible changes in the value assigned to each sentence involved in the process.

4.1 Arguments and Preferences

As we have already discussed, an argument is a piece of reasoning in support of a conclusion built from certain evidence. Arguments thus considered exhibit an internal structure in which certain substructures can be found that share the same general argumentative layout; these substructures, that are argument themselves, are called sub-arguments. Clearly, each subargument is in support a certain conclusion which in turn can be regarded an intermediate conclusion of the original argument. A common property which is enforced for structured arguments is consistency, expressed as the requirement that an argument does not have two subarguments which support contradictory conclusions; this is equivalent to asking that the set of formulas appearing in the reasoning used to infer the conclusion must be consistent [29, 24, 8, 10]. This property eliminates from consideration self-conflicting arguments, and enforces that an argument consistently supports its conclusion.

The definition of argument that will be introduced is similar to the definition of argument structure presented in [29], which has been used in many formalizations of structured argumentation systems (see [4, 9, 27]). Thus, in the context of stratified belief bases, an argument for a conclusion α is a set of consistent formulas that are part of the stratified belief base Σ which is $\subseteq$-minimal and that entails the conclusion. If we consider that an agent keeps its beliefs in the form of a stratified belief base $\Sigma = (\Sigma_0, \ldots, \Sigma_n)$, where Σ_0 contains the beliefs with the highest-value, and these values will be decreasing for each successive stratum; therefore, it is assumed that this stratum represents a set of indisputable beliefs which are the foundation for the other accepted beliefs. We formalize the discussion above in the definitions and remarks that will be introduced next, presenting examples to complete the presentation.

Definition 8 (Argument [14]). *Let $\Sigma = (\Sigma_0, \ldots, \Sigma_n)$ be a stratified belief base and α a sentence. A set of sentences A is an argument for α from Σ, denoted $\langle A, \alpha \rangle_\Sigma$ or simply $\langle A, \alpha \rangle$, if:*

1. $A \subseteq \Sigma_0 \cup \ldots \cup \Sigma_n$.
2. $A \cup \Sigma_0$ *is consistent.*
3. $\alpha \in Cn(A \cup \Sigma_0)$.
4. *There is no $X \subset A$ that satisfies conditions (2) and (3).*

Given the argument $\langle A, \alpha \rangle_\Sigma$, the sentence α is called the conclusion *of the argument, and the set A is called the* support *of α. Sometimes, when no confusion could arise, we will simplify the notation for arguments as $\langle A, \alpha \rangle$.*

Definition 9 (Subargument [14]). *Let Σ be a stratified belief base. An argument $\langle B, \beta \rangle_\Sigma$ is a sub-argument of $\langle A, \alpha \rangle_\Sigma$, if $B \subseteq A$.*

Observation 3 *Given an argument $\langle A, \alpha \rangle_\Sigma$ built from a stratified belief base $\Sigma = (\Sigma_0, \ldots, \Sigma_n)$, the condition (2) of Definition 8 determines that $\neg \alpha \notin Cn(\Sigma_0)$.*

Observation 4 Let Σ be a stratified belief base, and let $\langle A, \alpha \rangle_\Sigma$ be an argument from Σ with sub-arguments $\langle B, \beta \rangle_\Sigma$, and $\langle C, \gamma \rangle_\Sigma$. Since A is consistent (Def. 8), $B \cup C$ is a consistent set of formulas.

Example 3. Consider the stratified belief base $\Sigma = (\Sigma_0, \Sigma_1, \Sigma_2, \Sigma_3, \Sigma_4)$, where:

$$\Sigma_0 = \begin{Bmatrix} w \\ u \\ u \to v \\ \neg b \to a \\ b \to \neg c \end{Bmatrix} \quad \Sigma_1 = \begin{Bmatrix} q \to r \\ s \to \neg q \\ u \to q \\ w \to \neg t \\ d \end{Bmatrix} \quad \Sigma_2 = \begin{Bmatrix} s \\ p \to q \\ t \to \neg r \\ u \to t \\ \neg e \to \neg f \end{Bmatrix} \quad \Sigma_3 = \begin{Bmatrix} p \\ v \\ b \to c \end{Bmatrix} \quad \Sigma_4 = \{b \to a\}$$

Note that for any sentence α in Σ such that $\{\alpha\} \cup \Sigma_0$ is consistent there exists an argument $\langle \{\alpha\}, \alpha \rangle$ (e.g., $\langle \{w\}, w \rangle$, and $\langle \{q \to r\}, q \to r \rangle$). Some of the arguments that can be obtained from Σ follow and we will use them in other examples below:

$\langle A_0, \neg s \rangle$ where $A_0 = \{p, p \to q, s \to \neg q\}$
$\langle A_1, r \rangle$ where $A_1 = \{p, p \to q, q \to r\}$
$\langle A_2, q \rangle$ where $A_2 = \{p, p \to q\}$
$\langle A_3, \neg q \rangle$ where $A_3 = \{s, s \to \neg q\}$
$\langle A_4, \neg r \rangle$ where $A_4 = \{u, u \to t, t \to \neg r\}$
$\langle A_5, t \rangle$ where $A_5 = \{u, u \to t\}$
$\langle A_6, q \rangle$ where $A_6 = \{u, u \to q\}$
$\langle A_7, \neg t \rangle$ where $A_7 = \{w, w \to \neg t\}$

As an example, note that $\langle A_2, q \rangle$ is a sub-argument of both $\langle A_1, r \rangle$ and $\langle A_0, \neg s \rangle$, and that $\langle A_5, t \rangle$ is a sub-argument of $\langle A_4, \neg r \rangle$.

Since arguments can use sentences of different strata in their structure, the comparison of arguments will consider such information. The value of a set of sentences will be obtained from the strata the sentences belong to; arguments will have a higher value if they are built with sentences of a stratum with a lower index, where Σ_0 contains the highest valued sentences. Since a sentence can belong to more of one stratum, the stratum with lowest-index (thus with the greatest value) should be chosen. Therefore, the *index* of a sentence β in Σ, denoted $index_\Sigma(\beta)$, is the index corresponding to the lowest indexed stratum –and hence most valued– in Σ to which β belongs to.

Definition 10 (Stratum [14]). Let $\Sigma = (\Sigma_0, \ldots, \Sigma_n)$ be a stratified belief base such that $\beta \in (\Sigma_0 \cup \ldots \cup \Sigma_n)$. Then, $index_\Sigma(\beta) = i$ if and only if $\beta \in \Sigma_i$ and there is no Σ_j $(j < i)$ such that $\beta \in \Sigma_j$. We say that i is the stratum of β.

Therefore, using the weakest-link principle, the stratum of an argument is defined by the index of its weakest sentence.

Definition 11 (Stratum of a set [14]). Let Σ be a stratified belief base and $A = \{x_1, x_2, \ldots, x_m\}$ be a set of sentences in Σ. The stratum of the set A is

$$str_\Sigma(A) = \max\{index_\Sigma(x_1), index_\Sigma(x_2), \ldots, index_\Sigma(x_m)\}$$

Example 4. Consider Σ of Example 3, then it holds that $index_\Sigma(w) = 0$, $index_\Sigma(v) = 3$, and $index_\Sigma(u \to q) = 1$. It also holds that

$str_\Sigma(\langle\{w\}, w\rangle) = 0,$ $str_\Sigma(\langle\{u\}, u\rangle) = 0,$ $str_\Sigma(\langle\{s\}, s\rangle) = 2,$
$str_\Sigma(\langle\{v\}, v\rangle) = 3,$ $str_\Sigma(\langle\{u, u \to v\}, v\rangle) = 0,$ $str_\Sigma(\langle A_1, r\rangle) = 3,$
$str_\Sigma(\langle A_2, q\rangle) = 3,$ $str_\Sigma(\langle A_3, \neg q\rangle) = 2,$ $str_\Sigma(\langle A_4, \neg r\rangle) = 2,$
$str_\Sigma(\langle A_5, t\rangle) = 2,$ $str_\Sigma(\langle A_6, q\rangle) = 1,$ $str_\Sigma(\langle A_7, \neg t\rangle) = 1$

We will introduce now the order "$\succeq_\Sigma$" over the set of arguments that can be built from Σ; this order is based on $str_\Sigma(\cdot)$. Note that, as before, a lower number represents a greater value.

Definition 12 (Argument Comparison [14]). *Let Σ be a stratified belief base and let $\langle B, \beta\rangle_\Sigma$ and $\langle A, \alpha\rangle_\Sigma$ be two arguments from Σ. We will say that $\langle B, \beta\rangle_\Sigma$ is as least as good than $\langle A, \alpha\rangle_\Sigma$, denoted as $\langle B, \beta\rangle_\Sigma \succeq_\Sigma \langle A, \alpha\rangle_\Sigma$, if and only if $str_\Sigma(B) \leq str_\Sigma(A)$.*

The result bellow reflects the fact that a subargument, for its own nature of being a part of a bigger structure, cannot be weaker than any of its super-arguments.

Proposition 1. *Let Σ be a stratified belief base, if $\langle B, \beta\rangle_\Sigma$ is a sub-argument of $\langle A, \alpha\rangle_\Sigma$ then $\langle B, \beta\rangle_\Sigma \succeq_\Sigma \langle A, \alpha\rangle_\Sigma$.*

4.2 Defeat among Arguments

When a pair of arguments have conclusions that are contradictory these arguments interfere which each other, *i.e.*, they are in a relation of *conflict*. In the Example 3, the arguments $\langle A_2, q\rangle$ and $\langle A_3, \neg q\rangle$ are in conflict since their conclusions are contradictory. Moreover, given two conflicting arguments A and B, such as $B \succeq_\Sigma A$, it is only natural that the interference of B over A be regarded as successful; that is to say, argument B prevails over A, or in other words B *defeats* A, introducing the *defeat* relation which is a derived relation based on the existence of conflict and takes advantage of the order $\succeq_\Sigma$.

The defeat relation can take different forms since defeat can be achieved by different means. We have discussed a particular type of defeat to introduce the concept; in that intuitive discussion, the attack was aimed at the conclusion, this form of defeat is called *rebuttal*, but other kinds of defeat exists leading to a richer relation. It is possible to effect the attack indirectly by directing it to an intermediate step in the reasoning, this form of defeat is called *undercut*. In the framework presented, the undercut defeaters will be achieved by an argument attacking a sub-argument of another argument, *i.e.*, an undercut is a rebuttal over a sub-argument of the argument that receives the attack. There exists another form of attack, called *assumption attack*, where the attacker's aim is one or more of the assumptions made by the argument that constitutes the attacker's target, but since we will not use assumptions in our formalism, this latter form of attack will not appear in our presentation. The conflict and defeat formal definitions of conflict and defeat appear below.

Definition 13 (Argument Conflict [14]). *Let Σ be a stratified belief base. Two arguments $\langle A, \alpha \rangle_\Sigma$ and $\langle B, \beta \rangle_\Sigma$ are in conflict, if $\Sigma_0 \cup \{\alpha, \beta\}$ is an inconsistent set.*

It is important to examine the intuition behind making the stratum Σ_0 the context over which the inconsistency of the conclusions of two arguments is analyzed. The stratum Σ_0, is a set of consistent beliefs (as all strata are when considered individually) with the particular property of having maximum value of all strata attached to its sentences; therefore, as Σ_0 is a consistent set, any argument that is built, by the constraints imposed on its construction, should be consistent with it. Moreover, since each conclusion by itself is consistent with Σ_0, the conclusions of conflicting arguments must be contradictory with each other, possibly through the use of an appropriate subset of the beliefs in Σ_0. The example below shows the general case.

Example 5. Consider the stratified belief base $\Sigma' = (\Sigma'_0, \Sigma'_1)$, where:

$$\Sigma'_0 = \begin{Bmatrix} x \\ y \\ k \to z \\ h \to \neg z \end{Bmatrix} \quad \Sigma'_1 = \begin{Bmatrix} x \to k \\ y \to h \\ \neg x \end{Bmatrix}$$

The arguments $\langle A, k \rangle$ with $A = \{x, x \to k\}$ and $\langle B, h \rangle$ with $B = \{y, y \to h\}$ are in conflict even though their conclusions are not in direct contradiction.

Definition 14 (Argument Defeat [14]). *Let Σ be a stratified belief base. An argument $\langle B, \beta \rangle_\Sigma$ defeats $\langle A, \alpha \rangle_\Sigma$, if there exists a sub-argument $\langle C, \gamma \rangle_\Sigma$ of $\langle A, \alpha \rangle_\Sigma$, such that the following two conditions hold:*

- *$\langle B, \beta \rangle_\Sigma$ and $\langle C, \gamma \rangle_\Sigma$ are in conflict, and*
- *$\langle B, \beta \rangle_\Sigma \succeq_\Sigma \langle C, \gamma \rangle_\Sigma$.*

We will say that $\langle B, \beta \rangle_\Sigma$ defeats $\langle A, \alpha \rangle_\Sigma$ at sub-argument $\langle C, \gamma \rangle_\Sigma$.

Example 6. Consider the stratified belief base of Example 3. Argument $\langle A_3, \neg q \rangle$ defeats $\langle A_1, r \rangle$ (at the sub-argument $\langle A_2, q \rangle$), and $\langle A_6, q \rangle$ defeats $\langle A_3, \neg q \rangle$. Argument $\langle A_4, \neg r \rangle$ defeats $\langle A_1, r \rangle$ and $\langle A_7, \neg t \rangle$ defeats $\langle A_4, \neg r \rangle$ (at the sub-argument $\langle A_5, t \rangle$).

Consider the arguments $\langle C, \gamma \rangle_\Sigma$ and $\langle A, \alpha \rangle_\Sigma$ are in conflict and suppose that $str_\Sigma(C) = str_\Sigma(A)$. Then it holds that $\langle C, \gamma \rangle_\Sigma \succeq_\Sigma \langle A, \alpha \rangle_\Sigma$ and also that $\langle A, \alpha \rangle_\Sigma \succeq_\Sigma \langle C, \gamma \rangle_\Sigma$. Hence, in this particular case, it holds that $\langle C, \gamma \rangle_\Sigma$ defeats $\langle A, \alpha \rangle_\Sigma$ and $\langle A, \alpha \rangle_\Sigma$ defeats $\langle C, \gamma \rangle_\Sigma$.

As a consequence of the way the comparison criterion "$\succeq_\Sigma$" is defined, if B defeats A, then B will also defeat every super-argument of A, that is, B also defeats every argument C such that A is a sub-argument of C (see Proposition 1).

Observation 5 *Let Σ be a stratified belief base. If $\langle B, \beta \rangle_\Sigma$ is a sub-argument of $\langle A, \alpha \rangle_\Sigma$, and $\langle C, \gamma \rangle_\Sigma$ is a sub-argument of $\langle A, \alpha \rangle_\Sigma$, then $\langle B, \beta \rangle_\Sigma$ cannot defeat $\langle C, \gamma \rangle_\Sigma$ and vice versa. (see Observation 4).*

Proposition 2. Let $\langle A, \alpha \rangle_\Sigma$ be an argument. If $A \subseteq \Sigma_0$ then there is no argument $\langle B, \beta \rangle$ such that $\langle B, \beta \rangle_\Sigma$ defeats $\langle A, \alpha \rangle_\Sigma$.

Corollary 2. Given an argument $\langle A, \alpha \rangle_\Sigma$. If $str_\Sigma(A) = 0$ then there exists no argument that can defeat $\langle A, \alpha \rangle_\Sigma$.

So far, we have introduced an argumentation framework that provides an argument-based inferential mechanism that will obtain sets of beliefs supported from the stratified belief base, including the implicit beliefs. The set of beliefs supported from the SBB is the set of warranted arguments, also known as warranted beliefs; and, for obtaining them, we will use the semantics defined for abstract argumentation frameworks.

An abstract argumentation framework –as introduced in [13]– is described by a pair $\langle \mathfrak{Args}, \mathbf{R} \rangle$, where $\mathfrak{Args}$ is a finite set of arguments and $\mathbf{R}$ is a binary relation between arguments such that $\mathbf{R} \subseteq \mathfrak{Args} \times \mathfrak{Args}$. In the original presentation in [13], the notation $(A, B) \in \mathbf{R}$ (or, equivalently, $A \mathbf{R} B$) is described by saying that A *attacks* B; this relation corresponds to the notion of defeat introduced in this work, i.e., in Dung's formalism every attack was successful effecting defeat, therefore we will consistently use defeat instead of attack.

A key concept in abstract argumentation is that of argumentation semantics which defines sets of arguments that are acceptable together in the context of the attack relation. Different semantics have been defined in the original work of Dung, and others have been later introduced in the literature. The arguments contained in the sets produced by a particular semantics will be considered as being able to support their conclusions. Two different approaches to find these sets have been investigated: extension-based and labeling-based argumentation semantics; the extension-based is a declarative definition, whereas the labelling-based is procedural. An extension is a subset of arguments contained in the framework, and the extension-based approach specifies how to obtain the subsets that form the set of extensions. The labeling-based approach assigns a label from the set $\{in, out, undecided\}$ to each argument in the framework; the assignment of labels yields a set of labelings that correspond to the extensions found through the extension-based declarative method.

As we mentioned Dung [13] introduces several argumentation semantics which permit to evaluate the status of the arguments in the framework building extensions, e.g., *complete*, *grounded*, *stable*, and *preferred* semantics. Other semantics have been proposed after the initial definition, e.g., *stage*, *semi-stable*, *ideal*, *CF2*, and *prudent* semantics among others, see [2] for detailed exploration of this issue.

From a stratified belief base Σ, by applying the argumentation formalism defined above, it is possible to define an argumentation framework $\langle \mathfrak{Args}, \mathbf{R} \rangle$; to obtain the warranted conclusions one of the semantics mentioned can be applied. In particular, we will favor a skeptical approach to finding the warranted arguments. An argument will be skeptically warranted if and only if it is warranted in all the extensions produced by the chosen semantics. Since grounded semantics only produces one extension, it provides an example of skeptical semantics. We will use this semantics in the example below.

Going back to Example 3, we can define the argumentation framework $\langle \mathfrak{Args}, \mathbf{R} \rangle$, where $\{A_0, A_1, A_2, A_3, A_4, A_5, A_6, A_7\} \subseteq \mathfrak{Args}$ and $\mathbf{R}$ contains, for instance, the following tuples:

$$\{(A_3, A_2), (A_3, A_1), (A_4, A_1), (A_7, A_4), (A_7, A_5), (A_6, A_3)\}$$

Using grounded semantics [13], the arguments A_1, A_2, A_6, and A_7 are warranted; therefore, b, c, and $\neg f$ can be inferred from our stratified belief base, among others.

The choice of the particular semantics used to obtain the warranted arguments is a design decision that affects the inferential behavior of the system; for more details regarding different aspects of skeptical semantics in abstract argumentation frameworks see [1, 3]. The reader may obtain a general perspective on argumentation in general, and argumentation semantics in particular, in [4], or in [8, 27] where deeper and detailed accounts are presented.

5 Revision on Stratified Belief Bases

In this section we will introduce the definition of a revision operator on stratified belief bases and the postulates characterizing it [14]. This operator will allow for a multi-level reuse of discarded beliefs from the revision of each stratum.

After having presented the necessary elements to handle belief change by a set of sentences on a single belief base in Section 3 we will now define expansion and revision operators on stratified belief bases.

Definition 15. *Let $\Sigma = (\Sigma_0, \ldots, \Sigma_n)$ be a stratified belief base and let A be a consistent set. Then the expansion of Σ with respect to A at level i, $0 \leqslant i \leqslant n$, noted by $\Sigma + (A, i)$, is equal to:*

$$\Sigma' = (\Sigma_0, \ldots, \Sigma_{i-1}, \Sigma_i \cup A, \Sigma_{i+1}, \ldots, \Sigma_n)$$

Note that when $\Sigma'_i = \Sigma_i \cup A$ is inconsistent the resulting Σ' cannot be a stratified belief base because it will not satisfy the conditions of its definition, i.e., the expansion of a SBB *might not be a* SBB.

As stated in Section 2, now we are able to define a revision operator on stratified belief bases. The discarded beliefs from the revision of each stratum will be reused and, as a result of this, dynamically changing the value of beliefs.

Definition 16. *Let $*^j$ be a prioritized revision operator (either kernel or partial meet) on every $\Sigma_j \in \Sigma$, for all $0 \leqslant j \leqslant n$. This introduces a family of revision operators of the same type: kernel or partial meet. Let A be a consistent set of beliefs, accordingly with the type, the kernel or partial meet revision of Σ with respect to A at level i, $0 \leqslant i$, noted by $\Sigma * (A, i)$, is defined as:*

- *If $i > n$ then $\Sigma * (A, i) = \Sigma$.*
- *If $i \leqslant n$ then $\Sigma * (A, i) = \Sigma' = (\Sigma_0, \Sigma'_1, \ldots, \Sigma'_m)$ where $m = n$ or $m = n+1$ and:*
  - *$\Sigma'_j = \Sigma_j$ for all $j < i$.*

- $\Sigma'_i = \Sigma_i *^i A$ and $R_{i+1} = \Sigma_i \setminus \Sigma'_i$, and then:
  a) $\Sigma'_{i+k} = \Sigma_{i+k} *^{i+k} R_{i+k}$ and $R_{i+k+1} = \Sigma_{i+k} \setminus \Sigma'_{i+k}$ for $1 \leq k \leq n-i$.
  b) If $R_{n+1} \neq \emptyset$ then a new stratum is created as $\Sigma'_{n+1} = R_{n+1}$.

That is, our multiple revision operator can be defined by means of an iterative/recursive algorithm. We suppose that each $*^i$ is a prioritized multiple revision operator for Σ_i. The result of $\Sigma *(A, i)$ (the revision of Σ with respect to A at the level i) is $\Sigma' = (\Sigma'_1, \ldots, \Sigma'_m)$, such that in general, for all $j, 1 \leq j \leq n$:

$$\Sigma'_j = \Sigma_j *^j R_j,$$

where the R_j's are defined as follows:

- $R_j = \emptyset$ for $0 \leq j < i$, so $\Sigma'_j = \Sigma_j$ for all $j < i$;
- $R_j = A$ for $j = i$;
- if $\Sigma_j \setminus \Sigma'_j \neq \emptyset$ then $R_{j+1} = \Sigma_j \setminus \Sigma'_j$.

The revised stratified belief base can have an additional layer when R_{n+1} is not empty. Since every layer Σ_i ($1 \leq i \leq n$) is consistent, then every subset of it is also consistent. Therefore, $R_{i+1} \subseteq \Sigma_i$ is always consistent and the revision of every layer is well defined.

We have assumed that all revision operators applied on the layers satisfy (at least) Inclusion, Consistency, Weak Success, Vacuity 1 and 2, Uniformity 1 (and 2), and Core-Retainment, according to Theorems 1 and 2. This work generalizes the proposal in [16] where only the part of the represented knowledge corresponding to non-defeasible rules could be demoted transforming them in defeasible rules; here it is possible that any sentence could be demoted and stored at a lower level. We will illustrate this revision operator below with some examples.

Note that rejection of the new information is not only justified when such information is internally inconsistent: it would seem a natural behavior that an agent when faced with (consistent) information of extremely poor quality, may reasonably opt for *flat-out rejection*, without incorporating it to its stratified belief base at any layer. An interesting aspect of this approach is that the priority given to new information is relative to its value, in that manner it is possible to incorporate it in the proper place without disturbing the information stored in higher value levels.

6 A New Set of Operators

In [14] a revision operator is defined which allows the addition of sentences to a stratum in a stratified belief base. Here we propose a new set of operators for stratified belief bases.

Let Σ be a stratified belief base, A a consistent set of sentences, and i a stratum of Σ. We identify the following family of change operators:

Addition of Beliefs. We identify just one case of change: expansion. The expansion was presented in Definition 15 and it can be applied in some special circumstances, since every expansion could not be a stratified belief base.

Deletion of Beliefs. We identify two kinds of change:
- **Remove beliefs** (noted by $\Sigma \div (i, A)$): This operation removes the set of sentences of A from Σ at the stratum i. Remove can be defined as *multiple package contraction*, as it was suggested in [18] and adapted for belief bases in [17].
- **Clean beliefs** (noted by $\Sigma \div A$): this operation remove the set of sentences of A throughout all the strata of Σ. *Clean* operation can be defined in terms of a set of *Remove* operations.

Change of Beliefs. We identify three kinds of change:
- **Revise Beliefs**. Revision was presented in Definition 16.
- **Decrease Beliefs** (noted by $\Sigma \Downarrow (i, A)$): This operation takes place when a set of sentences A at the stratum i loses credibility or support.
- **Increase Beliefs** (noted by $\Sigma \Uparrow (i, A)$): this operation takes place when a set of sentences A at the stratum i turns more credible.

The *Decrease* and *Increase* operators can be defined in terms of *Revise* and *Remove* operators.

In the next subsections we will present the motivations of each new operator and the possible constructions for them.

6.1 The Remove Operator

The remove operator allows the elimination of a set of beliefs from a stratum in a stratified belief base. When should this type of operation occur? The answer to this question could be: "in those cases where it is discovered that such beliefs are no longer supported or do not contribute to the construction of *undefeated arguments*".

The elimination or contraction of beliefs has been widely debated in the belief change theory. In general, it is considered that there are no direct examples of contraction in real examples, except in cases of forgetfulness or memory loss. However, there is a consensus that the contraction could be a suboperation of a more complex operation such as revision.

Therefore, elimination of multiple beliefs on a stratum can be modelled by means of a multiple contraction operator. Fuhrmann and Hansson [18] propose two ways of multiple contraction on belief sets: *package contraction*, where all sentences of a set are removed from a belief set, and *choice contraction*, where at least one sentence of a set of sentences is removed from a belief set. Fermé et al [17] propose either package and choice multiple kernel contraction for belief bases. We will adopt multiple package kernel contraction for our model.

The following concepts were extracted from [17] where representation theorems for package multiple kernel contraction for belief bases are presented.

Definition 17. Let Σ_i, A be set in $\mathcal{L}$. The package kernel set of Σ_i by A, noted by $\Sigma_i \perp\!\!\!\perp_P A$, is the set such that $X \in \Sigma_i \perp\!\!\!\perp_P A$ if and only if:

1. $X \subseteq \Sigma_i$,

2. $A \cap Cn(X) \neq \emptyset$,

3. If $Y \subset X$ then $A \cap Cn(Y) = \emptyset$.

Definition 18. *A function σ is an* incision function *for Σ_i if and only if it satisfies the following conditions for every set A:*

1. $\sigma(\Sigma_i \perp\!\!\!\perp_P A) \subseteq \bigcup(\Sigma_i \perp\!\!\!\perp_P A)$,

2. *If $\emptyset \subset X \in \Sigma_i \perp\!\!\!\perp_P A$ then $X \cap (\Sigma_i \perp\!\!\!\perp_P A) \neq \emptyset$.*

Definition 19. *Let σ be an incision function for Σ_i. The* multiple package kernel contraction $\div$ *for Σ_i is defined as follows for every set A of $\mathcal{L}$.:*

$$\Sigma_i \div A = S \setminus \sigma(\Sigma_i \perp\!\!\!\perp_P A)$$

Theorem 3. *An operator $\div$ for a set Σ_i is an operator of* multiple package kernel contraction *if and only if it satisfies the following conditions:*

Package Success: *If $A \cap Cn(\emptyset) = \emptyset$ then $A \cap Cn(\Sigma_i \div A) = \emptyset$.*

Package Inclusion: $\Sigma_i \div A \subseteq S$.

Package Core Retainment: *If $\beta \in \Sigma_i$ and $\beta \notin \Sigma_i \div A$ then there exists a set B such that $B \subseteq \Sigma_i$, $A \cap Cn(B) = \emptyset$ but $A \cap Cn(B \cup \{\beta\}) \neq \emptyset$.*

Package Uniformity: *If every subset X of Σ_i implies some element of A if and only if implies some element of B then $\Sigma_i \div A = \Sigma_i \div B$.*

The above operator can be defined in terms package remainder sets [18], generating a multiple package partial meet contraction operator, but this will not included in current paper.

Definition 20. *Let $\Sigma = (\Sigma_0, \ldots, \Sigma_n)$ be a stratified belief base, A be a set of sentences and $\div$ be a multiple package kernel contraction operator for the stratum i of Σ. The* remove operator *of Σ with respect A at level i is defined as:*

$$\Sigma \div (i, A) = (\Sigma'_0, \ldots, \Sigma'_n)$$

where $\Sigma'_j = \Sigma_j$ for all $j \neq i$ and $\Sigma'_i = \Sigma_i \div A$.

6.2 The Clean Operator

The clean operator allows the elimination of a set of beliefs from every stratum in a stratified belief base. When should this type of operation occur? The answer to this question could be: "in those cases where it is discovered that such beliefs are false and they do not contribute to the construction of good *arguments*". That is, clean operator is the most drastic operator of elimination of beliefs in a stratified belief base. The clean operator can be defined for a stratified belief base Σ by a multiple package contraction operator for every stratum of the Σ.

Definition 21. *Let $\Sigma = (\Sigma_0, \ldots, \Sigma_n)$ be a stratified belief base and let $\div_i$ be a remove operator for each stratum i (that is, $\div_0$ is a remove operator for stratum 0, $\div_1$ is a remove operator for stratum 1, ..., $\div_n$ is a remove operator for stratum n. The* clean operator *of Σ with respect to A is defined as:*

$$\Sigma \div A = (\Sigma'_0, \ldots, \Sigma'_n)$$

such that $\Sigma'_i = \Sigma \div_i (i, A)$ for all $0 \leq i \leq n$.

6.3 Decrease and Increase Operators

If we have multiple sources of information, in many cases they win or lose credibility. A good formalism to treat with credibilities is presented in [30]. Here we propose different ways of changing credibility of beliefs, allowing some beliefs become more (or less) credible after a change process.

The way we propose to decrease a set of sentences A from a stratified belief base Σ at the level i, for $i \leqslant n$, is as follows:

1. Eliminate A from Σ_i.
2. Let A' be such that $A' = A \cap Cn(\Sigma_i)$. Then, we have two cases:
   (a) If $i = n$ then add a new stratum $n+1$ composed from A.
   (b) If $i < n$ then the new stratified belief bases is equal to the revision of the resulting stratified belief base with respect to A' at the level $i+1$.

Since every stratum of Σ is consistent, A' is a consistent subset of A. This filtering of A is to ensure that we are decreasing beliefs that actually are in the stratum i. Formally, we define the decrease operator as follows.

Definition 22. *Let $\Sigma = (\Sigma_0, \ldots, \Sigma_n)$ be a stratified belief base, $\div$ be a (kernel/partial meet) remove operator for Σ and $*$ be a (kernel/partial meet) revision operator for Σ. Let A' be such that $A' = A \cap Cn(\Sigma_i)$. Then, the decrease operator for Σ with respect to A at the level i for $i \leqslant n$ is equal to $\Sigma \Downarrow (i, A) = \Sigma'''$ such that:*

- $\Sigma' = \Sigma \div (i, A)$.
- *If $i = n$ then $\Sigma''' = (\Sigma'_0, \ldots, \Sigma'_n, A)$.*
- *If $i < n$ then $\Sigma''' = \Sigma' * (A', i+1)$.*

In a similar way, we propose to increase a set of sentences A from a stratified belief base Σ at the level i, for $0 < i$, as follows:

1. Eliminate A from Σ_i.
2. Let A' be such that $A' = A \cap Cn(\Sigma_i)$. The new stratified belief bases is equal to the revision of the resulting stratified belief base with respect to A' at the level $i-1$.

Again, since every stratum of Σ is consistent, A' is a consistent subset of A. This filtering of A is to ensure that we are increasing beliefs that actually are in the stratum i. Formally, we define the decrease operator as follows.

Definition 23. *Let $\Sigma = (\Sigma_0, \ldots, \Sigma_n)$ be a stratified belief base, $\div$ be a (kernel/partial meet) remove operator for Σ and $*$ be a (kernel/partial meet) revision operator for Σ. Let A' be such that $A' = A \cap Cn(\Sigma_i)$. Then, the increase operator for Σ with respect to A at the level i for $0 < i$ is equal to $\Sigma \Uparrow (i, A) = \Sigma'''$ such that:*

- $\Sigma' = \Sigma \div (i, A)$.
- $\Sigma''' = \Sigma' * (A', i-1)$.

Example 7. Consider the stratified belief base $\Sigma = (\Sigma_0, \Sigma_1, \Sigma_2)$, where:

$$\Sigma_0 = \begin{cases} a \\ a \to b \\ \neg c \to \neg b \\ \neg q \end{cases} \quad \Sigma_1 = \begin{cases} p \\ p \to q \\ \neg r \end{cases} \quad \Sigma_2 = \begin{cases} s \\ a \to \neg b \\ \neg p \\ \neg q \end{cases}$$

If we want to increase $\{p, q\}$ in Σ at level 1 we first remove p and q from the stratum 1, generating $\Sigma' = (\Sigma'_0, \Sigma'_1, \Sigma'_2)$ where:

$$\Sigma'_0 = \begin{cases} a \\ a \to b \\ \neg c \to \neg b \\ \neg q \end{cases} \quad \Sigma'_1 = \{\neg r\} \quad \Sigma'_2 = \begin{cases} s \\ a \to \neg b \\ \neg p \\ \neg q \end{cases}$$

After that, we obtain $\Sigma'' = (\Sigma''_0, \Sigma''_1, \Sigma''_2)$ where:

$$\Sigma''_0 = \begin{cases} a \\ a \to b \\ \neg c \to \neg b \\ p \\ q \end{cases} \quad \Sigma''_1 = \begin{cases} \neg q \\ \neg r \end{cases} \quad \Sigma''_2 = \begin{cases} s \\ a \to \neg b \\ \neg p \\ \neg q \end{cases}$$

Example 8. Consider the stratified belief base Σ of the Example 7. If we want to decrease $\{p, q, t\}$ in Σ at level 1 we first remove p, q and t from the stratum 1, generating the same Σ' of the above example. After that, we obtain $\Sigma'' = (\Sigma''_0, \Sigma''_1, \Sigma''_2, \Sigma''_3)$ where:

$$\Sigma''_0 = \begin{cases} a \\ a \to b \\ \neg c \to \neg b \\ \neg q \end{cases} \quad \Sigma''_1 = \{\neg r\} \quad \Sigma''_2 = \begin{cases} s \\ a \to \neg b \\ p \\ q \end{cases} \quad \Sigma''_3 = \begin{cases} \neg p \\ \neg r \end{cases}$$

Note that t is not added to the stratum 2 because t was not a consequence of stratum 1.

7 Conclusions

In this article we have presented a new set of operators for stratified belief bases. Taking as starting point previous works, we have proposed the following new operators: *remove*, which allows the elimination of a set of sentences from a stratum; *clean*, which allows the elimination of a set of sentences throughout all strata of a stratified belief base; *decrease*, which allows to diminish the credibility of a set of sentences; and *increase*, which allows to raise the credibility of a set of sentences.

It is important to note that new operators emerged from the practical use of reasoning systems in which stratified belief bases are used to represent knowledge and argumentative reasoning is used as inference mechanism. These operators are necessary for several reasons, among which it is possible to mention:

- Allow changing the credibility of beliefs.

- Remove "junk" beliefs that are left as a result of multiple processes of change but do not contribute to the construction of undefeated arguments or block some "reasonable" arguments.
- Clean a stratified belief base from "wrong" beliefs.

Clearly, the application of the new operators will require a deeper analysis aimed to decide which beliefs to remove and/or how to modify their credibility.

References

1. Pietro Baroni and Massimiliano Giacomin. Comparing argumentation semantics with respect to skepticism. In *9th European Conference on Symbolic and Quantitative Approaches to Reasoning with Uncertainty, (ECSQARU 2007), Hammamet, Tunisia*, pages 210–221, 2007.
2. Pietro Baroni and Massimiliano Giacomin. Semantics of abstract argument systems. In Iyad Rahwan and Guillermo R. Simari, editors, *Argumentation in Artificial Intelligence*, pages 24–44. Springer, 2009.
3. Pietro Baroni and Massimiliano Giacomin. Skepticism relations for comparing argumentation semantics. *International Journal Approximate Reasoning*, 50(6):854–866, 2009.
4. Trevor J. M. Bench-Capon and Paul E. Dunne. Argumentation in artificial intelligence. *Artificial Intelligence*, 171(10-15):619–641, 2007.
5. S. Benferhat, D. Dubois, and H. Prade. Argumentative inference in uncertain and inconsistent knowledge bases. In D. Heckerman and E. H. Mamdani, editors, *Proceedings of Ninth Annual Conference on Uncertainty in Artificial Intelligence, UAI 1993*, pages 411–419. Morgan Kaufmann, 1993.
6. Salem Benferhat, Didier Dubois, and Henri Prade. How to infer from inconsistent beliefs without revising. In *Proceedings of IJCAI 1995*, pages 1449–1455, 1995.
7. Salem Benferhat, Souhila Kaci, Daniel Le Berre, and Mary-Anne Williams. Weakening conflicting information for iterated revision and knowledge integration. *Artificial Intelligence Journal*, 153(1-2):339–371, 2004.
8. Philippe Besnard and Anthony Hunter. A logic-based theory of deductive arguments. *Artificial Intelligence*, 128(1-2):203–235, 2001.
9. Philippe Besnard and Anthony Hunter. *Elements of Argumentation*. MIT Press, 2008.
10. Philippe Besnard and Anthony Hunter. Argumentation based on classical logic. In Iyad Rahwan and Guillermo R. Simari, editors, *Argumentation in Artificial Intelligence*, pages 133–152. Springer, 2009.
11. Gerhard Brewka. Preferred subtheories: An extended logical framework for default reasoning. In *Inf Proceedings of Eleventh Joint Conference on Artificial Intelligence, (IJCAI 1989)*, volume 2, pages 1043–1048, San Mateo, Ca., 1989.
12. James P. Delgrande, Torsten Schaub, Hans Tompits, and Stefan Woltran. Belief revision of logic programs under answer set semantics. In Gerhard Brewka and Jérôme Lang, editors, *KR 2008*, pages 411–421. AAAI Press, 2008.
13. Phan Minh Dung. On the Acceptability of Arguments and its Fundamental Role in Nonmonotonic Reasoning, Logic Programming and n-Person Games. *Artificial Intelligence*, 77(2):321–358, 1995.
14. Marcelo A. Falappa, Alejandro J. García, Gabriele Kern-Isberner, and Guillermo R. Simari. Stratified belief bases revision with argumentative inference. *The Journal of Philosophical Logic*, 42(1):161–193, 2013.

15. Marcelo A. Falappa, Gabriele Kern-Isberner, Maurício D.L̃. Reis, and Guillermo R. Simari. Prioritized and non-prioritized multiple change on belief bases. *The Journal of Philosophical Logic*, 41(1):77–113, 2012.
16. Marcelo A. Falappa, Gabriele Kern-Isberner, and Guillermo R. Simari. Belief Revision, Explanations and Defeasible Reasoning. *Artificial Intelligence Journal*, 141:1–28, 2002.
17. Eduardo L. Fermé, Karina Saez, and Pablo Sanz. Multiple Kernel Contraction. *Studia Logica*, 73(2):183–195, 2003.
18. André Fuhrmann and Sven Ove Hansson. A Survey of Multiple Contractions. *The Journal of Logic, Language and Information*, 3:39–76, 1994.
19. Peter Gärdenfors. *Knowledge in Flux: Modelling the Dynamics of Epistemic States*. The MIT Press, Bradford Books, Cambridge, Massachusetts, 1988.
20. Peter Gärdenfors and David Makinson. Revisions of Knowledge Systems using Epistemic Entrenchment. *Second Conference on Theoretical Aspects of Reasoning About Knowledge*, pages 83–95, 1988.
21. Sven Ove Hansson. *A Textbook of Belief Dymanics: Theory Change and Database Updating*. Kluwer Academic Publishers, 1999.
22. Sten Lindström and Wlodzimierz Rabinowicz. Epistemic Entrenchment with Incomparabilities and Relational Belief Revision. In André Fuhrmann and Michael Morreau, editors, *The Logic of Theory Change*, volume 465 of *Lecture Notes in Computer Science*, pages 93–126. Springer, 1989.
23. Pavlos Peppas. The Limit Assumption and Multiple Revision. *Journal of Logic and Computation*, 14(3):355–371, 2004.
24. John L. Pollock. Justification and defeat. *Artificial Intelligence*, 67:377–407, 1994.
25. David Poole. A logical framework for default reasoning. *Artif. Intell.*, 36(1):27–47, 1988.
26. David Poole, Romas Aleliunas, and Randy Goebel. Theorist: A Logical Reasoning System for Defaults and Diagnosis. Technical report, University of Waterloo, Department of Computer Science, Waterloo, Canada, 1985.
27. Iyad Rahwan and Guillermo R. Simari. *Argumentation in Artificial Intelligence*. Springer, 2009.
28. Hans Rott. Preferential Belief Change using Generalized Epistemic Entrenchment. *The Journal of Logic, Language and Information*, 1:45–78, 1992.
29. Guillermo R. Simari and Ron P. Loui. A mathematical treatment of defeasible reasoning and its implementation. *Artificial Intelligence*, 53:125–157, 1992.
30. Luciano H. Tamargo, Alejandro J. García, Marcelo A. Falappa, and Guillermo R. Simari. On the revision of informant credibility orders. *Artificial Intelligence*, 212:36–58, 2014.

On the Iteration of KM-Update

Eduardo Fermé[1], Sara Gonçalves[2]

Abstract. In this paper we present a model for iteration of Katsuno and Mendelzon Update, inspired in those developed for iteration in AGM belief revision. We adapt Darwiche and Pearls' postulates of belief revision to update (as well as the *independence postulate*) and show two families of such operators, based in natural and lexicographic revision. In all cases, we provide a possible worlds semantics of the models.

Keywords: Belief Change, KM-Update, Iteration, Possible World Semantic.

1 Introduction

Belief change is the process through which an agent should modify its beliefs upon the arising of new information and it is present in our day to day lives. Even though we are not aware, changes occur everywhere, every time. These changes lead to a constant renewal of our beliefs. We end up accepting new beliefs, revising old ones and even eliminating some of them. Formalizing this, however, is not a simple problem, since there is not an unique way to perform these changes.

Belief change, when referred to in its wide sense, has been a subject of interest for almost as long as man itself exists. In recent years, its study was addressed in two large research traditions: philosophy and computer science. Numerous contributions in the field of artificial intelligence were given [5, 26, 9, 35, 32] (some of this models are summarized in [17]). In 1983, Fagin, Ullman and Vardi pointed out the necessity of defining the dynamics of the process of update in a database [12]. In philosophy the first steps in modern belief change were done with the works of Levi [25], Harper [14] and von Wright [34]. Through the study of the mechanisms by which scientific theories are developed, philosophers proposed criteria of rationality for revisions of probability assignments, providing much of the basic formal framework of belief change.

The most important model of belief change is now known as the AGM model of belief change, was proposed in 1985 by Alchourrón, Gärdenfors and Makinson in [1]. The AGM model is a formal framework to characterize the dynamics and state of belief of a rational agent. In the AGM framework, the beliefs are represented ideally by belief sets, which are deductively closed sets of sentences. A change consists in adding or removing a specific sentence from/to a belief set to obtain a new (consistent) belief set.

---

[1] University of Madeira and NOVA Laboratory of Computer Science and Informatics, Portugal, `ferme@uma.pt`

[2] University of Madeira, Portugal, `saraxrg@gmail.com`

In the last 30 years, the AGM model has acquired the status of a standard model of belief change. The AGM model inspired many researchers to propose extensions and generalizations as well as applications and connections with other fields (for an overview of these proposals see [13]). Among this extensions we can mention iterated change: A drawback of AGM definition of revision is that the conditions for the iteration of the process are very weak, and this is caused by the lack of expressive power of belief sets. In order to ensure good properties for the iteration of the revision process, a more complex structure is needed. So shifting from belief sets to epistemic states was proposed by Darwiche and Pearl in [10] to accommodate iteration. In this framework, it is possible to define interesting iterated revision operators. There exists several ways to introduce iteration in the AGM operators, for an overview see [30].

In 1992, Katsuno and Mendelzon presented a type of operator of change that they called update [18]. Whereas AGM operators are suited to capture changes that reflect evolving knowledge about a static situation, the KM-update operators are intended to represent changes in beliefs that result from changes in the objects of belief. The difference was pointed out for the first time by Keller and Winslett [19] (in the context of relational databases) and is captured in the following example [35]:

> Initially the agent knows that there is either a book on the table (α) or a magazine on the table (β), but not both.
> *Case 1:* The agent is told that there is a book on the table. She concludes that there is no magazine on the table. This is revision.
> *Case 2:* The agent is told that subsequently a book has been put on the table. In this case she should not conclude that there is no magazine on the table. This is update.

In this paper we propose to analyze iteration in the context of update, inspired in AGM revision. An important difference between AGM revision and KM-update is that belief revision is a local operation in the sense that a revision function is defined just for the current belief set and the language. KM-update, on the other hand is a global operator defined for the set of all possible belief sets and the language[3]. Since update is defined as a global operator, an operator $\diamond$ is defined for all the possible belief sets φ. Consequently, in a first view, iteration is not a problem since $(\varphi \diamond \alpha) \diamond \beta$ is well defined. However, what happens if we want to make changes in our preferences as a consequence of updating by α? In that case we can define a new operator $\blacklozenge_\alpha$ to reflect these changes. This imply that, $(\varphi \diamond \alpha) \diamond \beta$ and $\varphi \blacklozenge_\alpha \beta$ are different operations (in the discussion section we will get back to this topic to explain this difference).

In the paper we adapt the Darwiche and Pearls postulates of belief revision to update (as well as the *Independence postulate* proposed in [4, 16]) and we

---

[3] With *global* we refer that KM-update is defined for the set of all possible belief sets, whereas AGM is just defined for a *fixed* belief set. As pointed out by a reviewer, there is an opposite view, namely that AGM works globally on all models of a theory as a whole, while KM works locally on each model of the theory as such.

present two families of such operators, based in natural [7] and lexicographic revision [27, 29].

2 Background

2.1 Formal preliminaries

We denote by $\mathcal{L}$ the set of formulas of a propositional language built over a finite set of propositional variables $\mathcal{P}$. The elements of $\mathcal{L}$ are denoted by lower case Greek letters $\alpha, \beta, \ldots$ (possibly with subscripts). The set of valuation functions from the set of propositional variables into the boolean set $\{0,1\}$ (false, true) is denoted $\mathcal{V}$. As usual, we write $\omega \models \alpha$ when a valuation $\omega \in \mathcal{V}$ satisfies a formula α, i.e. when ω is a model of α. The set of models (or possible worlds) of a formula α is denoted by $[\![\alpha]\!]$. W is the set of all possible worlds. If M is a set of models we denote by α_M a formula such that $[\![\alpha_M]\!] = M$. We often omit the braces, by writing, e.g., $\alpha_{\omega,\omega'}$ instead of $\alpha_{\{\omega,\omega'\}}$. If $\leqslant$ is a total pre-order (a total and transitive relation), then $\simeq$ is a notation for the associated equivalence relation ($a \simeq b$ iff $a \leqslant b$ and $b \leqslant a$), and $<$ is the notation for the associated strict order ($a < b$ iff $a \leqslant b$ and $b \not\leqslant a$).

2.2 AGM model

In the classical AGM theory beliefs are represented by belief sets K (logically closed sets of sentences) in a language $\mathcal{L}$, and a change consists on adding or removing a sentence from a belief set in order to obtain a new belief set.

AGM recognizes three change operations:

- Expansion: a sentence is added to the belief set and nothing is removed (represented as $+\alpha$ or $+\neg\alpha$);
- Contraction: a sentence is removed from the belief set and nothing is added (represented as $-\alpha$ or $-\neg\alpha$);
- Revision: a new sentence is added to the belief set and at the same time other sentences are removed if necessary to ensure the consistency of the revised set (represented as $*\alpha$ or $*\neg\alpha$).

If the language $\mathcal{L}$ is finite, we can define a belief set as a propositional sentence φ, such that $K = Cn(\varphi)$. Since we assumed $\mathcal{L}$ be finite, in the rest of the paper we will denote a belief set by φ. We will say that a formula φ is *complete* if it is consistent and for any propositional formula α it follows that $\varphi \vdash \alpha$ or $\varphi \vdash \neg\alpha$.

Alchourron, Gardenfors, and Makinson have proposed a set of postulates to govern the process of belief revision [1]. Katsuno and Mendelzon rephrased these postulates for a finite language [17].

(R1) $\varphi * \alpha \vdash \alpha$
(R2) If $\varphi \wedge \alpha \not\vdash \bot$ then $\varphi * \alpha \equiv \varphi \wedge \alpha$

(R3) If $\alpha \not\vdash \bot$ then $\varphi * \alpha \not\vdash \bot$
(R4) If $\varphi_1 \equiv \varphi_2$ and $\alpha_1 \equiv \alpha_2$ then $\varphi_1 * \alpha_1 \equiv \varphi_2 * \alpha_2$
(R5) $(\varphi * \alpha) \wedge \beta \vdash \varphi * (\alpha \wedge \beta)$
(R6) If $(\varphi * \alpha) \wedge \beta \not\vdash \bot$ then $\varphi * (\alpha \wedge \beta) \vdash (\varphi * \alpha) \wedge \beta$

Along with this definition Katsuno and Mendelzon provided a representation theorem that shows an equivalence between the postulates and a revision mechanism based on total pre-orders where these are defined as:

Definition 1. *Let W be the set of all worlds (or interpretations) of a propositional language $\mathcal{L}$. A function that maps each sentence φ in $\mathcal{L}$ to a total pre-order $\leqslant_\varphi$ on worlds W is called a faithful assignment if and only if*[4]*:*

(1) $\omega_1, \omega_2 \models \varphi$ *only if* $\omega_1 =_\varphi \omega_2$
(2) $\omega_1 \models \varphi$ *and* $\omega_2 \not\models \varphi$ *only if* $\omega_1 <_\varphi \omega_2$
(3) $\varphi \equiv \psi$ *only if* $\leqslant_\varphi = \leqslant_\psi$

Their representation theorem shows that a revision operator is equivalent to a faithful assignment where the result of a revision $\varphi * \alpha$ is determined by α and the total pre-order assigned to φ:

Proposition 1. *[17] A revision operator $*$ satisfies postulates (R1) - (R6) precisely when there exists a faithful assignment that maps each sentence φ into a total pre-order $\leqslant_\varphi$ such that*

$$[\![\varphi * \alpha]\!] = min([\![\alpha]\!], \leqslant_\varphi)$$

where $[\![\alpha]\!]$ is the set of all worlds satisfying α and $min([\![\alpha]\!], \leqslant_\varphi)$ contains all worlds that are minimal in $[\![\alpha]\!]$ according to the total pre-order $\leqslant_\varphi$, i.e. all the worlds that include α and are closer to φ.

The Figure 1 provides a diagram of the possible worlds approach[5].

2.3 Iteration for Belief Revision

Although the AGM paper was of great importance to the development of the area of belief revision, their postulates were not sufficient to ensure the rational preservation of conditional beliefs. The AGM postulates are not able to capture the dynamics of the structure used to encode one-step revision policies and

---

[4] $\omega_1 <_\varphi \omega_2$ is defined as $\omega_1 \leqslant_\varphi \omega_2$ and $\omega_2 \not\leqslant_\varphi \omega_1$ and $\omega_1 =_\varphi \omega_2$ is defined as $\omega_1 \leqslant_\varphi \omega_2$ and $\omega_2 \leqslant_\varphi \omega_1$.
[5] This kind of diagram is due to Konieczny and Pino Pérez.

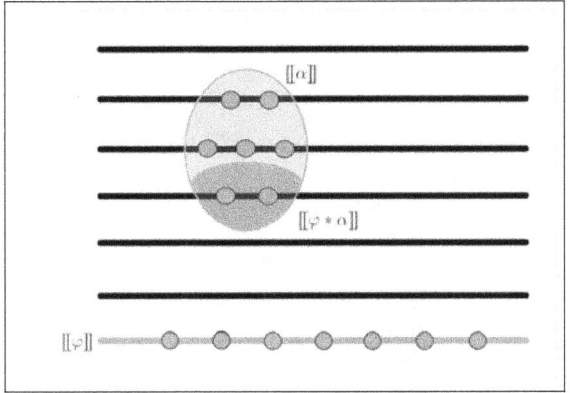

Fig. 1. Belief Revision.

are, therefore, too weak to properly regulate iterated belief revision, that is, the sequential revision of beliefs in response to a string of propositions. Figure 2 illustrates this problem.

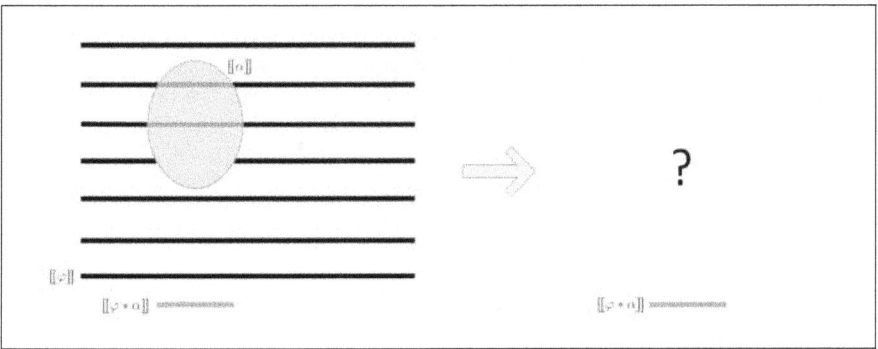

Fig. 2. AGM doesn't provide insight to iteration.

Darwiche and Pearl [10] propose, in order to define iteration, that revision functions must operate on epistemic states instead of belief sets. While belief sets characterize the set of propositions that an agent holds, epistemic states also contains a more complex structure which command the belief changes. The formal definition of an epistemic state in the Darwiche and Pearl meaning is:

Definition 2. *An epistemic state Ψ is an object to which we associate a consistent propositional formula $B(\Psi)$ that denotes the current beliefs of the agent in the epistemic state Ψ. Let us denote by $\mathcal{E}$ the set of epistemic states.*

Darwiche and Pearl modified the (R1)-(R6) postulates to work in the more general framework of epistemic states.

(R*1) $B(\Psi * \alpha) \vdash \alpha$
(R*2) If $B(\Psi) \wedge \alpha \not\vdash \bot$ then $B(\Psi * \alpha) \equiv B(\Psi) \wedge \alpha$
(R*3) If $\alpha \not\vdash \bot$ then $B(\Psi * \alpha) \not\vdash \bot$
(R*4) If $\Psi_1 = \Psi_2$ and $\alpha_1 \equiv \alpha_2$ then $B(\Psi_1 * \alpha_1) \equiv B(\Psi_2 * \alpha_2)$
(R*5) $B(\Psi * \alpha) \wedge \beta \vdash B(\Psi * (\alpha \wedge \beta))$
(R*6) If $B(\Psi * \alpha) \wedge \beta \not\vdash \bot$ then $B(\Psi * (\alpha \wedge \beta)) \vdash B(\Psi * \alpha) \wedge \beta$

Note that this modification is achieved by a weakening of postulate (R4) allowing belief revision to be a function of an epistemic state. The new postulate (R*4) requires the epistemic states to be identical in order to let them lead to equivalent belief states when revised by equivalent evidence.
The four additional postulates proposed by these authors, and which are known as DP-postulates, are the following:

(C1) If $\alpha \vdash \mu$ then $B((\Psi * \mu) * \alpha) \equiv B(\Psi * \alpha)$
(C2) If $\alpha \vdash \neg\mu$, then $B((\Psi * \mu) * \alpha) \equiv B(\Psi * \alpha)$
(C3) If $B(\Psi * \alpha) \vdash \mu$, then $B((\Psi * \mu) * \alpha) \vdash \mu$
(C4) If $B(\Psi * \alpha) \not\vdash \neg\mu$, then $B((\Psi * \mu) * \alpha) \not\vdash \neg\mu$

Postulate (C1) states that the later evidence α cannot discredit the previous evidence μ because α entails μ or, in other words, evidence α alone can yield the same belief set, making evidence μ redundant. Postulate (C2) allows the later evidence α to discredit the previous evidence μ due to the fact that α logically contradicts μ and, as before, evidence α alone would yield the same belief set. Postulate (C3) on the other hand, retains evidence μ after accommodating the more recent evidence α given that α implies μ given current beliefs. Lastly, postulate (C4) stipulates that if μ is not contradicted after seeing α then it should remain uncontradicted when α is preceded by μ itself.

Postulates (C1)-(C4) have a correspondence in terms of total preorders, where the definition of faithful assignment must be adapted to belief states (i.e., by using $\leqslant_\Psi$ instead of $\leqslant_\varphi$):

Proposition 2. *[10, Theorem 13] Suppose that a revision operator satisfies postulates (R*1)-(R*6). The operator satisfies postulates (C1)-(C4) iff the operator and its corresponding faithful assignment satisfy:*

(CR1) If $\omega_1 \models \mu$ and $\omega_2 \models \mu$, then $\omega_1 \leqslant_\Psi \omega_2$ iff $\omega_1 \leqslant_{\Psi * \mu} \omega_2$
(CR2) If $\omega_1 \models \neg\mu$ and $\omega_2 \models \neg\mu$, then $\omega_1 \leqslant_\Psi \omega_2$ iff $\omega_1 \leqslant_{\Psi * \mu} \omega_2$
(CR3) If $\omega_1 \models \mu$ and $\omega_2 \models \neg\mu$, then $\omega_1 <_\Psi \omega_2$ only if $\omega_1 <_{\Psi * \mu} \omega_2$
(CR4) If $\omega_1 \models \mu$ and $\omega_2 \models \neg\mu$, then $\omega_1 \leqslant_\Psi \omega_2$ only if $\omega_1 \leqslant_{\Psi * \mu} \omega_2$

According to (CR1), the order among the μ-worlds remains unchanged after revision by μ. According to (CR2) the order among the $\neg\mu$-worlds remains unchanged after revision by μ. (CR3) says that if a μ-world is strictly preferred to a $\neg\mu$-world, then that strict preference is maintained after revision by μ. (CR4) says that if a μ-world is weakly preferred to a $\neg\mu$-world, then that weak preference is maintained after revision by μ.

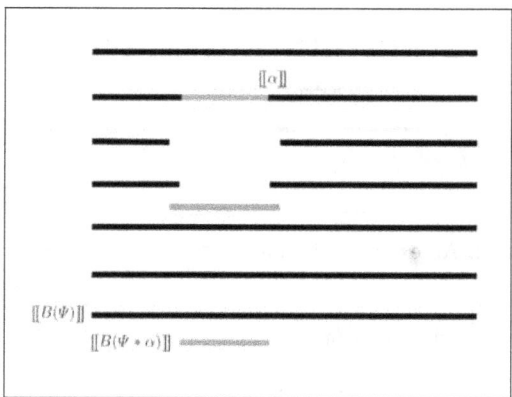

Fig. 3. An example of a belief revision function satisfying (CR1)-(CR4).

(C1)-(C4) have become the benchmark for iterated revision, and new proposals are almost invariably compared to them. However, Booth and Meyer [4] and Jin and Thielscher [16] have pointed out that these postulates are too permissive since they do not rule out operators by which all newly acquired information is given up as soon as an agent learns a fact that contradicts some of its current beliefs. To avoid this they proposed the following additional condition (known as the *independence postulate*):

(Ind) If $B(\Psi) * \alpha \not\vdash \neg\mu$, then $B((\Psi * \mu) * \alpha) \vdash \mu$

It is easily seen that (Ind) is clearly stronger than (C3) and (C4). (C1), (C2) and (Ind) were considered characteristic of a family of operators called admissible revision operators [4]. In total preorders, (Ind) corresponds to the following postulate:

(CRInd) For $\omega_1 \models \mu$ and $\omega_2 \models \neg\mu$, if $\omega_1 \leq_\Psi \omega_2$ then $\omega_1 <_{\Psi*\mu} \omega_2$

In the literature, the two more cited revision operators that satisfy the DP-postulates (C1)-(C4) are:

Conservative revision, originally called natural revision, has been studied by Boutilier [6, 7] and Rott [31]. This operation is conservative in the sense that it only makes the minimal changes of the preorder that are needed to accept the input. In revision by μ, the minimal μ-worlds are moved to the bottom of the preorder which is otherwise left unchanged. The main

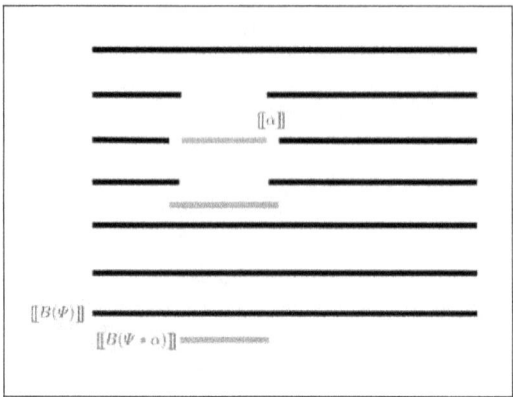

Fig. 4. An example of an admissible revision function.

characteristic of this operator is:

(Nat) If $B(\Psi * \mu) \vdash \neg\alpha$, then $B((\Psi * \mu) * \alpha) = B(\Psi * \alpha)$

(CRNat) If $\omega_1 \notin [[\Psi * \mu]]$ and $\omega_2 \notin [[\Psi * \mu]]$, then $\omega_1 \leq_\Psi \omega_2 \Leftrightarrow \omega_1 \leq_{\Psi*\mu} \omega_2$

Lexicographic revision was proposed by Nayak in [28] and deeply studied by Nayak, Pagnucco and Peppas [29]. When revising by μ it rearranges the preorder by placing the μ-worlds at bottom (but conserving their relative order) and the $\neg\mu$-worlds at top (but conserving their relative order). It has the following property.

(Lex) If $\alpha \not\vdash \neg\mu$, then $B((\Psi * \mu) * \alpha) \vdash \mu$

(CRLex) If $\omega_1 \models \mu$ and $\omega_2 \models \neg\mu$, then $\omega_1 <_{\Psi*\mu} \omega_2$

2.4 KM-Update

As mentioned in the introduction, Katsuno and Mendelzon proposed an alternative model to represent changes in beliefs that result from changes in the objects of belief. The difference between belief revision and update is more evident in the possible world approach [19, 35]. Instead of a preorder and a global distance to the worlds where the input is verified, Katsuno and Mendelzon pointed out that, since the agent believe that one of the current possible worlds is the real world (even if he didn't know which of them), then when the world changes, an update function should look to each one of the possible world and finds the minimal way of changing each of them in order to accommodate the input. In the formal presentation Katsuno and Mendelzon assume that $\mathcal{L}$ is finite.

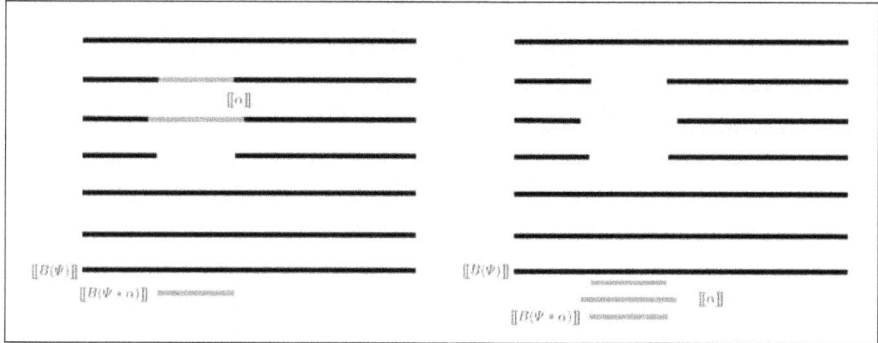

Fig. 5. An example of Natural Revision (left) and Lexicographic revision (right).

The postulates for update are the following:

Definition 3. *Let φ and α be sentences of $\mathcal{L}$. Then $\varphi \diamond \alpha$ is the update of φ by α if and only if it satisfies:*

(U1) $\varphi \diamond \alpha \vdash \alpha$
(U2) *If* $\varphi \vdash \alpha$, *then* $\varphi \diamond \alpha \equiv \varphi$
(U3) *If* $\varphi \not\vdash \bot$ *and* $\alpha \not\vdash \bot$, *then* $\varphi \diamond \alpha \not\vdash \bot$
(U4) *If* $\varphi_1 \equiv \varphi_2$ *and* $\alpha_1 \equiv \alpha_2$ *then* $\varphi_1 \diamond \alpha_1 \equiv \varphi_2 \diamond \alpha_2$
(U5) $\varphi \diamond \alpha \wedge \beta \vdash \varphi \diamond (\alpha \wedge \beta)$
(U6) *If* $\varphi \diamond \alpha \vdash \beta$ *and* $\varphi \diamond \beta \vdash \alpha$, *then* $\varphi \diamond \alpha \equiv \varphi \diamond \beta$
(U7) *If* φ *is a complete formula, then* $(\varphi \diamond \alpha) \wedge (\varphi \diamond \beta) \vdash \varphi \diamond (\alpha \vee \beta)$
(U8) $(\varphi \vee \varphi) \diamond \alpha \equiv (\varphi \diamond \alpha) \vee (\varphi \diamond \alpha)$

If we compare update postulates with AGM (R1)-(R6), we can note some interesting formal differences. In particular, postulate (R2) (vacuity) does not hold for update. Update and its relation with revision has been further studied by Becher [3] and others. Katsuno and Mendelzon characterized update in terms of possible worlds:

Definition 4. *Let W be the set of all worlds. A faithful assignment is a function mapping each world ω to a total pre-order $\leq_\omega$ such that if $\omega \neq \omega'$, then $\omega <_\omega \omega'$.*

Proposition 3. [18] *$\diamond$ is an update operator if and only if there exists a faithful assignment that maps each possible world ω to a partial pre-order $<_\omega$ such that:*

$$[\![\varphi \diamond \alpha]\!] = \bigcup_{\omega \models \varphi} \min([\![\alpha]\!], \leq_\omega)$$

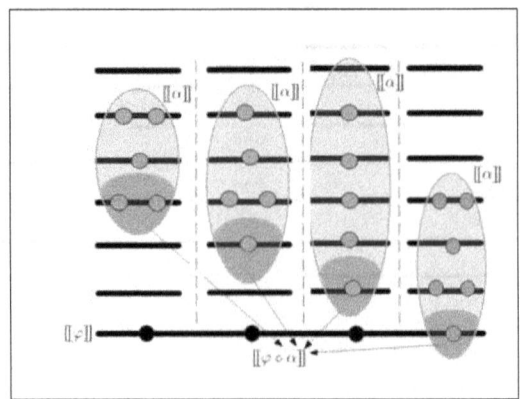

Fig. 6. An example of an update operator.

3 Iteration for KM-update

As we mentioned, a KM-update operator $\diamond$ is defined for all the possible belief sets φ. Consequently, in a first view, iteration does not require a special attention, since $(\varphi \diamond \alpha) \diamond \beta$ is well defined. However, what happens if we want to make changes in our preferences as a consequence of updating by α? In that case we need to define a new kind of operator $\diamond_\alpha$ to reflect these changes. This implies that $(\varphi \diamond \alpha) \diamond \beta$ and $\varphi \diamond_\alpha \beta$ will be, in general, different operations.

In order to differentiate these new operators from the original KM operators we will use the notation $\blacklozenge_s$, where $s = \alpha_1, \ldots, \alpha_n$ is a sequence of updates. If s is an empty sequence, we will denote $\blacklozenge$, if $s = \alpha$ we will denote $\blacklozenge_\alpha$. (U1)-(U8) will remain unchanged for the new class of operators.

In order to define iteration of update we start by adapting the (C1)-(C4) postulates:

(CU1) If $\alpha \vdash \mu$ then $\varphi \blacklozenge_\mu \alpha \equiv \varphi \blacklozenge \alpha$
(CU2) If $\alpha \vdash \neg\mu$ then $\varphi \blacklozenge_\mu \alpha \equiv \varphi \blacklozenge \alpha$
(CU3) If $\varphi \blacklozenge \alpha \vdash \mu$ then $\varphi \blacklozenge_\mu \alpha \vdash \mu$
(CU4) If $\varphi \blacklozenge \alpha \not\vdash \neg\mu$ then $\varphi \blacklozenge_\mu \alpha \not\vdash \neg\mu$

Basically, the only change needed is to replace the revision operator by update and notation. In this case, all the additional structure that was needed in belief revision, resides in the operator $\blacklozenge$. Consequently, a belief state can be defined by the pair $(\varphi, \blacklozenge)$.

In terms of faithful assignment, the corresponding (CR) properties, where the definition of faithful assignment must be adapted to belief states (i.e., by using $\leqslant_{\{\bullet,\omega\}}$ instead of $\leqslant_\omega$), are:[6]

(CRU1) If $\omega_1 \models \mu$ and $\omega_2 \models \mu$, then $\omega_1 \leqslant_{\{\bullet,\omega\}} \omega_2 \Leftrightarrow \omega_1 \leqslant_{\{\bullet_\mu,\omega\}} \omega_2$
(CRU2) If $\omega_1 \models \neg\mu$ and $\omega_2 \models \neg\mu$, then $\omega_1 \leqslant_{\{\bullet,\omega\}} \omega_2 \Leftrightarrow \omega_1 \leqslant_{\{\bullet_\mu,\omega\}} \omega_2$
(CRU3) If $\omega_1 \models \mu$ and $\omega_2 \models \neg\mu$, then $\omega_1 <_{\{\bullet,\omega\}} \omega_2$ implies $\omega_1 <_{\{\bullet_\mu,\omega\}} \omega_2$

(CRU4) If $\omega_1 \models \mu$ and $\omega_2 \models \neg\mu$, then $\omega_1 \leqslant_{\{\bullet,\omega\}} \omega_2$ implies $\omega_1 \leqslant_{\{\bullet_\mu,\omega\}} \omega_2$

(CRU1)-(CRU2) require that the order among μ-worlds and the order among the $\neg\mu$-worlds remains unchanged after update by μ in all of the pre-orders defined for each ω_i. In the same way, (CRU3) says that if a μ-world is strictly preferred to a $\neg\mu$-world, then that strict preference is maintained after updating by μ in all of the preorders defined for each ω_i and finally (CRU4) says that if a μ-world is weakly preferred to a $\neg\mu$-world, then that weak preference is maintained after revision by μ in all of the preorders defined for each ω_i.

The next theorem proves our last assertion.

Proposition 4. $\bullet_s$ *is an update operator if and only if there exists a faithful assignment that maps each possible world ω to a partial pre-order $<_{\{\bullet_s,\omega\}}$ such that:*

$$[\![\varphi \bullet_s \alpha]\!] = \bigcup_{\omega \models \varphi} \min([\![\alpha]\!], \leqslant_{\{\bullet_s,\omega\}})$$

Theorem 1. *Let $\bullet_s$ be an update operator. Then:*

1. *$\bullet_s$ satisfies (CU1) iff its corresponding faithful assignment satisfies (CRU1)*
2. *$\bullet_s$ satisfies (CU2) iff its corresponding faithful assignment satisfies (CRU2)*
3. *$\bullet_s$ satisfies (CU3) iff its corresponding faithful assignment satisfies (CRU3)*
4. *$\bullet_s$ satisfies (CU4) iff its corresponding faithful assignment satisfies (CRU4)*

At this point, an important difference between iteration of revision and iteration of update appears. In belief revision, AGM, without the iteration postulates, didn't provide any insight about the new preorder. On the other hand, $\bullet_\alpha$, without any change w.r.t. $\bullet$ is well defined and satisfies (CU1)-(CU4):

Proposition 5. *Let $\bullet_s$ be an update operator and let $\bullet_\alpha = \bullet$. Then $\bullet_\alpha$ satisfies (CU1)-(CU4).*

For that reason, the claims pointed out by Booth and Meyer [4] and Jin and Thielscher [16] gain a new importance in update. The corresponding *independence postulate* for iterated update is

---

[6] $\leqslant_{\{\bullet,\omega\}}$ represents the total preorder centered in ω for the update operator $\bullet$, and $\leqslant_{\{\bullet_\alpha,\omega\}}$ represents the total preorder centered in ω after updating by α using $\bullet$.

(U-Ind) If φ is a complete formula and $\varphi \bullet \alpha \not\vdash \neg\mu$, then $\varphi \bullet_\mu \alpha \vdash \mu$

which corresponds, in terms of faithful assignments, to the following property:

(CRUInd) If $\omega_1 \models \mu$ and $\omega_2 \models \neg\mu$, then $\omega_1 \leqslant_{\{\bullet,\omega\}} \omega_2 \Rightarrow \omega_1 <_{\{\bullet_\mu,\omega\}} \omega_2$

Theorem 2. *Let $\bullet_s$ be an update operator. Then $\bullet_s$ satisfies(U-Ind) iff its corresponding faithful assignment satisfies (CRUInd).*

Proposition 6. *Let $\bullet$ be an update operator satisfying (C-Ind). Then, in general, $\bullet_\alpha \neq \bullet$.*

The following example illustrate the claim of Proposition 6, i..e, that $(\varphi \bullet \alpha) \bullet \beta$ can differ to $\varphi \bullet_\alpha \beta$:

Example 1. Let $\mathcal{P} = \{\alpha, \beta\}$ and the correspondent possible worlds $W = \{\omega_1, \omega_2, \omega_3, \omega_4\}$ such that

$[\![\{\alpha, \beta\}]\!] = \{\omega_1\}$
$[\![\{\alpha, \neg\beta\}]\!] = \{\omega_2\}$
$[\![\{\neg\alpha, \beta\}]\!] = \{\omega_3\}$
$[\![\{\neg\alpha, \neg\beta\}]\!] = \{\omega_4\}$

Let $\bullet$ be an update operator satisfying (CU1)-(CU4) and (U-Ind). Let $\varphi \equiv \neg\beta$, i.e., $[\![\varphi]\!] = \{\omega_2, \omega_4\}$. Consider the total preorders represented in Figure 7. Then:

$[\![\varphi \bullet \alpha]\!] = \{\omega_2\}$.
$[\![(\varphi \bullet \alpha) \bullet \beta]\!] = \{\omega_1, \omega_3\}$.
$[\![\varphi \bullet_\alpha \beta]\!] = \{\omega_1\}$.

Explanation: On Figure 7 we see the distribution of the possible worlds for φ. In the first case (see Figure 8 top left and top right, marked with $\bullet$), $[\![(\varphi \bullet \alpha) \bullet \beta]\!]$ takes the nearest β-worlds using the order defined by worlds $\bullet$ for $[\![\varphi \bullet \alpha]\!] = \{\omega_2\}$. In the second case (see Figure 8 bottom left and bottom right, marked with $\bullet_\alpha$), ω_1 is improved in $\leqslant_{\{\bullet_\alpha,\omega_2\}}$, since is an α-world. Hence, the outcome of $[\![\varphi \bullet_\alpha \beta]\!]$ is $\{\omega_1\}$.

3.1 Two families of KM-update

In this subsection we show how to define (as in belief revision) natural and lexicographic update. First we need to define (adapt) the postulates:

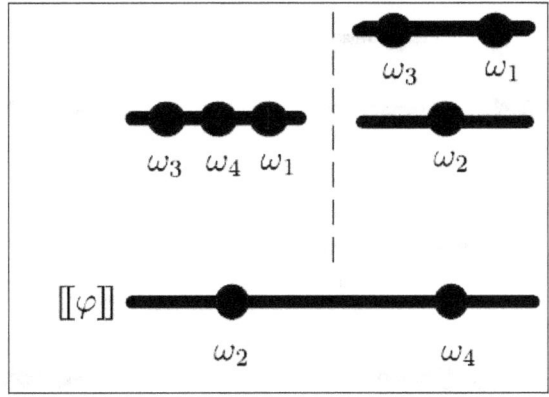

Fig. 7. The total preorders associated to Example 1.

(U-Nat) If φ is a complete formula and $\varphi \diamond \mu \vdash \neg\alpha$, then $\varphi \diamond_\mu \alpha \equiv \varphi \diamond \alpha$

(U-Lex) If $\varphi \vdash \neg\alpha$ and $\alpha \nvdash \neg\mu$, then $\varphi \diamond_\mu \alpha \vdash \mu$

and then provide their correspondent postulates in terms of possible worlds:

(CRUNat) If $\omega_1, \omega_2 \models \neg(\varphi \diamond \mu)$, then $\omega_1 \leqslant_{\{\diamond,\omega\}} \omega_2 \Leftrightarrow \omega_1 \leqslant_{\{\diamond_\mu,\omega\}} \omega_2$

(CRULex) If $\omega_1 \models \mu$ and $\omega_2 \models \neg\mu$, then $\omega_1 <_{\{\diamond_\mu,\omega\}} \omega_2$ for all $\omega \neq \omega_2$

The following representation theorem shows the equivalences:

Theorem 3. *Let $\diamond$ be an update operator. Let f be its corresponding faithful assignment, i.e., such that*

$$[\![\varphi \diamond \alpha]\!] = \bigcup_{\omega \models \varphi} \min([\![\alpha]\!], \leqslant_{\{\diamond,\omega\}})$$

Then:
1. *$\diamond$ satisfies(U-Nat) iff the faithful assignment satisfies (CRUNat)*
2. *$\diamond$ satisfies(U-Lex) iff the faithful assignment satisfies (CRULex)*

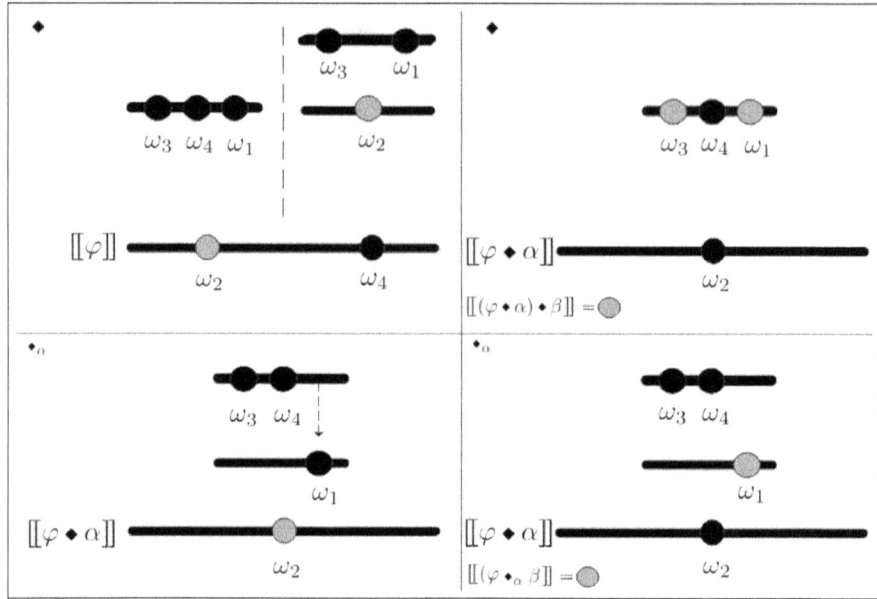

Fig. 8. Example 1 (cont.).

4 Conclusions, related and future works

We have defined the basis for iteration of update. According to our definitions it is possible to define a sequence of update, where the sequence of update plays a role in how the new belief state will be updated. This is different to applying the original update operator to the updated belief set as we have seen on Proposition 6. The postulates that we presented (CRU1)-(CRU4), (CRUInd), (CRUNat) and (CRULex) allow to define large families of update operators. New postulates can be added in order to specify a new operator. For instance, we can define an update operator $\blacklozenge$ that satisfies (CRU1)-(CRU2), (CRUInd) and

(CRUSoft) If $\omega_1 \models \mu, \omega_2 \models \neg\mu$ then $\omega_2 <_{\{\blacklozenge,\omega\}} \omega_1 \Rightarrow \omega_2 \leqslant_{\{\blacklozenge_\mu,\omega\}} \omega_1$

This postulate allows only little (soft) changes in the preorder (this property comes from improvement operators [24].). If a model of $\neg\mu$ is just a little more plausible than a model of μ regarding ω, then after an update the two models will have the same plausibility.

The use of different postulates can help to define different behaviors. For instance, the previous update function satisfies $\blacklozenge_{\{\alpha,\beta\}} = \blacklozenge_{\{\beta,\alpha\}}$. In future works we will analyze additional properties.

There are other approaches for update (for a good overview see [15]) that were not explored in this paper. There are also similar works in the literature, in particular in the area of logic programming (sequences of updates) [2, 11, 33], and in iterative alignment of ontologies [8] and dynamics of ontologies [33]. Gabriele Kern Isberner has several works in the area of iteration of conditionals in nonmonotonic reasoning and belief revision. In particular, in [20] revision operators have been proposed for multiple revision by sets of propositions and by sets of conditionals. The principle of conditional preservation proposed by Darwiche and Pearl is elaborated in detail as an invariance property, which is further developed in [22]. Iterated revision and update is studied in [23]. In [21], the framework of c-revisions is adapted to multiple propositional revision.

The aim of this preliminary report was only to define iteration for Katsuno and Mendelzon's update in the spirit of Darwiche and Pearl postulates. We have worked only with functions and postulates adapted from belief revision to KM-update. However, due to the intrinsic differences between update and revision, an interesting future point is to explore new proper iterated update functions. The relation between our proposal and the related works mentioned in the above paragraph require a deep study and will be studied in future works.

5 Acknowledgements

We wish to thank the reviewers for their valuable comments which have contributed to improve this paper. We also want to thank Maurício Reis and Juan Mikalef for providing generous comments, suggestions and corrections on the earlier versions of this paper.

Personal note from Eduardo Fermé: I want to congratulate Gabriele in her Festschrift and thank her for more than ten years of camaraderie and friendship. It is always a pleasure and an honor to interact with her at the professional and at the personal level.

Appendix: Proofs

Proof of Theorem 1. [7]
(CU1) $\Leftrightarrow$ (CRU1)

($\Rightarrow$) Assume (CU1) holds and let $\omega_1 \models \mu$ and $\omega_2 \models \mu$. Let $\alpha \equiv \alpha_{\{\omega_1,\omega_2\}}$ Then $\alpha \vdash \mu$ and due to (CU1) $\varphi \bullet_\mu \alpha \equiv \varphi \bullet \alpha$. Hence $\min(\{\omega_1,\omega_2\}, \leqslant_{\{\bullet,\omega\}}) = \min(\{\omega_1,\omega_2\}, \leqslant_{\{\bullet_\mu,\omega\}})$ from which it follows that $\omega_1 \leqslant_{\{\bullet,\omega\}} \omega_2 \Leftrightarrow \omega_1 \leqslant_{\{\bullet_\mu,\omega\}} \omega_2$.

($\Leftarrow$) Assume (CRU1) holds and let $\alpha \vdash \mu$. We want to show that $\varphi \bullet_\mu \alpha \equiv \varphi \bullet \alpha$. Condition $(CRU1)$ implies that $\leqslant_{\{\bullet,\omega\}}$ and $\leqslant_{\{\bullet_\mu,\omega\}}$ are equivalent for all $\omega' \in [\![\alpha]\!]$ since $[\![\alpha]\!] \subseteq [\![\mu]\!]$. Hence:

---

[7] Adapted from [10, Proof of Theorem 13]. For the sake of simplicity, we will use in the proofs $\bullet$ instead of $\bullet_s$, since the subscript is not necessary for the proofs.

$$[\![\varphi \bullet \alpha]\!] = \bigcup_{\omega \models \varphi} \min([\![\alpha]\!], \leqslant_{\{\bullet,\omega\}})$$
$$[\![\varphi \bullet \alpha]\!] = \bigcup_{\omega \models \varphi} \min([\![\alpha]\!], \leqslant_{\{\bullet_\mu,\omega\}})$$
$$[\![\varphi \bullet \alpha]\!] = [\![\varphi \bullet_\mu \alpha]\!]$$

(CU2) $\Leftrightarrow$ **(CRU2)**. The proof is symmetric with the one above.

(CU3) $\Leftrightarrow$ **(CRU3)**

($\Rightarrow$) Assume (CU3) holds and let $\omega_1 \models \mu$, $\omega_2 \models \neg\mu$ and $\omega_1 <_{\{\bullet,\omega\}} \omega_2$. Let $\alpha \equiv \alpha_{\{\omega_1,\omega_2\}}$. Then $[\![\alpha_\omega \bullet \alpha]\!] = \min([\![\alpha]\!], \leqslant_{\{\bullet,\omega\}}) = \{\omega_1\}$, from which it follows that $\alpha_\omega \bullet \alpha \vdash \mu$. By (CU3) $\alpha_\omega \bullet_\mu \alpha \vdash \mu$, from which it follows that $[\![\alpha_\omega \bullet_\mu \alpha]\!] = \min([\![\alpha]\!], \leqslant_{\{\bullet_\mu,\omega\}}) \subseteq [\![\mu]\!]$, hence $[\![\alpha_\omega \bullet_\mu \alpha]\!] = \{\omega_1\}$, from which we can conclude that $\omega_1 <_{\{\bullet_\mu,\omega\}} \omega_2$.

($\Leftarrow$) Assume (CRU3) holds and let $\varphi \bullet \alpha \vdash \mu$. From $[\![\varphi \bullet \alpha]\!] = \bigcup_{\omega \models \varphi} \min([\![\alpha]\!], \leqslant_{\{\bullet,\omega\}})$ it follows that for all $\omega \models \varphi$ if $\omega' \in \min([\![\alpha]\!], \leqslant_{\{\bullet,\omega\}})$ implies that $\omega' \models \alpha \wedge \mu$ and for all $\omega'' \models \alpha \wedge \neg\mu$ it follows that $\omega' <_{\{\bullet,\omega\}} \omega''$. (CRU3) yields $\omega' <_{\{\bullet_\mu,\omega\}} \omega''$ for all $\omega'' \models \alpha \wedge \neg\mu$, hence $\omega'' \notin \min([\![\alpha]\!], \leqslant_{\{\bullet_\mu,\omega\}})$. Since this is valid for all $\omega \models \varphi$ we can conclude that $\varphi \bullet_\mu \alpha \vdash \mu$.

(CU4) $\Leftrightarrow$ **(CRU4)**

($\Rightarrow$) Assume (CU4) holds and let $\omega_1 \models \mu$, $\omega_2 \models \neg\mu$ and $\omega_1 \leqslant_{\{\bullet,\omega\}} \omega_2$. Let $\alpha \equiv \alpha_{\{\omega_1,\omega_2\}}$. Then $\omega_1 \in [\![\alpha_\omega \bullet \alpha]\!] = \min([\![\alpha]\!], \leqslant_{\{\bullet,\omega\}})$, from which it follows that $\alpha_\omega \bullet \alpha \nvdash \neg\mu$. By (CU4) $\alpha_\omega \bullet_\mu \alpha \nvdash \neg\mu$, from which it follows that $[\![\alpha_\omega \bullet_\mu \alpha]\!] \cap [\![\mu]\!] \neq \emptyset$, i.e., $\min([\![\alpha]\!], \leqslant_{\{\bullet_\mu,\omega\}}) \cap [\![\mu]\!] \neq \emptyset$, hence $\omega_1 \in [\![\alpha_\omega \bullet_\mu \alpha]\!]$, from which we can conclude that $\omega_1 \leqslant_{\{\bullet_\mu,\omega\}} \omega_2$.

($\Leftarrow$) Assume (CRU4) holds and let $\varphi \bullet \alpha \nvdash \neg\mu$. From $[\![\varphi \bullet \alpha]\!] = \bigcup_{\omega \models \varphi} \min([\![\alpha]\!], \leqslant_{\{\bullet,\omega\}})$ it follows that there exists some v such that $v \models \varphi$ and for some $\omega' \in \min([\![\alpha]\!], \leqslant_{\{\bullet,v\}})$ it holds that $\omega' \models \alpha \wedge \mu$ and for all $\omega'' \models \alpha \wedge \neg\mu$ it follows that $\omega' \leqslant_{\{\bullet,v\}} \omega''$. (CRU4) yields $\omega' \leqslant_{\{\bullet_\mu,v\}} \omega''$ for all $\omega'' \models \alpha \wedge \neg\mu$, hence $\omega' \in \min([\![\alpha]\!], \leqslant_{\{\bullet_\mu,v\}})$, from which it follows that $\omega' \in \bigcup_{\omega \models \varphi} \min([\![\alpha]\!], \leqslant_{\{\bullet_\mu,\omega\}})$. Hence $\varphi \bullet_\mu \alpha \nvdash \neg\mu$. ∎

Proof of Theorem 2.

(U-Ind) $\Leftrightarrow$ **(CRUInd)**

($\Rightarrow$) Assume (U-Ind) holds and let $\omega_1 \models \mu$, $\omega_2 \models \neg\mu$ and $\omega_1 \leqslant_{\{\bullet,\omega\}} \omega_2$. Let $\alpha \equiv \alpha_{\{\omega_1,\omega_2\}}$. Then $\omega_1 \in [\![\alpha_\omega \bullet \alpha]\!] = \min([\![\alpha]\!], \leqslant_{\{\bullet,\omega\}})$, from which it follows that $\alpha_\omega \bullet \alpha \nvdash \neg\mu$. By (U-Ind) $\alpha_\omega \bullet_\mu \alpha \vdash \mu$, from which it follows that $[\![\alpha_\omega \bullet_\mu \alpha]\!] = \min([\![\alpha]\!], \leqslant_{\{\bullet_\mu,\omega\}}) \subseteq [\![\mu]\!]$, hence $[\![\alpha_\omega \bullet_\mu \alpha]\!] = \{\omega_1\}$, from which we can conclude that $\omega_1 <_{\{\bullet_\mu,\omega\}} \omega_2$.

($\Leftarrow$) Assume (CRInd) holds and and let φ be a complete formula such that $\varphi \bullet \alpha \nvdash \neg\mu$. Since φ is complete, there exists some v such that $\varphi \equiv \alpha_v$.

Then $[\![\varphi \bullet \alpha]\!] = \min([\![\alpha]\!], \leqslant_{\{\bullet,v\}})$. Due to $\varphi \bullet \alpha \not\vdash \neg\mu$ it follows that there exists $\omega' \in \min([\![\alpha]\!], \leqslant_{\{\bullet,\omega\}})$ such that $\omega' \models \alpha \wedge \mu$ and for all ω'' such that $\omega'' \models \alpha \wedge \neg\mu$ it follows that $\omega' \leqslant_{\{\bullet,v\}} \omega''$. (CRUInd) yields $\omega' <_{\{\bullet_\mu,v\}} \omega''$ for all $\omega'' \models \alpha \wedge \neg\mu$, hence $\omega' \in \min([\![\alpha]\!], \leqslant_{\{\bullet_\mu,v\}})$ and there ω'' not exists such that $\omega'' \models \alpha \wedge \neg\mu$ and $\omega'' \in \min([\![\alpha]\!], \leqslant_{\{\bullet_\mu,v\}})$. Hence $\varphi \bullet_\mu \alpha \vdash \mu$. ∎

Proof of Theorem 3.

(U-Nat) ⇔ (CRUNat)

(⇒) Assume (U-Nat) holds and let $\omega_1, \omega_2 \models \neg(\varphi \bullet \mu)$. Let $\alpha \equiv \alpha_{\{\omega_1,\omega_2\}}$. Then $\alpha \vdash \neg\varphi \bullet \mu$ from which it follows that $\varphi \bullet \mu \vdash \neg\alpha$ and (by U-Nat) $\varphi \bullet_\mu \alpha \equiv \varphi \bullet \alpha$. Since φ is a complete formula, there exists v such that $\alpha_v \equiv \varphi$. Hence $\min(\{\omega_1, \omega_2\}, \leqslant_{\{\bullet,v\}}) = \min(\{\omega_1, \omega_2\}, \leqslant_{\{\bullet_\mu,v\}})$ from which it follows that $\omega_1 \leqslant_{\{\bullet,v\}} \omega_2 \Leftrightarrow \omega_1 \leqslant_{\{\bullet_\mu,v\}} \omega_2$.

(⇐) Assume (CRUNat) holds and and let φ be a complete formula such that $\varphi \bullet \mu \vdash \neg\alpha$. Since φ is complete there exists v such that $\varphi \equiv \alpha_v$. Then $[\![\varphi \bullet \alpha]\!] = \min([\![\alpha]\!], \leqslant_{\{\bullet,v\}})$. We have that $[\![\alpha]\!] \models \neg(\varphi \bullet \mu)$. Condition (CRUNat) implies that $\leqslant_{\{\bullet,v\}}$ and $\leqslant_{\{\bullet_\mu,v\}}$ are equivalent for all $\omega' \in [\![\alpha]\!]$. Hence:

$$[\![\varphi \bullet \alpha]\!] = \min([\![\alpha]\!], \leqslant_{\{\bullet,v\}})$$
$$[\![\varphi \bullet \alpha]\!] = \min([\![\alpha]\!], \leqslant_{\{\bullet_\mu,v\}})$$
$$[\![\varphi \bullet \alpha]\!] = [\![\varphi \bullet_\mu \alpha]\!]$$

(U-Lex) ⇔ (CRULex)

(⇒) Assume (U-Lex) holds and let $\omega_1 \models \mu$ and $\omega_2 \models \neg\mu$. Let $\alpha \equiv \alpha_{\{\omega_1,\omega_2\}}$. Then $\alpha \not\vdash \neg\mu$. Let φ be a complete formula such that $\varphi \vdash \neg\alpha$. Then $\varphi \not\equiv \alpha_{\omega_2}$. Then it follows by (U-Lex) that $\varphi \bullet_\mu \alpha \vdash \mu$. Since φ is complete there exists $v \neq \omega_2$ such that $\varphi \equiv \alpha_v$. Then $[\![\varphi \bullet_\mu \alpha]\!] = \min([\![\alpha]\!], \leqslant_{\{\bullet_\mu,v\}})$. Since $[\![\varphi \bullet_\mu \alpha]\!] \subseteq [\![\mu]\!]$ it follows that $\min([\![\alpha]\!], \leqslant_{\{\bullet_\mu,v\}}) = \{\omega_1\}$. Hence $\omega_1 <_{\{\bullet_\mu,v\}} \omega_2$.

(⇐) Assume (CRULex) holds and let φ, α and μ such that $\varphi \vdash \neg\alpha$ and $\alpha \not\vdash \neg\mu$. $\varphi \equiv \alpha_{\{\omega_1,\ldots,\omega_n\}}$, for $\omega_1, \ldots, \omega_n \in W$. Then $\{\omega_1, \ldots, \omega_n\} \subseteq [\![\neg\alpha]\!]$. Then $\omega_i \notin \min([\![\alpha]\!], \leqslant_{\{\bullet_\mu,\omega_i\}})$, for $i = 1, \ldots, n$. It follows from (CRULex) that $\omega_j <_{\{\bullet_\mu,\omega_i\}} \omega_k$ for all $\omega_j \models \mu$ and $\omega_k \models \neg\mu$, $k \neq i$ and $i = 1, \ldots, n$. Then $\min([\![\alpha]\!], \leqslant_{\{\bullet_\mu,\omega_i\}}) \cap [\![\neg\mu]\!] = \emptyset$. Hence $\varphi \bullet_\mu \alpha \vdash \mu$. ∎

References

1. Carlos Alchourrón, Peter Gärdenfors, and David Makinson. On the logic of theory change: Partial meet contraction and revision functions. *Journal of Symbolic Logic*, 50:510–530, 1985.
2. Jose Julio Alferes, Joao Alexandre Leite, Luis Moniz Pereira, Halina Przymusinska, and Teodor C. Przymusinski. Dynamic updates of non-monotonic knowledge bases. *The Journal of Logic Programming*, 45(1):43–70, 2000.

3. Verónica Becher. *Binary Functions for Theory Change*. PhD thesis, University of Buenos Aires, June 1999.
4. Richard Booth and Thomas Meyer. Admissible and restrained revision. *Journal of Artificial Intelligence Research*, 26:127–151, 2006.
5. Alexander Borgida. Language features for flexible handling of exceptions in information systems. *ACM Trans. Database Syst.*, 10(4):565–603, 1985.
6. Craig Boutilier. Revision sequences and nested conditionals. In *Proc. 13th Int. Joint Conf. on Artificial Intelligence (IJCAI'93)*, pages 519–525, 1993.
7. Craig Boutilier. Iterated revision and minimal change of conditional beliefs. *Journal of Philosophical Logic*, 25:263–305, 1996.
8. Danny Chen, John Lastusky, James Starz, and Stephen Hookway. A user guided iterative alignment approach for ontology mapping. In *Proceedings of the International Conference on Semantic Web and Web Services. Las Vegas, USA*, pages 51–56, 2008.
9. Mukesh Dalal. Investigations into a theory of knowledge base revision: Preliminary report. In *Seventh National Conference on Artificial Intelligence, (AAAI-88)*, pages 475–479, St. Paul, 1988.
10. Adnan Darwiche and Judea Pearl. On the logic of iterated belief revision. *Artificial intelligence*, 89(1-2):1–29, 1997.
11. Thomas Eiter, Michael Fink, Giuliana Sabbatini, and Hans Tompits. On properties of update sequences based on causal rejection. *Theory and Practice of Logic programming*, 2(06):711–767, 2002.
12. Ronald Fagin, Jeffrey Ullman, and Moshe Vardi. On the semantics of updates in databases: Preliminary report. In *Proceedings of Second ACM SIGACT-SIGMOD Symposium on Principles of Database Systems*, pages 352–365, 1983.
13. Eduardo Fermé and Sven Ove Hansson. AGM 25 years: Twenty-five years of research in belief change. *Journal of Philosophical Logic*, 40:295–331, 2011.
14. William Harper. Rational conceptual change. In The University of Chicago Press, editor, *PSA: Proceedings of the Biennial Meeting of the Philosophy of Science Association, Volume Two: Symposia and Invited Papers*, pages 462–494, 1977.
15. Andreas Herzig and Omar Rifi. Propositional belief base update and minimal change. *Artificial Intelligence*, 115(1):107–138, 1999.
16. Yi Jin and Michael Thielscher. Iterated belief revision, revised. *Artificial Intelligence*, 171:1–18, 2007.
17. Hirofumi Katsuno and Alberto Mendelzon. Propositional knowledge base revision and minimal change. *Journal of Artificial Intelligence*, 52:263–294, 1991.
18. Hirofumi Katsuno and Alberto Mendelzon. On the difference between updating a knowledge base and revising it. In Peter Gärdenfors, editor, *Belief Revision*, number 29 in Cambridge Tracts in Theoretical Computer Science, pages 183–203. Cambridge University Press, 1992.
19. Arthur M. Keller and Marianne Winslett. On the use of an extended relational model to handle changing incomplete information. *IEEE Transactions on Software Engineering*, 11(7):620–633, 1985.
20. Gabriele Kern-Isberner. *Conditionals in nonmonotonic reasoning and belief revision: considering conditionals as agents*. Lecture Notes in Artificial Intelligence LNAI 2087. Springer-Verlag, 2001.
21. Gabriele Kern-Isberner and Daniela Huvermann. Multiple iterated belief revision without independence. In *Proceedings of the Twenty- Eighth International Florida Artificial Intelligence Research Society Conference, FLAIRS 2015, Hollywood, Florida*, pages 570–575, 2015.
22. Gabrielle Kern-Isberner. A thorough axiomatization of a principle of conditional preservation in belief revision. *Annals of Mathematics and Artificial Intelligence*, 40(1-2):127–164, 2004.

23. Gabrielle Kern-Isberner. Linking iterated belief change operations to nonmonotonic reasoning. In G. Brewka and J. Lang, editors, *Proceedings 11th International Conference on Knowledge Representation and Reasoning, KR'2008*, pages 166–176, Menlo Park, CA, 2008. AAAI Press.
24. Sebastien Konieczny and Ramón Pino Perez. Improvement operators. In *Eleventh International Conference on Principles of Knowledge Representation and Reasoning (KR'08)*, pages 177–186, 2008.
25. Isaac Levi. Subjunctives, dispositions, and chances. *Synthèse*, 34:423–455, 1977.
26. João Martins and Stuart Shapiro. A model for belief revision. *Artificial Intelligence*, 35:25–79, 1988.
27. Abhaya Nayak. Foundational belief change. *Journal of Philosophical Logic*, 23:495–533, 1994.
28. Abhaya Nayak. Iterated belief change based on epistemic entrenchment. *Erkenntnis*, 41:353–390, 1994.
29. Abhaya Nayak, Maurice Pagnucco, and Pavlos Peppas. Dynamic belief revision operators. *Artificial Intelligence*, 146:2,:193–228, 2003.
30. Pavlos Peppas. A panorama of iterated revision. In Sven Ove Hansson, editor, *David Makinson on Classical Methods for Non-Classical Problems*, volume 3 of *Outstanding Contributions to Logic*, pages 71–94. Springer Netherlands, 2014.
31. Hans Rott. Coherence and conservatism in the dynamics of belief. part II: Iterated belief change without dispositional coherence. *Journal of Logic and Computation*, 13:111–145, 2003.
32. Ken Satoh. Nonmonotonic reasoning by minimal belief revision. In *Proceedings of the International Conference on Fifth Generation Computer Systems (FGCS'88)*, pages 455–462, 1988.
33. Martin Slota, João Leite, and Theresa Swift. On updates of hybrid knowledge bases composed of ontologies and rules. *Artificial Intelligence*, 229:33–104, 2015.
34. G. H. von Wright. *Explanation and Understanding*. Cornell University Press, 1971.
35. Marianne Winslett. Reasoning about action using a possible models approach. In *AAAI*, pages 89–93, 1988.

Part V
Uncertain Reasoning

On de Finetti on Iterated Conditionals

Igor Douven[1]

Abstract. De Finetti's semantics for conditionals has much to recommend it. However, this paper argues that it is materially inadequate because it gets the truth conditions and probabilities of nested conditionals badly wrong.

1 Introduction

Conditionals are sentences of the form "If φ, [then] ψ" or "ψ if φ," with φ being called "the antecedent" and ψ "the consequent."[2] As has frequently been remarked, conditionals play a pivotal role in human reasoning, so it is small wonder that they have attracted attention from philosophers, psychologists, linguists, and computer scientists alike. It might be thought that these different research communities are interested in mostly different questions concerning conditionals, but that is not so: by far the most research on conditionals is focused on a very limited number of questions, most notably the question of what the truth conditions of conditionals are, and the question of what their probabilities are.

On the other hand, there are marked contrasts between the *answers* to the aforementioned questions that enjoy, or have been enjoying, popularity in the different research communities. For instance, the mental models theory of Johnson-Laird and Byrne [2002] has long influenced theorizing about conditionals in psychology. Other fields have largely ignored this approach to the semantics of conditionals, which to a large extent may be due to its lack of formal precision.[3] In linguistics, Kratzer's [1986] modal semantics for conditionals has been dominating the debate for some time now. Again, this semantics has gone mostly unnoticed in other fields (though some philosophers have paid attention to it; see Yalcin [2010], Égré and Cozic [2011], and MacFarlane [2014]). As to philosophy, it is probably correct to say that, for some decades, there has been *no* account of conditionals that has enjoyed any popularity; virtually every philosopher working on conditionals has his or her own particular account (see Douven [2015b, Ch. 1]).

---

[1] Sciences, Normes, Décision (CNRS), Paris-Sorbonne University, igor.douven@paris-sorbonne.fr

[2] This paper is exclusively concerned with indicative conditionals—roughly, conditionals whose auxiliary verb is in the indicative mood—which throughout will simply be referred to as "conditionals," unless otherwise stated.

[3] This is most clearly illustrated by the fact that the notion of modulation, which is crucial to the mental models theory of conditionals, never receives a formal definition; see Douven [2015a]. There are other problems with the account; see Evans and Over [2004], and also Baratgin et al. [2015], on the most recent version of the mental models account of conditionals.

In computer science, especially among AI researchers, the preferences are again different. For many years, computer science was the only field in which de Finetti's semantics for conditionals had gained serious traction (see, e.g., Calabrese [1987], [1990], [2005], Goodman and Nguyen [1988], Goodman, Nguyen, and Walker [1991], Benferhat, Dubois, and Prade [1997], Kern-Isberner [2001], and Coletti and Scozzafava [2002]). It is only since very recently that psychologists have also become interested in this semantics (Politzer, Over, and Baratgin [2010], Baratgin, Over, and Politzer [2013], [2014]). The lack of interest from other researchers is surprising, given that de Finetti's semantics has much to recommend it. Among other things, it attributes truth conditions to conditionals that seem to accord well with people's evaluations of the truth values of such sentences (Baratgin, Over, and Politzer [2013]). Moreover, it allows one to maintain a relationship between the probabilities of conditionals and conditional probabilities that has seemed perfectly natural to most, and that is also supported by a vast amount of experimental results, but that is under pressure from Lewis' [1976] so-called triviality results.

Amongst its most notable features is the fact that de Finetti's semantics, unlike any other known semantics that renders the triviality results inapplicable, lets us swiftly combine conditionals with other parts of the language and in particular allows for unlimited nesting of conditionals. However, this note argues that the fact that de Finetti's semantics makes it unproblematic to attribute truth conditions to nested conditionals, and also to assign probabilities to them, is of no avail, for the semantics gets those truth conditions and probabilities *wrong*. In other words, the argument will be that de Finetti's semantics is materially inadequate.

This would be an extremely unkind ending for a contribution to a Festschrift for Gabriele Kern-Isberner, whose work on conditionals I have always admired, and learned much from, but which has de Finetti's semantics as one important pillar. So, instead, I will conclude by pointing out a recent alternative approach to conditionals that, while not de Finettian, is nevertheless in line with the basic idea underlying Kern-Isberner's work, to wit, that "conditionals are used to describe plausible relationships between antecedent and consequent" (Kern-Isberner [2001:27]); this approach will be seen to have no difficulty with the problem cases for de Finetti's semantics.

2 The problem of nested conditionals

To show the attraction of de Finetti's semantics for conditionals, we begin by briefly recapitulating the gist of Lewis' [1976] previously mentioned triviality results. With these results Lewis seemed to have presented a dilemma for any semantics of conditionals, by proving, from fairly weak assumptions, that no such semantics could both be propositional (i.e., take conditionals to express propositions) *and* validate what is often called "the Equation," that is, the claim that the probability of a conditional equals the probability of the conditional's consequent given its antecedent.[4] This was generally regarded to be bad

---

[4] I am saying that the triviality results *seemed* to present a dilemma for semantics of conditionals because it is debatable whether the assumptions underlying these

news because, on the one hand, anything less than a propositional semantics of conditionals seemed to make it impossible to explain how conditionals can combine with other parts of the language—for instance, how they can occur as component parts of conjunctions—or how there can be meaningful nested conditionals; and on the other hand, the Equation appeared just too obviously correct to abandon—as van Fraassen [1976:272 f] put it: "What is the probability that I throw a six if I throw an even number, if not the probability that: if I throw an even number, it will be a six?" In fact, if we grant that conditionals do combine with the rest of our language, then from a present-day perspective Lewis' results are even more unsettling, for in the meantime evidence has piled up that people do tend to evaluate the probability of a conditional as the corresponding conditional probability.[5]

There is actually a surprising number of researchers who have taken the first horn of this apparent dilemma, including Adams ([1975], [1998]), Bennett ([2003]), Edgington ([1986], [1995]), and Gibbard ([1981]). According to these researchers, conditionals do not express propositions. They acknowledge that this makes it impossible for conditionals to be embedded in constructions with the propositional operators of conjunction, disjunction, and negation, but then they go on to deny that conditionals in actual speech and thinking are ever embedded in such constructions. What might appear to be a conjunction of two conditionals—such as "If Jack cooks dinner, we'll have lamb, and if Jill cooks dinner, we'll have pork"—is really the *juxtaposition* of two conditionals; the "and" should be thought of as merely serving stylistic purposes here. According to the same authors, we have no real use for disjunctions of conditionals—"If Jack cooks dinner, we'll have lamb, or if Jill cooks dinner, we'll have pork" does not seem to convey anything different from the corresponding conjunction, which, as stated, we are supposed to think of as the juxtaposition of the two conditionals. And conditionals are ever only *seemingly* in the scope of a negation operator, in that we routinely interpret "It's not the case that if φ, ψ" as "If φ, then it's not the case that ψ."

The point about negated conditionals is more or less generally accepted, and the other two claims at least do not appear completely implausible. Nested conditionals, however, pose more of a problem for those denying that conditionals express propositions. First consider some examples of such conditionals:

1. If you forget your umbrella, you'll get soaked if it starts raining.
2. If this material becomes soft if it gets hot, it is not suited for our purposes.
3. If your mother gets angry if you come home with a B, then she'll get furious if you come home with a C.

---

results are really weak enough; see van Fraassen [1976], Douven and Verbrugge [2013], and Douven [2015b, Ch. 3]. However, this issue need not detain us here.

[5] See Hadjichristidis et al. [2001], Evans, Handley, and Over [2003], Oberauer and Wilhelm [2003], Over and Evans [2003], Evans and Over [2004], Weidenfeld, Oberauer, and Hornig [2005], Evans et al. [2007], Oaksford and Chater [2007], Oberauer, Weidenfeld, and Fischer [2007], Over et al. [2007], Gauffroy and Barrioullet [2009], Douven and Verbrugge [2010], [2013], Pfeifer and Kleiter [2010], Politzer, Over, and Baratgin [2010], Fugard et al. [2011], and Over, Douven, and Verbrugge [2013]. However, see also Skovgaard-Olsen, Singmann, and Klauer [2015].

4. If the glass breaks if it's dropped on the floor, then it breaks if it's smashed against the wall.

Of these, (1) is a so-called right-nested conditional, (2) a left-nested conditional, and (3) and (4) are both left- and right-nested.

To see why nested conditionals create trouble for the "non-propositional" account of conditionals, at least if the point is to uphold the Equation, note that conditional probability is defined in terms of conjunction. Thus, to maintain that the probability of "If φ, ψ" equals the probability of ψ given φ, is by the definition of conditional probability to maintain that the former equals the probability of the conjunction of φ and ψ, divided by the probability of φ. And the problem is that in the definition of conditional probability the conjunction operator does *not* play a merely stylistic role, nor is it dispensable for other reasons. Hence, if there is no conjunction of φ and ψ—as will be the case on the current account if φ or ψ are conditionals—then a fortiori there will be no *probability* of the conjunction of φ and ψ. Consequently, the conditional probability of ψ given φ will be undefined, and so will be the probability of "If φ, ψ." But that is not at all how it appears for (1)–(4); we have no difficulty imagining situations in which we would deem these conditionals to have very definite probabilities.

What advocates of the non-propositional account have said in response is that insofar as nested conditionals can be understood at all, they can be rephrased as simple (i.e., non-nested) conditionals. Unfortunately, non-propositionalists have done little to show how the rephrasing is supposed to go. For right-nested conditionals, such as (1), they have offered the so-called Import–Export principle, according to which "If φ, then if ψ, χ" is equivalent to "If φ and ψ, then χ," provided none of φ, ψ, and χ is itself conditional in form (McGee [1985]). This principle is generally found to be pre-theoretically plausible.[6] Granting Import–Export, what are the simple conditionals equivalent to (2)–(4), and to similar left- and left-and-right-nested conditionals that are as readily interpretable as (1) is? While no one has produced an argument to the effect that there *could not be* simple-conditional equivalents of those nested conditionals, the onus is surely on the non-propositionalists to show that, whenever we can make sense of a nested conditional, there is some simple conditional or other that is equivalent to that nested conditional. Surprisingly, they have had nothing informative to say on this matter.[7]

3 De Finetti's semantics for conditionals

There is a way to slip between the horns of the dilemma presented by the triviality results. For these results to apply, it is not enough that conditionals have truth conditions. They must have *binary* truth conditions, meaning that,

---

[6] Although the few empirical results concerning the principle that we presently have show a somewhat mixed picture; see Douven and Verbrugge [2013] and van Wijnbergen, Elqayam, and Over [2015].

[7] For a discussion of other problems facing non-propositionalism, see Douven [2015b], Chs. 1 and 2].

figuratively speaking, in each possible world they are either true or false; or, as we put it, they must express propositions. Long before Lewis presented his results, however, de Finetti had come to endorse a *trivalent* semantics of conditionals. By considering, on the one hand, the relation between truth and bets and, on the other hand, the betting conditions for conditionals, he proposed that besides truth and falsehood, a conditional can have an indeterminate truth value, corresponding to the case in which a bet on a conditional is deemed void, namely, if the antecedent fails to obtain. In full, de Finetti's semantics of the conditional, which will henceforth be symbolized by "$\to$," is given by this truth table (the "I" stands for "indeterminate"):

	$\varphi \to \psi$	φ		
		T	I	F
	T	T	I	I
ψ	I	I	I	I
	F	F	I	I

De Finetti does not leave it at a truth table for the conditional. Rather, he provides trivalent truth tables for conjunction, disjunction, and negation as well, thereby making it unproblematic for conditionals to occur in conjunctions and disjunctions, or be negated. These further truth tables are now generally known as "the strong Kleene tables" (they were independently proposed by Kleene [1952]).[8] It can be shown that, from the combination of the de Finetti truth table for the conditional and the strong Kleene tables, it follows that conjunctions and disjunctions of conditionals can be reduced to single simple conditionals, and that the negation operator still takes narrow scope under the conditional, in accordance with what we said to be the generally accepted view on this matter.[9]

It is not just that de Finetti's semantics *permits* one to uphold the Equation; given this semantics, it is actually possible to *justify* the Equation a priori, via a Dutch book argument.[10] To appreciate the significance of this, note that earlier proponents of the Equation had considered it to be an eminently reasonable principle—see again the passage from van Fraassen [1976] cited above—but had been unable to give it a normative underpinning. Indeed, insofar as these proponents wanted to stick to a propositional semantics of conditionals,[11] the triviality results showed that they could not possibly have succeeded in that regard. Note also that thereby de Finetti's semantics automatically receives strong empirical support from the wealth of evidence for the Equation (see note 5).

---

[8] According to these truth tables, a conjunction is true iff both conjuncts are true, false iff at least one conjunct is false, and indeterminate in the remaining cases; a disjunction is true iff at least one disjunct is true, false iff both disjuncts are false, and indeterminate in the remaining cases; and a negated sentence is true iff the unnegated sentence is false, and it is indeterminate otherwise.

[9] See Goodman, Nguyen, and Walker [1991:63] for a proof.

[10] See de Finetti [1937/1964:68 f]. More exactly, the Dutch book argument applies to all conditionals whose antecedents have positive probability.

[11] And still ignoring the concern mentioned in note 4.

In fact, not only is there empirical backing for the probabilistic implications of the semantics, also the predictions it makes about truth assignments have been recently subjected to testing and have been found to hold water (Baratgin, Over, and Politzer [2013]). And at a more fundamental level, there is evidence supporting the specific supposition de Finetti makes concerning the relationship between conditionals and bets, which is the basic motivation for his semantics (Politzer, Over, and Baratgin [2010]).

These are all important points in favor of de Finetti's semantics: The semantics explains how conditionals can be embedded in conjunctions, and so on; it explains why the probabilities of conditionals equal the corresponding conditional probabilities; and it enjoys considerable empirical support. How about nested conditionals, which we saw to be so problematic for the non-propositional account? Such conditionals did not figure among the materials that were used to test de Finetti's semantics. Since de Finetti's semantics is able to account for nested conditionals, it is clearly a candidate for such tests. Nested conditionals were problematic for non-propositionalism because, on that account, conditionals cannot be embedded in conjunctions. But, as stated, on de Finetti's semantics they can. Nor is there anything else in this semantics to prevent one from nesting conditionals endlessly. What is more, while for de Finetti there is no *need* to reduce nested conditionals to simple ones, it turns out that, given his semantics, every nested conditional is, *as a matter of fact*, equivalent to a simple conditional.

To prove this reducibility claim, we need some definitions. First, we shall assume a propositional language $\mathcal{L}$ equipped with a conjunction ($\wedge$), disjunction ($\vee$), and negation ($\neg$) operator as well as with the conditional operator, all as defined by de Finetti's truth tables. Furthermore, we use Roman capitals as variables for *factive* sentences in $\mathcal{L}$, where a sentence is factive if it contains at most the conjunction, disjunction, and negation operator; we use small Greek letters as variables for factive and non-factive sentences alike. Further note that we can, without loss of generality, define nested conditionals as follows:

Definition 1. *The* nested conditionals *among the sentences of $\mathcal{L}$ are the smallest set satisfying the following clauses:*

- $A \rightarrow (B \rightarrow C)$ and $(A \rightarrow B) \rightarrow C$ are nested conditionals, for any factive A, B, and C;
- if φ and ψ are nested conditionals, then so are $A \rightarrow \varphi$, $\varphi \rightarrow A$, and $\varphi \rightarrow \psi$, for any factive A.

That with this definition we are not loosing generality follows from the above-mentioned reducibility of conjunctions and disjunctions of conditionals to simple conditionals and the fact that conditionals never genuinely occur in the scope of the negation operator. For in view of this, conditionals of such forms as $\big(\neg(\varphi \rightarrow \psi) \vee (\chi \rightarrow \rho)\big) \rightarrow \neg(\sigma \rightarrow \tau)$ and $\varphi \rightarrow \big((\psi \rightarrow \chi) \wedge \neg(\rho \rightarrow \sigma)\big)$, which are nested in the ordinary sense of the word, will be equivalent to conditionals that are nested in the sense of Definition 1.

We will state the reducibility claim in terms of a nested conditional's factive components and a designated element (the factive consequent) of those components.

Definition 2. *For all $A, \varphi \in \mathcal{L}$, A is a* factive component *of φ iff $A \in$ COMP(φ), where*

$$\text{COMP}(\varphi) := \begin{cases} \{\varphi\} & \text{if } \varphi \text{ is factive;} \\ \text{COMP}(\psi) \cup \text{COMP}(\chi) & \text{if } \varphi = \psi \to \chi \text{ for some } \psi, \chi \in \mathcal{L}. \end{cases}$$

For example, COMP$(([\neg A \wedge B] \to [C \to D]) \to (E \vee F)) = \{\neg A \wedge B, C, D, E \vee F\}$.

Definition 3. *For all $\varphi, \psi \in \mathcal{L}$, the* factive consequent *of $\varphi \to \psi$, CONS$(\varphi \to \psi)$, is defined as follows:*

$$\text{CONS}(\varphi \to \psi) := \begin{cases} \psi & \text{if } \psi \text{ is factive;} \\ \text{CONS}(\chi \to \rho) & \text{if } \psi = \chi \to \rho \text{ for some } \chi, \rho \in \mathcal{L}. \end{cases}$$

So, for instance, CONS$((A \to B) \to C) = C$ and CONS$(A \to (B \to [C \to D])) = D$.

With the new terminology in place, our previous claim concerning nested conditionals in de Finetti's semantics can be stated as follows (here $\equiv$ means "have the same de Finetti truth table"):

Proposition 1 *For all $\varphi, \psi \in \mathcal{L}$,*

$$\varphi \to \psi \equiv \bigwedge \{\text{COMP}(\varphi \to \psi) \setminus \{\text{CONS}(\varphi \to \psi)\}\} \to \text{CONS}(\varphi \to \psi).$$

Note that the right-hand side of the equivalence in Proposition 1 is a simple conditional. So, this proposition not only tells us that any conditional in de Finetti's semantics—no matter how complex it is—*is* equivalent to (i.e., has the same truth table as) some simple conditional; it also offers an *effective procedure* for arriving at a simple conditional that is equivalent to the complex conditional: take all the factive components of a given nested conditional; single out the factive consequent among those components; form a simple conditional by letting the conjunction of all factive components minus the factive consequent be its antecedent and the factive consequent its consequent.[12]

To illustrate, Proposition 1 tells us that $((A \to B) \to C) \to (D \to (E \to F))$ is equivalent to $(A \wedge B \wedge C \wedge D \wedge E) \to F$. And indeed, one verifies that the former conditional is true iff $(A \to B) \to C$ and $D \to (E \to F)$ are both true, which is the case iff, first, $A \to B$ and C are both true, and second, D and $E \to F$ are both true, where the first is the case iff A, B, and C are all true, and the second, iff D, E, and F are all true. But those are precisely the conditions under which $(A \wedge B \wedge C \wedge D \wedge E) \to F$ is true. On the other hand, $((A \to B) \to C) \to (D \to (E \to F))$ is false iff $(A \to B) \to C$ is true and $D \to (E \to F)$ is false. We just saw that the former is the case iff A, B, and C are true, and one easily checks that $D \to (E \to F)$ is false iff D is true and $E \to F$ is false, the latter being the case iff E is true and F false. In short, $((A \to B) \to C) \to (D \to (E \to F))$ is false iff all of A, B, C, D,

---

[12] Note that COMP and CONS are both decidable.

and E are true while F is false—which are exactly the conditions under which $(A \wedge B \wedge C \wedge D \wedge E) \to F$ is false.

To go beyond a mere illustration, and to prove the proposition, we need one more definition:

Definition 4. *For all $\varphi \in \mathcal{L}$, the rank of φ, $\mathrm{rk}(\varphi)$ is defined as follows:*

$$\mathrm{rk}(\varphi) := \begin{cases} 0 & \text{if } \varphi \text{ is factive;} \\ \max\{\mathrm{rk}(\psi), \mathrm{rk}(\chi)\} + 1 & \text{if } \varphi = \psi \to \chi \text{ for some } \psi, \chi \in \mathcal{L}. \end{cases}$$

So $\mathrm{rk}(\varphi)$ tells us how deeply nested φ is. For instance, $\mathrm{rk}((A \to [B \to C]) \to (D \to E)) = 3$.

The proof of Proposition 1 is now by induction on the rank of conditional sentences, as follows:

Proof of Proposition 1: For the base case, with $\mathrm{rk}(\varphi \to \psi) = 1$ (and thus φ and ψ both factive), it suffices to note that $\bigwedge\{\mathrm{COMP}(\varphi \to \psi) \setminus \{\mathrm{CONS}(\varphi \to \psi)\}\} = \varphi$ and $\mathrm{CONS}(\varphi \to \psi) = \psi$. Then assume that the proposition holds for all conditionals of rank n (Induction Hypothesis). It is to be shown that, under this assumption, it also holds for conditionals of rank $n+1$. Suppose $\mathrm{rk}(\varphi \to \psi) = n+1$. Then both $\mathrm{rk}(\varphi) \leqslant n$ and $\mathrm{rk}(\psi) \leqslant n$. Without loss of generality, let $\varphi = \chi \to \rho$, with $\mathrm{COMP}(\chi \to \rho) = \{A_1, \ldots, A_k\}$ and $\mathrm{CONS}(\chi \to \rho) = A_k$, and let $\psi = \sigma \to \tau$, with $\mathrm{COMP}(\sigma \to \tau) = \{B_1, \ldots, B_l\}$ and $\mathrm{CONS}(\sigma \to \tau) = B_l$. Thus, by the Induction Hypothesis,

$$\varphi \equiv (A_1 \wedge \cdots \wedge A_{k-1}) \to A_k \quad \text{and} \quad \psi \equiv (B_1 \wedge \cdots \wedge B_{l-1}) \to B_l.$$

To see that then also

$$\varphi \to \psi \equiv ((A_1 \wedge \cdots \wedge A_{k-1}) \to A_k) \to ((B_1 \wedge \cdots \wedge B_{l-1}) \to B_l), \quad (*)$$

note that the left-hand side of $(*)$ is true iff both φ and ψ are true, iff both $(A_1 \wedge \cdots \wedge A_{k-1}) \to A_k$ and $(B_1 \wedge \cdots \wedge B_{l-1}) \to B_l$ are true, iff all of $\{A_i\}_{1 \leqslant i \leqslant k}$ as well as all of $\{B_i\}_{1 \leqslant i \leqslant l}$ are true, iff the right-hand side of $(*)$ is true. Further note that the left-hand side of $(*)$ is false iff φ is true and ψ is false, iff $(A_1 \wedge \cdots \wedge A_{k-1}) \to A_k$ is true and $(B_1 \wedge \cdots \wedge B_{l-1}) \to B_l$ is false, iff all of $\{A_i\}_{1 \leqslant i \leqslant k}$ and all of $\{B_i\}_{1 \leqslant i \leqslant l-1}$ are true but B_l is false, iff the right-hand side of $(*)$ is false. (In all other cases, both sides of $(*)$ are indeterminate.) Consequently, both sides of $(*)$ have the same truth conditions. It follows that Proposition 1 holds for conditionals of arbitrary rank. ∎

Clearly, the earlier-mentioned Import–Export principle is a straightforward corollary of Proposition 1. While this may be a welcome result in view of the intuitive support the principle enjoys, it equally follows from Proposition 1 that $(A \to B) \to C \equiv (A \wedge B) \to C$, for all A, B, and C, which appears dubious: our earlier example (2) simply does not seem to mean the same as

(2*) If this material becomes soft and it gets hot, it is not suited for our purposes.

Be this as it may, to demonstrate the material inadequacy of de Finetti's semantics, it will suffice to focus on the consequences of Proposition 1 for left-*and*-right-nested conditionals, such as (3) and (4). By Proposition 1, (3) is equivalent to

(3*) If you come home with a B and your mother gets angry and you come home with a C, then your mother will get furious.

and (4) is equivalent to

(4*) If the glass is dropped on the floor and it breaks and it is smashed against the wall, then it breaks.

Supposing (3) to be asserted in anticipation of your result for some test paper, the antecedent of (3*) is inconsistent: you may not come home either with a B or with a C (you may do better, or worse still), but you *cannot* come home with *both* a B *and* a C. This means that, on de Finetti's semantics, (3*) is indeterminate, and even *necessarily* indeterminate. Given that, in view of Proposition 1, it is equivalent to (3), the latter should be indeterminate as well, and necessarily so. Pre-theoretically, however, that conditional can be both true and false, presumably depending on the character of the mother. And (4*) sounds both nonsensical and trivial, in sharp contrast with (4). It might be thought that, at least probabilistically, de Finetti is getting things right here, for what other probability than 1 could (4*) have? And that may not be too far off from the probability one would like to assign to (4). But consider the following variant of (4):

(5) If the glass breaks if it's dropped on the floor, then it breaks if it's dropped on a pillow.

This is actually rather unlikely, yet

(5*) If the glass is dropped on the floor and it breaks and it is dropped on a pillow, then the glass breaks.

would again seem to have a probability of 1.[13]

The predictions that de Finetti's semantics makes regarding the truth conditions and probabilities of nested conditionals deviate so starkly from what common sense suggests that one need not even conduct an experiment to know that the semantics' predictive accuracy vanishes when we turn to nested conditionals. To see that this constitutes a *major* problem for de Finetti, note that the above counterexamples are not consequences of some principle in the periphery of his account, perhaps specially added to handle nested conditionals. Rather, they follow directly from the truth table for the conditional, which forms the very core of the account.[14]

---

[13] I have not been able to find anything in de Finetti's writings that provides the means for determining the probabilities of conditionals with inconsistent antecedents, such as (3*). Some authors who have aimed to further develop de Finetti's semantics simply exclude such conditionals from the outset (e.g., Calabrese [1987]). This makes it impossible for me to say how the probabilities of (3) and (3*) compare.

[14] One might try to explain away the counterexamples by arguing that (3) and (4) are not really nested conditionals but are to be interpreted as simple conditionals

4 The inferentialist alternative

As stated at the outset, I do not want to end my contribution on the negative note that formed the conclusion of the previous section ("*Now let Gabi sort it out!*"). For a more cheerful ending, I would like to draw attention to an alternative approach to conditionals that appears to have no difficulty with nestings and has other important virtues besides. Versions of this alternative semantics, which has been termed "inferentialism," have recently been proposed in Krzyżanowska, Wenmackers, and Douven [2014], Douven [2015b, Ch. 2], [2015c], Krzyżanowska [2015], and Skovgaard-Olsen [2015a], [2015b]. In this type of semantics, the presence of an inferential connection between antecedent and consequent is essential to the truth of a conditional. This is not a new idea, but where previous proposals in this vein had required the inferential connection to be deductive, inferentialism allows the connection to be of a weaker variety, like inductive or abductive, or to consist of a number of steps, some of which may be deductively valid, others inductively valid, and yet others abductively valid. More exactly, according to inferentialism a conditional is true in a given context iff its consequent follows from its antecedent plus (possibly) contextually relevant background knowledge, with "follows" being understood in a broad sense so as to encompass all of the designated types of inference as well as combinations thereof. Informally, the position can be said to require for a conditional's truth the antecedent together with background premises to provide a strong enough argument for, or a good enough reason for believing, the consequent.

---

relating two claims about dispositions (or mechanisms, or something similar). For instance, it might be held that (4) is really the claim that if the glass has the disposition of breaking when dropped on the floor, then it has the disposition of breaking when slammed against the wall. But first note that this response is helpful only if we can have an analysis of dispositions that does not rely on an account of conditionals or something even less well understood than conditionals. And many believe that, at the moment, our best theory of dispositions is still the one to be found in Lewis [1997], according to which an object has the disposition to Φ under such-and-such circumstances iff it has some intrinsic property such that, if the designated circumstances were to obtain, and if the object were to retain the intrinsic property long enough, the circumstances and intrinsic property together would cause the object to Φ—an analysis presupposing an account of subjunctive conditionals, which are certainly not better understood than indicative conditionals. Second, even if one disagrees with Lewis' view on dispositions and is hopeful that, some day, we shall have a theory of dispositions (or mechanisms, etc.) that does not presuppose an account of conditionals or anything as complicated, one still has to explain how it can happen that, although "If the glass is dropped on the floor, it will break" and "If the glass is slammed against the wall, it will break" *look* like bona fide conditionals, and when not embedded presumably also *are*, they stop being conditionals once they are embedded. For as soon as one agrees that they *are* conditionals (whether or not these then express dispositions, or mechanisms, or what have you), de Finetti's truth table applies, and from that we derive that (4) is equivalent to (4*). (Thanks to Nicole Cruz and David Over for pressing this objection.)

As the aforementioned authors argue, inferentialism is intuitively attractive, explaining in the most straightforward manner the broadly felt idea that a conditional makes no sense unless its antecedent and consequent *have something to do with each other*, that, to repeat Kern-Isberner, "conditionals are used to describe plausible relationships between antecedent and consequent"—an idea that most other semantics of conditionals leave to pragmatics to account for (invariably without specifying *how* the pragmatic explanation is supposed to go; see Douven [2015c]). The same authors have also shown that inferentialism has many theoretical virtues, for instance, in that it helps explain away some long-standing puzzles regarding conditionals (such as Gibbard's [1981] famous riverboat puzzle).

Here, I would like to highlight a feature that so far has only been noted in Skovgaard-Olsen [2015b], namely, that inferentialism naturally handles nested conditionals, and in the exact same manner in which it handles simple conditionals. I am unable to imagine a context in which I would deem (3) true in which your mother getting angry if you come home with a B is not a good reason for believing that she will get furious if you come home with a C. Similarly, imagining (4) to be about the glass standing on my desk now, I judge the conditional to be true; and I also deem the glass breaking if dropped on the floor a good reason for believing that it will break as well if smashed against the wall. This is not essentially different from holding the antecedent of (4) to be true because the glass dropping on the floor is a good reason for believing that it will break, nor from holding the consequent of the same nested conditional true because the glass being smashed against the wall is a good reason for believing that it will break, or from holding that the antecedent of (3) is true because your coming home with a B is a good reason for believing that your mother will get angry, and so on.

Inferentialism is also particularly well poised to respond to a point that non-propositionalists like to make in relation to nested conditionals, to wit, that *no* semantics of conditionals will be able to get much dialectical mileage from nested conditionals; as David Over put it to me, with regard to such conditionals, we are all between a rock and a hard place. Specifically, the problem is supposed to be that while for non-propositionalists *no* nested conditional should make sense, from the perspective of any semantics that does attribute truth conditions to conditionals, *all* nested conditionals should make sense. But surely there are some nested conditionals that do make sense and others that do not.

I should begin by noting that I disagree with this assessment of the dialectics for reasons that are unrelated to inferentialism. Theorists holding that we can attribute truth conditions to nested conditionals are no more committed to holding that all such conditionals must make sense than theorists holding that we can attribute truth conditions to conjunctions, disjunctions, and negations are committed to holding that *all* sentences make sense which do not contain the conditional operator. A sentence may fail to make sense for a variety of reasons; for instance, it may be too grammatically complex for us to parse, or we may have difficulty comprehending how the words constituting the sentence combine to make a meaningful whole.

Given inferentialism, one very concrete way in which a conditional may not make sense to us is if we just cannot figure out how its antecedent could be inferentially connected to its consequent. Consider Gibbard's [1981:235] well-known example

(6) If Kripke was there if Strawson was there, then Anscombe was there.

According to non-propositionalists, we do not really understand this sentence. Not assuming any contextual background information, I am inclined to agree with their judgment. However, I would suggest that we do not understand it precisely for the aforementioned reason: we cannot comprehend how Kripke's having been present if Strawson was could possibly be a good reason for believing that Anscombe was present.

Of course, there is an obvious litmus test for this suggestion: specifying a context which does provide the so-far-missing reason relations should make the conditional interpretable. And the following, I believe, is just such a context: Yesterday you overheard a discussion between a group of philosophers (including Anscombe, Kripke, and Strawson) who were talking about the plan to split up in groups of three (one group would be Anscombe, Kripke, and Strawson) that are supposed to stay together all day long for the rest of the week, with each group having the goal of trying to solve a different open problem in philosophy. You heard that most members of the group liked this plan a lot, though you then had to head home and could not hear what they decided in the end. We meet the day after you overheard this discussion and I tell you that I *think* I saw Strawson in the pub. Then you tell me what you heard in the pub the other day. We both know that Kripke and Strawson are not really that fond of each other and would not hang out with each other for social reasons. Then if you say, "So if Kripke was there if Strawson was there, then Anscombe was there," I *would* understand what you were saying.

We have seen that, unlike non-propositionalism, de Finetti's semantics has no difficulty accounting for embedded (including nested) conditionals. In itself, that is a virtue, and it is not its only one, as we have also seen. However, on closer inspection it appeared that the truth conditions as well as the probabilities that the semantics predicts nested conditionals have can be completely out of step with our pre-theoretical judgments of those truth conditions and probabilities. To be sure, that is not a knock-down argument against de Finetti's semantics. In defense, one could question various of the assumptions underlying my argument, including even probability theory or the notion of logical equivalence involved. That way, however, the package we are being offered may well become so revisionary that many will want to try some simpler approach to conditionals first.[15] I recommend inferentialism as such a simpler approach.

---

[15] Especially in view of the fact that, even if simplicity is not a guide to truth in general, it may very well be in the linguistic domain; see Douven [2006:450].

Acknowledgments.

This paper was inspired by discussions I had with Jean Baratgin, Nicole Cruz De Echeverria Loebell, David Over, and Guy Politzer, for which I thank them. I am also greatly indebted to Niels Skovgaard-Olsen, Christopher von Bülow, and two anonymous referees for valuable comments on previous versions.

References

Adams, E. W. [1975] *The Logic of Conditionals*, Dordrecht: Reidel.

Adams, E. W. [1998] *A Primer of Probability Logic*, Stanford: CSLI Publications.

Baratgin, J., Douven, I., Evans, J. St. B. T., Oaksford, M., Over, D. E., and Politzer, G. [2015] "The New Paradigm and Mental Models," *Trends in Cognitive Sciences* 19:547–548.

Baratgin, J., Over, D. E., and Politzer, G. [2013] "Uncertainty and the de Finetti Tables," *Thinking and Reasoning* 19:308–328.

Baratgin, J., Over, D. E., and Politzer, G. [2014] "New Psychological Paradigm for Conditionals: Its Philosophy and Psychology," *Mind and Language* 29:73–84.

Benferhat, S., Dubois, D., and Prade, H. [1997] "Nonmonotonic Reasoning, Conditional Objects and Possibility Theory," *Artificial Intelligence* 92:259–276.

Bennett, J. [2003] *A Philosophical Guide to Conditionals*, Oxford: Oxford University Press.

Calabrese, P. G. [1987] "An Algebraic Synthesis of the Foundations of Logic and Probability," *Information Sciences* 42:187–237.

Calabrese, P. G. [1990] "Reasoning with Uncertainty Using Conditional Logic and Probability," *Proceedings of the First International Symposium on Uncertainty Modeling and Analysis*, IEEE Computer Science, pp. 682–688.

Calabrese, P. G. [2005] "Reflections on Logic and Probability in the Context of Conditionals," in G. Kern-Isberner, W. Rödder, and F. Kulmann (eds.) *Conditionals, Information, and Inference*, Berlin: Springer, pp. 12–37.

Coletti, G. and Scozzafava, R. [2002] *Probabilistic Logic in a Coherent Setting*, Dordrecht: Kluwer.

de Finetti, B. [1937/1964] "Foresight: Its Logical Laws, its Subjective Sources," in H. E. Kyburg and H. E. Smokler (eds.) *Studies in Subjective Probability*, New York: Wiley, pp. 55–118.

Douven, I. [2006] "Assertion, Knowledge, and Rational Credibility," *Philosophical Review* 115:449–485.

Douven, I. [2015a] "Experimental Approaches to Conditionals," in W. Buckwalter and J. Sytsma (eds.) *The Blackwell Companion to Experimental Philosophy*, Oxford: Wiley-Blackwell, in press.

Douven, I. [2015b] *The Epistemology of Indicative Conditionals*, Cambridge: Cambridge University Press.

Douven, I. [2015c] "How to Account for the Oddness of Missing Link Conditionals," *Synthese*, in press.

Douven, I. and Verbrugge, S. [2010] "The Adams Family," *Cognition* 117:302–318.

Douven, I. and Verbrugge, S. [2013] "The Probabilities of Conditionals Revisited," *Cognitive Science* 37:711–730.

Edgington, D. [1986] "Do Conditionals Have Truth-Conditions?" *Crítica* 18:3–30.

Edgington, D. [1995] "On Conditionals," *Mind* 104:235–329.

Égré, P. and Cozic, M. [2011] "If-Clauses and Probability Operators," *Topoi* 30:17–29.

Evans, J. St. B. T., Handley, S. J., Neilens, H., and Over, D. E. [2007] "Thinking about Conditionals: A Study of Individual Differences," *Memory and Cognition* 35:1759–1771.

Evans, J. St. B. T., Handley, S. J., and Over, D. E. [2003] "Conditionals and Conditional Probability," *Journal of Experimental Psychology: Learning, Memory, and Cognition* 29:321–355.

Evans, J. St. B. T. and Over, D. E. [2004] *If*, Oxford: Oxford University Press.

Fugard, A., Pfeifer, N., Mayerhofer, B., and Kleiter, G. [2011] "How People Interpret Conditionals: Shifts toward the Conditional Event," *Journal of Experimental Psychology: Learning, Memory, and Cognition* 37:635–648.

Gauffroy, C. and Barrouillet, P. [2009] "Heuristic and Analytic Processes in Mental Models for Conditionals: An Integrative Developmental Theory," *Developmental Review* 29:249–282.

Gibbard, A. [1981] "Two Recent Theories of Conditionals," in W. L. Harper, R. Stalnaker, and G. Pearce (eds.) *Ifs*, Dordrecht: Reidel, pp. 211–247.

Goodman, I. R. and Nguyen, H. T. [1988] "Conditional Objects and the Modelling of Uncertainties," in M. M. Gupta and T. Yamakawa (eds.) *Fuzzy Computing: Theory, Hardware, and Applications*, Amsterdam: North-Holland, pp. 119–138.

Goodman, I. R., Nguyen, H. T., and Walker, E. A. [1991] *Conditional Inference and Logic for Intelligent Systems*, Amsterdam: North-Holland.

Hadjichristidis, C., Stevenson, R. J., Over, D. E., Sloman, S. A., Evans, J. St. B. T., and Feeney, A.[2001] "On the Evaluation of 'if p then q' Conditionals," in J. D. Moore and K. Stenning (eds.) *Proceedings of the 23rd Annual Meeting of the Cognitive Science Society*, Edinburgh, pp. 381–386.

Johnson-Laird, P. N. and Byrne, R. M. J. [2002] "Conditionals: A Theory of Meaning, Pragmatics, and Inference," *Psychological Review* 19:646–678.

Kern-Isberner, G. [2001] *Conditionals in Nonmonotonic Reasoning and Belief Revision*, Berlin: Springer.

Kleene, S. C. [1952] *Introduction to Metamathematics*, Amsterdam: North-Holland.

Kratzer, A. [1986] "Conditionals," in A. M. Farley, P. Farley, and K. E. McCollough (eds.) *Papers from the Parasession on Pragmatics and Grammatical Theory*, Chicago: Chicago Linguistics Society, pp. 115–135.

Krzyżanowska, K. H. [2015] *Between "If" and "Then"*, PhD dissertation, University of Groningen.

Krzyżanowska, K. H., Wenmackers, S., and Douven, I. [2014] "Rethinking Gibbard's Riverboat Argument," *Studia Logica* 102:771–792.

Lewis, D. K. [1976] "Probabilities of Conditionals and Conditional Probabilities," *Philosophical Review* 85:297–315.

Lewis, D. K. [1997] "Finkish Dispositions," *Philosophical Quarterly* 47:143–158.

MacFarlane, J. [2014] *Assessment Sensitivity: Relative Truth and Its Applications*, Oxford: Oxford University Press.

McGee, V. [1985] "A Counterexample to Modus Ponens," *Journal of Philosophy* 82:462–471.

Oaksford, M. and Chater, N. [2007] *Bayesian Rationality*, Oxford: Oxford University Press.

Oberauer, K., Geiger, S. M., Fischer, K., and Weidenfeld, A. [2007] "Two Meanings of 'If'? Individual Differences in the Interpretation of Conditionals," *Quarterly Journal of Experimental Psychology* 60:790–819.

Oberauer, K., Weidenfeld, A., and Fischer, K. [2007] "What Makes Us Believe a Conditional? The Roles of Covariation and Causality," *Thinking and Reasoning* 13:340–369.

Oberauer, K. and Wilhelm, O. [2003] "The Meaning(s) of Conditionals: Conditional Probabilities, Mental Models and Personal Utilities," *Journal of Experimental Psychology: Learning, Memory and Cognition* 29:688–693.

Over, D. E., Douven, I., and Verbrugge, S. [2013] "Scope Ambiguities and Conditionals," *Thinking and Reasoning* 19:284–307.

Over, D. E. and Evans, J. St. B. T. [2003] "The Probability of Conditionals: The Psychological Evidence," *Mind and Language* 18:340–358.

Over, D. E., Hadjichristidis, C., Evans, J. St. B. T., Handley, S. J., and Sloman, S. A. [2007] "The Probability of Causal Conditionals," *Cognitive Psychology* 54:62–97.

Pfeifer, N. and Kleiter, G. D. [2010] "The Conditional in Mental Probability Logic," in M. Oaksford and N. Chater (eds.) *Cognition and Conditionals*, Oxford: Oxford University Press, pp. 153–173.

Politzer, G., Over, D. E., and Baratgin, J. [2010] "Betting on Conditionals," *Thinking and Reasoning* 16:172–197.

Skovgaard-Olsen, N. [2015a] "Ranking Theory and Conditional Reasoning," *Cognitive Science*, in press.

Skovgaard-Olsen, N. [2015b] "Motivating the Relevance Approach to Conditionals," *Mind and Language*, in press.

Skovgaard-Olsen, N., Singmann, H., and Klauer, K. C. [2015] "The Relevance Effect and Conditionals," manuscript.

van Fraassen, B. C. [1976] "Probabilities of Conditionals," in W. L. Harper and C. A. Hooker (eds.) *Foundations of Probability Theory, Statistical Inference, and Statistical Theories of Science* (Vol. I), Dordrecht: Reidel, pp. 261–301.

van Wijnbergen-Huitink, J., Elqayam, S., and Over, D. E. [2015] "The Probability of Iterated Conditionals," *Cognitive Science* 39:788–803.

Yalcin, S. [2010] "Probability Operators," *Philosophy Compass* 5:916–937.

Qualitative and Semi-Quantitative Modeling of Uncertain Knowledge - A Discussion

Didier Dubois[1], Henri Prade[2]

Abstract. Belief revision mechanisms and nonmonotonic reasoning rely on implicit or explicit orderings that may be conveniently represented by means of possibility distributions or of ordinal conditional functions. The paper provides a comparative discussion of the two representation settings. Possibility distributions may range on finite qualitative scales, or use more quantitative scales that may be mapped on the integer-valued scale of ordinal conditional functions. The use of a scale with finite lower and upper bounds in possibility theory makes their use natural as possibility degrees assigned to particular states that are respectively impossible and fully possible, while $+\infty$, the upper bound of the ordinal conditional function scale expressing impossibility, is never assigned to a particular state. Moreover, possibility theory is naturally associated with four set functions reflecting (graded) modalities, two of them corresponding to a particular system of lower and upper probabilities when the scale is taken as the unit real interval. On its side, the setting of ordinal conditional functions essentially exploits a disbelief function (even if a belief function can be defined as the disbelief of the negation of the event), which may be viewed as the exponent of an infinitesimal probability.

1 Introduction

An assignment of weights in a set of mutually exclusive possible worlds provides a convenient way for rank-ordering these worlds according to their plausibility. Such an assignment may be valued in a quantitative way as in probability theory, or may have a more qualitative flavor. Two knowledge representation settings are based on assignment functions that are ranging on scales that are less quantitative than the real unit interval [0, 1] used in probability theory. These two settings are i) qualitative possibility theory [68, 31], where possibility degrees may belong to a finite, or a denumerable scale, or to [0, 1] then viewed as an ordinal scale, and ii) ordinal conditional functions, also named ranking functions [60, 64], which take their values on $\mathbb{N} \cup \{0, \infty\}$. Using such scales is of interest when plausibility can be only assessed qualitatively, or when we are just interested in specifying a well-ordered partition of the worlds.

Qualitative possibility theory and ordinal conditional functions have been extensively used in nonmonotonic reasoning and in belief revision [28, 22, 45]. The two uncertainty theory settings can be closely related, but have some distinctive features that are worth discussing, especially in the perspective of

---

[1] IRIT, CNRS and University of Toulouse, 31062 Toulouse Cedex 9, France
[2] IRIT, CNRS and University of Toulouse, 31062 Toulouse Cedex 9, France

the two above-mentioned artificial intelligence fields of research. This is the topic of this overview and discussion paper.

The paper is organized as follows. The next section introduces possibility theory and then provides a summary presentation of the ordinal conditional function setting, before discussing the linkage between the two frameworks through different variants of possibility theory according to the scale which is used. Section 3 compares the two theories in belief revision and then in nonmonotonic reasoning. The concluding remarks point out issues in relation with other reasoning problems such as fusion or decision under risk or uncertainty.

This paper is a small tribute to our dear friend and colleague Gabriele Kern-Isberner for all the rich discussions, exchanges and collaborations we had with her over two decades, at the occasion of her 60th birthday anniversary.

2 Frameworks for plausibility ranking

Possibility theory was proposed by L. A. Zadeh in the late 1970's for representing uncertain pieces of information expressed by fuzzy linguistic statements [68], and later developed in an artificial intelligence perspective [38, 31, 33]. Formally speaking, the proposal is quite similar to the one made twenty-five years before by the economist G. L. S. Shackle [56, 57], who had been advocating and developing a non probabilist view of uncertainty based on the idea of degree of surprise, which corresponds to a degree of impossibility. About a decade later, in the late 1980's, W. Spohn [60] introduced the notion of ordinal conditional functions as a basis for a dynamic theory of epistemic states.

We first restate these two settings, which both represent an epistemic state by means of a function defined on a referential U, before pointing out their link [24] and discussing different scales for grading possibility. The set U is to be understood as a set of mutually exclusive possible worlds, or configurations, or possible states of affairs (or descriptions thereof), or states for short. This set can be the domain of an attribute (numerical or categorical), the Cartesian product of attribute domains, the set of interpretations of a propositional language, etc.

2.1 Possibility theory

The basic building block of possibility theory is the notion of a possibility distribution to which Zadeh [68] associates a possibility measure. A possibility distribution is a mapping π from U to a totally ordered scale $\mathcal{S}$, with top denoted by 1 and bottom by 0. The possibility scale may be any finite chain $\mathcal{S} = \{1 = \lambda_1 > \cdots > \lambda_n > \lambda_{n+1} = 0\}$, the unit interval as suggested by Zadeh, or more generally any totally ordered chain. The function π represents the state of knowledge of an agent (about the actual state of affairs), also called an *epistemic state* distinguishing what is plausible from what is less plausible, what is the normal course of things from what is not, what is surprising from what is expected, with the following conventions:

i) $\pi(u) = 0$ means that state u is rejected as impossible;

ii) $\pi(u) = 1$ means that state u is totally possible (= plausible);
iii) the larger $\pi(u)$, the more plausible the state u is.

If the universe U is exhaustive, at least one of the elements in S should be the actual world, so that $\exists u, \pi(u) = 1$ (normalization). This condition expresses the consistency of the epistemic state described by π. Distinct values may simultaneously have a degree of possibility equal to 1.

In the $\{0, 1\}$-valued case, π is just the characteristic function of a subset $E \subseteq U$ of mutually exclusive states, ruling out all those states outside E considered as impossible. This represents incomplete information when E is not a singleton. Complete ignorance corresponds to the possibility distribution $\pi(u) = 1, \forall u \in U$ (all states are possible). A possibility distribution π is said to be at least as specific as another π' if and only if for each state of affairs u: $\pi(u) \leq \pi'(u)$. Then, π is at least as restrictive and informative as π', since it rules out at least as many states with at least as much strength.

Possibility theory is driven by the *principle of minimal specificity*. It states that *any hypothesis not known to be impossible cannot be ruled out*. It is a minimal commitment, cautious, information principle. Basically, we must always try to maximize possibility degrees, taking constraints into account. Given a piece of information in the form x is F where F is a fuzzy set restricting the values of the ill-known quantity x, it leads to represent the knowledge by the inequality $\pi \leq \mu_F$, the membership function of F. The minimal specificity principle enforces possibility distribution $\pi = \mu_F$, if no other piece of knowledge is available. Given several pieces of knowledge of the form x is F_i, for $i = 1, \ldots, n$, each of them translates into the constraint $\pi \leq \mu_{F_i}$, which leads to the inequality $\pi \leq \min_{i=1}^n \mu_{F_i}$ and on behalf of the minimal specificity principle, to the possibility distribution $\pi = \min_{i=1}^n \mu_{F_i}$. It justifies the use of the minimum operation for combining information items. This fully agrees with classical logic, since a set of a classical logic formulas is equivalent to the logical conjunction of the logical formulas that belong to the the set, and its models is obtained by intersecting the sets of models of its formulas. Moreover, asserting a logical proposition p amounts to declaring that any interpretation (state) that makes p false is impossible.

Possibility and necessity functions Given a simple query of the form "does event A occur?", where A is a subset of states, the response to the query can be obtained by computing the degree of possibility of A [68]

$$\Pi(A) = \sup_{u \in A} \pi(u).$$

$\Pi(A)$ evaluates to what extent A is consistent with π ($\Pi(A) = 1$ if and only if $\{u \in U | \pi(u) = 1\} \cap A \neq \emptyset$). In particular, $\Pi(U) = 1$ (since π is normalized) and $\Pi(\emptyset) = 0$ by convention. Besides the possibility of the event A^c opposite to A, modeled by the complement A^c, $\Pi(A^c)$ can be easily related to the idea of certainty of A. Indeed, the less $\Pi(A^c)$, the more A^c is impossible and the more certain is A. If the possibility scale $\mathcal{S}$ is equipped with an order-reversing map denoted by $1 - (\cdot)$ (if $\mathcal{S}$ is finite, $1 - \lambda_i = \lambda_{n+2-i}$), it enables a degree of necessity (certainty) of A to be defined in the form

$$N(A) = 1 - \Pi(A^c),$$

which expresses the well-known duality between possibility and necessity [23]. $N(A) = \inf_{u \notin A} 1 - \pi(u)$ evaluates to what extent A is certainly implied by π ($N(A) = 1$ if and only if $\{u \in U | \pi(u) > 0\} \subseteq A$). Clearly, $N(U) = 1$ and $N(\emptyset) = 0$.

Possibility measures satisfy the characteristic "maxitivity" property

$$\Pi(A \cup B) = \max(\Pi(A), \Pi(B)).$$

Necessity measures satisfy an axiom dual to that of possibility measures:

$$N(A \cap B) = \min(N(A), N(B)).$$

On infinite spaces, these axioms must hold for infinite families of sets. Since π is normalized, $\min(N(A), N(A^c)) = 0$ and $\max(\Pi(A), \Pi(A^c)) = 1$, or equivalently $\Pi(A) = 1$ whenever $N(A) > 0$, which totally fits the intuition behind this formalism, namely that something somewhat certain should be first fully possible, i.e. fully consistent with the available information. Moreover, one cannot be somewhat certain of both A and A^c, without being inconsistent. Note also that we only have $N(A \cup B) \geq \max(N(A), N(B))$. This property is in agreement with with the idea that one may be certain about the event $A \cup B$, without being really certain about more specific events such as A and B.

Certainty qualification Human knowledge is often expressed in a declarative way using statements to which belief degrees are attached. Certainty-qualified pieces of information of the form "A is certain to degree α" can be modeled by the constraint $N(A) \geq \alpha$. It represents a family of possible epistemic states π that obey this constraint. The least specific possibility distribution among them exists and is given by [38]:

$$\pi_{(A,\alpha)}(u) = \begin{cases} 1 & \text{if } u \in A \\ 1 - \alpha & \text{otherwise.} \end{cases}$$

If $\alpha = 1$ we get the characteristic function of A. If $\alpha = 0$, we get total ignorance. This possibility distribution is a key building-block to construct possibility distributions from several pieces of uncertain knowledge. It is instrumental in possibilistic logic semantics. Indeed, e.g. in the finite case, any possibility distribution can be viewed as a collection of nested certainty-qualified statements. Let $E_i = \{u \mid \pi(u) \geq \lambda_i \in S\}$ be the λ_i-cut of π. Then it can be checked that $\pi(u) = \min_{i: u \notin E_i} \lambda_{i+1}$ and $N(E_i) = 1 - \lambda_{i+1}$.

We can also consider possibility-qualified statements of the form $\Pi(A) \geq \beta$; however, the least specific epistemic state compatible with this constraint is trivial and expresses total ignorance.

Conditioning Conditional possibility can be defined similarly to probability theory using a Bayesian-like equation of the form

$$\Pi(B \cap A) = \Pi(B \mid A) \star \Pi(A)$$

where $\Pi(A) > 0$ and $\star$ may be the minimum or the product; moreover we have $N(B \mid A) = 1 - \Pi(B^c \mid A)$. The above equation makes little sense for necessity measures, as it becomes trivial when $N(A) = 0$, that is under lack of certainty, while in the above definition, the equation becomes problematic only if $\Pi(A) = 0$, which is natural as then A is considered impossible (see [14] for the handling of this situation). If operation $\star$ is the minimum, the equation $\Pi(B \cap A) = \min(\Pi(B \mid A), \Pi(A))$ fails to characterize $\Pi(B \mid A)$, and we must resort to the minimal specificity principle to come up with a qualitative conditioning of possibility [38]:

$$\Pi(B \mid A) = \begin{cases} 1 \text{ if } \Pi(B \cap A) = \Pi(A) > 0, \\ \Pi(B \cap A) \text{ otherwise.} \end{cases}$$

It is clear that $N(B \mid A) > 0$ if and only if $\Pi(B \cap A) > \Pi(B^c \cap A)$. Note also that $N(B \mid A) = N(A^c \cup B)$ if $N(B \mid A) > 0$. Moreover, if $\Pi(B \mid A) > \Pi(B)$ then $\Pi(B \mid A) = 1$, which points out the limited expressiveness of this qualitative notion (no gradual positive reinforcement of possibility). However, it is possible to have that $N(B) > 0, N(B^c \mid A_1) > 0, N(B \mid A_1 \cap A_2) > 0$ (i.e., oscillating beliefs).

In the numerical setting, we must choose $\star = $ product that preserves continuity, so that

$$\Pi(B \mid A) = \frac{\Pi(B \cap A)}{\Pi(A)}$$

which makes possibilistic and probabilistic conditionings very similar [16]. Gradual positive reinforcement of possibility is then allowed. Thus, in *quantitative* possibility theory where the scale $[0, 1]$ is used, conditioning is based on product, while in *qualitative* possibility theory it is based on minimum. Note that the Bayesian-like equation with $\star = \min$ already appears in [57] without resulting in a proper definition of conditioning due to the lack of a principle like the minimum specificity principle.

Two other measures The information conveyed by a possibility distribution π is qualitatively summarized wrt to the pair of opposite events (A, A^c) by computing the maximum of π over A and A^c. This leads to the measures $\Pi(A)$ and $N(A)$. But the minimum of π over A and A^c is also of interest and provides other pieces of information; this leads to consider two other set functions. Thus a measure of *guaranteed possibility* or *strong* possibility can be defined, that differs from the functions Π (*weak* possibility) and N (*strong* necessity) [26, 20]:

$$\Delta(A) = \inf_{u \in A} \pi(u).$$

It estimates to what extent *all* states in A are actually possible according to evidence. $\Delta(A)$ can be used as a degree of evidential support for A. Of course, this function possesses a dual conjugate ∇ such that $\nabla(A) = 1 - \Delta(A^c) = \sup_{u \notin A} 1 - \pi(u)$. Function $\nabla(A)$ evaluates the degree of potential or *weak* necessity of A, as it is 1 only if some state s out of A is impossible. It follows that the functions Δ and ∇ are *decreasing* with respect to set inclusion, and that they satisfy the characteristic properties $\Delta(A \cup B) = \min(\Delta(A), \Delta(B))$ and $\nabla(A \cap B) = \max(\nabla(A), \nabla(B))$ respectively.

Uncertain statements of the form "A is possible to degree β" often mean that any realization of A are possible to degree β (e.g. "it is possible that the museum is open this afternoon"). They can then be modeled by a constraint of the form $\Delta(A) \geq \beta$. It corresponds to the idea of observed evidence. This type of information is better exploited by assuming an informational principle opposite to the one of minimal specificity, namely, *any situation not yet observed is tentatively considered as potentially impossible*. This is similar to the closed-world assumption. The *most specific* distribution $\delta_{(A,\beta)}$ in agreement with $\Delta(A) \geq \beta$ is:

$$\pi_{[A,\beta]}(u) = \begin{cases} \beta & \text{if } u \in A \\ 0 & \text{otherwise.} \end{cases}$$

Note that while possibility distributions induced from certainty-qualified pieces of knowledge combine conjunctively, by discarding possible states, evidential support distributions induced by possibility-qualified pieces of evidence combine disjunctively, by accumulating possible states. Given several pieces of knowledge of the form *x is F_i is possible* (in the sense of guaranteed or strong possibility), for $i = 1, \ldots, n$, each of them translates into the constraint $\pi \geq \mu_{F_i}$; hence, several constraints lead to the inequality $\pi \geq \max_{i=1}^{n} \mu_{F_i}$ and on behalf of the closed-world assumption-like principle based on maximal specificity, expressed by the possibility distribution

$$\pi = \max_{i=1}^{n} \pi_i$$

where π_i represents the information item *x is F_i is possible*. This principle justifies the use of the maximum for combining evidential support functions. Acquiring pieces of possibility-qualified evidence leads to updating $\pi_{[A,\beta]}$ into some wider distribution $\pi > \pi_{[A,\beta]}$. Any possibility distribution can be represented as a collection of nested possibility-qualified statements of the form $(E_i, \Delta(E_i))$, with $E_i = \{u | \pi(u) \geq \lambda_i\}$, since $\pi(u) = \max_{i: u \in E_i} \Delta(E_i)$, dually to the case of certainty-qualified statements.

As a conclusion, possibility theory appears to be a very rich representation setting for handling epistemic states, where we can distinguish between the disbelief in an event A (modeled by $1 - \Pi(A)$) and the belief in an event A (modeled by $N(A)$). This setting has a logical counterpart (the reader is referred to [21, 33] for detailed presentations of possibilistic logic), and a strong modal logic flavor. Indeed generalized possibilistic logic, which accommodates both certainty-qualified and possibility-qualified statements is a fragment of a graded version of the modal logic KD [36]. Moreover, the counterpart of the two other set functions can be also handled in logical settings [20, 37].

2.2 Ordinal conditional functions

Ordinal conditional functions (OCF for short) (also simply named *ranking functions* have been proposed by Spohn [60, 61, 64]. They offer a representation framework for uncertainty that is quite close to the one of possibility theory, as we shall see. As stated by Shenoy [59], "the main ingredients of Spohn's theory are (1) a functional representation of an epistemic state called a natural (or ordinal) conditional function and (2) a rule for revising this function in light of new information." Let us restate them.

A distinctive feature of OCFs is the use of the scale $\mathbb{N} \cup \{0, +\infty\}$, in place of $[0, 1]$ (or a subset of it) in possibility theory. Indeed each interpretation (also called configuration) u is associated with a degree $\kappa(u)$, which is an integer or $+\infty$. The smaller $\kappa(u)$, the more possible u. $\kappa(u) = +\infty$ means that u is impossible, while $\kappa(u) = 0$ expresses that there is absolutely nothing against that u be the real state of the world. Note that as long as the function κ takes only integer values, nothing is fully impossible with such a representation. Besides, a counterpart to possibilistic normalization is supposed to hold, namely $\exists u, \kappa(u) = 0$, i.e., some state is not at all impossible.

Disbelief and belief set functions A set function, also usually denoted by κ, is associated with the "kappa" distribution or ranking function:

$$\kappa(A) = \min_{u \in A} \kappa(u) \text{ and } \kappa(\emptyset) = +\infty.$$

Note that $\kappa(A \cup B) = \min(\kappa(A), \kappa(B))$ (this extends to infinite disjunction), and that due to normalization, we have $\min(\kappa(A), \kappa(A^c)) = 0$. This set function thus behaves as a *disbelief* function [59]. Indeed, the larger $\kappa(A)$, the more impossible any configuration u that makes the event A true, so the more disbelieved A.

After a suggestion made by Spohn [61], Shenoy [59] introduces an associated Spohnian belief function β defined as follows:

$$\beta(A) = \begin{cases} -\kappa(A) & \text{if } \kappa(A) > 0 \\ \kappa(A^c) & \text{if } \kappa(A) = 0. \end{cases}$$

Note that the range of β is $\mathbb{Z} \cup \{-\infty, +\infty\}$ (and not $\mathbb{N} \cup \{0, +\infty\}$), and that

$$\beta(A) = -\beta(A^c),$$

which expresses that the more A is believed, the less the opposite event A^c is believed. This relation is reminiscent from the relationship existing between the measure of belief $MB(H, E) \in [0, 1]$ in an hypothesis H given an evidence E and the measure of disbelief $MD(H, E) = MB(H^c, E)$, introduced for the MYCIN expert system [13], whose difference $MB(H, E) - MD(H, E) \in [-1, 1]$ defines the notion of a certainty factor. This kind of relation expresses a duality similar to the one between necessity and (im)possibility, and the disbelief function can be retrieved from the belief function:

$$\kappa(A) = \begin{cases} 0 & \text{if } \beta(A) \geq 0 \\ -\beta(A) & \text{if } \beta(A) < 0. \end{cases}$$

Positive and negative ranking functions More recently, a simpler way to handle the above duality has been advocated by Spohn [62] with the idea of positive and negative ranking functions. The set function κ which expresses a grading of disbelief, which is something negative, is called a *negative* ranking function. Then the set function λ, defined by

$$\lambda(A) = \kappa(A^c),$$

which is such that $\lambda(\emptyset) = 0$, $\lambda(U) = +\infty$, $\lambda(A \cap B) = \min(\lambda(A), \lambda(B))$, expresses a grading of belief (in A) and is called a *positive* ranking function. The larger $\lambda(A)$, the more A is believed. The set function λ is very similar to a necessity measure, except that the range of λ is $\mathbb{N} \cup \{0, +\infty\}$ rather than a bounded scale such as the unit interval, while the set function $\kappa(A)$ is similar to $N(A^c) = 1 - \Pi(A)$.

Lastly, Spohn [62] considers the *two-sided ranking* function

$$\tau(A) = \lambda(A) - \kappa(A) = \kappa(A^c) - \kappa(A),$$

which insures that $\tau(A^c) = -\tau(A)$ (as for the set function β, although τ is more like a certainty factor). The range of τ is clearly $\mathbb{Z} \cup \{-\infty, +\infty\}$. However, "the formal behavior [of τ] is awkward", as stated in [62]. Note also that in possibility theory $N(A) - N(A^c) = N(A) - (1 - \Pi(A)) = N(A) \in [0, 1]$ if $N(A) > 0$ (the certainty factor reduces to the non zero necessity degree), while $N(A) - N(A^c) = \Pi(A) - 1 \in [-1, 0]$ if $N(A) = 0$ (it reduces to the opposite of impossibility).

Conditioning Conditioning is a key notion in ranking function theory. It could be defined in the following way:

$$\kappa(u \mid B) = \begin{cases} \kappa(u) - \kappa(B) & \text{if } u \in B \\ +\infty & \text{otherwise.} \end{cases}$$

Letting $\kappa(u \mid B) = +\infty$ if $u \in B^c$ amounts to saying that any value in B^c is fully impossible after conditioning by B. In fact, Spohn [60] only defines $\kappa(u \mid B)$ for $u \in B$. He prefers to keep the result of conditioning finite in any case by completing the definition when $u \in B^c$ as follows, using a parameter $\alpha \in \mathbb{N}$:

$$\kappa_{B,n}(u) = \begin{cases} \kappa(u \mid B) & \text{if } u \in B \\ n + \kappa(u \mid B^c) & \text{if } u \in B^c. \end{cases}$$

Since $\kappa_{B,n}(B) = 0$ and $\kappa_{B,n}(B^c) = n$, the parameter n can be understood as the "firmness" with which B is believed in $\kappa_{B,n}$. Darwiche and Pearl [15] have proposed a special case of the above revision operator (where $n = \kappa(B^c) + 1$) which ensures that the revision by B always strengthens the belief in B:

$$\kappa_B^{DP}(u) = \begin{cases} \kappa(u) - \kappa(B) & \text{if } u \in B \\ \kappa(u) + 1 & \text{if } u \in B^c. \end{cases}$$

Relation with probability theory Spohn [60] suggests to understand $\kappa(u)$ as the exponent of an infinitesimal probability, which fully agrees with the fact that $\kappa(A \cup B) = \min(\kappa(A), \kappa(B))$, while $\kappa(u \mid B)$ is then the exponent of the infinitesimal conditional probability $P(u \mid B)$. See also [41] for another viewpoint.

Relation with possibility theory It is easy to change a ranking function into a possibility distribution by means of the following transformations [24] :

$$\pi_\kappa(u) = 2^{-\kappa(u)}, \Pi_\kappa(A) = 2^{-\kappa(A)}.$$

Then π_κ et Π_κ take their values on a subset of rationals of $[0,1]$. Π_κ is indeed a possibility measure : $\Pi_\kappa(A \cup B) = 2^{-\min(\kappa(A),\kappa(B))} = \max(\Pi_\kappa(A), \Pi_\kappa(B))$.

Moreover, for the conditional, we have $\forall u$:

$$\pi_\kappa(u|B) = \begin{cases} 2^{-\kappa(u)-\kappa(B)} = \frac{2^{-\kappa(u)}}{2^{-\kappa(B)}} = \frac{\pi_\kappa(u)}{\Pi_\kappa(B)} & \text{if } u \in B, \\ 0 \text{ if } u \in B^c, \end{cases}$$

which is the possibilistic conditioning based on product. Similarly, $\pi_{\kappa_B,n}(u) = \pi_\kappa(u|B)$ if $u \in B$, and $\pi_{\kappa_B,n}(u) = 2^{-n} \cdot \pi_\kappa(u|B^c)$ if $u \in B^c$.

The converse transformation only applies when $\kappa(u) = -log_2(\pi(u))$ takes integer values.

As can be seen, OCF theory and possibility theory are very close. Interestingly enough, from the beginning Spohn [60, 61, 64] has acknowledged the close relationship of his proposal with Shackle's theory of potential surprise (and its defense by Levi [52–54]), while he seems to not to be aware of possibility theory when he introduces ordinal conditional functions [60, 61], and more recently, he seems to base his critique of it [64] on a partial and limited knowledge of the corresponding literature.

For Spohn [60], "The essential point is that Shackle has no precise and workable account of conditional degrees of potential surprise, of changes of FPSs, etc." This is no longer true for possibility theory! See [32, 30, 8, 9].

The other point of difference is Spohn's reluctance to state the full impossibility of a configuration u by assigning $+\infty$ to it, while it may be convenient (but it is not compulsory!) in possibility theory to state that some u is fully impossible by letting $\pi(u) = 0$, or that the potential surprise (or the impossibility) is maximal (i.e., $1 - \pi(u) = 1$) as Shackle's view allows.

2.3 Different scales for graded possibility

The above discussion about the linkage between OCF theory and possibility theory illustrates the fact that there are several representations of epistemic states in agreement with the idea of graded possibility, such as well ordered partitions of U [60] or Lewis' systems of spheres [43], OCF's [60], or numerical possibilities. But all these representations of epistemic states do not have the same expressive power, and range from purely qualitative to quantitative possibility distributions, using weak orders, qualitative scales, integers, or reals. In fact we can distinguish several representation settings according to the expressiveness of the scale used [9]:

1. The pure ordinal finite setting, where an epistemic state on a set of possible worlds is simply encoded by means of a total preorder $\geq$, telling which worlds are more normal, less surprising than other ones. The quotient set $U/\sim$, built from the equivalence relation $\sim$ extracted from a relation $\geq$

modeling "more possible than", forms a *well-ordered partition* $U_0, \ldots, U_k$ such that $\Pi(U_0) = 1 > \cdots > \Pi(U_k)$, where Π is the possibility measure induced by any possibility distribution π that represents $\geq$ (i.e., $\pi(u_1) > \pi(u_2)$ if and only if $u_1 > u_2$, $>$ being the strict part of $\geq$). The less $\pi(u)$, the less plausible u, or the less likely it is the real world. This is the setting used by Grove [43], and Gärdenfors [39] when modeling belief revision. However, note that this purely ordinal representation is less expressive than the qualitative encoding of a possibility distribution on a totally ordered scale as the former cannot express impossibility.

2. The qualitative finite setting, with possibility degrees in a finite totally ordered scale : $L = \{\alpha_0 = 1 > \alpha_1 > \cdots > \alpha_{m-1} > 0\}$. This setting is used in possibilistic logic [21].

3. The denumerable setting, using a scale $L = \{\alpha^0 = 1 > \alpha^1 > \cdots > \alpha^i > \cdots > 0\}$, for some $\alpha \in (0,1)$. This is isomorphic, as recalled above, to the use of integers in κ-functions by Spohn [61], where the set of natural integers is used as a disbelief scale. Note that this framework is not really qualitative; it is (minimally) numerical as it allows for sums of exponents.

4. The dense ordinal setting using $L = [0,1]$, seen as an ordinal scale. In this case, the possibility distribution π is defined up to any monotone increasing transformation $f : [0,1] \to [0,1], f(0) = 0, f(1) = 1$. This setting is also used in possibilistic logic [21].

5. The dense absolute setting, where $L = [0,1]$, seen as a genuine numerical scale equipped with product. In this case, a possibility measure can be viewed as special case of a Shafer [58] plausibility function, actually a consonant plausibility function, or as an upper probability [65]. One clear advantage of using the scale $[0,1]$ is also the capability of introducing as many intermediary possibility levels as it is needed.

The reader is referred to [34] for a detailed description of existing elicitation methods of possibility distributions for these different types of scales.

3 Belief revision and default reasoning

In the following we do not intend a general discussion on the revision of epistemic states, but only deal with one form of revision when a body of prior uncertain evidence is to be revised by the input information B acting at the same level as the prior information modeled by the plausibility ranking [18], in the possibility theory and OCF settings. The minimum-based and the product-based conditioning in possibility theory gives birth to qualitative and quantitative revision operators. We have already indicated that $\kappa(u \mid B)$ is the OCF counterpart of $\pi_\kappa(u|B) = \frac{\pi_\kappa(u)}{\Pi_\kappa(B)}$ if $u \in B$ (and $\pi_\kappa(u|B) = 0$ otherwise) for the product-based conditioning. Similarly, the minimum-based conditioning

$$\pi(u|_{min} B) = \begin{cases} 1 & \text{if } \pi(\omega) = \Pi(B) \text{ and } u \in B \\ \pi(u) & \text{if } \pi(u) < \Pi(B) \text{ and } u \in B \\ 0 & \text{if } u \in B^c. \end{cases}$$

has the following form

$$\kappa(u|_{min}B) = \begin{cases} 0 & \text{if } \kappa(u) = \kappa(B) \text{ and } u \in B \\ \kappa(u) & \text{if } \kappa(u) > \kappa(B) \text{ and } u \in B \\ +\infty & \text{if } u \in B^c. \end{cases}$$

in the OCF framework [22].

These semantic revision operators have been studied axiomatically and their syntactic counterpart provided in the setting of possibilistic logic (when the possibility distribution to be revised is the semantics of a possibilistic logic base [21]) [25, 6, 8].

Belief revision with uncertain inputs Besides, the possibilistic revision operators have been extended to uncertain inputs [30, 22, 9]. Uncertain inputs are then viewed as pieces of information of the form $N(B) \geq \alpha$; revision may then either enforce uncertainty (the revision result should satisfy the input information viewed as a constraint), or may treat the uncertainty-qualified input as poorly informed if the initial epistemic state is already more certain that B is true than the input (in this case no change takes place).

We only consider the first option in the following, which is expressed in possibility theory by the following revision rule [25, 30]:

$$\pi(u \mid (B, \alpha)) = \begin{cases} \pi(u \mid_\star B) & \text{if } u \in B \\ (1 - \alpha) \star \pi(u \mid_\star B^c) & \text{if } u \in B^c. \end{cases}$$

where $\star$ stands for minimum or product ($\pi(u \mid_\star B)$) refers to the minimum or product-based conditioning respectively. This revision rule can be viewed as a counterpart of Jeffrey's rule in probability theory, changing the convex sum into a qualitative mixture ($P(A \mid (B, \alpha)) = \alpha \cdot P(A \mid B) + (1 - \alpha) \cdot P(A \mid B^c)$) [44]).

In case operation $\star$ is the product, we recognize Spohn's definition of $\kappa_{B,n}(u)$ up to the transformation from a possibility distribution to an OCF (with $\alpha = 1 - 2^{-n}$). Formally speaking, this means that Spohn's rule is implicitly a revision under an uncertain input (where the uncertainty is directly related to the firmness n). When $\star = \min$, it can be shown [22] that what is obtained corresponds exactly to a qualitative transmutation [66] called *adjustment*, introduced by M.-A. Williams [67] in the Spohnian setting.

Belief revision with conditionals A generalization of Spohn's rule of conditioning, called *c-revision*, has been proposed by G. Kern-Isberner [46, 45], where an OCF is now revised by a conditional statement modeled by a conditional event [17]. A conditional event $B \mid A$ is a tri-valued entity which is true if A and B are true, which is false if B is false and A is true, and which is not applicable if A is false; conditional events provide a simple modeling of non monotonic consequence relations [27]. C-revision is defined as

$$\kappa_{B|A,n}(u) = \begin{cases} \kappa(u) - \kappa(B \mid A) & \text{if } u \in A \cap B \\ \kappa(u) - \kappa(B^c \mid A) + n & \text{if } u \in A \cap B^c \\ \kappa(u) & \text{if } u \in A^c. \end{cases}$$

where $\kappa(B \mid A) = \kappa(A \cap B) - \kappa(A)$. Spohn's revision rule is retrieved for the conditional event $B \mid U$. This rule has been extended to multiple revision [48]. The c-revision rule may be used in a default reasoning perspective, since belief revision and non monotonic reasoning are closely related [40]. We first restate the possibilistic approach to non-monotonic reasoning.

Default reasoning Nonmonotonic reasoning has been extensively studied in AI in relation with the problem of reasoning under incomplete information with rules having potential exceptions. The possibilistic approach [5, 7] provides a faithful representation of the postulate-based approach proposed by Kraus, Lehmann and Magidor [49, 51]. A default rule "if a then b, generally" linking two propositions a and b, is then understood formally as the constraint

$$\Pi(a \wedge b) > \Pi(a \wedge \neg b)$$

on a possibility measure Π describing the semantics of the available knowledge. It expresses that in the context where a is true, there exists situations where having b true is strictly more satisfactory than any situations where b is false in the same context. As already said, such a constraint is equivalently expressed with conditional necessity as $N(b \mid a) > 0$.

The method then consists in turning each default $p_i \rightsquigarrow q_i$ in the knowledge base into a possibilistic clause $(\neg p_i \vee q_i, N(\neg p_i \vee q_i))$, where the proposition $\neg p_i \vee q_i$ is associated with the priority level $N(\neg p_i \vee q_i)$. N is the necessity measure defined from the greatest possibility distribution π induced by the set of constraints $\Pi(p_i \wedge q_i) > \Pi(p_i \wedge \neg q_i)$ encoding to the default knowledge base. We thus obtain a possibilistic logic base K. This encodes the generic knowledge embedded in the default rules. The ranking of the defaults induced by the degrees $N(\neg p_i \vee q_i)$ is the same as the Z-ranking introduced by Pearl [55]. Then we apply the possibilistic logic inference for reasoning with the formulas in K encoding the defaults, together with the available factual knowledge encoded as fully certain possibilistic formulas in a base F.

Such an approach has been proved to be in full agreement with the Kraus-Lehmann-Magidor postulates-based approach to nonmonotonic reasoning [4]. More precisely, two nonmonotonic entailments can be defined in the possibilistic setting, the one presented above, based on the least specific possibility distribution compatible with the constraints encoding the set of defaults, and another one more cautious, where one considers that b can be deduced in the situation where all we know is $F = \{a\}$ if and only if the inequality $\Pi(a \wedge b) > \Pi(a \wedge \neg b)$ holds true for *all* Π functions compatible with the constraints encoding the set of defaults. The first entailment coincides with the rational closure inference [51], while the later corresponds to the (cautious) preferential entailment [4].

While the consequences obtained with the preferential entailment are hardly debatable, the ones derived with the rational closure are more adventurous. However these latter consequences can be always modified if necessary by the *addition* of further defaults [5, 7]. These added defaults may express independence information of the type "in context c, the truth or the falsity of a has no influence on the truth of b" [19, 1].

Thus, a default rule "if a then b, generally", corresponding to the conditional event $b \mid a$, is encoded by a possibility distribution equal to $1 - \alpha$ for the interpretations of $a\neg b$ (they falsify the rule), and equal to 1 for the other interpretations. Besides, possibilistic inference forgets all formulas with weights under the level of inconsistency of $K \cup F$, even when they are not involved in the inconsistency, which may prevent to draw some desirable conclusions. This is avoided in the more sophisticated approach to default reasoning, inspired from [10], developed by G. Kern-Isberner [45] in the OCF setting, and which can be expressed in possibility theory-based representations [47]. In this approach, put in possibilistic terms, the interpretations that falsifies a still have a possibility degree equal to 1, $a\neg b$ receives a degree $1 - \alpha^-$, and interpretation ab now receives a degree $1 - \alpha^+$ (instead of 1, in the previous approach). This is a consequence of the c-revision point of view in terms of conditional events. The merits of this approach to default reasoning have still to be compared in detail with approaches presented in [5], and in [10, 3]. This is a topic for further research.

4 Concluding remarks

Generally speaking, possibility theory and OCF theory are two frameworks for plausibility ranking that are quite similar. Still the use of different scales for grading (im)plausibility makes their expressive powers somewhat different. While one may refrain to assess $+\infty$ as the implausibility of some configuration, the use of a bounded scale such as $[0, 1]$ makes it easy to introduce graded counterparts to modal operators. It is also worth noticing that the connections of the two frameworks with probabilities are different, numerical possibilities encoding imprecise probabilities while ranking functions are inspired by infinitesimal probabilities.

In this paper, beyond comparing the set functions defined in each setting, the presentation has mainly focused on the capability to translate conditioning and associated revision operators from one theory to the other one as well as their handling of default reasoning. In that respect, other areas of applications such as information fusion [2] [50], decision making under risk and uncertainty [29, 35] [42], or causality ascription [11, 12] [63] would be worth discussing as well.

References

1. N. Ben Amor, K. Mellouli, S. Benferhat, D. Dubois, and H. Prade. A theoretical framework for possibilistic independence in a weakly ordered setting. *Int. J. of Uncertainty, Fuzziness and Knowledge-Based Systems*, 10(2):117–155, 2002.
2. S. Benferhat, D. Dubois, S. Kaci, and H. Prade. Possibilistic merging and distance-based fusion of propositional information. *Annals of Mathematics and Artificial Intelligence*, 34(1-3):217–252, 2002.
3. S. Benferhat, D. Dubois, J. Lang, H. Prade, A. Saffiotti, and Ph. Smets. Reasoning under inconsistency based on implicitly-specified partial qualitative probability relations: a unified framework. In *Proc. 15th National Conf. on Artificial Intelligence (AAAI-98), Madison*, pages 121–126, Menlo Park, 1998. AAAI Press.
4. S. Benferhat, D. Dubois, and H. Prade. Nonmonotonic reasoning, conditional objects and possibility theory. *Artificial Intelligence*, 92:259–276, 1997.

5. S. Benferhat, D. Dubois, and H. Prade. Practical handling of exception-tainted rules and independence information in possibilistic logic. *Appl. Intell.*, 9(2):101–127, 1998.
6. S. Benferhat, D. Dubois, and H. Prade. A computational model for belief change and fusing ordered belief bases. In M.-A. Williams and H. Rott, editors, *Frontiers in Belief Revision*, pages 109–134. Kluwer Acad. Publ., 2001.
7. S. Benferhat, D. Dubois, and H. Prade. The possibilistic handling of irrelevance in exception-tolerant reasoning. *Ann. Math. Artif. Intell.*, 35(1-4):29–61, 2002.
8. S. Benferhat, D. Dubois, H. Prade, and M -A. Williams. A practical approach to revising prioritized knowledge bases. *Studia Logica*, 70(1):105–130, 2002.
9. S. Benferhat, D. Dubois, H. Prade, and M.-A. Williams. A framework for iterated belief revision using possibilistic counterparts to Jeffrey's rule. *Fundam. Inform.*, 99(2):147–168, 2010.
10. S. Benferhat, A. Saffiotti, and Ph. Smets. Belief functions and default reasoning. *Artif. Intell.*, 122(1-2):1–69, 2000.
11. J.-F. Bonnefon, R. Da Silva Neves, D. Dubois, and H. Prade. Predicting causality ascriptions from background knowledge: Model and experimental validation. *Int. J. of Approximate Reasoning*, 48:752–765, 2008.
12. J.-F. Bonnefon, R. Da Silva Neves, D. Dubois, and H. Prade. Qualitative and quantitative conditions for the transitivity of perceived causation: - Theoretical and experimental results. *Annals of Mathematics and Artificial Intelligence*, 64(2-3):311–333, 2012.
13. B. G. Buchanan and E. H. Shortliffe. *Rule-Based Expert Systems*. Addison-Wesley, Reading, Mass., 1984.
14. G. Coletti and B. Vantaggi. T-conditional possibilities: Coherence and inference. *Fuzzy Sets and Systems*, 160(3):306–324, 2009.
15. A. Darwiche and J. Pearl. On the logic of iterated belief revision. *Artificial Intelligence*, 89(1-2):1–29, 1997.
16. B. De Baets, E. Tsiporkova, and R. Mesiar. Conditioning in possibility with strict order norms. *Fuzzy Sets and Systems*, 106:221–229, 1999.
17. B. De Finetti. La logique des probabilités. In *Congrès International de Philosophie Scientifique*, pages 1–9, Paris, 1936. Hermann et Cie.
18. D. Dubois. Three scenarios for the revision of epistemic states. *J. Log. Comput.*, 18(5):721–738, 2008.
19. D. Dubois, L. Fariñas del Cerro, A. Herzig, and H. Prade. A roadmap of qualitative independence. In D. Dubois, H. Prade, and E. P. Klement, editors, *Fuzzy Sets, Logics and Reasoning about Knowledge*, volume 15 of *Applied Logic series*, pages 325–350. Kluwer Acad. Publ., Dordrecht, 1999.
20. D. Dubois, P. Hajek, and H. Prade. Knowledge-driven versus data-driven logics. *J. Logic, Language, and Information*, 9:65–89, 2000.
21. D. Dubois, J. Lang, and H. Prade. Possibilistic logic. In D. M. Gabbay, C. J. Hogger, J. A. Robinson, and D. Nute, editors, *Handbook of Logic in Artificial Intelligence and Logic Programming, Vol. 3*, pages 439–513. Oxford University Press, 1994.
22. D. Dubois, S. Moral, and H. Prade. Belief change rules in ordinal and numerical uncertainty theories. In D. Dubois and H. Prade, editors, *Belief Change*, Gabbay, D. M. and Smets, Ph., series editors, Handbook of Defeasible Reasoning and Uncertainty Management Systems, pages 311–392. Kluwer, Dordrecht, 1998.
23. D. Dubois and H. Prade. *Fuzzy Sets and Systems - Theory and Applications*. Academic Press, New York, 1980.
24. D. Dubois and H. Prade. Epistemic entrenchment and possibilistic logic. *Artif. Intell.*, 50(2):223–239, 1991.

25. D. Dubois and H. Prade. Belief change and possibility theory. In P. Gärdenfors, editor, *Belief Revision*, pages 142–182. Cambridge University Press, 1992.
26. D. Dubois and H. Prade. Possibility theory as a basis for preference propagation in automated reasoning. In *Proc. 1st IEEE Inter. Conf. on Fuzzy Systems (FUZZ-IEEE'92), San Diego, Ca., March 8-12*, pages 821–832, 1992.
27. D. Dubois and H. Prade. Conditional objects as nonmonotonic consequence relationships. *IEEE Trans. on Systems, Man and Cybernetics*, 24(12):1724–1740, 1994.
28. D. Dubois and H. Prade. Conditional objects, possibiliy theory and default rules. In G. Crocco, L. Fariñas del Cerro, and A. Herzig, editors, *Conditionals : From Philosophy to Computer Sciences*, pages 301–336. Oxford University Press, 1995.
29. D. Dubois and H. Prade. Possibility theory as a basis for qualitative decision theory. In *Proc. 14th Int. Joint Conf. on Artificial Intelligence (IJCAI'95), Montréal, Aug. 20-25*, pages 1924–1930. Morgan Kaufmann, 1995.
30. D. Dubois and H. Prade. A synthetic view of belief revision with uncertain inputs in the framework of possibility theory. *Int. J. Approx. Reasoning*, 17(2-3):295–324, 1997.
31. D. Dubois and H. Prade. Possibility theory: Qualitative and quantitative aspects. In D. M. Gabbay and Ph. Smets, editors, *Quantified Representation of Uncertainty and Imprecision*, volume 1 of *Handbook of Defeasible Reasoning and Uncertainty Management Systems*, pages 169–226. Kluwer Acad. Publ., 1998.
32. D. Dubois and H. Prade. Accepted beliefs, revision and bipolarity in the possibilistic framework. In F. Huber and C. Schmidt-Petri, editors, *Degrees of Belief*, pages 161–184. Oxford University Press, 2009.
33. D. Dubois and H. Prade. Possibilistic logic – An overview. In J. H. Siekmann, editor, *Computational Logic*, volume 9 of *Handbook of the History of Logic*, pages 283–342. Elsevier, 2014.
34. D. Dubois and H. Prade. Practical methods for constructing possibility distributions. *Int. J. of Intelligent Systems*, DOI: 10.1002/int.21782, 2015.
35. D. Dubois, H. Prade, and R. Sabbadin. Decision-theoretic foundations of qualitative possibility theory. *Europ. J. of Operational Research*, 128:459–478, 2001.
36. D. Dubois, H. Prade, and S. Schockaert. Stable models in generalized possibilistic logic. In *Proc.13th Inter. Conf. on Principles of Knowledge Representation and Reasoning, Rome, June 10-14*, pages 519–529, 2012.
37. D. Dubois, H. Prade, and S. Schockaert. Reasoning about uncertainty and explicit ignorance in generalized possibilistic logic. In T. Schaub, G. Friedrich, and B. O'Sullivan, editors, *Proc. 21st Europ. Conf. on Artificial Intelligence ECAI'14, Aug. 18-22, Prague*, volume 263 of *Frontiers in Artificial Intelligence and Applications*, pages 261–266. IOS Press, 2014.
38. Didier Dubois and Henri Prade. *Possibility Theory: An Approach to Computerized Processing of Uncertainty, (with the collaboration of H. Farreny, R. Martin-Clouaire, and C. Testemale)*. Plenum Press, 1988.
39. P. Gärdenfors. *Knowledge in Flux: Modeling the Dynamics of Epistemic States*. Bradford Books. MIT Press, Cambridge, 1988.
40. P. Gärdenfors. Belief revision and nonmonotonic logic: Two sides of the same coin? In L. Aiello, editor, *Proc. 9th Europ. Conf. on Artificial Intelligence (ECAI'90)*, pages 768–773, London / Boston, 1990. Pitman.
41. P. H. Giang and P. P. Shenoy. On transformations between probability and Spolinian disbelief functions. In K .B. Laskey and H. Prade, editors, *Proc. 15th Conf. on Uncertainty in Artificial Intelligence (UAI'99), Stockholm, July 30 - Aug. 1*, pages 236–244. Morgan Kaufmann, 1999.

42. P. H. Giang and P. P. Shenoy. A qualitative linear utility theory for Spohn's theory of epistemic beliefs. In C. Boutilier and M. Goldszmidt, editors, *Proc. 16th Conf. in Uncertainty in Artificial Intelligence (UAI'00), Stanford, June 30 - July 3*, pages 220–229. Morgan Kaufmann, 2000.
43. A. Grove. Two modellings for theory change. *J. of Philosophical Logic*, 17(157-180), 1988.
44. R. Jeffrey. *The Logic of Decision*. McGraw-Hill, New York, 1965.
45. G. Kern-Isberner. *Conditionals in Nonmonotonic Reasoning and Belief Revision - Considering Conditionals as Agents*, volume 2087 of *LNCS*. Springer, 2001.
46. G. Kern-Isberner. Handling conditionals adequately in uncertain reasoning. In S. Benferhat and Ph. Besnard, editors, *Proc. 6th Europ. Conf. Symbolic and Quantitative Approaches to Reasoning with Uncertainty (ECSQARU'01), Toulouse, Sept. 19-21*, volume 2143 of *LNCS*, pages 604–615. Springer, 2001.
47. G. Kern-Isberner. Representing and learning conditional information in possibility theory. In B. Reusch, editor, *Computational Intelligence, Theory and Applications, Proc. Int. Conf. 7th Fuzzy Days, Dortmund, Oct. 1-3*, volume 2206 of *LNCS*, pages 194–217. Springer, 2001.
48. G. Kern-Isberner and D. Huvermann. Multiple iterated belief revision without independence. In I. Russell and W. Eberle, editors, *Proc. 28th Int. Florida Artificial Intelligence Research Society Conf. (FLAIRS'15), Hollywood, Fl. May 18-20*, pages 570–575. AAAI Press, 2015.
49. S. Kraus, D. Lehmann, and M. Magidor. Nonmonotonic reasoning, preferential models and cumulative logics. *Artificial Intelligence*, 44:167–207, 1990.
50. N. Laverny and J. Lang. From knowledge-based programs to graded belief-based programs, part I: on-line reasoning*. *Synthese*, 147(2):277–321, 2005.
51. D. Lehmann and M. Magidor. What does a conditional knowledge base entail? *Artificial Intelligence*, 55:1–60, 1992.
52. I. Levi. On potential surprise. *Ratio*, 8:17–129, 1966.
53. I. Levi. *Gambling with Truth, chapters VIII and IX*. Knopf, New York, 1967.
54. I. Levi. Support and surprise: L. j. cohen's view of inductive probability. *Brit. J. Phil. Sci.*, 30:279–292, 1979.
55. J. Pearl. System Z : A natural ordering of defaults with tractable applications for default reasoning. In *Proc. of Theoretical Aspects of Reasoning about Knowledge (TARK'90)*, pages 121–135, 1990.
56. G. L. S. Shackle. *Expectation in Economics*. Cambridge University Press, UK, 2nd edition, 1952, 1949.
57. G. L. S. Shackle. *Decision, Order and Time in Human Affairs*. Cambridge University Press, 1961.
58. G. Shafer. Belief functions and possibility measures. In J. C. Bezdek, editor, *Analysis of Fuzzy Information, Vol. I: Mathematics and Logic*, pages 51–84. CRC Press, Boca Raton, 1987.
59. P. P. Shenoy. On Spohn's rule for revision of beliefs. *Int. J. Approx. Reasoning*, 5(2):149–181, 1991.
60. W. Spohn. Ordinal conditional functions: a dynamic theory of epistemic states. In W. L. Harper and B. Skyrms, editors, *Causation in Decision, Belief Change, and Statistics*, volume 2, pages 105–134. Kluwer, 1988.
61. W. Spohn. A general non-probabilistic theory of inductive reasoning. In R. D. Shachter, T. S. Levitt, L. N. Kanal, and J. F. Lemmer, editors, *Uncertainty in Artificial Intelligence*, volume 4, pages 149–158. Elsevier Sci. Publ. (North-Holland), 1990.
62. W. Spohn. A survey of ranking theory. In F. Huber and C. Schmidt-Petri, editors, *Degrees of Belief*, volume 342 of *Synthese Library*, pages 161–184. Springer, 2009.

63. W. Spohn. *Causation, Coherence and Concepts. A Collection of Essays*, volume 256 of *Boston Studies in the Philosophy of Science*. Springer, 2009.
64. W. Spohn. *The Laws of Belief: Ranking Theory and Its Philosophical Applications*. Oxford Univ. Press, 2012.
65. P. Walley. Measures of uncertainty in expert systems. *Artif. Intell.*, 83(1):1–58, 1996.
66. M.-A. Williams. Transmutations of knowledge systems. In J. Doyle, E. Sandewall, and P. Torasso, editors, *Proc. 4th Int. Conf. on Principles of Knowledge Representation and Reasoning (KR'94). Bonn, May 24-27*, pages 619–629, 1994.
67. Mary-Anne Williams. On the logic of theory base change. In C. MacNish, D. Pearce, and L. M. Pereira, editors, *Logics in Artificial Intelligence, European Workshop, JELIA '94, York, UK, September 5-8, 1994, Proceedings*, volume 838 of *LNCS*, pages 86–105. Springer, 1994.
68. L. A. Zadeh. Fuzzy sets as a basis for a theory of possibility. *Fuzzy Sets and Systems*, 1:3–28, 1978.

On the Relationship Between Aggregating Semantics and FO-PCL Grounding Semantics for Relational Probabilistic Conditionals

Marc Finthammer[1], Christoph Beierle[2]

Abstract. Aggregating semantics and FO-PCL grounding semantics are two approaches to extend the principle of maximum entropy to probabilistic conditionals containing free variables. As they use different variants of conditionals, we use the logic PCI, covering both approaches as special cases. Using the PCI framework, we state explicitly the relationships among various classes of models, including the class of models induced by a parametrically uniform knowledge base. In particular, we show that the models coincide if both maximum entropy and parametric uniformity are taken into account.

1 Introduction

Relational probabilistic conditionals of the from *if A, then B with probability d*, formally denoted by $(B \mid A)[d]$, are a powerful means for knowledge representation and reasoning when uncertainty is involved. As an illustration, consider the following example, adapted from [6], modelling the relationships among elephants in a zoo and their keepers. Elephants usually like their keepers, except for keeper Fred. But elephant Clyde gets along with everyone, and therefore he also likes Fred. The knowledge base $\mathcal{R}_{ek}$ consists of the following conditionals:

$ek_1 : (likes(E,K) \mid elephant(E), keeper(K))[0.9]$

$ek_2 : (likes(E, fred) \mid elephant(E), keeper(fred))[0.05]$

$ek_3 : (likes(clyde, fred) \mid elephant(clyde), keeper(fred)[0.85]$

Conditional ek_1 models statistical knowledge about the general relationship between elephants and their keepers (*"elephants like their keeper with probability 0.9"*), whereas conditional ek_2 represents knowledge about the exceptional keeper Fred and his relationship to elephants in general (*"elephants like keeper Fred only with probability 0.05"*). Conditional ek_3 models subjective belief about the relationship between the elephant Clyde and keeper Fred (*"elephant Clyde likes keeper Fred with probability 0.85"*). From a common-sense point of view, the knowledge base $\mathcal{R}_{ek}$ makes perfect sense: conditional ek_2 is an exception of ek_1, and ek_3 is an exception of ek_2.

However, assigning a formal semantics to $\mathcal{R}_{ek}$ is not straightforward. In the propositional case, the situation is easier. For a ground probabilistic conditional

---

[1] Faculty of Mathematics and Computer Science, University of Hagen, 58084 Hagen, Germany
[2] Faculty of Mathematics and Computer Science, University of Hagen, 58084 Hagen, Germany

$(B \mid A)[d]$, we can say that a probability distribution P over the possible worlds satisfies $(B \mid A)[d]$ iff $P(A) > 0$ and $P(B \mid A) = d$, i.e., iff the *conditional probability* of $(B \mid A)$ under P is d (see e.g. [21]). When extending this to the relational case with free variables as in $\mathcal{R}_{ek}$, the exact role of the variables has to be specified. While there are various approaches dealing with a combination of probabilities with a first-order language (e.g. [18, 17, 24, 27, 25, 37], for a comparison and evaluation of some approaches see e.g. [1, 23]), in this paper we focus on two semantics for probabilistic relational conditionals, the *aggregating semantics* [25] proposed by Kern-Isberner and the grounding semantics employed in the logic FO-PCL [15].

Both aggregating semantics and FO-PCL employ the principle of *maximum entropy* (*ME principle*). The ME principle is a well-established concept for choosing the uniquely determined model of a knowledge base $\mathcal{R}$ having maximum entropy; this model is the most unbiased model of $\mathcal{R}$ in the sense that it completes the knowledge given by $\mathcal{R}$ inductively, but adds as little additional information as possible [19, 36, 29, 30, 20, 31, 28, 21, 24, 32]. While for a set of propositional conditionals there is a general agreement about its ME model, the situation changes when the conditionals are built over a first-order language.

While the two approaches realized by aggregating semantics and in FO-PCL are related in the sense that they refer to a set of constants when interpreting the variables in the conditionals, there is also a major difference. FO-PCL requires all groundings of a conditional to have the same probability d given in the conditional, and in general, FO-PCL needs to restrict the possible instantiations for the variables occurring in a conditional by providing constraint formulas like $U \neq V$ or $U \neq a$ in order to avoid inconsistencies. On the other hand, under aggregating semantics the grounded instances may have distinct probabilities as long as they aggregate to the given probability d, and aggregating semantics as defined in [25] considers only conditionals without constraint formulas.

In [9], the logical framework PCI (*p*robabilistic *c*onditionals with *i*nstantiation restrictions) extending aggregating semantics to conditionals with instantiation restrictions and also providing a grounding semantics as in FO-PCL is proposed; in [4], PCI is used to compare the instances of relational conditionals under the ME models emerging in aggregating and FO-PCL grounding semantics. From a knowledge representation point of view, PCI provides greater flexibility e.g. when expressing knowledge about individuals known to be exceptional with respect to some relationship. The purpose of this paper is to present a brief introduction both to FO-PCL and to aggregating semantics and to study the formal relationship between the models of these two semantic approaches within the PCI framework. In contrast to [38] where aggregating semantics could not be applied directly to an FO-PCL knowledge base having instantiation restrictions, the use of the PCI framework allows us to precisely state equalities and differences among the models of a knowledge base $\mathcal{R}$ under FO-PCL grounding semantics and under aggregating semantics extended to conditionals with instantiation restrictions.

The rest of this paper is organized as follows. After briefly presenting the required background on FO-PCL and aggregating semantics in Section 2, we recall the formal definition of PCI introduced in [9] and prove some first results about the relationship between the model classes under FO-PCL grounding semantics and under aggregating semantics in Section 3. Section 4 focusses on applying the ME principle to both semantics and demonstrates that the resulting ME models are different in general. In Section 5, the subclass of parametrically uniform knowledge bases [14, 15, 3] is considered. It is proven that also for parametrically uniform knowledge bases, which have been shown to allow for a simplified ME model computation in FO-PCL, the model classes are still different under the two semantics. In Section 6, based on a result stated in [8], we prove that the models under FO-PCL grounding semantics and under aggregating semantics coincide if both concepts considered in the two previous sections are employed, i.e., when applying the ME principle to a knowledge base that is parametrically uniform. Finally, in Section 7 we conclude and point out further work.

2 Background: FO-PCL and Aggregating Semantics

Simply grounding a relational knowledge base $\mathcal{R}$ easily leads to an inconsistency. For instance, a straightforward complete grounding of $\mathcal{R}_{ek}$ yields a grounded knowledge base that can be viewed as a propositional knowledge base. However, this grounded knowledge is inconsistent since it contains both

$\qquad$ (likes(clyde, fred) | elephant(clyde), keeper(fred)) [0.9]
and $\quad$ (likes(clyde, fred) | elephant(clyde), keeper(fred)) [0.05],

which arise from ek_1 and ek_2, respectively, but no probability distribution P can satisfy both

$\qquad$ $P($ likes(clyde, fred) | elephant(clyde), keeper(fred) $) = 0.9$
and $\quad$ $P($ likes(clyde, fred) | elephant(clyde), keeper(fred) $) = 0.05$

simultaneously. Therefore, the logic FO-PCL [14, 15] employs instantiation restrictions for the free variables of a conditional. An FO-PCL conditional has additionally a constraint formula determining the *admissible instantiations* of free variables, and the FO-PCL grounding semantic requires that all admissible ground instances of a conditional r have the probability given by r.

Example 1 (Elephant Keeper with instantiation restrictions). In FO-PCL, adding $K \neq fred$ to conditional ek_1 and $E \neq clyde$ to conditional ek_2 in $\mathcal{R}_{ek}$ yields the knowledge base $\mathcal{R}'_{ek}$ with:

ek'_1 : $\langle (likes(E, K) | elephant(E), keeper(K))[0.9], K \neq fred \rangle$
ek'_2 : $\langle (likes(E, fred) | elephant(E), keeper(fred))[0.05], E \neq clyde \rangle$
ek'_3 : $\langle (likes(clyde, fred) | elephant(clyde), keeper(fred))[0.85], \top \rangle$

Note that e.g. (likes(clyde, fred) | elephant(clyde), keeper(fred)) [0.05] is the ground instance of conditional ek'_2 that is not admissible, and that the set of admissible ground instances of $\mathcal{R}'_{ek}$ is indeed consistent under probabilistic semantics for propositional knowledge bases as considered e.g. in [35, 21]. ◇

Thus, under FO-PCL grounding semantics, $\mathcal{R}'_{ek}$ is consistent, where a probability distribution P satisfies an FO-PCL conditional r iff all admissible ground instances of r have the probability specified by r.

In contrast, the aggregating semantics, as given in [25], does not consider instantiation restrictions, since its satisfaction relation (in this paper denoted by $\models_\odot^{orig}$ to indicate the original definition of aggregating semantics without instantiation restrictions), is less strict with respect to probabilities of ground instances: $P \models_\odot^{orig} (B \mid A) [d]$ iff the quotient of the sum of all probabilities $P(B_l \wedge A_l)$ and the sum of $P(A_l)$ is d, where $(B_1 \mid A_1), \ldots, (B_G \mid A_G)$ are the ground instances of $(B \mid A)$. In this way, the aggregating semantics is capable of balancing the probabilities of ground instances, resulting in greater flexibility and higher tolerance with respect to consistency issues. Provided that there are enough individuals so that the corresponding aggregating over all probabilities is possible, the knowledge base $\mathcal{R}_{ek}$ that is inconsistent under FO-PCL grounding semantics is consistent under aggregating semantics.

3 PCI Logic

The logical framework PCI uses probabilistic conditionals with and without instantiation restrictions and provides different options for a satisfaction relation. The syntax of PCI given in [9] uses the syntax of FO-PCL [14, 15]. In the following, we will precisely state the formal relationship among aggregating semantics, FO-PCL grounding semantics, and the satisfaction relations offered by PCI.

As FO-PCL, PCI uses function-free, sorted signatures of the form $\Sigma = (Sort, Const, Pred)$. In a PCI-signature Σ, $Sort = \{s_1, \ldots, s_k\}$ is a set of sort names or just sorts. The set $Const$ is a finite set of constants symbols where each $d \in Const$ has a unique sort $s \in Sort$. With $Const^{(s)}$ we denote the set of all constants having sort s; thus $Const = \bigcup_{s \in Sort} Const^{(s)}$ is a set being the union of (disjoint) sets of sorted constant symbols. $Pred$ is a set of predicate symbols, each having a particular number of arguments. If $p \in Pred$ is a predicate taking n arguments, each argument position i must be filled with a constant or variable of a specific sort s_i. Thus, each $p \in Pred$ comes with an arity of the form $s_1 \times \ldots \times s_n \in Sort^n$ indicating the required sorts for the arguments. Variables Var also have a unique sort, and all formulas and variable substitutions must obey the obvious sort restrictions. In the following, we will adopt the unique names assumption, i.e., different constants denote different elements. The set of all terms is defined as $Term_\Sigma := Var \cup Const$. Let $\mathcal{L}_\Sigma$ be the set of quantifier-free first-order formulas defined over Σ and Var in the usual way.

Definition 1 (Instantiation Restriction). *An* instantiation restriction *is a conjunction of inequality atoms of the from* $t_1 \neq t_2$ *with* $t_1, t_2 \in Term_\Sigma$. *The set of all instantiation restriction is denoted by* $\mathcal{C}_\Sigma$.

Since an instantiation restriction may be a conjunction of inequality atoms, we can express that a conditional has multiple restrictions, e.g., by stating $E \neq clyde \wedge K \neq fred$.

Definition 2 (q-, p-, r-Conditional). *Let $A, B \in \mathcal{L}_\Sigma$ be quantifier-free first-order formulas over Σ and Var.*

1. *(B | A) is called a* qualitative conditional *(or just* q-conditional*). A is the antecedent and B the consequence of the qualitative conditional. The set of all qualitative conditionals over $\mathcal{L}_\Sigma$ is denoted by $(\mathcal{L}_\Sigma | \mathcal{L}_\Sigma)$.*

2. *Let $(B \mid A) \in (\mathcal{L}_\Sigma | \mathcal{L}_\Sigma)$ be a qualitative conditional and let $d \in [0,1]$ be a real value. $(B \mid A) [d]$ is called a* probabilistic conditional *(or just* p-conditional*) with probability d. The set of all probabilistic conditionals over $\mathcal{L}_\Sigma$ is denoted by $(\mathcal{L}_\Sigma | \mathcal{L}_\Sigma)^{prob}$.*

3. *Let $(B \mid A) [d] \in (\mathcal{L}_\Sigma | \mathcal{L}_\Sigma)^{prob}$ be a probabilistic conditional and let $C \in \mathcal{C}_\Sigma$ be an instantiation restriction. $\langle (B \mid A) [d], C \rangle$ is called an* instantiation restricted conditional *(or just* r-conditional*). The set of all instantiation restricted conditionals over $\mathcal{L}_\Sigma$ is denoted by $(\mathcal{L}_\Sigma | \mathcal{L}_\Sigma)^{prob}_{\mathcal{C}_\Sigma}$.*

Instantiation restricted qualitative conditionals are defined analogously. If it is clear from the context, we may omit *qualitative*, *probabilistic*, and *instantiation restricted* and just use the term *conditional*.

Definition 3 (PCI Knowledge Base). *A pair $(\Sigma, \mathcal{R})$ consisting of a PCI signature $\Sigma = (Sort, Const, Pred)$ and a set of instantiation restricted conditionals $\mathcal{R} = \{r_1, \ldots, r_m\}$ with $r_i \in (\mathcal{L}_\Sigma | \mathcal{L}_\Sigma)^{prob}_{\mathcal{C}_\Sigma}$ is called a* PCI knowledge base*.*

For an instantiation restricted conditional $r = \langle (B \mid A) [d], C \rangle$, $\Theta_\Sigma(r)$ denotes the set of all ground substitutions with respect to the variables in r. A ground substitution $\theta \in \Theta_\Sigma(r)$ is applied to the formulas A, B and C in the usual way, i.e., each variable is replaced by a certain constant according to the mapping $\theta = \{v_1/c_1, \ldots, v_s/c_s\}$ with $v_k \in Var$, $c_k \in Const$, $1 \leqslant k \leqslant s$. So $\theta(A)$, $\theta(B)$, and $\theta(C)$ are ground formulas and we have $\theta((B \mid A)) := (\theta(B) \mid \theta(A))$.

Given a ground substitution θ over the variables occurring in an instantiation restriction $C \in \mathcal{C}_\Sigma$, the evaluation of C under θ, denoted by $[\![C]\!]_\theta$, yields true iff $\theta(t_1)$ and $\theta(t_2)$ are different constants for all $t_1 \neq t_2 \in C$.

Definition 4 (Admissible Ground Substitutions and Instances). *Let $\Sigma = (Sort, Const, Pred)$ be a sorted signature and let $r = \langle (B \mid A) [d], C \rangle \in (\mathcal{L}_\Sigma | \mathcal{L}_\Sigma)^{prob}_{\mathcal{C}_\Sigma}$ be an instantiation restricted conditional. The set of* admissible ground substitutions *of r is defined as:*

$$\Theta^{adm}_\Sigma(r) := \{\theta \in \Theta_\Sigma(r) \mid [\![C]\!]_\theta = true\}$$

The set of admissible ground instances *of r is defined as:*

$$gnd^{adm}_\Sigma(r) := \{\theta((B \mid A))[d] \mid \theta \in \Theta^{adm}_\Sigma(r)\}$$

In the following, when we talk about the ground instances of a conditional, we will always refer to its admissible ground instances.

Without loss of generality, let $\text{var}(r) = \{X_1, \ldots, X_n\}$ be the set of variables appearing in $r = (B \mid A)$ and let $\theta \in \Theta_\Sigma^{\text{adm}}(r)$ with

$$\theta = \{X_1/a_1, \ldots, X_n/a_n\}$$

be a ground substitution of r. Then we also write r as

$$r = (B(\boldsymbol{X}) \mid A(\boldsymbol{X}))$$

where $\boldsymbol{X} = (X_1, \ldots, X_n)$ is a tuple which contains the variables of r in arbitrary but fixed order. That way, we can write a ground instance $\theta(r)$ more compactly by defining

$$r(\boldsymbol{a}) := \theta(r)$$
$$(B(\boldsymbol{a}) \mid A(\boldsymbol{a})) := (\theta(B(\boldsymbol{X})) \mid \theta(A(\boldsymbol{X})))$$

where $\boldsymbol{a} = (a_1, \ldots, a_n)$ is the respective tuple which contains the constants of θ in the appropriate order with respect to $\boldsymbol{X}$.

With G denoting the number of ground instances of r, we can enumerate these ground instances as follows:

$$\text{gnd}_\Sigma^{\text{adm}}(r) = \{r(\boldsymbol{a}^{(1)}), \ldots, r(\boldsymbol{a}^{(G)})\} \tag{1}$$

As for an FO-PCL knowledge base [14], for a PCI knowledge base $(\Sigma, \mathcal{R})$ we define the *Herbrand base* $\mathcal{H}(\mathcal{R})$ as the set of all ground atoms in all $\text{gnd}_\Sigma^{\text{adm}}(r_i)$ with $r_i \in \mathcal{R}$. Every subset $\omega \subseteq \mathcal{H}(\mathcal{R})$ is a *Herbrand interpretation*, defining a logical semantics for $\mathcal{R}$. The set $\Omega_\Sigma := \{\omega \mid \omega \subseteq \mathcal{H}(\mathcal{R})\}$ denotes the set of all Herbrand interpretations. Herbrand interpretations are also called *possible worlds*.

Definition 5 (PCI Interpretation). *The probabilistic semantics of $(\Sigma, \mathcal{R})$ is a possible world semantics [18] where the ground atoms in $\mathcal{H}(\mathcal{R})$ are binary random variables. A PCI interpretation P of a knowledge base $(\Sigma, \mathcal{R})$ is thus a probability distribution $P : \Omega_\Sigma \to [0, 1]$. The set of all probability distributions over Ω_Σ is denoted by Prob_Σ.*

The PCI framework offers two different satisfaction relations: $\models_\triangle^{\text{pci}}$ is based on grounding as in FO-PCL, and $\models_\odot^{\text{pci}}$ extends aggregating semantics to r-conditionals.

Definition 6 (PCI Satisfaction Relations). *Let $P \in \text{Prob}_\Sigma$ and let*

$$r = \langle (B(\boldsymbol{X}) \mid A(\boldsymbol{X}))[d], C \rangle \in (\mathcal{L}_\Sigma \mid \mathcal{L}_\Sigma)_{\mathcal{C}_\Sigma}^{\text{prob}}$$

with

$$\text{gnd}_\Sigma^{\text{adm}}(r) = \{(B(\boldsymbol{a}^{(1)}) \mid A(\boldsymbol{a}^{(1)}))[d], \ldots, (B(\boldsymbol{a}^{(G)}) \mid A(\boldsymbol{a}^{(G)}))[d]\}$$

be an r-conditional. The two PCI satisfaction relations $\models_\triangle^{\text{pci}}$ and $\models_\odot^{\text{pci}}$ are defined by

$$P \models_\triangle^{\text{pci}} r \quad \text{iff} \quad P\left(A(\boldsymbol{a}^{(l)})\right) > 0 \text{ for } 1 \leqslant l \leqslant G \quad \text{and}$$

$$\frac{P\left(A(\boldsymbol{a}^{(1)})B(\boldsymbol{a}^{(1)})\right)}{P\left(A(\boldsymbol{a}^{(1)})\right)} = d \quad \text{and}$$

$$\vdots \quad \vdots \quad (2)$$

$$\text{and}$$

$$\frac{P\left(A(\boldsymbol{a}^{(G)})B(\boldsymbol{a}^{(G)})\right)}{P\left(A(\boldsymbol{a}^{(G)})\right)} = d$$

and by

$$P \models_\odot^{\text{pci}} r \quad \text{iff} \quad \sum_{r(\boldsymbol{a}^{(l)}) \in \text{gnd}_\Sigma^{\text{adm}}(r)} P\left(A(\boldsymbol{a}^{(l)})\right) > 0 \quad \text{and}$$

$$\frac{\sum_{r(\boldsymbol{a}^{(l)}) \in \text{gnd}_\Sigma^{\text{adm}}(r)} P\left(A(\boldsymbol{a}^{(l)})B(\boldsymbol{a}^{(l)})\right)}{\sum_{r(\boldsymbol{a}^{(l)}) \in \text{gnd}_\Sigma^{\text{adm}}(r)} P\left(A(\boldsymbol{a}^{(l)})\right)} = d \quad (3)$$

As usual, the satisfaction relations $\models_\star^{\text{pci}}$ with $\star \in \{\triangle, \odot\}$ are extended to a set of conditionals $\mathcal{R}$ by defining

$$P \models_\star^{\text{pci}} \mathcal{R} \quad \text{iff} \quad P \models_\star^{\text{pci}} r_i \text{ for all } r_i \in \mathcal{R}.$$

The following proposition shows that PCI not only captures the grounding semantics of FO-PCL [14], but also the aggregating semantics $\models_\odot^{\text{orig}}$ of [25].

Proposition 1 (PCI Captures Aggregating Semantics [9]). Let $(B(\boldsymbol{X}) \mid A(\boldsymbol{X}))[d]$ be a p-conditional. Then we have:

$$P \models_\odot^{\text{pci}} \langle (B(\boldsymbol{X}) \mid A(\boldsymbol{X}))[d], \top \rangle \quad \text{iff} \quad P \models_\odot^{\text{orig}} (B(\boldsymbol{X}) \mid A(\boldsymbol{X}))[d] \quad (4)$$

Thus, while $\models_\triangle^{\text{pci}}$ coincides with the grounding semantics of FO-PCL by definition, $\models_\odot^{\text{pci}}$ properly generalizes the original aggregating semantics [25] to conditional knowledge bases with instantiation restrictions. Hence, PCI logic provides a suitable framework for investigating the formal relationship between these two semantics.

Notation: In the rest of this paper, we will focus on the two semantics induced by $\models_\triangle^{\text{pci}}$ and $\models_\odot^{\text{pci}}$. In order to simplify our presentation, we will call the semantics induced by $\models_\triangle^{\text{pci}}$ just *grounding semantics* and the semantics induced by $\models_\odot^{\text{pci}}$ just *aggregating semantics*.

Definition 7 ($Mod_\triangle(\mathcal{R}), Mod_\odot(\mathcal{R})$). *For a knowledge base $\mathcal{R}$ and $\star \in \{\triangle, \odot\}$, the set of grounding and aggregating models, respectively, is given by:*

$$Mod_\star(\mathcal{R}) = \{P \in Prob_\Sigma \mid P \models_\star^{pci} \mathcal{R}\}$$

As our first result regarding the formal relationship between grounding and aggregating semantics we observe that every grounding model is also an aggregating model.

Proposition 2. *For every knowledge base $\mathcal{R}$, the following holds:*

$$Mod_\triangle(\mathcal{R}) \subseteq Mod_\odot(\mathcal{R}) \tag{5}$$

Proof. We will show that for every conditional $r = \langle (B(\boldsymbol{X}) \mid A(\boldsymbol{X}))[d], C \rangle \in \mathcal{R}$ and every $P \in Mod_\triangle(\mathcal{R})$, the relationship $P \models_\odot^{pci} r$ holds. If $P \models_\triangle^{pci} r$, we can multiply each of the equations in (2) by its denominator and obtain, using the notation as in Definition 6, the system of equations:

$$P\left(A(\boldsymbol{a}^{(1)})B(\boldsymbol{a}^{(1)})\right) = d\,P\left(A(\boldsymbol{a}^{(1)})\right)$$
$$\vdots \tag{6}$$
$$P\left(A(\boldsymbol{a}^{(G)})B(\boldsymbol{a}^{(G)})\right) = d\,P\left(A(\boldsymbol{a}^{(G)})\right)$$

Adding all equations in (6) yields

$$\sum_{1 \leq l \leq G} P\left(A(\boldsymbol{a}^{(l)})B(\boldsymbol{a}^{(l)})\right) = d \sum_{1 \leq l \leq G} P\left(A(\boldsymbol{a}^{(l)})\right) \tag{7}$$

and deviding (7) by the sum on the right hand side yields (3). Thus, $P \models_\odot^{pci} r$, implying $Mod_\triangle(\mathcal{R}) \subseteq Mod_\odot(\mathcal{R})$. □

We will now show that the relation in (5) is a proper subset relationship.

Proposition 3. *In general, for a knowledge base $\mathcal{R}$, the following holds:*

$$Mod_\triangle(\mathcal{R}) \subsetneq Mod_\odot(\mathcal{R}) \tag{8}$$

Proof. Consider the PCI knowledge base $(\Sigma, \mathcal{R})$ with $\Sigma = (Sort, Const, Pred)$, $Sort = \{s\}$, $Const^{(s)} = \{a, b\}$, $Pred = \{q\}$ where q is a unary predicate taking an argument of sort s, and $\mathcal{R} = \{r\}$ with $r = \langle (q(X) \mid \top)[0.8], \top \rangle$. So we have:

$$gnd_\Sigma^{adm}(r) = \{(q(a) \mid \top)[0.8], (q(b) \mid \top)[0.8]\}$$

Let P be the probability distribution over the possible worlds over Σ given by:

$$P(\{\}) = 0.05$$
$$P(\{q(a)\}) = 0.20$$
$$P(\{q(b)\}) = 0.10$$
$$P(\{q(a), q(b)\}) = 0.65$$

P is an aggregating model of $(\Sigma, \mathcal{R})$, because we have (cf. (3)):

$$\frac{P(q(a) \wedge \top) + P(q(b) \wedge \top)}{P(\top) + P(\top)} = \frac{0.20 + 0.65 + 0.1 + 0.65}{1+1} = \frac{1.6}{2} = 0.8$$

On the other hand, P is not a grounding model of $(\Sigma, \mathcal{R})$, because we have (cf. (2)):

$$\frac{P(q(a) \wedge \top)}{P(\top)} = \frac{0.20 + 0.65}{1} = 0.85 \neq 0.8$$

$$\frac{P(q(b) \wedge \top)}{P(\top)} = \frac{0.10 + 0.65}{1} = 0.75 \neq 0.8$$

Thus, we have $P \models_{\odot}^{pci} r$, but $P \not\models_{\triangle}^{pci} r$, implying $Mod_{\triangle}(\mathcal{R}) \neq Mod_{\odot}(\mathcal{R})$ and hence, together with Proposition 2, $Mod_{\triangle}(\mathcal{R}) \subsetneq Mod_{\odot}(\mathcal{R})$. □

4 PCI Logic and Maximum Entropy Semantics

If a knowledge base $\mathcal{R}$ is consistent, there are usually many different models satisfying $\mathcal{R}$. The principle of maximum entropy chooses the unique distribution which has maximum entropy among all distributions satisfying a knowledge base $\mathcal{R}$ [29, 21]. Applying this principle to the PCI satisfaction relations $\models_{\triangle}^{pci}$ and $\models_{\odot}^{pci}$ yields

$$\text{ME}_{\triangle}(\mathcal{R}) := \arg\max_{P \in Mod_{\triangle}(\mathcal{R})} H(P) \tag{9}$$

$$\text{ME}_{\odot}(\mathcal{R}) := \arg\max_{P \in Mod_{\odot}(\mathcal{R})} H(P) \tag{10}$$

where

$$H(P) = -\sum_{\omega \in \Omega} P(\omega) \log P(\omega)$$

is the *entropy* of a probability distribution P. Thus, for instance, $\text{ME}_{\triangle}(\mathcal{R})(\omega)$ is the probability of a world ω under the ME distribution of $\mathcal{R}$ with respect to grounding semantics, and $\text{ME}_{\odot}(\mathcal{R})(r)$ denotes the probability of a conditional r under the ME distribution of $\mathcal{R}$ with respect to aggregating semantics.

Example 2 (Misanthrope). The knowledge base $\mathcal{R}_{mi}$ adapted from [14] models friendship relations within a group of people with one exceptional member, a misanthrope. In general, if a person V likes another person U, then it is very likely that U likes V, too. But there is one person, the misanthrope a, who generally does not like other people:

$$\mathcal{R}_{mi} := \left\{ \begin{array}{l} r_1 \colon \langle (\,likes(U,V) \mid likes(V,U)\,)\,[0.9], U \neq V \rangle \\ r_2 \colon \langle (\,likes(a,V) \mid \top\,)\,[0.05], V \neq a \rangle \end{array} \right\}$$

Within the PCI framework, consider $\mathcal{R}_{mi}$ together with constants $\mathit{Const} = \{a, b, c\}$. So the ground instances are:

$$\mathrm{gnd}_{\Sigma}^{\mathrm{adm}}(\mathcal{R}_{mi}) := \begin{cases} r_1^{(1)}: \langle (\mathit{likes}(a,b) \mid \mathit{likes}(b,a))\ [0.9], \top \rangle \\ r_1^{(2)}: \langle (\mathit{likes}(a,c) \mid \mathit{likes}(c,a))\ [0.9], \top \rangle \\ r_1^{(3)}: \langle (\mathit{likes}(b,a) \mid \mathit{likes}(a,b))\ [0.9], \top \rangle \\ r_1^{(4)}: \langle (\mathit{likes}(b,c) \mid \mathit{likes}(c,b))\ [0.9], \top \rangle \\ r_1^{(5)}: \langle (\mathit{likes}(c,a) \mid \mathit{likes}(a,c))\ [0.9], \top \rangle \\ r_1^{(6)}: \langle (\mathit{likes}(c,b) \mid \mathit{likes}(b,c))\ [0.9], \top \rangle \\ r_2^{(1)}: \langle (\mathit{likes}(a,b) \mid \top)\ [0.05], \top \rangle \\ r_2^{(2)}: \langle (\mathit{likes}(a,c) \mid \top)\ [0.05], \top \rangle \end{cases}$$

Let $\mathrm{ME}_{\triangle}(\mathcal{R}_{mi})$ and $\mathrm{ME}_{\odot}(\mathcal{R}_{mi})$ denote the corresponding ME distributions under PCI-grounding and PCI-aggregating semantics. The following table shows the qualitative parts of the ground instances in $\mathrm{gnd}_{\Sigma}^{\mathrm{adm}}(\mathcal{R}_{mi})$, i.e., the corresponding qualitative conditionals, as well as the probabilities these conditionals have under the ME distributions $\mathrm{ME}_{\triangle}(\mathcal{R}_{mi})$ and $\mathrm{ME}_{\odot}(\mathcal{R}_{mi})$:

qualitative part of $r_i^{(l)}$	$\mathrm{ME}_{\triangle}(\mathcal{R}_{mi})(r_i^{(l)})$	$\mathrm{ME}_{\odot}(\mathcal{R}_{mi})(r_i^{(l)})$
$r_1^{(1)}: (\mathit{likes}(a,b) \mid \mathit{likes}(b,a))$	0.900	0.460
$r_1^{(2)}: (\mathit{likes}(a,c) \mid \mathit{likes}(c,a))$	0.900	0.460
$r_1^{(3)}: (\mathit{likes}(b,a) \mid \mathit{likes}(a,b))$	0.900	0.967
$r_1^{(4)}: (\mathit{likes}(b,c) \mid \mathit{likes}(c,b))$	0.900	0.967
$r_1^{(5)}: (\mathit{likes}(c,a) \mid \mathit{likes}(a,c))$	0.900	0.967
$r_1^{(6)}: (\mathit{likes}(c,b) \mid \mathit{likes}(b,c))$	0.900	0.967
$r_2^{(1)}: (\mathit{likes}(a,b) \mid \top)$	0.050	0.050
$r_2^{(2)}: (\mathit{likes}(a,c) \mid \top)$	0.050	0.050

Under the distribution $\mathrm{ME}_{\triangle}(\mathcal{R}_{mi})$, all six ground conditionals emerging from r_1 have probability 0.9. However, under $\mathrm{ME}_{\odot}(\mathcal{R}_{mi})$, two of the ground conditionals resulting from r_1 have different probabilities than the four others. ◇

The knowledge base $\mathcal{R}_{mi}$ in Example 2 proves that the maximum entropy models under grounding and under aggregating semantics may differ:

Proposition 4. *In general, for a knowledge base $\mathcal{R}$, the following holds:*

$$\mathrm{ME}_{\triangle}(\mathcal{R}) \neq \mathrm{ME}_{\odot}(\mathcal{R}) \tag{11}$$

5 PCI Logic and Parametric Uniformity

As elaborated in [14], the computation of the uniquely determined maximum entropy model $\mathrm{ME}_{\triangle}(\mathcal{R})$ under grounding semantics can be done by solving an optimization problem whose solution $\mathrm{ME}_\triangle(\mathcal{R})$ can be represented by a Gibbs distribution [16]

$$\mathrm{ME}_\triangle(\mathcal{R})(\omega) = \frac{1}{Z} \exp\left(\sum_{r_i \in \mathcal{R}} \sum_{r_i(\boldsymbol{a}_i^{(l)}) \in \mathrm{gnd}_\Sigma^{\mathrm{adm}}(r_i)} \lambda_i^{(l)} f_{\triangle,i}^{(l)}(\omega) \right) \quad (12)$$

where $f_{\triangle,i}^{(l)}$ is the feature function determined by $r_i(\boldsymbol{a}_i^{(l)})$, $\lambda_i^{(l)}$ is a Lagrange multiplier [5] and Z is a normalization constant. We will not elaborate on the details of Equation (12) as they are not important for the rest of this work (see [14] for a detailed explanation). What is important to note is that according to (12), one optimization parameter $\lambda_i^{(l)}$ has to be determined for *each single ground instance* $r_i(\boldsymbol{a}_i^{(l)})$ of each conditional $r_i \in \mathcal{R}$. This readily yields a computationally infeasible optimization problem for larger knowledge bases because there might be just too many ground instances.

However, there may be instances of a conditional sharing the same entropy-optimal parameter. The concept of *parametric uniformity* proposed in [14] means that for each conditional all its ground instances share the same entropy-optimal parameter value under grounding semantics. The advantage of a parametrically uniform knowledge base $\mathcal{R}^{\mathcal{PU}}$ is that just *one* optimization parameter λ_i per conditional r_i has to be computed instead of one parameter per ground instance [14]:

$$\mathrm{ME}_\triangle(\mathcal{R}^{\mathcal{PU}})(\omega) = \frac{1}{Z} \exp\left(\sum_{r_i \in \mathcal{R}^{\mathcal{PU}}} \lambda_i \sum_{r_i(\boldsymbol{a}_i^{(l)}) \in \mathrm{gnd}_\Sigma^{\mathrm{adm}}(r_i)} f_{\triangle,i}^{(l)}(\omega) \right) \quad (13)$$

That is, just $|\mathcal{R}^{\mathcal{PU}}|$ optimization parameters have to be computed for a parametrically uniform knowledge base in (13), compared to $|\mathrm{gnd}_\Sigma^{\mathrm{adm}}(\mathcal{R})|$ optimization parameters for a general knowledge base in (12). Note that in particular, the number of ground instances and hence the number of optimization parameters $\lambda_i^{(l)}$ grows with the number of constants in Σ in (12), but the number of optimization parameters λ_i is independent of the number of constants in (13).

Luckily, each knowledge base $\mathcal{R}$ that is not parametrically uniform can be transformed into a knowledge base that is equivalent to $\mathcal{R}$ regarding its ME model with respect to $\models_\triangle^{\mathrm{pci}}$ and that is parametrically uniform. This is achieved by the set of transformations rules $\mathcal{PU}$ developed in [3] where the reasons causing $\mathcal{R}$ not to be parametrically uniform are investigated in detail. A central observation is that for any parametrically uniform $\mathcal{R}$, the sets of ground instances of atoms and pairs of different atoms occurring in the admissible ground instances of two conditionals must be either disjoint or identical (otherwise there is an *imbalanced sharing* [3]). For instance, $\mathcal{R}_{mi}$ from Ex. 2 is not parametrically uniform since the sets of admissible ground atoms originating from $likes(U, V)$ of conditional r_1 and from $likes(a, V)$ of conditional r_2 are neither equal nor disjoint. For a single conditional it is required that each admissible ground instance of an atom p occurs the same number of times in the ground instances of the conditional as any other admissible ground instance of the atom p; an analogous requirement holds for each pair of different atoms p', p'' (otherwise there is an *imbalanced use* [3]). The syntactic criterion of *inter-rule* and *intra-rule interactions* [3] identifies these constellations by analysing the syntactic structure of the conditionals in $\mathcal{R}$; no grounding operation is required for this criterion.

Example 3 (Inter-Rule Interaction in $\mathcal{R}_{mi}$). As indicated above, the knowledge base $\mathcal{R}_{mi}$ of Ex. 2 is not parametrically uniform. This is detected by an inter-rule interaction between the rules r_1 and r_2 denoted by $r_2 \leftarrow \langle likes \rangle_{U,a} \rightarrow r_1$. For removing this type of inter-rule interaction, $\mathcal{PU}$ contains the transformation rule

$$(TE_1) \quad \frac{\mathcal{R} \cup \{r_1, r_2\}}{\mathcal{R} \cup \{r_1\} \cup \nu\{\sigma(r_2), \overline{\sigma}(r_2)\}} \quad \begin{array}{l} r_1 \leftarrow \langle P \rangle_{V,c} \rightarrow r_2, \\ \sigma = \{V/c\} \end{array}$$

where $\sigma(r)$ is the result of applying the variable substitution $\sigma = \{V/c\}$ to r, and $\bar{\sigma}(r)$ is the result of adding the constraint $V \neq c$ to the constraint formula of r. The operator ν transforms a conditional in constraint normal form, a normal form required for the recognition of interactions.

The application of (TE_1) to $\mathcal{R}_{mi}$ removes the interaction $r_2 \leftarrow \langle likes \rangle_{U,a} \rightarrow r_1$ and replaces the conditional r_1 with the two new conditionals

$$r_{1'}: \langle (\,likes(a, V) \mid likes(V, a)\,)\,[0.9], V \neq a \rangle$$
$$r_{1''}: \langle (\,likes(U, V) \mid \top\,)\,[0.9], U \neq V \wedge U \neq a \rangle$$

Note that the atom $likes(U, V)$ of r_1 that caused the interaction becomes $likes(a, V)$ in $r_{1'}$ and its set of ground atoms is now identical to the set of ground atoms from $likes(a, V)$ of r_2. The added constraint $U \neq a$ in $r_{1''}$ leads to disjoint sets of ground atoms of the discussed atoms.

The importance of inter- and intra-rule interactions is that they fully capture the reasons for a knowledge base not to be parametrically uniform. For each of the six different types of interactions there is a corresponding transformation rule analogously to (TE_1) [3]. The following proposition states that $\mathcal{PU}$ is correct and complete.

Proposition 5 ($\mathcal{PU}(\mathcal{R})$ [3]). *Exhaustively applying $\mathcal{PU}$ to a knowledge base $\mathcal{R}$ yields a knowledge base $\mathcal{PU}(\mathcal{R})$ such that $\mathcal{R}$ and $\mathcal{PU}(\mathcal{R})$ have the same maximum entropy model under grounding semantics and $\mathcal{PU}(\mathcal{R})$ is parametrically uniform, i.e.:*

$$\mathrm{ME}_\triangle(\mathcal{PU}(\mathcal{R})) = \mathrm{ME}_\triangle(\mathcal{R}) \tag{14}$$

Example 4 ($\mathcal{PU}(\mathcal{R}_{mi})$). By applying $\mathcal{PU}$ exhaustively to $\mathcal{R}_{mi}$ from Example 2, the conditional r_1 is replaced by three new conditionals and the resulting set $\mathcal{R}_{mi}^{\mathcal{PU}} := \mathcal{PU}(\mathcal{R}_{mi})$ is parametrically uniform and has the same maximum entropy model $\mathrm{ME}_\triangle(\mathcal{R}_{mi}^{\mathcal{PU}}) = \mathrm{ME}_\triangle(\mathcal{R}_{mi})$ under grounding semantics as $\mathcal{R}_{mi}$:

$$\mathcal{R}_{mi}^{\mathcal{PU}} := \left\{ \begin{array}{l} r_{1.1}: \langle (\,likes(U, V) \mid likes(V, U)\,)\,[0.9], U \neq V \wedge U \neq a \wedge V \neq a \rangle \\ r_{1.2}: \langle (\,likes(a, V) \mid likes(V, a)\,)\,[0.9], V \neq a \rangle \\ r_{1.3}: \langle (\,likes(U, a) \mid likes(a, U)\,)\,[0.9], U \neq a \rangle \\ r_2\ \ : \langle (\,likes(a, V) \mid \top\,)\,[0.05], V \neq a \rangle \end{array} \right\}$$

Considering $\mathcal{R}_{mi}^{\mathcal{PU}}$ again together with the constants $Const = \{a,b,c\}$, we get the same nine ground instances as in $\text{gnd}_\Sigma^{\text{adm}}(\mathcal{R}_{mi})$ (cf. Example 2):

$$\text{gnd}_\Sigma^{\text{adm}}(\mathcal{R}_{mi}^{\mathcal{PU}}) := \left\{ \begin{array}{l} r_{1.1}^{(1)}: \langle(\ likes(b,c)\ |\ likes(c,b)\ )\ [0.9], \top\rangle \\ r_{1.1}^{(2)}: \langle(\ likes(c,b)\ |\ likes(b,c)\ )\ [0.9], \top\rangle \\ r_{1.2}^{(1)}: \langle(\ likes(a,b)\ |\ likes(b,a)\ )\ [0.9], \top\rangle \\ r_{1.2}^{(2)}: \langle(\ likes(a,c)\ |\ likes(c,a)\ )\ [0.9], \top\rangle \\ r_{1.3}^{(1)}: \langle(\ likes(b,a)\ |\ likes(a,b)\ )\ [0.9], \top\rangle \\ r_{1.3}^{(2)}: \langle(\ likes(c,a)\ |\ likes(a,c)\ )\ [0.9], \top\rangle \\ r_2^{(1)}: \langle(\ likes(a,b)\ |\ \top\ )\ [0.05], \top\rangle \\ r_2^{(2)}: \langle(\ likes(a,c)\ |\ \top\ )\ [0.05], \top\rangle \end{array} \right\} \qquad \diamond$$

A generalization of Proposition 5 is that the $\mathcal{PU}$ transformation process not only leaves the ME grounding model unchanged, but also the full class of grounding models since the set of all ground instances of a knowledge base remains the same as illustrated in Example 4. On the other hand, this does not hold for aggregating semantics:

Proposition 6. *For every knowledge base $\mathcal{R}$, we have*

$$Mod_\triangle(\mathcal{PU}(\mathcal{R})) = Mod_\triangle(\mathcal{R}) \qquad (15)$$

and thus, in particular, $ME_\triangle(\mathcal{PU}(\mathcal{R})) = ME_\triangle(\mathcal{R})$ as in (14). However, for aggregating semantics, in general we have:

$$ME_\odot(\mathcal{PU}(\mathcal{R})) \neq ME_\odot(\mathcal{R}) \qquad (16)$$

Proof. Equation (15) is a consequence of [3, Prop. 9]. For the knowledge base $\mathcal{R}_{mi}$ from Example 2 it is shown in [4] that $ME_\odot(\mathcal{PU}(\mathcal{R}_{mi})) \neq ME_\odot(\mathcal{R}_{mi})$, implying (16). □

It has been shown that parametric uniformity simplifies the computation of $ME_\triangle(\mathcal{R})$ [8, 2]. When focussing only on the important subclass of parametrically uniform knowledge bases, $Mod_\triangle(\mathcal{R}^{\mathcal{PU}}) \subseteq Mod_\odot(\mathcal{R}^{\mathcal{PU}})$ still holds for every parametrically uniform knowledge base $\mathcal{R}^{\mathcal{PU}}$ due to Proposition 2, but the set of models under grounding and under aggregating semantics remain different:

Proposition 7. *In general, for a parametrically uniform knowledge base $\mathcal{R}^{\mathcal{PU}}$, the following holds:*

$$Mod_\triangle(\mathcal{R}^{\mathcal{PU}}) \subsetneq Mod_\odot(\mathcal{R}^{\mathcal{PU}}) \qquad (17)$$

Proof. Since the knowledge base $\mathcal{R}$ used in the proof of Proposition 3 is parametrically uniform, (17) can be shown by the same argumentation as in the proof of Proposition 3. □

Taking into account the observations about the relationship between grounding and aggregating models made in Propositions 2 – 7, it is not obvious under which circumstances the two semantics coincide. In the next section, we will show that if both maximum entropy and parametric uniformity are taken into account, the two semantics are identical.

6 Maximum Entropy Meets Parametric Uniformity

We will show that solving the optimizations problems (9) and (10) defining $\text{ME}_\triangle(\mathcal{R})$ and $\text{ME}_\odot(\mathcal{R})$, respectively, can be reduced to solving the same optimization problem if $\mathcal{R} = \mathcal{R}^{\mathcal{PU}}$ is parametrically uniform, implying $\text{ME}_\triangle(\mathcal{R}^{\mathcal{PU}}) = \text{ME}_\odot(\mathcal{R}^{\mathcal{PU}})$.

6.1 Aggregating Semantics: Satisfaction Relation via Feature Functions

For a conditional $r_i = \langle (B_i(\boldsymbol{X}) \mid A_i(\boldsymbol{X})) [d_i], C_i \rangle$, let G_i denote its number of ground instances as in (1). The *counting functions* of r_i, proposed by Kern-Isberner in [22] and used e.g. in [26, 7, 10, 34, 33], determine for each world ω the number of ground instances of r_i verified and falsified by ω, respectively. Here, we use the following variant of the counting functions focussing on the applicability of r_i:

$$\text{app}_i(\omega) := |\{r_i(\boldsymbol{a}_i^{(l)}) \in \text{gnd}_\Sigma^{\text{adm}}(r_i) \mid \omega \models A_i(\boldsymbol{a}_i^{(l)})\}| \tag{18}$$

$$\text{napp}_i(\omega) := |\{r_i(\boldsymbol{a}_i^{(l)}) \in \text{gnd}_\Sigma^{\text{adm}}(r_i) \mid \omega \models \overline{A_i(\boldsymbol{a}_i^{(l)})}\}| = G_i - \text{app}_i(\omega) \tag{19}$$

$$\text{ver}_i(\omega) := |\{r_i(\boldsymbol{a}_i^{(l)}) \in \text{gnd}_\Sigma^{\text{adm}}(r_i) \mid \omega \models A_i(\boldsymbol{a}_i^{(l)}) B_i(\boldsymbol{a}_i^{(l)})\}| \tag{20}$$

That is, $\text{app}_i(\omega)$ is the number of ground instances of r_i which are applicable with respect to a world ω, whereas $\text{napp}_i(\omega)$ is the corresponding number of ground instances which are not applicable. Similarly, $\text{ver}_i(\omega)$ is the number of ground instances of r_i which are verified by ω.

The satisfaction relation $\models_\odot^{orig}$ of the original aggregating semantics can also be expressed by employing a feature function based on counting functions; in [7, 10], this is exploited to develop algorithms computing the ME model with respect to $\models_\odot^{orig}$. Expressing $\models_\odot^{orig}$ by a feature function directly translates to the satisfaction relation $\models_\odot^{pci}$ for PCI-aggregating semantics, which also takes instantiation restrictions into account. The corresponding feature function is given by:

$$f_{\odot,i}(\omega) := \text{ver}_i(\omega) - \text{app}_i(\omega) \cdot d_i \tag{21}$$

Proposition 8 ($\models_\odot^{pci}$ **via Feature Function**). *Let $P \in \text{Prob}_\Sigma$ and let r_i be an r-conditional with $\sum_{r_i(\boldsymbol{a}_i^{(l)}) \in \text{gnd}_\Sigma^{\text{adm}}(r_i)} P\left(A_i(\boldsymbol{a}_i^{(l)})\right) > 0$. Then we have:*

$$P \models_\odot^{pci} r_i \quad \text{iff} \quad \sum_{\omega \in \Omega} P(\omega) \cdot f_{\odot,i}(\omega) = 0 \tag{22}$$

For a knowledge base $\mathcal{R} = \{r_1, \ldots, r_m\}$ consisting of m r-conditionals, we define the following equation system:

$$EQ_\mathcal{R}^\odot := \begin{bmatrix} \sum_{\omega \in \Omega} P(\omega) \cdot f_{\odot,1}(\omega) = 0 \\ \vdots \\ \sum_{\omega \in \Omega} P(\omega) \cdot f_{\odot,m}(\omega) = 0 \end{bmatrix} \tag{23}$$

Thus, according to (22), for every probability distribution $P \in \text{Prob}_\Sigma$, the solution set of $EQ_\mathcal{R}^\odot$ is equal to $\text{Mod}_\odot(\mathcal{R})$, i.e., we have:

$$P \text{ is a solution of } EQ_\mathcal{R}^\odot \quad \text{iff} \quad P \in \text{Mod}_\odot(\mathcal{R}) \qquad (24)$$

6.2 Grounding Semantics: Determining $\text{ME}_\triangle(\mathcal{R}^{\mathcal{PU}})$ More Efficiently

In general, determining $\text{ME}_\triangle(\mathcal{R})$ requires solving an equation system consisting of $|\text{gnd}_\Sigma^{\text{adm}}(\mathcal{R})| = \sum_{r_i \in \mathcal{R}} G_i$ linear constraints. In [8], we presented an approach significantly reducing the number of linear constraints which have to be considered for a parametrically uniform knowledge base $\mathcal{R}^{\mathcal{PU}} = \{r_1, \ldots, r_m\}$ consisting of m r-conditionals. By employing the counting functions (19) and (20), we define the feature function $F_{\triangle, i}$ for $r_i \in \mathcal{R}^{\mathcal{PU}}$ by:

$$F_{\triangle, i}(\omega) := \text{ver}_i(\omega) + \text{napp}_i(\omega) \cdot d_i \qquad (25)$$

Note that the feature function $F_{\triangle,i}$ is defined for each conditional $r_i \in \mathcal{R}^{\mathcal{PU}}$ directly, whereas the feature functions $f_{\triangle,i}^{(l)}$ mentioned in Section 5 are defined for each ground instance $r_i(\boldsymbol{a}_i^{(l)}) \in \text{gnd}_\Sigma^{\text{adm}}(r_i)$ of each conditional r_i.

Based on the feature functions $F_{\triangle,i}$, we define the following equation system consisting of m linear constraints:

$$EQ_\mathcal{R}^{\text{sum}F} := \begin{bmatrix} \sum_{\omega \in \Omega} P(\omega) \cdot F_{\triangle,1}(\omega) = G_1 \cdot d_1 \\ \vdots \\ \sum_{\omega \in \Omega} P(\omega) \cdot F_{\triangle,m}(\omega) = G_m \cdot d_m \end{bmatrix} \qquad (26)$$

Furthermore, with

$$\text{Sol}_\mathcal{R}^{\text{sum}F} := \{P \in \text{Prob}_\Sigma \mid P \text{ is a solution of } EQ_\mathcal{R}^{\text{sum}F}\}$$

denoting the set of probability distributions solving the equation system $EQ_\mathcal{R}^{\text{sum}F}$, we recall the following result stated in [8]:

Proposition 9 ([8]). *For a parametrically uniform knowledge base $\mathcal{R}^{\mathcal{PU}}$, the following holds:*

$$\text{ME}_\triangle(\mathcal{R}^{\mathcal{PU}}) = \underset{P \in \text{Sol}_\mathcal{R}^{\text{sum}F}}{\arg\max}\, H(P) \qquad (27)$$

That is, the ME model $\text{ME}_\triangle(\mathcal{R}^{\mathcal{PU}})$ under grounding semantics corresponds to the solution of ME optimization problem induced by the equation system $EQ_\mathcal{R}^{\text{sum}F}$. Note that $EQ_\mathcal{R}^{\text{sum}F}$ consists of merely m linear constraints, i.e., one linear constraint for each conditional in $\mathcal{R}^{\mathcal{PU}}$. So the number of linear constraints does not depend on the number of ground instances in $\text{gnd}_\Sigma^{\text{adm}}(\mathcal{R}^{\mathcal{PU}})$ and is thereby independent of the number of constants in Σ.

6.3 ME Models Under Grounding and Aggregating Semantics Coincide for Parametrically Uniform Knowledge Bases

Based on the results stated in Sections 6.1 and 6.2, we can now prove that the ME models under grounding and aggregating semantics coincide for a parametrically uniform knowledge base.

Proposition 10. *For a parametrically uniform knowledge base $\mathcal{R}^{\mathcal{PU}}$, the following holds:*

$$\mathrm{ME}_{\triangle}(\mathcal{R}^{\mathcal{PU}}) = \mathrm{ME}_{\odot}(\mathcal{R}^{\mathcal{PU}}) \tag{28}$$

Proof. For a conditional $r_i \in \mathcal{R}^{\mathcal{PU}}$, let

$$EQ_{r_i}^{\mathrm{sum}F} := \sum_{\omega \in \Omega} P(\omega) \cdot F_{\triangle,i}(\omega) = G_i \cdot d_i \tag{29}$$

and

$$EQ_{r_i}^{\odot} := \sum_{\omega \in \Omega} P(\omega) f_{\odot,i}(\omega) = 0 \tag{30}$$

denote the corresponding equations in the equation system in (26) and (23), respectively[3]. Now we will show that the equations (29) and (30) are equivalent:

$$
\begin{array}{ll}
& EQ_{r_i}^{\mathrm{sum}F} \\
\text{iff} & \sum_{\omega \in \Omega} P(\omega) \cdot F_{\triangle,i}(\omega) = G_i \cdot d_i \\
\text{iff} & \sum_{\omega \in \Omega} P(\omega) \cdot (\mathrm{ver}_i(\omega) + \mathrm{napp}_i(\omega) \cdot d_i) = G_i \cdot d_i \\
\text{iff} & \sum_{\omega \in \Omega} P(\omega) \cdot (\mathrm{ver}_i(\omega) + (G_i - \mathrm{app}_i(\omega)) \cdot d_i) = G_i \cdot d_i \\
\text{iff} & \sum_{\omega \in \Omega} P(\omega) \cdot (\mathrm{ver}_i(\omega) + G_i \cdot d_i - \mathrm{app}_i(\omega) \cdot d_i) = G_i \cdot d_i \\
\text{iff} & \sum_{\omega \in \Omega} P(\omega)(G_i \cdot d_i) + \sum_{\omega \in \Omega} P(\omega) \cdot (\mathrm{ver}_i(\omega) - \mathrm{app}_i(\omega) \cdot d_i) = G_i \cdot d_i \\
\text{iff} & G_i \cdot d_i + \sum_{\omega \in \Omega} P(\omega) \cdot (\mathrm{ver}_i(\omega) - \mathrm{app}_i(\omega) \cdot d_i) = G_i \cdot d_i \\
\text{iff} & \sum_{\omega \in \Omega} P(\omega) \cdot (\mathrm{ver}_i(\omega) - \mathrm{app}_i(\omega) \cdot d_i) = 0 \\
\text{iff} & \sum_{\omega \in \Omega} P(\omega) f_{\odot,i}(\omega) = 0 \\
\text{iff} & EQ_{r_i}^{\odot}
\end{array}
$$

By performing the above transformations on the equations of all $r_i \in \mathcal{R}^{\mathcal{PU}}$ in parallel, we get:

$$EQ_{\mathcal{R}}^{\mathrm{sum}F} \equiv EQ_{\mathcal{R}}^{\odot} \tag{31}$$

---

[3] For ease of readability, we denote $\mathcal{R}^{\mathcal{PU}}$ by just $\mathcal{R}$ in the following subscripts, i.e. we omit the superscript $\mathcal{PU}$.

Since the equation systems are equivalent, their solution sets must coincide, and with (24) we have:

$$Sol_{\mathcal{R}}^{\text{sum}F} = Mod_{\mathcal{R}}^{\odot} \qquad (32)$$

By considering (27) and (10) concerning $\text{ME}_{\triangle}(\mathcal{R}^{\mathcal{PU}})$ and $\text{ME}_{\odot}(\mathcal{R}^{\mathcal{PU}})$, respectively, (32) yields:

$$\text{ME}_{\triangle}(\mathcal{R}^{\mathcal{PU}}) = \underset{P \in Sol_{\mathcal{R}}^{\text{sum}F}}{\arg\max}\, H(P) = \underset{P \in Mod_{\mathcal{R}}^{\odot}}{\arg\max}\, H(P) = \text{ME}_{\odot}(\mathcal{R}^{\mathcal{PU}})$$

□

Thus, while in general $\text{ME}_{\triangle}(\mathcal{R}) \neq \text{ME}_{\odot}(\mathcal{R})$ holds, a parametrically uniform knowledge base $\mathcal{R}^{\mathcal{PU}}$ ensures that we have $\text{ME}_{\triangle}(\mathcal{R}^{\mathcal{PU}}) = \text{ME}_{\odot}(\mathcal{R}^{\mathcal{PU}})$.

7 Conclusions and Further Work

The purpose of this paper was to provide a brief introduction to aggregating and grounding semantics of probabilistic relational conditionals, and to study the relationships among the corresponding model classes. We stated various explicit conditions where the models differ, and using a result of [8], we proved that the maximum entropy models of the two semantics are identical when considering only knowledge bases that are parametrically uniform.

Implementations for aggregating and grounding semantics have been developed within the KREATOR environment[4], an integrated development environment for relational probabilistic logic [13], and various reasoning problems both for aggregating and grounding semantics are studied in [33]. Our current work includes the design of an ME model computation that allows for a simpler determination of the equivalence classes induced by the conditional structures [21] of a knowledge base and to employ this for both types of semantics; some steps in this direction are given in [11,12].

References

1. C. Beierle, M. Finthammer, G. Kern-Isberner, and M. Thimm. Evaluation and comparison criteria for approaches to probabilistic relational knowledge representation. In J. Bach and S. Edelkamp, editors, *KI 2011*, volume 7006 of *LNCS*, pages 63–74. Springer, 2011.
2. C. Beierle, M. Höhnerbach, and M. Marto. Implementation of a transformation system for relational probabilistic knowledge bases simplifying the maximum entropy model computation. In *Proc. FLAIRS 2014*, pages 486–489, Menlo Park, CA, 2014. AAAI Press.
3. C. Beierle and A. Krämer. Achieving parametric uniformity for knowledge bases in a relational probabilistic conditional logic with maximum entropy semantics. *Annals of Mathematics and Artificial Intelligence*, 73(1-2):5–45, 2015.

---

[4] KREATOR can be found at http://kreator-ide.sourceforge.net/

4. Christoph Beierle, Marc Finthammer, and Gabriele Kern-Isberner. Relational probabilistic conditionals and their instantiations under maximum entropy semantics for first-order knowledge bases. *Entropy*, 17(2):852–865, 2015.
5. S. Boyd and L. Vandenberghe. *Convex Optimization*. Cambridge University Press, New York, NY, USA, 2004.
6. J. Delgrande. On first-order conditional logics. *Artificial Intelligence*, 105:105–137, 1998.
7. M. Finthammer. An iterative scaling algorithm for maximum entropy reasoning in relational probabilistic conditional logic. In E. Hüllermeier, S. Link, T. Fober, and B. Seeger, editors, *Scalable Uncertainty Management, 6th International Conference, Proceedings*, volume 7520 of *LNAI*, pages 351–364. Springer, 2012.
8. M. Finthammer and C. Beierle. How to exploit parametric uniformity for maximum entropy reasoning in a relational probabilistic logic. In L. Fariñas del Cerro, A. Herzig, and J. Mengin, editors, *JELIA 2012*, volume 7519 of *LNAI*, pages 189–201. Springer, 2012.
9. M. Finthammer and C. Beierle. Instantiation restrictions for relational probabilistic conditionals. In E. Hüllermeier, S. Link, T. Fober, and B. Seeger, editors, *Scalable Uncertainty Management, 6th International Conference, Proceedings*, volume 7520 of *LNAI*, pages 598–605. Springer, 2012.
10. M. Finthammer and C. Beierle. Using equivalences of worlds for aggregation semantics of relational conditionals. In B. Glimm and A. Krüger, editors, *KI 2012*, LNAI 7526, pages 49–60. Springer, 2012.
11. M. Finthammer and C. Beierle. A two-level approach to maximum entropy model computation for relational probabilistic logic based on weighted conditional impacts. In U. Straccia and A. Calì, editors, *SUM 2014*, volume 8720 of *LNAI*, pages 162–175. Springer, 2014.
12. M. Finthammer and C. Beierle. Towards a more efficient computation of weighted conditional impacts for relational probabilistic knowledge bases under maximum entropy semantics. In Steffen Hölldobler, Markus Krötzsch, Rafael Peñaloza, and Sebastian Rudolph, editors, *KI 2015*, volume 9324 of *LNAI*, pages 72–86. Springer, 2015.
13. M. Finthammer and M. Thimm. An integrated development environment for probabilistic relational reasoning. *Logic Journal of the IGPL*, 20(5):831–871, 2012.
14. J. Fisseler. *Learning and Modeling with Probabilistic Conditional Logic*, volume 328 of *Dissertations in Artificial Intelligence*. IOS Press, Amsterdam, 2010.
15. J. Fisseler. First-order probabilistic conditional logic and maximum entropy. *Logic Journal of the IGPL*, 20(5):796–830, 2012.
16. S. Geman and D. Geman. Stochastic relaxation, gibbs distributions, and the Bayesian restoration of images. *IEEE Trans. on Pattern Analysis and Machine Intelligence*, 6:721–741, 1984.
17. L. Getoor and B. Taskar, editors. *Introduction to Statistical Relational Learning*. MIT Press, 2007.
18. J.Y. Halpern. *Reasoning About Uncertainty*. MIT Press, 2005.
19. E.T. Jaynes. *Papers on Probability, Statistics and Statistical Physics*. D. Reidel Publishing Company, Dordrecht, Holland, 1983.
20. G. Kern-Isberner. Characterizing the principle of minimum cross-entropy within a conditional-logical framework. *Artificial Intelligence*, 98:169–208, 1998.
21. G. Kern-Isberner. *Conditionals in nonmonotonic reasoning and belief revision*. Springer, Lecture Notes in Artificial Intelligence LNAI 2087, 2001.
22. G. Kern-Isberner. Relational probabilistic reasoning at maximum entropy. KReate technical project paper, June 2009.

23. G. Kern-Isberner, C. Beierle, M. Finthammer, and M. Thimm. Comparing and evaluating approaches to probabilistic reasoning: Theory, implementation, and applications. *Transactions on Large-Scale Data- and Knowledge-Centered Systems*, 6:31–75, 2012.
24. G. Kern-Isberner and T. Lukasiewicz. Combining probabilistic logic programming with the power of maximum entropy. *Artificial Intelligence, Special Issue on Nonmonotonic Reasoning*, 157(1-2):139–202, 2004.
25. G. Kern-Isberner and M. Thimm. Novel semantical approaches to relational probabilistic conditionals. In Fangzhen Lin, Ulrike Sattler, and Miroslaw Truszczynski, editors, *Proceedings Twelfth International Conference on the Principles of Knowledge Representation and Reasoning, KR'2010*, pages 382–391. AAAI Press, 2010.
26. G. Kern-Isberner and M. Thimm. A ranking semantics for first-order conditionals. In *ECAI-2012*, pages 456–461. IOS Press, 2012.
27. S. Loh, M. Thimm, and G. Kern-Isberner. On the problem of grounding a relational probabilistic conditional knowledge base. In T. Meyer and E. Ternovska, editors, *Proceedings 13th International Workshop on Nonmonotonic Reasoning NMR'2010, Subworkshop on NMR and Uncertainty*, 2010.
28. T. Lukasiewicz and G. Kern-Isberner. Probabilistic logic programming under maximum entropy. In *Proceedings ECSQARU-99*, volume 1638, pages 279–292. Springer Lecture Notes in Artificial Intelligence, 1999.
29. J.B. Paris. *The uncertain reasoner's companion – A mathematical perspective*. Cambridge University Press, 1994.
30. J.B. Paris and A. Vencovska. In defence of the maximum entropy inference process. *International Journal of Approximate Reasoning*, 17(1):77–103, 1997.
31. Jeff Paris. Common sense and maximum entropy. *Synthese*, 117:75–93, 1999.
32. Jeff B. Paris. What you see is what you get. *Entropy*, 16(11):6186–6194, 2014.
33. N. Potyka. Solving reasoning problems for probabilistic conditional logics with consistent and inconsistent information, 2015. Dissertation, University of Hagen.
34. N. Potyka, C. Beierle, and G. Kern-Isberner. A concept for the evolution of relational probabilistic belief states and the computation of their changes under optimum entropy semantics. *Journal of Applied Logic*, 13:414–440, 2015.
35. W. Rödder, E. Reucher, and F. Kulmann. Features of the expert-system-shell SPIRIT. *Logic Journal of the IGPL*, 14(3):483–500, 2006.
36. J.E. Shore and R.W. Johnson. Axiomatic derivation of the principle of maximum entropy and the principle of minimum cross-entropy. *IEEE Transactions on Information Theory*, IT-26:26–37, 1980.
37. M. Thimm. Probabilistic reasoning with incomplete and inconsistent beliefs, 2011. Dissertation, TU Dortmund.
38. M. Thimm, G. Kern-Isberner, and J. Fisseler. Relational probabilistic conditional reasoning at maximum entropy. In *ECSQARU-11*, LNCS 6717, pages 447–458. Springer, 2011.

The Finite Values Property

Elizabeth Howarth[1], Jeff Paris[2]

Affectionately dedicated to Gabriele Kern-Isberner

Abstract. We argue that the simplicity condition on a probability function on sentences of a predicate language L that it takes only finitely many values on the sentences of any finite sublanguage of L can be viewed as rational. We then go on to investigate consequences of this condition, linking it to the model theoretic notion of quantifier elimination.

Introduction

A much studied problem over the last 35 years in AI is how one should choose a particular probability function to satisfy a given set of (satisfiable) constraints. Obviously in all but the most trivial cases such a choice has to go beyond what is actually stated in the constraints per se and as a result various methodologies have been proposed on the basis of what it seems reasonable to additionally assume in the circumstances. For example if the constraints actually apply to an objective probability function then an approach based on some sort of averaging of all the possible candidate probability functions might seem appropriate whilst if one is seeking a subjective, common sense, probability function then one might opt for minimizing the information content beyond what is already there in the constraints, for example by maximizing entropy (relevant to the context of this paper and volume see [1],[12],[13],[14],[15],[21],[22],[23]).

In this paper we shall consider a property of the chosen probability function which, in the context of a predicate language, seems attractive on the basis of both pragmatic and rational considerations, namely that in a sense to be explained shortly the chosen probability function only takes finitely many values on the sentences of any particular finite sublanguage. 'Pragmatic' because such a property can simplify predictions and 'rational' in the sense of Occam's Razor, that it is rational to adopt as simple an explanation as possible.

Firstly however we need to introduce the particular context in which we shall be working, namely Pure Inductive Logic, see for example [17], [18], [19].

---

[1] School of Mathematics, The University of Manchester, Manchester M13 9PL, lizhowarth@outlook.com, supported by a UK Engineering and Physical Sciences Research Council (EPSRC) Studentship

[2] School of Mathematics, The University of Manchester, Manchester M13 9PL, jeff.paris@manchester.ac.uk, supported by a UK Engineering and Physical Sciences Research Council Research Grant

Context

Let L be the first order language with constant symbols a_n, $n \in \mathbb{N}^+ = \{1, 2, 3, \ldots\}$ and relation symbols $R_1, R_2, \ldots, R_q$ of arities $r_1, r_2, \ldots, r_q$ respectively but no function symbols nor equality.

Let $SL, QFSL$ denote the sentences and quantifier free sentences of L and let $SL^{(n)}$ denote those sentences of L which do not mention any constants a_r with $r > n$. In other words sentences of the finite sublanguage of L with just the relation symbols $R_1, R_2, \ldots, R_q$ and the constant symbols $a_1, a_2, \ldots, a_n$.

We say that a function $w : SL \to [0, 1]$ is a *probability function on SL* if for all $\theta, \varphi, \exists x \, \psi(x) \in SL$

(P1) $\models \theta \implies w(\theta) = 1$.

(P2) $\theta \models \neg\varphi \implies w(\theta \vee \varphi) = w(\theta) + w(\varphi)$.

(P3) $w(\exists x \, \psi(x)) = \lim_{n \to \infty} w(\psi(a_1) \vee \psi(a_2) \vee \ldots \vee \psi(a_n))$.

With this definition all the expected properties of probabilities hold, in particular because of their significance to this paper, if $\theta, \varphi \in SL$ and $\theta \models \varphi$ then $w(\theta) \leq w(\varphi)$ and if $\models \theta \leftrightarrow \varphi$ then $w(\theta) = w(\varphi)$ (see for example [19, Chapter 3] for more details).

Let $\mathcal{T}L$ be the set of structures M for L with universe the interpretations of the a_n (also denoted a_n). So for $M \in \mathcal{T}L$ every element in the universe of M is named by a constant.

In our view the central question which Pure Inductive Logic aims to investigate is:

Question: Given an agent $\mathcal{A}$ inhabiting an unknown structure $M \in \mathcal{T}L$ and $\theta \in SL$ what probability $w(\theta)$ should $\mathcal{A}$ *rationally*, or *logically*, assign to θ?

– or more generally given that we obviously expect $\mathcal{A}$'s answers to be mutually consistent:

Question: Given an agent $\mathcal{A}$ inhabiting an unknown structure $M \in \mathcal{T}L$, *rationally* or *logically*, what probability function w should $\mathcal{A}$ adopt?

It is important to appreciate here that by 'probability' we mean subjective probability (i.e. degree of belief or willingness to bet) and that $\mathcal{A}$ should know nothing more about M, so have no intended interpretation in mind for the constant and relation symbols.

The key obstacle in answering this question is of course what we mean by 'rational'. In the absence of any precise definition of this term the main approach (since Carnap essentially founded the subject in the 1940's, see for example [2], [3], [4], [5], [6]) is to postulate possible properties of w which are 'rational' in some intuitive sense and then investigate what consequences these entail. Of these properties the most widely accepted is that w should satisfy:

Constant Exchangeability, Ex:

For $\varphi(a_{i_1}, a_{i_2}, \ldots, a_{i_m}) \in SL$,[3]

$$w(\varphi(a_1, a_2, \ldots, a_m)) = w(\varphi(a_{i_1}, a_{i_2}, \ldots, a_{i_m})).$$

The rational justification here is that the agent has no knowledge about any of the a_i so it would be irrational to treat them differently when assigning probabilities.

Constant Exchangeability, Ex, is so widely accepted in this area that we will henceforth *assume it throughout for all the probability functions we consider*.

The purpose of this paper is to investigate a further ostensibly rational principle (the Finite Values Property, FVP) which is based on the putative idea that simplicity is a facet of rationality.

The Finite Values Property

Let w be a probability function on SL satisfying Ex. We say that w satisfies the *Finite Values Property*, FVP, if

$$\{w(\theta) \mid \theta \in SL^{(n)}\}$$

is finite for each $n \in \mathbb{N}$.

The Finite Values Property may seem rather surprising, since although at each 'level' n, the number of constants in each sentence of $SL^{(n)}$ is bounded, no such restriction is placed on the length or complexity of these sentences. FVP is the formal version of the pragmatically and rationally desirable 'finiteness property' alluded to in the introduction.[4]

We shall say that w satisfies FVP_n if

$$\{w(\theta) \mid \theta \in SL^{(n)}\}$$

is finite. So FVP amounts to $\forall n \in \mathbb{N}$, FVP_n. Clearly if w satisfies FVP_n then it also satisfies FVP_m for $m < n$.

The aim of this paper is to initiate an investigation into which probability functions satisfy this property and what further structure they must have. Our first result is that FVP always holds if the language L is unary (i.e. each $r_i = 1$), a standard assumption in fact in Inductive Logic up to the 21st century.

---

[3] The convention is that when a sentence φ is written in this form it is assumed (unless otherwise stated) that the displayed constants are distinct and include all the constants actually occurring in φ.

[4] We would point out that in the context of Pure Inductive Logic our current favoured choice of probability functions on grounds of rationality are the so called *homogeneous* probability functions, see [19, Chapter 30], and these do indeed additionally satisfy this further pragmatic requirement of FVP, see [9].

Theorem 1 *If L is unary and w is a probability function on SL then w satisfies FVP.*

Proof. Suppose L is unary and let $\alpha_i(x)$ for $i = 1, 2, \ldots, 2^q$ enumerate the *atoms* of L, that is the formulae of L of the form

$$R_1(x)^{\epsilon_1} \wedge R_2(x)^{\epsilon_2} \wedge \ldots \wedge R_q(x)^{\epsilon_q}$$

where the $\epsilon_i \in \{0, 1\}$ and for a formula φ, $\varphi^1 = \varphi$, $\varphi^0 = \neg \varphi$.

Let $\theta(a_1, \ldots, a_n) \in SL^{(n)}$. It is well known, see for example [16, Theorem 4], that θ is logically equivalent to a sentence θ' of the form

$$\bigvee_{k=1}^{l} \left(\bigwedge_{j=1}^{2^q} (\exists x\, \alpha_j(x))^{\epsilon_{k_j}} \wedge \bigwedge_{i=1}^{n} \alpha_{f_{k_i}}(a_i) \right),$$

where each $\epsilon_k \in \{0,1\}^{2^q}$ and the disjuncts are disjoint and satisfiable.

Let B be the set of satisfiable disjuncts (up to logical equivalence)

$$\bigwedge_{j=1}^{2^q} (\exists x\, \alpha_j(x))^{\epsilon_j} \wedge \bigwedge_{i=1}^{n} \alpha_{f_i}(a_i). \tag{1}$$

Since there are 2 choices for ϵ_j for each $j = 1, \ldots, 2^q$, and at most 2^q choices for α_{f_i} for each $i = 1, \ldots, n$, this gives

$$|B| \leqslant 2^{2^q + qn}.$$

Since $\theta(a_1, \ldots, a_n)$ is logically equivalent to the disjunction of some subset of B, the size of $SL^{(n)}$ up to logical equivalence is bounded by the number $2^{|B|}$ of distinct subsets of B. Since logically equivalent sentences must get the same probability FVP_n, and hence FVP, follows.

It turns out that Theorem 1 is a special, though important, case of a much more general result. Before we can give that result however it will be useful to introduce the concept of *ions*.

Ions

The following characterization of FVP_n shows that the set of sentences B in Theorem 1 has a counterpart wherever FVP_n occurs.

Theorem 2 *A probability function w on SL satisfies FVP_n just if there is a finite set of sentences*

$$B = \{\varphi_1, \ldots, \varphi_g\} \subset SL^{(n)}$$

such that

- $w(\varphi_i \wedge \varphi_j) = 0$ for any $1 \leqslant i < j \leqslant g$,
- $\sum_{i=1}^{g} w(\varphi_i) = 1$,
- for any $\theta \in SL^{(n)}$ there is a subset B_θ of B such that

$$w\left(\theta \leftrightarrow \bigvee_{\varphi \in B_\theta} \varphi\right) = 1.$$

Proof. From left to right, suppose that w satisfies FVP_n. Initially let $B' = \{\top\}$ where $\top \in SL^{(0)}$ (the set of sentences of L mentioning no constants) is a tautology. Now repeatedly 'split' sentences in B' as follows. If possible pick $\varphi \in B'$ for which there exists $\theta \in SL^{(n)}$ such that

$$0 < w(\varphi \wedge \theta), \, w(\varphi \wedge \neg \theta) < w(\varphi),$$

and replace φ in B' by $\varphi \wedge \theta$ and $\varphi \wedge \neg \theta$. Repeat this step until no such φ remains. Note that at each stage of this process

$$w\left(\bigvee_{\varphi \in B'} \varphi\right) = 1 \tag{2}$$

and for any distinct $\varphi, \eta \in B'$, $w(\varphi \wedge \eta) = 0$.

To show that this process must halt after a finite number of steps suppose on the contrary that it did not. Then the $w(\varphi)$ for φ appearing in B' at some stage cannot be bounded away from 0 since otherwise by (2) the $|B'|$ would also have to be uniformly bounded at all stages, which clearly is not the case. But if the $w(\varphi)$ for φ appearing in a B' at some stage are not bounded away from 0 then this contradicts FVP_n.

Let
$$B = \{\varphi_1, \varphi_2, \ldots, \varphi_g\} \subset SL^{(n)}$$
be the halting B', so for any $\theta \in SL^{(n)}$ and any $\varphi_j \in B$,

$$w(\theta \wedge \varphi_j) \in \{0, w(\varphi_j)\}.$$

For $\theta \in SL^{(n)}$ let $B_\theta = \{\varphi_j \in B \mid w(\theta \wedge \varphi_j) = w(\varphi_j)\}$. Then from (2)

$$w(\theta) = w\left(\theta \wedge \bigvee_{\varphi \in B} \varphi\right) = \sum_{\varphi \in B} w(\theta \wedge \varphi) = \sum_{\varphi \in B_\theta} w(\varphi) \tag{3}$$

since $w(\theta \wedge \varphi) = 0$ for $\varphi \in B - B_\theta$.

Furthermore, since
$$w(\varphi_j) = w(\varphi_j \wedge \theta) + w(\varphi_j \wedge \neg \theta)$$
we have
$$B_{\neg \theta} = B - B_\theta. \tag{4}$$

Hence from (3) for $\neg \theta$,

$$w(\neg \theta \wedge \bigvee_{\varphi \in B_\theta} \varphi) = w(\bigvee_{\psi \in B_{\neg \theta}} \psi \wedge \bigvee_{\varphi \in B_\theta} \varphi) = 0$$

since by (4) for $\varphi \in B_\theta, \psi \in B_{\neg \theta}$, $\varphi \neq \psi$ so $w(\varphi \wedge \psi) = 0$. Together with (3) this forces that

$$w(\theta \leftrightarrow \bigvee_{\varphi \in B_\theta} \varphi) = 1.$$

In the other direction, it is clear that if $B = \{\varphi_1, \ldots, \varphi_g\} \subset SL^{(n)}$ is as described in the statement of the result, then for any $\theta \in SL^{(n)}$

$$w(\theta) = \sum_{\varphi \in B_\theta} w(\varphi),$$

and since the number of possible subsets B_θ of B is finite, then so is the range of $w \restriction SL^{(n)}$.

We will call such a set $B \subset SL^{(n)}$ with the properties given in Theorem 2 a set of *n-ions* for w. Note from the above result that when φ is an n-ion for w and $\theta \in SL^{(n)}$, $B_{\theta \wedge \varphi}$ is either equal to $\{\varphi\}$ or to $\emptyset$, so that

$$w(\theta \wedge \varphi) \in \{0, w(\varphi)\}. \tag{5}$$

Theorem 2 suggests a connection between FVP and the well known and very significant notion of *Quantifier Elimination* in Model Theory, a connection which we shall investigate in the next section.

Generalized Quantifier Elimination

Let T be a theory in $L^{(0)}$, meaning that $T \subset SL^{(0)}$ and T is closed under logical consequence. We shall say that T satisfies *Generalized Quantifier Elimination*, GQE, if there is a finite set of formulae of $L^{(0)}$,

$$\{\zeta_i(x_1, x_2, \ldots, x_k) \mid i = 1, 2, \ldots, m\}$$

such that for each formula $\theta(x_1, x_2, \ldots, x_n)$ of $L^{(0)}$ there is a Boolean combination $\psi(x_1, x_2, \ldots, x_n)$ of formulae $\zeta_i(y_1, y_2, \ldots, y_k)$ where $\{y_1, y_2, \ldots, y_k\} \subseteq \{x_1, x_2, \ldots, x_n\}$ and $i \in \{1, 2, \ldots, m\}$, such that $T \models \theta \leftrightarrow \psi$.

Note that this reduces to the standard notion of simply *Quantifier Elimination* in the case where the $\zeta_i(x_1, x_2, \ldots, x_k)$ are just the $R_i(x_1, x_2, \ldots, x_{r_i})$ for R_i a relation symbol of L and $k = \max\{r_i\}$.

Theorem 3 *Let w be a probability function on SL satisfying Ex and such that*

$$Th(w) = \{\varphi \in SL^{(0)} \mid w(\varphi) = 1\}$$

satisfies GQE. Then w satisfies FVP.

Proof. Let w be as given with the $\zeta_i(x_1, x_2, \ldots, x_k)$ for $i = 1, 2, \ldots, m$ as in the definition of GQE. Then for $\theta(x_1, x_2, \ldots, x_n)$ a formula of $L^{(0)}$ there is a Boolean combination $\psi(x_1, x_2, \ldots, x_n)$ of formulae $\zeta_i(y_1, y_2, \ldots, y_k)$ where $\{y_1, y_2, \ldots, y_k\} \subseteq \{x_1, x_2, \ldots, x_n\}$ and $i \in \{1, 2, \ldots, m\}$, such that $Th(w) \models \theta \leftrightarrow \psi$. Hence $\theta \leftrightarrow \psi$ is a logical consequence of some finite subset of $Th(w)$ and since the members of $Th(w)$ are all sentences it further follows that

$$\theta(a_1, \ldots, a_n) \leftrightarrow \psi(a_1, \ldots, a_n)$$

is also a logical consequence of this finite subset. Therefore, since $w(\varphi) = 1$ for each $\varphi \in Th(w)$,

$$w(\theta(a_1, \ldots, a_n) \leftrightarrow \psi(a_1, \ldots, a_n)) = 1$$

and consequently

$$w(\theta(a_1, \ldots, a_n)) = w(\psi(a_1, \ldots, a_n)).$$

But clearly here there are, up to logical equivalence, only finitely many choices for $\psi(a_1, \ldots, a_n)$ and hence only finitely many choices for $w(\theta)$ when $\theta \in SL^{(n)}$.

To date all the examples we have of probability functions satisfying FVP actually also have theories which satisfy GQE. For example in the unary case we see that we can take the ζ to be the possible disjuncts in (1). Other examples are the t-heterogeneous and homogeneous probability functions (see [9]) and the probability functions $°w^\Psi$ (see [19, page 184]) when Ψ is standard.

For later use we observe that for L a not purely unary language there are probability functions on SL which fail even FVP_0. For example, in the notation of [19, page 217] let

$$\bar{p}_n = \langle 0, n^{-1}, n^{-1}, \ldots, n^{-1}, 0, 0, \ldots \rangle$$

for $n \in \mathbb{N}^+$ where there are n copies of n^{-1}. Then the probability function

$$w = \sum_{n=1}^{\infty} 2^{-n} v^{\bar{p}_n, L} \qquad (6)$$

satisfies Ex. However for each $m \in \mathbb{N}^+$ there is a sentence $\theta_m \in SL^{(0)}$ such that

$$v^{\bar{p}_n, L}(\theta_m) = \begin{cases} 1 & \text{if } m = n, \\ 0 & \text{if } m \neq n, \end{cases}$$

so $w(\theta_m) = 2^{-m}$. (With apologies for the terseness of this example we refer the reader to [9].)

This observation may suggest that the converse to Theorem 3 holds, that if w satisfies FVP then $Th(w)$ satisfies GQE. We now show that this does indeed hold in cases where the n-ions of w for some large enough n have a certain property.

Theorem 4 *Let w satisfy FVP and suppose that there is some k such that if*

$$\{\zeta_1(a_1, a_2, \ldots, a_k), \ldots, \zeta_m(a_1, a_2, \ldots, a_k)\}$$

is a set of k-ions for w then for $n \geqslant k$ the set of Boolean combinations $\psi(a_1, a_2, \ldots, a_n)$ of sentences $\zeta_i(b_1, b_2, \ldots, b_k)$, where $\{b_1, b_2, \ldots, b_k\} \subseteq \{a_1, a_2, \ldots, a_n\}$ and $i \in \{1, 2, \ldots, m\}$, includes a set of n-ions for w. Then $Th(w)$ satisfies GQE.

Proof. Let $\rho(x_1, x_2)$ be the formula

$$\bigwedge_{i=1}^{q} \bigwedge_{f=1}^{r_i} \forall z_1, \ldots, z_{f-1}, z_{f+1}, \ldots, z_{r_i}$$
$$(R_i(z_1, \ldots, z_{f-1}, x_1, z_{f+1}, \ldots, z_{r_i}) \leftrightarrow R_i(z_1, \ldots, z_{f-1}, x_2, z_{f+1}, \ldots, z_{r_i}))$$

which expresses that x_1 and x_2 are indistinguishable from each other as far as the relations $R_1, \ldots, R_q$ of L are concerned.

The formula $\rho(x_1, x_2)$ clearly acts like equality in that it satisfies the axioms of equality (see for example [16]), in particular satisfying that for each $i = 1, 2, \ldots, q$,

$$\models \left(\bigwedge_{f=1}^{r_i} \rho(x_f, x_{r_i + f}) \right) \rightarrow (R_i(x_1, x_2, \ldots, x_{r_i}) \leftrightarrow R_i(x_{r_i+1}, x_{r_i+2}, \ldots, x_{2r_i})).$$

Consequently we also have that for any formula $\varphi(x_1, x_2, \ldots, x_n)$ of $L^{(0)}$ (or even L) that

$$\models \left(\bigwedge_{f=1}^{n} \rho(x_f, x_{n+f}) \right) \rightarrow (\varphi(x_1, x_2, \ldots, x_n) \leftrightarrow \varphi(x_{n+1}, x_{n+2}, \ldots, x_{2n})). \quad (7)$$

Let w be as described in the statement of the theorem. Let $\theta(x_1, x_2, \ldots, x_n)$ be a formula of $L^{(0)}$ and (without loss of generality) let $n \geqslant k$. We claim that there is a Boolean combination $\psi(x_1, x_2, \ldots, x_n)$ of the $\rho(x_i, x_j)$ for $1 \leqslant i, j \leqslant n$ and the $\zeta_i(y_1, y_2, \ldots, y_k)$, where $\{y_1, \ldots, y_k\} \subseteq \{x_1, \ldots, x_n\}$, such that

$$w(\forall x_1, \ldots, x_n \, (\theta(x_1, \ldots, x_n) \leftrightarrow \psi(x_1, \ldots, x_n))) = 1,$$

which clearly gives the required result.

Let $\mathcal{H} = \{H_1, H_2, \ldots, H_e\}$ be a partition of $\{1, 2, \ldots, n\}$ and let $\eta_{\mathcal{H}}(x_1, \ldots, x_n)$ be the formula

$$\bigwedge_{i=1}^{e} \left(\bigwedge_{s,t \in H_i} \rho(x_s, x_t) \wedge \bigwedge_{1 \leqslant i < j \leqslant e} \bigwedge_{\substack{s \in H_i \\ t \in H_j}} \neg \rho(x_s, x_t) \right).$$

By the assumption of the theorem there is a Boolean combination $\psi_{\mathcal{H}}(x_1, x_2, \ldots, x_n)$ of the $\zeta_i(y_1, y_2, \ldots, y_k)$, where $\{y_1, y_2, \ldots, y_k\} \subseteq \{x_1, x_2, \ldots, x_n\}$ and $i \in \{1, 2, \ldots, m\}$, such that

$$w(\psi_{\mathcal{H}}(a_1, \ldots, a_n) \leftrightarrow (\theta(a_1, \ldots, a_n) \wedge \eta_{\mathcal{H}}(a_1, \ldots, a_n))) = 1$$

and hence

$$w((\psi_{\mathcal{H}}(a_1, \ldots, a_n) \wedge \eta_{\mathcal{H}}(a_1, \ldots, a_n)) \leftrightarrow (\theta(a_1, \ldots, a_n) \wedge \eta_{\mathcal{H}}(a_1, \ldots, a_n))) = 1. \tag{8}$$

We can now refine (8) to give[5]

$$w((\psi_{\mathcal{H}}(a_{i_1}, \ldots, a_{i_n}) \wedge \eta_{\mathcal{H}}(a_{i_1}, \ldots, a_{i_n})) \leftrightarrow (\theta(a_{i_1}, \ldots, a_{i_n}) \wedge \eta_{\mathcal{H}}(a_{i_1}, \ldots, a_{i_n}))) = 1 \tag{9}$$

for any $i_1, i_2, \ldots, i_n$, not necessarily distinct. To see this let $\mathcal{G} = \{G_1, G_2, \ldots, G_c\}$ be the partition of $\{1, 2, \ldots, n\}$ such that k, j are in the same class just if $i_k = i_j$. If $\mathcal{G}$ is not a refinement of $\mathcal{H}$ then both of

$$\psi_{\mathcal{H}}(a_{i_1}, \ldots, a_{i_n}) \wedge \eta_{\mathcal{H}}(a_{i_1}, \ldots, a_{i_n}) \quad \text{and} \quad \theta(a_{i_1}, \ldots, a_{i_n}) \wedge \eta_{\mathcal{H}}(a_{i_1}, \ldots, a_{i_n})$$

are inconsistent so get probability 0 and the required conclusion (9) holds.

On the other hand if $\mathcal{G}$ is a refinement of $\mathcal{H}$ then from the fact that ρ satisfies the axioms of equality (7) we have that

$$(\psi_{\mathcal{H}}(a_{i_1}, \ldots, a_{i_n}) \wedge \eta_{\mathcal{H}}(a_{i_1}, \ldots, a_{i_n})) \leftrightarrow (\psi_{\mathcal{H}}(a_{h_1}, \ldots, a_{h_n}) \wedge \eta_{\mathcal{H}}(a_{h_1}, \ldots, a_{h_n}))$$

gets probability 1 according to w, where h_t is the least i_j such that t, j are in the same equivalence class according to $\mathcal{H}$. Similarly we have that

$$(\theta(a_{i_1}, \ldots, a_{i_n}) \wedge \eta_{\mathcal{H}}(a_{i_1}, \ldots, a_{i_n})) \leftrightarrow (\theta(a_{h_1}, \ldots, a_{h_n}) \wedge \eta_{\mathcal{H}}(a_{h_1}, \ldots, a_{h_n}))$$

gets probability 1 according to w.

---

[5] One might have supposed that we could get a much less torturous proof here by noting that, by the assumption of the theorem, there is such a Boolean combination $\psi(a_1, \ldots, a_n)$ for which

$$w(\psi(a_1, \ldots, a_n) \leftrightarrow \theta(a_1, \ldots, a_n)) = 1$$

and then concluding that for any $i_1, i_2, \ldots, i_n$ we must also have

$$w(\psi(a_{i_1}, \ldots, a_{i_n}) \leftrightarrow \theta(a_{i_1}, \ldots, a_{i_n})) = 1.$$

Unfortunately this need not be the case. For example if L has just a single binary relation symbol R and w is the probability function defined by

$$w(R(a_i, a_j)) = \begin{cases} 1 & \text{if } i = j, \\ 0 & \text{if } i \neq j, \end{cases}$$

then w satisfies Ex and for $\bot$ a contradiction, $w(R(a_1, a_2) \leftrightarrow \bot) = 1$ but $w(R(a_1, a_1) \leftrightarrow \bot) = 0$.

Hence the sentence in (9) gets the same probability as

$$(\psi_{\mathcal{H}}(a_{h_1},\ldots,a_{h_n}) \wedge \eta_{\mathcal{H}}(a_{h_1},\ldots,a_{h_n})) \leftrightarrow (\theta(a_{h_1},\ldots,a_{h_n}) \wedge \eta_{\mathcal{H}}(a_{h_1},\ldots,a_{h_n})). \quad (10)$$

Applying the corresponding argument in the case of

$$\psi_{\mathcal{H}}(a_1,\ldots,a_n) \wedge \eta_{\mathcal{H}}(a_1,\ldots,a_n) \text{ and } \theta(a_1,\ldots,a_n) \wedge \eta_{\mathcal{H}}(a_1,\ldots,a_n),$$

and noticing that the the non-refinement situation cannot occur in this case, we see that

$$(\psi_{\mathcal{H}}(a_1,\ldots,a_n) \wedge \eta_{\mathcal{H}}(a_1,\ldots,a_n)) \leftrightarrow (\theta(a_1,\ldots,a_n) \wedge \eta_{\mathcal{H}}(a_1,\ldots,a_n)) \quad (11)$$

has the same probability, i.e. 1 by (8), as

$$(\psi_{\mathcal{H}}(a_{s_1},\ldots,a_{s_n}) \wedge \eta_{\mathcal{H}}(a_{s_1},\ldots,a_{s_n})) \leftrightarrow (\theta(a_{s_1},\ldots,a_{s_n}) \wedge \eta_{\mathcal{H}}(a_{s_1},\ldots,a_{s_n})) \quad (12)$$

where s_t is the least j such that t, j are in the same equivalence class according to $\mathcal{H}$. But clearly the sentences in (12) and (10), and consequently also (9), must get the same probability by Ex, to wit probability 1.

To complete the proof notice that $\theta(a_{i_1},\ldots,a_{i_n}))$ is logically equivalent to

$$\bigvee_{\mathcal{H}} (\theta(a_{i_1},\ldots,a_{i_n}) \wedge \eta_{\mathcal{H}}(a_{i_1},\ldots,a_{i_n}))$$

where the disjunction is over all partitions $\mathcal{H}$ of $\{1,2,\ldots,n\}$ and therefore by (9) to

$$\bigvee_{\mathcal{H}} (\psi_{\mathcal{H}}(a_{i_1},\ldots,a_{i_n}) \wedge \eta_{\mathcal{H}}(a_{i_1},\ldots,a_{i_n})).$$

Hence we now have that

$$\bigwedge_{i_1,\ldots,i_n \leq r} \left(\theta(a_{i_1},\ldots,a_{i_n}) \leftrightarrow \bigvee_{\mathcal{H}} (\psi_{\mathcal{H}}(a_{i_1},\ldots,a_{i_n}) \wedge \eta_{\mathcal{H}}(a_{i_1},\ldots,a_{i_n})) \right)$$

gets probability 1 according to w. Taking the limit $r \to \infty$ (and using the standard result [19, Lemma 3.8]) now gives as required that

$$w(\forall x_1,\ldots,x_n\,(\theta(x_1,\ldots,x_n) \leftrightarrow \bigvee_{\mathcal{H}} (\eta_{\mathcal{H}}(x_1,\ldots,x_n) \wedge \psi_{\mathcal{H}}(x_1,\ldots,x_n)))) = 1,$$

the disjunction over $\mathcal{H}$ being simply a Boolean combination of the $\rho(x_i, x_j)$ for $1 \leq i, j \leq n$ and the $\zeta_i(y_1, y_2, \ldots, y_k)$.

At this time we know of no example of a probability function w satisfying FVP which does not satisfy the requirement of Theorem 4. This might lead one to conjecture that FVP and GQE are essentially the same thing, a point we shall revisit later in the concluding section.

The Strong Finite Values Property

Given the suggested rationality of FVP it seems natural to take a further step in this direction and consider probability functions satisfying Ex for which even

$$\{w(\theta) \mid \theta \in SL\}$$

is finite. We say that such a probability function satisfies the *Strong Finite Values Property*, SFVP.

The plan now is to give a characterization (in fact several) of the probability functions satisfying SFVP. Before doing so however we need to introduce some notation.

A sentence $\Theta(a_{i_1}, a_{i_2}, \ldots, a_{i_n}) \in QFSL$ is a *state description*[6] for $a_{i_1}, a_{i_2}, \ldots, a_{i_n}$ if it is (up to the order of conjuncts) of the form

$$\bigwedge_{j=1}^{q} \bigwedge_{b_1,\ldots,b_{r_j}} \pm R_j(b_1, \ldots, b_{r_j})$$

where the $b_1, \ldots, b_{r_j} \in \{a_{i_1}, a_{i_2}, \ldots, a_{i_n}\}$ and $\pm R_j(\boldsymbol{b})$ stands for one of $R_j(\boldsymbol{b})$, $\neg R_j(\boldsymbol{b})$. We shall adopt the usual notation that state descriptions are designated by upper case letters Θ, Φ, Ψ etc..

Let I_n be the set of state descriptions of L for $a_1, \ldots, a_n$ which are invariant up to logical equivalence under any permutation of $a_1, \ldots, a_n$. That is:

$$I_n = \{\Phi(a_1, \ldots, a_n) \mid \Phi(a_1, \ldots, a_n) \equiv \Phi(a_{\sigma(1)}, \ldots, a_{\sigma(n)}) \quad \forall \sigma \in S_n\}, \quad (13)$$

where S_n is the set of permutations of $\{1, 2, \ldots, n\}$ (and where logically equivalent members are identified). Let r be at least as large as the arity of any relation in L, i.e. $r \geqslant \max\{r_j \mid j = 1, 2, \ldots, q\}$.

For $\Theta(a_1, \ldots, a_r) \in I_r$ we define a unique structure $M_\Theta \in TL$ as follows: For $j = 1, 2, \ldots, q$ and not necessarily distinct $i_1, i_2, \ldots, i_{r_j} \in \mathbb{N}^+$ set

$$M \models R_j(a_{i_1}, a_{i_2}, \ldots, a_{i_{r_j}}) \iff \Theta(a_1, \ldots, a_r) \models R_j(a_{\tau(i_1)}, a_{\tau(i_2)}, \ldots, a_{\tau(i_{r_j})})$$

where $\tau(i_t)$ is the least s such that $i_t = i_s$.

In this case, referring to $\langle \tau(i_1), \tau(i_2), \ldots, \tau(i_{r_j}) \rangle$ as the collapse of $\langle i_1, i_2, \ldots, i_{r_j} \rangle$, notice that if σ is a permutation of $\mathbb{N}$ and then $\langle \sigma(i_1), \sigma(i_2), \ldots, \sigma(i_{r_j}) \rangle$ has the identical collapse, and furthermore all its coordinates are in $\{1, 2, \ldots, r\}$. From this it follows that $M_\Theta \models \Theta$, for each $n \in \mathbb{N}^+$ M_Θ is a model of

$$\bigvee_{\Phi \in I_n} \Phi(a_1, \ldots, a_n) \quad (14)$$

---

[6] State descriptions are important in this subject because by a result of Gaifman [7] every probability function on SL is already determined by its values on the state descriptions.

for each $n \in \mathbb{N}^+$, and furthermore M_Θ is the unique structure in $\mathcal{TL}$ with these two properties. Hence

$$\{M \in \mathcal{TL} \,|\, M \models \bigvee_{\Phi \in I_n} \Phi(a_1, \ldots, a_n), \;\; \forall n \in \mathbb{N}^+\} = \{M_\Theta \,|\, \Theta \in I_r\}. \qquad (15)$$

Now define the probability function V_{M_Θ} on SL by:

$$V_{M_\Theta}(\varphi) = \begin{cases} 1 & \text{if } M_\Theta \models \varphi, \\ 0 & \text{if } M_\Theta \models \neg\varphi. \end{cases}$$

Note that by the construction of M_Θ, V_{M_Θ} satisfies Ex.

The following theorem characterizes those probability functions on SL satisfying SFVP (and Ex).

Theorem 5 *If w is a probability function on SL then the following statements are equivalent:*

1. w satisfies SFVP.
2. $w\left(\bigvee_{\Phi(a_1,\ldots,a_n) \in I_n} \Phi(a_1, \ldots, a_n)\right) = 1$ for each $n \in \mathbb{N}^+$.
3. w is a convex sum of the functions V_{M_Θ} for $\Theta \in I_r$.
4. For every $n \in \mathbb{N}$, $\theta \in SL^{(n)}$ and $\sigma \in S_n$,

$$w(\theta(a_1, \ldots, a_n) \leftrightarrow \theta(a_{\sigma(1)}, \ldots, a_{\sigma(n)})) = 1.$$

Since the proof of this result (as we know it) requires a digression into Nonstandard Analysis we shall refer the curious reader to [8, page 112].

Theorem 5 clearly shows that SFVP puts very strong constraints on a probability function, too strong in our view to be considered as a desirable simplicity condition.

FVP and Super Regularity

As far as practical applications are concerned there is a major question we have essentially ignored up to now, namely given a satisfiable $\theta \in SL^{(0)}$ is there a probability function w satisfying Ex + FVP and such that $w(\theta) = 1$? Equivalently given a satisfiable finite set of linear constraints

$$\sum_{i=1}^{n} \beta_{i,j} w(\theta_i) = 0, \quad j = 1, 2, \ldots, m$$

with the $\theta_j \in SL^{(0)}$ is there necessarily a probability function w both satisfying these constraints, FVP and Ex? Unfortunately at present we do not know the answer to this question though we would conjecture that it is yes.

Of course we would obtain an immediate affirmative answer to this conjecture if we could find a probability function satisfying Ex + FVP + SReg, where SReg, standing for *Super Regularity*, is the requirement that $w(\theta) > 0$ for all satisfiable $\theta \in SL$. Unfortunately this is not possible when L is not purely unary as the following theorem shows:

Theorem 6 *If L is not purely unary and w is a probability function on SL satisfying FVP_n for some $n \in \mathbb{N}$, then w does not satisfy SReg.*

Proof. Suppose L and w are as described. Then by Theorem 2, there is some set of n-ions for w
$$B = \{\varphi_1, \ldots, \varphi_g\} \subset SL^{(n)}$$
such that $w(\varphi_i \wedge \varphi_j) = 0$ for $1 \leqslant i < j \leqslant g$,
$$\sum_{i=1}^{g} w(\varphi_i) = 1,$$
and for every $\theta \in SL^{(n)}$ there is some $B_\theta \subseteq B$ such that
$$w\left(\theta \leftrightarrow \bigvee_{\varphi_i \in B_\theta} \varphi_i\right) = 1.$$

Suppose that
$$\models \bigvee_{i=1}^{g} \varphi_i$$
and for each $\theta \in SL^{(n)}$
$$\models \theta \leftrightarrow \bigvee_{\varphi_i \in B_\theta} \varphi_i.$$

Then by Theorem 2, every probability function on SL would satisfy FVP_n with n-ions B, contradicting our earlier observation that there are probability functions on SL, for example (6), for which FVP_0 (and hence FVP_n) fails. Therefore, either $\neg \bigvee_{i=1}^{g} \varphi_i$ is consistent, but assigned probability zero by w, or for some $\theta \in SL^{(n)}$, $\neg \left(\theta \leftrightarrow \bigvee_{\varphi_i \in B_\theta} \varphi_i\right)$ is consistent, but assigned probability zero by w. In either case, w fails to satisfy SReg.

FVP and FVP_n

Given our results so far a natural question to ask is whether FVP_n might actually imply FVP for sufficiently large n. The *simple* answer to this question is, perhaps unsurprisingly, no. We give here an outline to show that FVP_1 does not imply FVP (or even FVP_2). Similar examples can, with some effort, be constructed in general to show that FVP_n does not imply FVP_{n+1}, see [10].

Let L be the language with a single binary relation symbol R. Let M be the structure for L with universe $\mathbb{Z}$, R interpreted as immediate successor and for $n \in \mathbb{N}^+$ let a_n^M, the interpretation of a_n in M, be $n/2$ if n is even and $(1-n)/2$ if n is odd. Notice that $M \in \mathcal{TL}$.

For any $i, j \in \mathbb{Z}$ there is an isomorphism of M sending i to j so for $\theta(a_1) \in SL^{(1)}$,
$$M \models \theta(a_i) \iff M \models \theta(a_j). \tag{16}$$

Hence the a_i for which $\theta(a_i)$ holds in M is either all of them or none of them. This similarity however breaks down when we allow sentences from $SL^{(2)}$ since for $m \in \mathbb{N}$ we can clearly write down provably disjoint sentences $\psi_m(a_i, a_j)$ such that
$$M \models \psi_m(a_i, a_j) \iff |a_i^M - a_j^M| = m. \tag{17}$$

Now define a probability function V_M on SL by
$$V_M(\theta(a_1, a_2, \ldots, a_n)) = \begin{cases} 1 & \text{if } M \models \theta(a_1, a_2, \ldots, a_n), \\ 0 & \text{otherwise,} \end{cases}$$
and in turn a further function w on SL by
$$w(\theta(a_1, a_2, \ldots, a_n)) = \sum_\tau V_M(\theta(a_{\tau(1)}, a_{\tau(2)}, \ldots, a_{\tau(n)})) \cdot \prod_{i=1}^n 2^{-\tau(i)}$$
where τ runs over all maps from $\{1, 2, \ldots, n\}$ into $\mathbb{N}^+$. By a theorem of Gaifman, see [7] or [19, Chapter 26], w is a probability function on SL satisfying Ex.

By (16) all the $V_M(\theta(a_i))$ are 0 or all are 1 so $w(\theta(a_1))$ is either 0 or 1 and w satisfies FVP$_1$. However for $m \in \mathbb{N}$ the $w(\psi_m(a_1, a_2))$ are clearly non-zero and have sum at most 1 (because they are provably disjoint) so w must fail FVP$_2$.

We referred to this construction and the supplement at [10] as giving a 'simple answer' to the question of whether FVP$_n$ alone implies FVP. The reason for this qualifier is that in these examples as the n increases so does the largest arity of the relation symbols used in the language. As far as we currently know it is possible that if w satisfies FVP$_n$ on a language L with no relation symbols of arity greater than n then w must also satisfy FVP.

We would conjecture that this is indeed the case, and even more that once we have FVP$_n$ at this largest arity level then the j-ions beyond that are as described earlier just Boolean combinations of the n-ions.

Conclusion

Theorem 2 shows that if a function w satisfies FVP$_n$, its n-ions correspond to various 'possible worlds', in each of which w is able to 'decide' every $\theta \in SL^{(n)}$, so that the probability it assigns to any such θ is the sum of the probabilities assigned to those worlds where θ is decided positively. This demonstrates that there is an underlying simplicity to those functions which satisfy FVP, beyond the superficial simplicity evident in its definition, in that it entails a rather 'neat', and arguably natural, way of assigning probabilities.

Simplicity, as a feature of probability functions used to model rational belief, was endorsed by Kemeny in [11], and considered by Paris & Vencovská in [20], but seems otherwise to have received little attention in Inductive Logic. Kemeny is not explicit about what constitutes simplicity, and the notion discussed by Paris & Vencovská is rather different from that considered here in relation to FVP. With these and likely other different ideas of simplicity available it would be reckless to claim without qualification that simplicity is always a desirable feature of probability functions, in fact we reach the opposite conclusion in the case of the Strong Finite Values Property. However, the particular simplicity entailed by FVP and interpreted above in terms of systematic reasoning about 'possible worlds', seems to be an appealing and arguably a rational feature.

In the course of this paper we have made two somewhat rash conjectures:

- If $\theta \in SL$ is consistent then there is a probability function w satisfying FVP for which $w(\theta) = 1$.
- If w satisfies FVP_n when n is the largest arity of any relation symbol in the language L then w satisfies the FVP and furthermore the j-ions for $j > n$ are just Boolean combinations of the n-ions.

Clearly a positive answer to these conjectures would considerably strengthen the structural importance of FVP. In particular as shown in Theorem 4 a positive answer to this second bullet point would equate FVP with GQE.

References

1. Beierle, C., Finthammer, M. & Kern-Isberner, G., Relational Probabilistic Conditionals and Their Instantiations under Maximum Entropy Semantics for First-Order Knowledge Bases, *Entropy*, 2015, **17**(2):852-865.
2. Carnap, R., On Inductive Logic, *Philosophy of Science*, 1945, **12**(2):72-97.
3. Carnap, R., On the Application of Inductive Logic, *Philosophy and Phenomenology Research*, 1947, **8**:133-147.
4. Carnap, R., *Logical Foundations of Probability*, University of Chicago Press, Chicago, Routledge & Kegan Paul, London, 1950.
5. Carnap, R., *The Continuum of Inductive Methods*, University of Chicago Press, 1952.
6. Carnap, R., The Aim of Inductive Logic, in *Logic, Methodology and Philosophy of Science*, eds. E.Nagel, P.Suppes & A.Tarski, Stanford University Press, Stanford, California, 1962, pp303-318.
7. Gaifman, H., Concerning Measures on First Order Calculi, *Israel Journal of Mathematics*, 1964, **2**:1-18.
8. Howarth, E., New Rationality Principles in Pure Inductive Logic, Ph.D. Thesis, Manchester University, June 2015. Available at http://www.maths.manchester.ac.uk/~jeff/theses/lwthesis.pdf
9. Howarth, E. & Paris, J.B., The Theory of Spectrum Exchangeability, *Review of Symbolic Logic*, **8**(01):108-130, 2015.
10. Howarth, E. & Paris, J.B., A proof that FVP_n does not imply FVP_{n+1}. Available at http://www.maths.manchester.ac.uk/~jeff/papers/lw150729GKIsup.pdf
11. Kemeny, J.G., Carnap's Theory of Probability and Induction, in *The Philosophy of Rudolf Carnap*, ed. P.A.Schilpp, La Salle, Illinois, Open Court, 1963, pp711-738.
12. Kern-Isberner, G., Conditionals in Nonmonotonic Reasoning and Belief Revision - Considering Conditionals as Agents, in *Lecture Notes in Computer Science, 2087*, Springer, 2001, ISBN 3-540-42367-2

13. Kern-Isberner, G. & Lukasiewicz, T., Combining probabilistic logic programming with the power of maximum entropy, *Artificial Intelligence*, 2004, **157**(1-2):139-202.
14. Kern-Isberner, G. & Thimm, M., Novel Semantical Approaches to Relational Probabilistic Conditionals, in the *Proceedings of the Twelfth International Conference on the Principles of Knowledge Representation and Reasoning (KR'10)*, Eds. F.Lin, U.Sattler, M.Truszczynski, AAAI Press, Toronto, Canada, May 2010, pp382-392.
15. Landes, J. & Williamson, J., Justifying Objective Bayesianism on Predicate Languages, *Entropy [Online]*, 2015, **17**:2459-2543.
16. Paris, J.B., *A Short Course in Predicate Logic*. 2015, ISBN 978-87-403-0795-5. Available online at bookboon.com.
17. Paris, J.B., Pure Inductive Logic, in *The Continuum Companion to Philosophical Logic*, eds. L.Horsten & R.Pettigrew, Continuum International Publishing Group, London, 2011, pp428-449.
18. Paris, J.B., *Pure Inductive Logic: Workshop Notes for Progic 2015*. URL= <http://www.maths.manchester.ac.uk/~jeff/lecture-notes/Progic15.pdf>
19. Paris, J.B. & Vencovská, A., *Pure Inductive Logic*, in the Association of Symbolic Logic Perspectives in Mathematical Logic Series, Cambridge University Press, April 2015.
20. Paris, J.B. & Vencovská, A., The Twin Continua of Inductive Methods, in *Logic Without Borders*, eds. A.Hirvonen, J.Kontinen, R.Kossak, & A.Villaveces, Ontos Mathematical Logic, Frankfurt-Heusenstamm, Germany, 2015, pp355-366.
21. Paris, J.B. & Rad, S.R., Inference Processes for Quantified Predicate Knowledge, in *Logic, Language, Information and Computation*, WoLLIC, Edinburgh, 2008, eds. W.Hodges & R.deQueiroz, Springer LNAI 5110, pp249-259.
22. Paris, J.B. & Rad, S.R., A note on the least informative model of a theory, in *Programs, Proofs, Processes, CiE 2010*, eds. F.Ferreira, B.Löwe, E.Mayordomo, & L.Mendes Gomes, Springer LNCS 6158, pp342-351.
23. Williamson, J., *In Defence of Objective Bayesianism*, Oxford University Press, 2010.

Relationships Between Semantics for Relational Probabilistic Conditional Logics

Nico Potyka[1]

Abstract. We investigate the relationships between some relational probabilistic conditional logics by comparing their semantics. In order to do so, we will order the different semantics with respect to their strength. Subsequently, we will provide several results that allow drawing conclusions from reasoning results under particular semantics about the results under related semantics.

1 Introduction

In applications of knowledge representation and reasoning, classical logics are often insufficient because they cannot express uncertainty. Probabilistic logics address this problem by enriching classical logical formulas with probabilities, see, e.g., [Nilsson, 1986], [Paris and Vencovská, 1990], [Lukasiewicz, 1999], [Kern-Isberner, 2001] for some examples. The intuitive meaning of a propositional probabilistic conditional $(G \mid F)[l, u]$, with propositional formulas G, F and probabilities $l, u \in [0, 1], l \leq u$, is that the conditional probability of G given that F holds is between l and u.

For relational probabilistic logics, different semantics have been considered, because a relational conditional like $(Flies(X) \mid Bird(X))[0.9]$ can be understood in different ways. For example, it might be a statistical statement claiming that 90% of all birds fly. It might also express a degree of belief and could state that our belief that some arbitrary bird flies is 90% or that our average belief over all birds is 90% whereas our belief might deviate substantially for certain individuals. Our investigation will be restricted to semantics that are defined with respect to probability functions over Herbrand interpretations over a finite set of predicate symbols and constants. In the terminology of [Halpern, 1990], we consider only finite *type 2* probabilistic logics. In contrast, in *type 1* probabilistic logics, probabilities are not defined over possible worlds, but over the elements of the domain of the language. For instance, in [Grove et al., 1994], probabilities for formulas are obtained by computing fractions of the number of satisfying worlds over the number of all possible worlds while letting the number of constants go to infinity.

The most basic idea to define semantics for type 2 relational logics is to regard free conditionals (i.e., conditionals that contain variables) as templates for ground conditionals [Lukasiewicz, 1999], [Fisseler, 2012], [Loh et al., 2010]. This approach, however, can be too strong when applied naively because it basically assumes that all individuals are exchangeable. In order to account for

---

[1] Department of Computer Science, FernUniversität in Hagen, Germany

exceptional individuals, one can enrich the language with constraints over variables [Fisseler, 2012] or consider more sophisticated ways to build the ground instances of free conditionals [Loh et al., 2010].

In [Kern-Isberner and Thimm, 2010], two novel semantical approaches have been proposed. The rough idea is to cumulate the conditional probabilities of ground instances in some way instead of demanding that the probability of the free conditional holds for all its ground instances. Furthermore, there are different ways to handle the case that the probability of a conditional's condition is zero. Allowing zero probabilities for the condition has some technical and computational advantages that will be explained in Section 2. However, a serious drawback from a knowledge representation perspective is that the knowledge base $\{(G\,|\,F)[0,0], (G\,|\,F)[1,1]\}$ is consistent under this approach because it is satisfied by each probability function with $P(F) = 0$. On the other hand, demanding $P(F) > 0$ is technically difficult and is usually not supported by algorithms that solve our reasoning problems.

Given that some semantics are easier to handle than others, it is an interesting question how these semantics relate to each other. To the best of my knowledge, there exist only a few results in this direction. As noted in [Kern-Isberner and Thimm, 2010], the aggregating and averaging semantics coincide for probabilistic facts (the case in which the condition is tautological). In [Beierle et al., 2015], the authors noted that the maximum entropy model under the grounding semantics from [Fisseler, 2012] and the maximum entropy model under the aggregating semantics from [Kern-Isberner and Thimm, 2010] coincide under certain preconditions. The purpose of this work is to investigate such relationships between semantics in a more general setting without restricting to certain reasoning approaches like Maximum entropy reasoning. In particular, we are interested in whether reasoning results under one semantics can be used to draw conclusions about reasoning results in other semantics. In Section 2, we will provide some background information on the relational languages, the different semantics and the reasoning problems that we have in mind. In particular, we will quickly recap some results on different semantics that explain why some semantics are technically and computationally beneficial. We will then introduce an ordering on semantics in Section 3 and explain what conclusions we can draw from reasoning results in one semantics about the results in related semantics. In Section 4, we will provide some ordering results on the introduced semantics and explain the practical implications of these results. Finally, we will have a closer look at the different variations of handling conditions with probability zero in Section 5.

The contributions of this paper can be summarized as follows:

- We propose an ordering of semantics for relational probabilistic logics and explain what conclusions we can draw from the relationships between semantics about the relationships between results of the corresponding reasoning problems (Section 3).
- We order grounding, averaging and aggregating semantics and their variations with respect to their strength (Section 4).
- We provide a more fine-grained investigation of the relationships between reasoning results under strict semantics and their relaxations (Section 5).

2 Background

In this section, we recap the basics that are necessary to understand this work. The reader who is familiar with the topic might want to skip the section or restrict to particular subsections.

2.1 Relational Language and Semantics

Our basic building block is a simple relational language $\mathcal{L}$ build up over a signature $\Sigma = (Sorts, Const, Pred)$. *Sorts* is a finite set of sorts. Basically, a sort represents a type like *person* or *disease*. *Const* is a finite set of constants, where each constant is associated with a sort $s \in Sorts$. Finally, *Pred* is a finite set of relation symbols $R(s_1, \ldots, s_k)$, where $s_1, \ldots, s_k \in Sorts$ indicate the sort of the i-th argument. We also allow relation symbols R of arity 0 and call them *propositional variables*. If Σ contains only propositional variables, $\mathcal{L}$ is called *propositional*.

In order to build up $\mathcal{L}$, we also consider a set *Var* of sorted variables, that is, each variable is associated with a sort $s \in Sorts$. A *term of sort* $s \in Sorts$ is a constant of sort s or a variable of sort s. $\mathcal{L}$ is the smallest set that contains

1. $R(t_1, \ldots, t_k)$ for each $R(s_1, \ldots, s_k) \in Pred$ and terms t_i of sort s_i,
2. $\neg F$ for each $F \in \mathcal{L}$,
3. $F_1 \wedge F_2$ for all $F_1, F_2 \in \mathcal{L}$.

Formulas of the form $R(t_1, \ldots, t_k)$ are called *atoms*. A *ground formula* is a formula that contains no variables. We do not allow quantifiers. However, we allow disjunction, implication and coimplication by regarding them as metasymbols that abbreviate the following formulas [Ebbinghaus et al., 1996]:

1. $F \vee G = \neg(\neg F \wedge \neg G)$.
2. $F \rightarrow G = (\neg F \vee G)$.
3. $F \leftrightarrow G = ((F \rightarrow G) \wedge (G \rightarrow F))$.

A *possible world* is a truth assignment to the ground atoms in our language. We denote the set of all possible worlds by Ω. It suffices for our purposes to define the semantics of ground formulas. The semantics of non-ground formulas will be reduced to this case later on by considering the ground instances in different ways. A possible world $\omega \in \Omega$ satisfies a ground formula F, written as $\omega \models F$, iff

1. $F = R(c_1, \ldots, c_k)$ and $\omega(R(c_1, \ldots, c_k)) = 1$,
2. $F = \neg G$ and $\omega \not\models G$,
3. $F = F_1 \wedge F_2$ and both $\omega \models F_1$ and $\omega \models F_2$ are true.

2.2 Relational Probabilistic Conditional Language

In order to build up a probabilistic language over $\mathcal{L}$, we need one more ingredient. A *constraint formula* over Σ is a formula that is obtained from the terms over Σ using only the inequality relation $\neq$, which is interpreted by syntactical inequality, and the junctors as explained before. For instance, if X, Y are variables and c is a constant, the constraint formula $(X \neq Y) \wedge (X \neq c)$ expresses that the variables X and Y denote distinct objects and that X does not refer to c. We denote the set of all constraint formulas over Σ by $\mathcal{C}$. Now we build up a probabilistic conditional language

$$(\mathcal{L}|\mathcal{L}) = \{(G|F)[l, u]\langle C \rangle \mid F, G \in \mathcal{L}, l, u \in [0,1], l \leq u, C \in \mathcal{C}\}$$

over $(\mathcal{L}|\mathcal{L})$. Syntactically, our language is a combination of the languages proposed in [Lukasiewicz, 1999] and [Fisseler, 2012]. Roughly speaking, the conditional $(G|F)[l, u]\langle C \rangle$ expresses that the conditional probability of G given that F holds is between l and u. The precise semantics is explained by means of the ground instances of G and F that can be restricted by C. If $l = u$, we speak of *point probabilities* and if $l < u$, we speak of *interval probabilities*. Sometimes it will be convenient to use the following abbreviations of $(G|F)[l, u]\langle C \rangle$:

- If $l = u$, we write $(G|F)[u]\langle C \rangle$.
- If $F \equiv \top$, we write $(G)[l, u]\langle C \rangle$.
- If $C \equiv \top$, we write $(G|F)[l, u]$.

A *knowledge base* over $(\mathcal{L}|\mathcal{L})$ is a finite set $\mathcal{K} \subset (\mathcal{L}|\mathcal{L})$.

In order to define semantics of knowledge bases, we consider probability functions over possible worlds. Let $\mathcal{P} = \{P : \Omega \to [0,1] \mid \sum_{\omega \in \Omega} P(\omega) = 1\}$ denote the set of all such probability functions. For a ground formula F, we let

$$P(F) = \sum_{\omega \models F} P(\omega).$$

Probabilities of non-ground formulas will be defined by means of *grounding operators* that map formulas to the set of its ground instances. Formally, ground instances can be defined by means of *ground substitutions*. In general, a ground substitution is a mapping $\sigma : Var \to Const$ mapping variables to constants of the same sort. σ can be extended to a mapping that maps formulas to ground formulas and conditionals to ground conditionals recursively via

1. $\sigma(R(t_1, \ldots, t_k)) = R(\sigma(t_1), \ldots, \sigma(t_k))$.
2. $\sigma(\neg F) = \neg \sigma(F)$.
3. $\sigma(F \wedge G) = \sigma(F) \wedge \sigma(G)$.
4. $\sigma((G|F)[\rho]\langle C \rangle) = (\sigma(G)|\sigma(F))[\rho]\langle \sigma(C) \rangle$.

Following [Fisseler, 2012], we call a ground instance of a constrained conditional *admissible* iff $\sigma(C)$ evaluates to true.

Example 1. Consider the signature $\Sigma = (\{s\}, \{a, b, c\}, \{Knows(s, s)\})$. The conditional $(Knows(X, Y))[0.6]\langle X \neq Y \rangle$, has 9 possible ground instances, but only 6 of them are admissible. For instance, $(Knows(a, a))[0.6]\langle a \neq a \rangle$ is not admissible. $(Knows(a, b))[0.6]\langle a \neq b \rangle$ is an admissible ground instance.

A thorough discussion of grounding operators can be found in [Loh et al., 2010]. Here, we will just assume that some grounding operator gr is given and that

- $\mathrm{gr}(\theta) = \{\theta\}$ whenever $\theta \in (\mathcal{L}|\mathcal{L})$ is ground,

– gr depends only on the conditional (in contrast, Thimm defined some grounding operators that also depend on probability functions [Thimm, 2011]).

For concreteness, you can think of gr as the operator that maps a formula to the set of all admissible ground instances (more sophisticated grounding operators in [Loh et al., 2010] avoid conflicts between ground instances when grounding whole knowledge bases).

2.3 Relational Probabilistic Conditional Semantics

Grounding Semantics. The basic idea of grounding semantics is to regard conditionals that contain variables as templates for their ground instances [Lukasiewicz, 1999], [Fisseler, 2012], [Loh et al., 2010]. Given a conditional $\theta \in (\mathcal{L}|\mathcal{L})$, a grounding operator gr and a probability function P, we say that P satisfies θ with respect to the grounding operator gr under grounding semantics iff

$$l \cdot P(F) \leq P(G \wedge F) \leq u \cdot P(F) \text{ for all } (G|F)[l, u]\langle C\rangle \in \text{gr}(\theta).$$

If $P(F) > 0$, we can divide this inequality by $P(F)$ and obtain $l \leq \frac{P(G \wedge F)}{P(F)} \leq u$. Note that $\frac{P(G \wedge F)}{P(F)}$ corresponds to the conditional probability of G given that F holds. Sometimes $P(F) > 0$ is demanded for all $(G|F)[l, u]\langle C\rangle \in \text{gr}(\theta)$. We will discuss this semantical variation later on.

Averaging Semantics. Intuitively, a conditional θ holds with probability $[l, u]$ under averaging semantics iff the average conditional probability of its ground instances is between l and u [Kern-Isberner and Thimm, 2010]. Formally, given a conditional $\theta \in (\mathcal{L}|\mathcal{L})$, a grounding operator gr and a probability function P, we say that P satisfies θ with respect to the grounding operator gr under averaging semantics iff

$$l \leq \frac{1}{|\text{gr}(\theta)|} \sum_{(G|F)[l,u]\langle C\rangle \in \text{gr}(\theta)} \frac{P(G \wedge F)}{P(F)} \leq u.$$

Aggregating Semantics. The aggregating semantics can be motivated statistically, see [Kern-Isberner and Thimm, 2010] and [Thimm et al., 2011] for a thorough discussion. Given a conditional $\theta \in (\mathcal{L}|\mathcal{L})$, a grounding operator gr and a probability function P, we say that P satisfies θ with respect to the grounding operator gr under aggregating semantics iff

$$l \leq \frac{\sum_{(G|F)[l,u]\langle C\rangle \in \text{gr}(\theta)} P(G \wedge F)}{\sum_{(G|F)[l,u]\langle C\rangle \in \text{gr}(\theta)} P(F)} \leq u.$$

There is some ambiguity here because no explicit precondition on the denominator has been made in [Kern-Isberner and Thimm, 2010]. The implicit constraint is that $\sum_{(G|F)[l,u]\langle C\rangle \in \text{gr}(\theta)} P(F) > 0$. This corresponds to demanding that there exists a $(G|F)[l, u]\langle C\rangle \in \text{gr}(\theta)$ such that $P(F) > 0$. However, we could also demand that this holds for all ground instances similarly to the variation of the grounding semantics. Finally, we could again multiply the inequality by the denominator to obtain a relaxed definition without the need to demand positive probability for any condition. We will discuss all three variants in the following.

Overview of Semantics. We saw three basic semantics for probabilistic relational conditional logics. Each of these semantics depends on a grounding operator. Additionally, we can consider variations of the grounding and aggregating semantics. The first variant demands that $P(F) > 0$ for all $(G|F)[l,u]\langle C\rangle \in \mathrm{gr}(\theta)$. We call this variant the *strict version* and regard it as the basic version because we also need this precondition for the averaging semantics. The *first relaxation* demands $P(F) > 0$ for a single $(G|F)[l,u]\langle C\rangle \in \mathrm{gr}(\theta)$. Finally, the *second relaxation* does not make any demands for the probability of the condition. We will now define satisfaction relations for all semantics and their variations. We will indicate the strict version by a 0 the first relaxation by a 1 and the second relation by a 2.

Definition 1 (Satisfaction Relations). *Let $\theta \in (\mathcal{L}|\mathcal{L})$, let gr be a grounding operator gr and let P be a probability function. We say that P satisfies θ with respect to gr under*

- *the strict grounding semantics, $P \models^{\mathrm{gr}}_{gnd\text{-}0} \theta$, iff*

$$l \cdot P(F) \leq P(G \wedge F) \leq u \cdot P(F) \text{ for all } (G|F)[l,u]\langle C\rangle \in \mathrm{gr}(\theta)$$
$$\text{and } P(F) > 0 \text{ for all } (G|F)[l,u]\langle C\rangle \in \mathrm{gr}(\theta)$$

- *the first relaxation of the grounding semantics, $P \models^{\mathrm{gr}}_{gnd\text{-}1} \theta$, iff*

$$l \cdot P(F) \leq P(G \wedge F) \leq u \cdot P(F) \text{ for all } (G|F)[l,u]\langle C\rangle \in \mathrm{gr}(\theta)$$
$$\text{and } P(F) > 0 \text{ for at least one } (G|F)[l,u]\langle C\rangle \in \mathrm{gr}(\theta)$$

- *the second relaxation of the grounding semantics, $P \models^{\mathrm{gr}}_{gnd\text{-}2} \theta$, iff*

$$l \cdot P(F) \leq P(G \wedge F) \leq u \cdot P(F) \text{ for all } (G|F)[l,u]\langle C\rangle \in \mathrm{gr}(\theta)$$

- *the (strict) averaging semantics, $P \models^{\mathrm{gr}}_{avg\text{-}0} \theta$, iff*

$$l \leq \frac{1}{|\mathrm{gr}(\theta)|} \sum_{(G|F)[l,u]\langle C\rangle \in \mathrm{gr}(\theta)} \frac{P(G \wedge F)}{P(F)} \leq u.$$

- *the strict aggregating semantics, $P \models^{\mathrm{gr}}_{agg\text{-}0} \theta$, iff*

$$l \cdot \sum_{(G|F)[l,u]\langle C\rangle \in \mathrm{gr}(\theta)} P(F) \leq \sum_{(G|F)[l,u]\langle C\rangle \in \mathrm{gr}(\theta)} P(G \wedge F) \leq u \cdot \sum_{(G|F)[l,u]\langle C\rangle \in \mathrm{gr}(\theta)} P(F)$$
$$\text{and } P(F) > 0 \text{ for all } (G|F)[l,u]\langle C\rangle \in \mathrm{gr}(\theta).$$

- *the first relaxation of the aggregating semantics,* $P \models_{agg\text{-}1}^{gr} \theta$, *iff*

$$l \cdot \sum_{(G|F)[l,u]\langle C \rangle \in \text{gr}(\theta)} P(F) \leqslant \sum_{(G|F)[l,u]\langle C \rangle \in \text{gr}(\theta)} P(G \wedge F) \leqslant u \cdot \sum_{(G|F)[l,u]\langle C \rangle \in \text{gr}(\theta)} P(F)$$

and $P(F) > 0$ for at least one $(G|F)[l,u]\langle C \rangle \in \text{gr}(\theta)$.

- *the second relaxation of the aggregating semantics,* $P \models_{agg\text{-}2}^{gr} \theta$, *iff*

$$l \cdot \sum_{(G|F)[l,u]\langle C \rangle \in \text{gr}(\theta)} P(F) \leqslant \sum_{(G|F)[l,u]\langle C \rangle \in \text{gr}(\theta)} P(G \wedge F) \leqslant u \cdot \sum_{(G|F)[l,u]\langle C \rangle \in \text{gr}(\theta)} P(F).$$

Remark 1. Note that all strict (first/second relaxations) semantics coincide if $\text{gr}(\theta)$ contains only a single conditional. This is true whenever the conditional is ground and, in particular, if the conditional is propositional. In this special case, there is also no difference between the first and second relaxation.

Definition 2 (Models, Consistency, Inconsistency). *Let $\models_S$ denote the satisfaction relation of some semantics S and let $\mathcal{K}$ be a knowledge base. The set*

$$\text{Mod}_S(\mathcal{K}) = \{P \in \mathcal{P} \mid P \models_S \theta \text{ for all } \theta \in \mathcal{K}\}$$

is called the set of models *of $\mathcal{K}$ under S. If $\text{Mod}_S(\mathcal{K}) \neq \emptyset$, $\mathcal{K}$ is called* consistent *under S and* inconsistent *under S otherwise.*

2.4 Reasoning with Relational Probabilistic Logics

The reasoning problem that we have in mind here is to compute probabilities for queries. To keep things simple, we will only consider ground queries. Hence, a query in our framework is a ground conditional $(G \mid F)$ and we are interested in deriving a meaningful conditional probability for this query. There are two popular approaches to derive probabilities for $(G \mid F)$ if the knowledge base is consistent. The first approach is to select a best model with respect to some quality criterion like maximum entropy [Paris and Vencovská, 1990] or minimum relative entropy with respect to a prior [Kern-Isberner, 2001]. Then this best model can be used to compute the conditional probability of the query (of course, things are more involved if we also allow non-ground queries). The second approach is to derive tight lower and upper bounds on the conditional probability of the query among all models of our knowledge base, see, e.g., [Hailperin, 1986], [Lukasiewicz, 1999] for some examples.

Model Selection. To evaluate the quality of probability functions, we consider some utility function $U : \mathcal{P} \to \mathbb{R}$. For instance, U could measure the entropy or U could measure the negative distance to a prior P_0 with respect to some metric or relative entropy, see [Paris and Vencovská, 1990] and [Kern-Isberner, 2001] for a thorough discussion of such approaches.

Definition 3 (Model Selection Problem). *Given some semantics $\mathcal{S}$, a consistent knowledge base $\mathcal{K}$ and a utility function U, the model selection problem is to compute*

$$P^* = \arg\max_{P \in \text{Mod}_\mathcal{S}(\mathcal{K})} U(P).$$

We call P^* the best model (with respect to $\mathcal{S}$ and U).

In order to guarantee that P^* is well-defined, we have to assure that a unique maximum of the optimization problem exists. Basic results from numerical optimization guarantee that this is true if $\text{Mod}_\mathcal{S}(\mathcal{K})$ is compact and convex and the utility function is continuous and strictly concave (like entropy or negative relative entropy and negative euclidean distance with respect to some prior). If the solution is not unique, we call each optimal probability function a best model.

Given the best model P^* and some ground query $(G|F)$, we can then compute $P^*(G \wedge F)$ and $P^*(F)$ to return the conditional probability $\frac{P^*(G \wedge F)}{P^*(F)}$ if $P^*(F) > 0$. If $P^*(F) = 0$, we can return the complete probability interval $[0,1]$ (we can conclude arbitrary things from false assumptions) or an error.

Probabilistic Entailment. The problem of deriving lower and upper bounds is often called probabilistic entailment [Jaumard et al., 1991].

Definition 4 (Probabilistic Entailment Problem). *Given some semantics $\mathcal{S}$, a consistent knowledge base $\mathcal{K}$ and a ground query $(G|F)$, the probabilistic entailment problem [Nilsson, 1986], [Jaumard et al., 1991] is to compute*

$$\text{opt}_{P \in \text{Mod}_\mathcal{S}(\mathcal{K})} \frac{P(G \wedge F)}{P(F)} \quad (P(F) > 0),$$

where $\text{opt} \in \{\inf, \sup\}$. If l is the lower and u the upper bound, the interval $[l, u]$ is called the result *(of solving the probabilistic entailment problem).*

If feasible, the minimization problem yields a tight lower and the maximization problem a tight upper bound on the conditional probability of G given F. Since we assume that the knowledge base is consistent, the optimization problem cannot be infeasible unless there is no $P \in \text{Mod}_\mathcal{S}(\mathcal{K})$ with $P(F) > 0$. In the latter case, we can return the complete probability interval $[0,1]$ like when answering a query with the best model or the empty interval $[1,0]$ to distinguish from the case, where there exists a $P \in \text{Mod}_\mathcal{S}(\mathcal{K})$ with $P(F) > 0$ but the bounds are $[0,1]$ anyway.

The probabilistic entailment problem is commonly solved by applying fractional programming techniques [Charnes and Cooper, 1962], [Hailperin, 1986]. The basic idea is to replace the normalization constraint $\sum_{\omega \in \Omega} P(\omega) = 1$ with the constraint $P(F) = 1$. Hence, the functions P that are considered in the optimization problem are not necessarily probability functions (they violate the normalization constraint, unless F is tautological). However, they can easily be rescaled to be probability functions by multiplying the scalar $s = \frac{1}{\sum_{\omega \in \Omega} P(\omega)}$. This is not even necessary though because in the fraction $\frac{P(G \wedge F)}{P(F)}$ each scalar of P cancels out. If we demand $P(F) = 1$, the objective function becomes just $P(G \wedge F)$, which is a linear function of P. Since linear functions are continuous, there are functions in $\text{Mod}_\mathcal{S}(\mathcal{K})$ that take the optimal values whenever $\text{Mod}_\mathcal{S}(\mathcal{K})$ is compact. If $\text{Mod}_\mathcal{S}(\mathcal{K})$ is also convex, the optimization problems are convex and therefore do not suffer from non-global local optima. In fact, if $\text{Mod}_\mathcal{S}(\mathcal{K})$ can be described by linear constraints, the optimization problems are linear. The latter is true for the second relaxation of grounding [Lukasiewicz, 1999] and aggregating semantics [Potyka, 2015].

Difficulty of reasoning problems under different semantics. We are interested in two computational problems here. Given a consistent knowledge base and a semantics, the first problem is to compute a best model with respect to some utility function. The second problem is to compute tight upper and lower bounds on the probability of a query. Simple sufficient conditions for both problems to be well-defined are usually that the model sets are compact and convex. This is true for the second relaxation of grounding and aggregating semantics [Potyka, 2015] (actually, this was shown only for point probabilities but the same argumentation applies to interval probabilities). Example 6.7, in [Thimm, 2011], shows that the model sets under averaging semantics can be non-convex. Example, 3.53 in [Potyka, 2015] shows that the averaging semantics and the strict grounding and aggregating semantics as well as their first relaxation can yield non-compact model sets (the example is propositional and therefore applies to all these semantics, c.f., Remark 1).

An interesting empirical observation is that the optimization problems under aggregating semantics are often solved faster than the corresponding problems under grounding semantics. The reason is most likely that the number of constraints under aggregating semantics can be significantly smaller (the aggregating semantics yields one constraint for each conditional, whereas the grounding semantics yields one constraint for each ground instance of each conditional).

Taking these observations into account, it is interesting to investigate if relationships between different semantics can be exploited to reduce the reasoning problem under one semantics to another semantics or to approximate the solution of the reasoning problem under one semantics by using another semantics. In this direction, it was already noted in [Kern-Isberner and Thimm, 2010] that the aggregating and averaging semantics coincide for probabilistic facts and in [Beierle et al., 2015] that the maximum entropy model under strict grounding and aggregating semantics coincides under certain preconditions.

3 Ordering of Semantics and Implications

In this section, we will introduce an ordering on semantics and explain what the ordering can tell us about reasoning results.

Definition 5 (Stronger, Weaker Semantics). *Let* $\models_{\mathcal{S}_1}, \models_{\mathcal{S}_2}$ *denote the satisfaction relation of some semantics* $\mathcal{S}_1, \mathcal{S}_2$. *We say that* $\mathcal{S}_1$ *is stronger than* $\mathcal{S}_2$, *written as* $\mathcal{S}_1 \leq \mathcal{S}_2$ *iff* $\models_{\mathcal{S}_1} \subseteq \models_{\mathcal{S}_2}$. *If the subset relationship is proper,* $\models_{\mathcal{S}_1} \subset \models_{\mathcal{S}_2}$, we say that $\mathcal{S}_1$ is strictly stronger than $\mathcal{S}_2$ and write $\mathcal{S}_1 \prec \mathcal{S}_2$. Furthermore, if $\mathcal{S}_1 \preceq \mathcal{S}_2$ ($\mathcal{S}_1 \prec \mathcal{S}_2$), we say that $\mathcal{S}_2$ is (strictly) weaker than $\mathcal{S}_1$.

Remark 2. Note that $\models_{\mathcal{S}_1} \subseteq \models_{\mathcal{S}_2}$ means that $P \models_{\mathcal{S}_1} \mathcal{K}$ implies that $P \models_{\mathcal{S}_2} \mathcal{K}$. That is, if $\mathcal{K}$ is satisfied by P under the stronger semantics, it is also satisfied under the weaker semantics. In particular, if $\mathcal{S}_1 \preceq \mathcal{S}_2$ ($\mathcal{S}_1 \prec \mathcal{S}_2$), then $\mathrm{Mod}_{\mathcal{S}_1}(\mathcal{K}) \subseteq \mathrm{Mod}_{\mathcal{S}_2}(\mathcal{K})$ ($\mathrm{Mod}_{\mathcal{S}_1}(\mathcal{K}) \subset \mathrm{Mod}_{\mathcal{S}_2}(\mathcal{K})$).

We now state some relationships between stronger and weaker semantics. If we are interested in solving reasoning problems for different semantics that can be ordered, the following proposition shows that it suffices to test consistency for the strongest semantics.

Proposition 1 (Satisfiability under weaker semantics). *If $\mathcal{S}_1 \preceq \mathcal{S}_2$ and $\mathcal{K}$ is consistent under $\mathcal{S}_1$, then $\mathcal{K}$ is consistent under $\mathcal{S}_2$.*

Proof. The claim follows immediately from $\mathrm{Mod}_{\mathcal{S}_1}(\mathcal{K}) \subseteq \mathrm{Mod}_{\mathcal{S}_2}(\mathcal{K})$ (Remark 2). □

The next proposition shows that if we computed a best model under one semantics, and it happens to be also a model under a stronger semantics, than it is a best model under the stronger semantics as well.

Proposition 2 (Best model under stronger semantics). *Suppose that $\mathcal{S}_1 \preceq \mathcal{S}_2$ and that $\mathcal{K}$ is consistent under $\mathcal{S}_2$. Let P^* be a best model of $\mathcal{K}$ with respect to some utility function U under $\mathcal{S}_2$. If $P^* \models_{\mathcal{S}_1} \mathcal{K}$, then P^* is a best model of $\mathcal{K}$ with respect to U under $\mathcal{S}_1$.*

Proof. For the sake of contradiction, suppose that P^* is not a best model of $\mathcal{K}$ under $\mathcal{S}_1$. Then there is a better model $P_1 \in \mathrm{Mod}_{\mathcal{S}_1}(\mathcal{K})$. But since $\mathrm{Mod}_{\mathcal{S}_1}(\mathcal{K}) \subseteq \mathrm{Mod}_{\mathcal{S}_2}(\mathcal{K})$, also $P_1 \in \mathrm{Mod}_{\mathcal{S}_2}(\mathcal{K})$ contradicting optimality of P^* under $\mathcal{S}_2$. □

Remark 3. Note that we do not need to assume that the best model is unique. However, if it is unique under $\mathcal{S}_2$ and the assumption of the proposition applies, then it is also unique under $\mathcal{S}_1$ (since $\mathrm{Mod}_{\mathcal{S}_1}(\mathcal{K}) \subseteq \mathrm{Mod}_{\mathcal{S}_2}(\mathcal{K})$).

Finally, the following proposition states that probabilistic entailment results under one particular semantics bound the results for all stronger semantics.

Proposition 3 (Probabilistic entailment under stronger semantics). *Suppose that $\mathcal{S}_1 \preceq \mathcal{S}_2$ and that $\mathcal{K}$ is consistent under $\mathcal{S}_1$. Let q be some query for the probabilistic entailment problem and suppose that $[l_i, u_i]$ is the entailment result under $\mathcal{S}_i$, $i = 1, 2$. Then $[l_1, u_1] \subseteq [l_2, u_2]$.*

Proof. The claim follows immediately from $\mathrm{Mod}_{\mathcal{S}_1}(\mathcal{K}) \subseteq \mathrm{Mod}_{\mathcal{S}_2}(\mathcal{K})$ and the definition of the probabilistic entailment problem. □

Remark 4. Note that if we regard the entailment result under $\mathcal{S}_2$ as an approximation to the entailment result under $\mathcal{S}_1$, then the absolute error cannot be larger than $u_2 - l_2$ and the relative error cannot be larger than $\frac{u_2 - l_2}{l_2}$. Hence, we get upper bounds on the approximation error for free. In particular, if $\mathcal{S}_2$ yields a point probability ($l_2 = u_2$), the same probability has to be valid under $\mathcal{S}_1$ (this is often true if our knowledge base contains only point probabilities).

4 Some Ordering Results

In this section, we present some results on the order of the introduced semantics for relational probabilistic logics. As the name suggests, strict semantics are strictly stronger than their first relaxations, which, in turn, are strictly stronger than the second relaxations.

Proposition 4 (Strict < First Relaxation < Second Relaxation). *We denote the grounding semantics and the aggregating semantics with respect to some grounding operator gr by gnd_{gr} and agg_{gr}, respectively.*

1. $gnd_{gr}\text{-}0 < gnd_{gr}\text{-}1 < gnd_{gr}\text{-}2$
2. $agg_{gr}\text{-}0 < agg_{gr}\text{-}1 < agg_{gr}\text{-}2$

Proof. $gnd_{gr}\text{-}0 \leq gnd_{gr}\text{-}1 \leq gnd_{gr}\text{-}2$ and $agg_{gr}\text{-}0 \leq agg_{gr}\text{-}1 \leq agg_{gr}\text{-}2$ follow immediately from the fact that the strict semantics adds a stronger precondition (for all) than the first relaxation (exists) and that the second relaxation omits this precondition completely. To show that the relationships are strict, we consider two propositional examples.

Let $\Sigma_1 = (\emptyset, \emptyset, \{A, B\})$ and $\mathcal{K}_1 = \{(B|A)[0.5], (A)[0]\}$. Then $\mathcal{K}_1$ is inconsistent with respect to the first relaxation, but consistent with respect to the second relaxation of both the aggregating and grounding semantics (each probability function with $P(A) = 0$ satisfies $\mathcal{K}_1$ under the second relaxation). This shows that the second relationship is strict.

To show that the first relationship is also strict, consider the signature $\Sigma_2 = (\{s\}, \{a, b\}, \{Q(s), R(s)\})$ and let $\mathcal{K}_2 = \{(Q(X)|R(X))[0.5], (R(a))[0]\}$. Then $\mathcal{K}_2$ is inconsistent with respect to the strict semantics, but consistent with respect to the first relaxation (now we need $P(R(a)) = 0$, $P(R(b)) > 0$ and $\frac{P(Q(b) \wedge R(b))}{P(R(b))} = 0.5$). □

Taking into account the results from the last section, we can see that we can approximate the strict and first relaxation of grounding and aggregating semantics by using the computationally convenient second relaxation. We will further strengthen this consequence in the next section.

Next, we show that the grounding semantics is strictly stronger than both the aggregating semantics and the averaging semantics.

Proposition 5 (Grounding < Averaging, Grounding < Aggregating). *We denote the grounding semantics, the averaging semantics and the aggregating semantics with respect to some grounding operator gr by gnd_{gr}, avg_{gr} and agg_{gr}, respectively.*

1. $gnd_{gr}\text{-}0 < avg_{gr}\text{-}0$
2. $gnd_{gr}\text{-}0 < agg_{gr}\text{-}0$
3. $gnd_{gr}\text{-}1 < agg_{gr}\text{-}1$
4. $gnd_{gr}\text{-}2 < agg_{gr}\text{-}2$

Proof. We first prove the $\leq$-part and show afterwards that the relationship is strict. Suppose that $P \models^{gr}_{gnd\text{-}0} \theta$. That is,

$$l \cdot P(F) \leq P(G \wedge F) \leq u \cdot P(F) \text{ for all } (G|F)[l, u]\langle C \rangle \in \text{gr}(\theta)$$
$$\text{and } P(F) > 0 \text{ for all } (G|F)[l, u]\langle C \rangle \in \text{gr}(\theta)$$

Let us first look at the averaging semantics. Note that our assumption guarantees $l \leq \frac{P(G \wedge F)}{P(F)} \leq u$ for all $(G|F)[l,u]\langle C \rangle \in \text{gr}(\theta)$. Hence,

$$\frac{1}{|\text{gr}(\theta)|} \sum_{(G|F)[l,u]\langle C \rangle \in \text{gr}(\theta)} \frac{P(G \wedge F)}{P(F)} \leq \frac{1}{|\text{gr}(\theta)|} \sum_{(G|F)[l,u]\langle C \rangle \in \text{gr}(\theta)} u = u$$

and

$$\frac{1}{|\text{gr}(\theta)|} \sum_{(G|F)[l,u]\langle C \rangle \in \text{gr}(\theta)} \frac{P(G \wedge F)}{P(F)} \geq \frac{1}{|\text{gr}(\theta)|} \sum_{(G|F)[l,u]\langle C \rangle \in \text{gr}(\theta)} l = l.$$

Hence, $P \models^{\text{gr}}_{avg\text{-}0} \theta$, which proves the $\leq$-part of 1.

For the aggregating semantics, we obtain

$$\sum_{(G|F)[l,u]\langle C \rangle \in \text{gr}(\theta)} P(G \wedge F) \leq \sum_{(G|F)[l,u]\langle C \rangle \in \text{gr}(\theta)} u \cdot P(F) = u \cdot \sum_{(G|F)[l,u]\langle C \rangle \in \text{gr}(\theta)} P(F)$$

and

$$\sum_{(G|F)[l,u]\langle C \rangle \in \text{gr}(\theta)} P(G \wedge F) \geq \sum_{(G|F)[l,u]\langle C \rangle \in \text{gr}(\theta)} l \cdot P(F) = l \cdot \sum_{(G|F)[l,u]\langle C \rangle \in \text{gr}(\theta)} P(F).$$

Hence, $P \models^{\text{gr}}_{agg\text{-}0} \theta$, which proves the $\leq$-part of 2. The proofs of 3 and 4 are analogous and are therefore left out.

It remains to show that the relationships are strict. To see this, let $\Sigma = (\{s\}, \{a,b\}, \{Q(s)\})$ and $\mathcal{K} = \{(Q(x))[0.8], (Q(a))[0.9], (Q(b))[0.7]\}$. Then $\mathcal{K}$ is inconsistent with respect to each variant of the grounding semantics, but consistent with respect to the averaging semantics and each variant of the aggregating semantics. □

Again, note that this result means that we can use the aggregating semantics to approximate the grounding semantics as explained in the foregoing section. However, there is no point in using the averaging semantics for this purpose because it is the computationally most difficult one among our semantics. Though we still get the interesting fact that satisfiability under grounding semantics implies satisfiability under averaging semantics.

Figure 1 summarizes the ordering of our semantics.

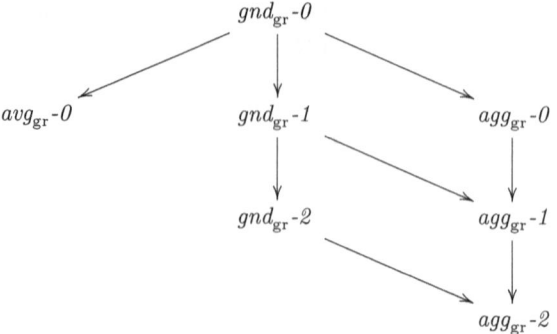

Fig. 1. Ordering of semantics. An edge (S_1, S_2) means that $S_1 \prec S_2$.

5 Relationships between strict and relaxed semantics

In this section, we will investigate the relationship between strict semantics and their relaxations in more detail. We will first show that the models under strict semantics and its first relaxation are topologically close to the models under the second relaxation. As we show afterwards, this implies that reasoning results under all three variations are equal under very mild preconditions.

In order to address these questions rigorously, we need some additional terminology. To combine probability functions, we will regard $\mathcal{P}$ as a subset of the vector space $\widehat{\mathcal{P}} = \{f : \Omega \to \mathbb{R}\}$, where $(f + g)(\omega) = f(\omega) + g(\omega)$ and $(c \cdot f)(\omega) = c \cdot f(\omega)$ for all $c \in \mathbb{R}, f, g \in \widehat{\mathcal{P}}$. It is easy to check that $\mathcal{P}$ is a convex subset of $\widehat{\mathcal{P}}$, i.e., it is closed under convex combinations. To talk about the distance between probability functions, it is convenient to employ the 1-norm, which is defined by $\|f\|_1 = \sum_{\omega \in \Omega} |f(\omega)|$. Using the 1-norm does not mean any loss of generality for our investigation because in finite-dimensional vector spaces all norms are equivalent in the sense that there is no difference in the induced neighborhoods. Using the 1-norm, we have

$$|P_1(F) - P_2(F)| \leq \sum_{\omega \models F} |P_1(\omega) - P_2(\omega)| \leq \|P_1 - P_2\|_1.$$

The interesting implication that we will use in the following is that if two semantics induce arbitrarily close models, they will also induce arbitrarily close reasoning results.

Let us quickly recall some topological definitions. An ϵ-ball in $S \subseteq \widehat{\mathcal{P}}$ centered at $f \in S$ is the set $\{g \in S \mid \|f - g\|_1 < \epsilon\}$. A neighborhood of f in S is a subset of S that is a superset of an ϵ-ball centered at f. $g \in \widehat{\mathcal{P}}$ is called a limit point of S if every neighborhood of g contains a point from S. The topological closure of S is obtained by adding all its limit points to S. Note that even though $\mathcal{P}$ is not a vector space, it is a topological space in its own right. In the following, all topological notions are meant with respect to $\mathcal{P}$.

How far are the models under strict semantics and their first relaxation from the models under the second relaxation? Not very far because, geometrically speaking, the strict and first relaxation exclude only functions on the boundary of $\mathcal{P}$ (probability functions with zero probabilities). This is made more precise in the following proposition.

Proposition 6 (Topological closure of strict and first relaxation corresponds to second relaxation). *Let $\mathcal{K}$ be a knowledge base and let $\mathcal{S} \in \{gnd_{gr}, agg_{gr}\}$.*

1. *If $\mathrm{Mod}_{\mathcal{S}\text{-}0}(\mathcal{K}) \neq \emptyset$, the topological closure of $\mathrm{Mod}_{\mathcal{S}\text{-}0}(\mathcal{K})$ is $\mathrm{Mod}_{\mathcal{S}\text{-}2}(\mathcal{K})$.*
2. *If $\mathrm{Mod}_{\mathcal{S}\text{-}1}(\mathcal{K}) \neq \emptyset$, the topological closure of $\mathrm{Mod}_{\mathcal{S}\text{-}1}(\mathcal{K})$ is $\mathrm{Mod}_{\mathcal{S}\text{-}2}(\mathcal{K})$.*

Proof. Let us pick $\mathrm{Mod}_{gnd_{gr}\text{-}0}(\mathcal{K})$ and show that its topological closure is equal to $\mathrm{Mod}_{gnd_{gr}\text{-}2}(\mathcal{K})$. First, let P be in the topological closure of $\mathrm{Mod}_{gnd_{gr}\text{-}0}(\mathcal{K})$. If $P \in \mathrm{Mod}_{gnd_{gr}\text{-}0}(\mathcal{K})$, then also $P \in \mathrm{Mod}_{gnd_{gr}\text{-}2}(\mathcal{K})$ and we are done. Otherwise, P is a limit point of $\mathrm{Mod}_{gnd_{gr}\text{-}0}(\mathcal{K})$. Hence, there is a sequence (P_n) in $\mathrm{Mod}_{gnd_{gr}\text{-}0}(\mathcal{K})$ that converges to P. But then (P_n) is also a sequence in $\mathrm{Mod}_{gnd_{gr}\text{-}2}(\mathcal{K})$ and since $\mathrm{Mod}_{gnd_{gr}\text{-}2}(\mathcal{K})$ is closed [Potyka, 2015], we have that $P \in \mathrm{Mod}_{gnd_{gr}\text{-}2}(\mathcal{K})$.

Conversely, suppose that $P \in \mathrm{Mod}_{gnd_{gr}\text{-}2}(\mathcal{K})$. Since $\mathrm{Mod}_{gnd_{gr}\text{-}0}(\mathcal{K}) \neq \emptyset$ by assumption, there is a $P' \in \mathrm{Mod}_{\mathcal{S}\text{-}0}(\mathcal{K})$. Also $P' \in \mathrm{Mod}_{\mathcal{S}\text{-}2}(\mathcal{K})$ and by convexity of $\mathrm{Mod}_{gnd_{gr}\text{-}2}(\mathcal{K})$ [Potyka, 2015], $P_t = t \cdot P + (1-t) \cdot P'$ is in $\mathrm{Mod}_{gnd_{gr}\text{-}2}(\mathcal{K})$ for all $t \in (0,1)$. In particular $P_t(\omega) > 0$ whenever $P'(\omega) > 0$. Hence, also $P_t \in \mathrm{Mod}_{gnd_{gr}\text{-}0}(\mathcal{K})$. But then $(P_{\frac{1}{n}})_{n \in \mathbb{N}}$ is a sequence in $\mathrm{Mod}_{gnd_{gr}\text{-}0}(\mathcal{K})$ that converges to P. Hence, P is a limit point of $\mathrm{Mod}_{gnd_{gr}\text{-}0}(\mathcal{K})$ and therefore is in the topological closure of $\mathrm{Mod}_{gnd_{gr}\text{-}0}(\mathcal{K})$.

This completes the proof for gnd_{gr}-0. The other cases can be proved analogously. □

Consequently, we should expect that entailment results under the second relaxation cannot be far from the entailment results under the first relaxation and strict semantics. The following proposition states that they are in fact equal in case that the condition of the query is not impossible with respect to our knowledge base.

Proposition 7 (Equal entailment results under strict, first and second relaxation). *Let $\mathcal{K}$ be a knowledge base, let $(G \mid F)$ be a ground query and let $\mathcal{S} \in \{gnd_{gr}, agg_{gr}\}$. Assume $\mathrm{Mod}_{\mathcal{S}\text{-}0}(\mathcal{K}) \neq \emptyset$ and that there is a $P \in \mathrm{Mod}_{\mathcal{S}\text{-}0}(\mathcal{K})$ such that $P(F) > 0$. For $i = 1, 2, 3$, let $[l_i, u_i]$ denote entailment results under $\mathcal{S}$-i. Then $[l_0, u_0] = [l_1, u_1] = [l_2, u_2]$.*

Proof. Let us pick $\mathrm{Mod}_{gnd_{gr}\text{-}0}(\mathcal{K})$. We already know that $[l_0, u_0] \subseteq [l_2, u_2]$ because gnd_{gr}-0 is stronger than gnd_{gr}-2. Hence, it remains to show that $[l_2, u_2] \subseteq [l_0, u_0]$. It suffices to show that l_2 (u_2) is the infimum (supremum) of the entailment problem under gnd_{gr}-0. We consider only l_2 (the argumentation for u_2 is analogously). Since $\mathrm{Mod}_{gnd_{gr}\text{-}2}(\mathcal{K})$ is closed, there is an optimal point taking the infimum, that is, a $P \in \mathrm{Mod}_{gnd_{gr}\text{-}2}(\mathcal{K})$ such that $P(F) > 0$ and $\frac{P(G \wedge F)}{P(F)} = l_2$. If also $P \in \mathrm{Mod}_{gnd_{gr}\text{-}0}(\mathcal{K})$, we are done. Otherwise, P must be a limit point of $\mathrm{Mod}_{gnd_{gr}\text{-}0}(\mathcal{K})$ by our foregoing discussion. Hence, there is a sequence (P_n) of probability functions in $\mathrm{Mod}_{gnd_{gr}\text{-}0}(\mathcal{K})$ that converges to P.

But then $P_n(G \wedge F)$ converges to $P(G \wedge F)$ and $P_n(F)$ converges to $P(F)$. Therefore, $\frac{P_n(G \wedge F)}{P_n(F)}$ converges to $\frac{P(G \wedge F)}{P(F)} = l_2$. We already know that $l_0 \geq l_2$ (otherwise $[l_0, u_0] \subseteq [l_2, u_2]$ cannot be true), hence l_2 must be the infimum of the entailment problem under gnd_{gr}-0.

This completes the proof for gnd_{gr}-0. The remaining cases can be proved analogously. □

What about the model selection problem? If we consider a continuous utility function, there are only two possible cases. Either the best probability function under the stronger semantics is not attained or it corresponds to the solution under the second relaxation. This is explained more thoroughly in the following proposition.

Proposition 8 (Model selection problem under strict, first and second relaxation). *Let $\mathcal{K}$ be a knowledge base, let U be a continuous utility function and let $\mathcal{S} \in \{gnd_{gr}, agg_{gr}\}$.*

1. *If $\mathrm{Mod}_{\mathcal{S}-0}(\mathcal{K}) \neq \emptyset$ and the model selection problem under $\mathcal{S}$-2 has a unique solution, then the solution under $\mathcal{S}$-0 either does not exist or corresponds to the solution under $\mathcal{S}$-2.*
2. *If $\mathrm{Mod}_{\mathcal{S}-1}(\mathcal{K}) \neq \emptyset$ and the model selection problem under $\mathcal{S}$-2 has a unique solution, then the solution under $\mathcal{S}$-1 either does not exist or corresponds to the solution under $\mathcal{S}$-2.*

Proof. Let us pick $\mathrm{Mod}_{gnd_{gr}-0}(\mathcal{K})$. If there is no best solution under $\mathcal{S}$-0 or if the best model under $\mathcal{S}$-2 is in $\mathrm{Mod}_{\mathcal{S}-0}(\mathcal{K})$, we are done. So let us suppose that the best solution P_0 under $\mathcal{S}$-0 does exist and does not coincide with the best solution P_2 under $\mathcal{S}$-2. Then $U(P_0) < U(P_2)$ by unique maximality of P_2. We know that P_2 is a limit point of $\mathrm{Mod}_{\mathcal{S}-0}(\mathcal{K})$. Hence, there is a sequence (P_n) in $\mathrm{Mod}_{\mathcal{S}-0}(\mathcal{K})$ that converges to P_2. By continuity of U, $(U(P_n))$ converges to to $U(P_2)$. Hence, there is a $m \in \mathbb{N}$ such that $U(P_2) - U(P_m) < U(P_2) - U(P_0)$, i.e., $U(P_m) > U(P_0)$ contradicting optimality of P_0. This completes the proof for gnd_{gr}-0. The other cases can be proved analogously. □

Note that, in particular, entropy, relative entropy with respect to a prior and p-norms with respect to a prior yield continuous utility functions.

6 Conclusions

We thoroughly investigated the relationships between the grounding, averaging and aggregating semantics and their variations. This investigation is arguably interesting in its own right, but there are also at least two practical implications. First, we can use the aggregating semantics to approximate the grounding semantics. This is particularly interesting for the probabilistic entailment problem because we get bounds on the approximation error for free. Second, we showed that there is often no difference in the reasoning results dependent on whether we use the strict, first or second relaxation of semantics. Hence, allowing zero probabilities for the condition in implementations (this is often unavoidable) is formally justified in many cases even when excluding them in the formal definition of the probabilistic logic.

References

[Beierle et al., 2015] Beierle, C., Finthammer, M., and Kern-Isberner, G. (2015). Relational probabilistic conditionals and their instantiations under maximum entropy semantics for first-order knowledge bases. *Entropy*, 17(2):852–865.

[Charnes and Cooper, 1962] Charnes, A. and Cooper, W. W. (1962). Programming with linear fractional functionals. *Naval Research logistics quarterly*, 9(3-4):181–186.

[Ebbinghaus et al., 1996] Ebbinghaus, H., Flum, J., and Thomas, W. (1996). *Mathematical Logic*. Undergraduate Texts in Mathematics. Springer New York.

[Fisseler, 2012] Fisseler, J. (2012). First-order probabilistic conditional logic and maximum entropy. *Logic Journal of the IGPL*, 20(5):796–830.

[Grove et al., 1994] Grove, A., Halpern, J., and Koller, D. (1994). Random worlds and maximum entropy. *J. of Artificial Intelligence Research*, 2:33–88.

[Hailperin, 1986] Hailperin, T. (1986). *Boole's Logic and Probability*, volume 85 of *Studies in Logic and the foundations of mathematics*. Elsevier, 2nd enlarged edition.

[Halpern, 1990] Halpern, J. Y. (1990). An analysis of first-order logics of probability. *Artificial Intelligence*, 46:311–350.

[Jaumard et al., 1991] Jaumard, B., Hansen, P., and Poggi, M. (1991). Column generation methods for probabilistic logic. *ORSA - Journal on Computing*, 3(2):135–148.

[Kern-Isberner, 2001] Kern-Isberner, G. (2001). *Conditionals in nonmonotonic reasoning and belief revision*. Springer, Lecture Notes in Artificial Intelligence LNAI 2087.

[Kern-Isberner and Thimm, 2010] Kern-Isberner, G. and Thimm, M. (2010). Novel semantical approaches to relational probabilistic conditionals. In Lin, F., Sattler, U., and Truszczynski, M., editors, *Proceedings Twelfth International Conference on the Principles of Knowledge Representation and Reasoning, KR'2010*, pages 382–391. AAAI Press.

[Loh et al., 2010] Loh, S., Thimm, M., and Kern-Isberner, G. (2010). On the problem of grounding a relational probabilistic conditional knowledge base. In *Proceedings of the 14th International Workshop on Non-Monotonic Reasoning (NMR'10)*, Toronto, Canada.

[Lukasiewicz, 1999] Lukasiewicz, T. (1999). Probabilistic deduction with conditional constraints over basic events. *Journal of Artificial Intelligence Research*, 10:380–391.

[Nilsson, 1986] Nilsson, N. J. (1986). Probabilistic logic. *Artificial Intelligence*, 28:71–88.

[Paris and Vencovská, 1990] Paris, J. and Vencovská, A. (1990). A note on the inevitability of maximum entropy. *International Journal of Approximate Reasoning*, 14:183–223.

[Potyka, 2015] Potyka, N. (2015). *Solving Reasoning Problems for Probabilistic Conditional Logics with Consistent and Inconsistent Information*. PhD thesis, Fernuniversität Hagen, Germany.

[Thimm, 2011] Thimm, M. (2011). *Probabilistic Reasoning with Incomplete and Inconsistent Beliefs*. PhD thesis, Technische Universität Dortmund, Germany.

[Thimm et al., 2011] Thimm, M., Kern-Isberner, G., and Fisseler, J. (2011). Relational probabilistic conditional reasoning at maximum entropy. In Liu, W., editor, *Proceedings 11th European Conference on Symbolic and Quantitative Approaches to Reasoning with Uncertainty, ECSQARU'11*, volume 6717 of *LNCS*, pages 447–458. Springer.

A New Rationality in Network Analysis - Status of Actors in a Conditional-logical Framework

Wilhelm Rödder[1], Friedhelm Kulmann[2], Andreas Dellnitz[3]

Abstract. An up-and-coming approach of Social Network Analysis considers conditional-logical rather than graphical structures. Links in a graph now become conditional propositions: if an actor has a message or an immaterial good, then also his neighbors will have it. Recent literature considers certain conditionals, only. This paper generalizes twofold. It shows that attenuated links all over the network are possible as long as the attenuation coefficient is constant. With varying coefficients the conditional structure might or might not be useful for the social fabric. In either case the Expert System Shell SPIRIT supports such analyses. Under this conditional-logical framework the statuses of actors experience new and meaningful interpretations. A middle size network illustrates the results.

1 Introduction

The Social Network (SN) as scientific object of investigation is much older than the internet. Sociologists studied such nets to find out why some groups of people in a society are successful and others are not. Why some companies dominate others and why networking perhaps is more important than knowledge, motivation, diligence or even all three of them. It was Émile Durkheim [4] who said that the whole is more than the sum of its parts. There is something intrinsic in a network which the members draw benefits from. And it was this "something" which made sociologist curious and inquisitive about this phenomenon. In the early 30s of the last century, Jakob Moreno introduced sociograms, i.e. graph theoretical representations of networks and paved the way for a more mathematical admittance to the subject, cf. [8].

A net is a set of actors and their relations. The actors then are nodes or vertices in a (hyper-)graph and the relations are (hyper-)edges or -links. For a good introduction into the concept see Scott [15] and for the more mathematical oriented reader we recommend Newman [9]. Both authors provide an overwhelming number of references in their bibliographies. The main purpose of such graphical representation of SN is the characterization of actors or groups of actors with respect to their centrality, embeddedness, betweeness or their roles in the social fabric, cf. again [15] and [9].

---

[1] FernUniversität in Hagen, Department of Operations Research, P.O. Box 940, D-58084 Hagen
[2] FernUniversität in Hagen, Department of Operations Research, P.O. Box 940, D-58084 Hagen
[3] FernUniversität in Hagen, Department of Operations Research, P.O. Box 940, D-58084 Hagen

In the early 1950s, Katz proposed a status index of actors in a Social Network, see [5]. If an actor receives choices from others he has a high status and this the more the more votes he gets. On the other hand there might be votes from more distant voters in the network and this makes their contribution less valuable. All in all the status index of an actor is a sum of attenuated votes from others all over the network. More precisely, an actor's status index is equal to a sum of all attenuated path distances from the other actors. The idea was improved by Bonacich [1] and finally by Bonacich and Lloyd [2], and modern Social Network Analysis (SNA) tools help the user to calculate such indices. Perhaps the equation most established today is the one developed by Bonacich and Lloyd, cf. again [2]. This equation reads

$$\mathbf{x} = (\mathbf{I} - \alpha \mathbf{A}^\mathsf{T})^{-1} \mathbb{1} \qquad (1)$$

Here $\mathbf{A}$ is the adjacency matrix of the net with $a_{ij} = 1$ iff actor i points to or chooses actor j, α is the attenuation coefficient, $\mathbb{1}^\mathsf{T} = (1,...,1)$ a unity vector, and finally is $\mathbf{x}$ the vector of all actors' status indices.

Katz in his paper says that such attenuation might be rethought as "the force of a probability of effectiveness of a single link", cf. [5, p. 41]. And he even questions whether these probabilities are the same on all links.

In recent years a new concept of analyzing networks is coming up. Rödder and Kulmann [12] presented a method to simulate cognitive processes in a group of actors and Rödder et al. [11] related on an entropy based evaluation of net structures. In either case the arrows of a directed graph are substituted by conditional propositions: if-then. In such a context if-then rules or conditionals represent transfers of attitudes, messages or (im-)material goods from actors to actors, rather than votes like in Katz's contribution. This immediately leads to the question: What is an actor's capability to relaunch a message throughtout the whole net or what is his capability to receive messages from others – diffusion and reception. The authors in [11] study this issue and show diffusion and reception to measure in the information theoretical unit bit. Unfortunately these authors only study certain conditionals; attenuation or uncertainty of a message or attitude to be launched to or to be received from others is not examined.

In this paper we generalize and permit probabilities – or p-attenuation –, respectively. With these probabilities we construct a global probability distribution on the set of all possible states of the network. Once such a distribution is available the analyst has a very rich instrument for evaluating the net. What is each actor's status: How strong is his diffusion and his reception potential, even if transfers are not certain. What is the probability that an actor receives the message from a particular sender. What are these probabilities if actor i or actor j launches a message, what if actor i sends but actor j denies it. The availability of the global distribution allows for answers to umpteen questions which mere graph theoretical analyses are not able to supply.

The scientific basis for such answers is a profound analysis of conditional logics under maximum entropy (MaxEnt) and minimum relative entropy (MinREnt), respectively. These principles find their axiomatic justifications in [3], [16] and [6]. Especially the last contribution of Kern-Isberner "Charaterizing the principle of minimum cross-entropy within a conditional-logical framework" opened the way for the implementation of an expert system shell, named SPIRIT [17]. The shell supports knowledge processing in such conditional-logical framework.

To present the new method we need probabilistic and information theoretical preliminaries given in section 2. Some effort must be taken to prove the existence of a global distribution on the net meeting our requirements; that is what section 3 is about. Section 4 then shows how this distribution helps to analyze networks. In 4.1 we do this for little nets of 2, 3 or 4 actors just to get familiar with the new method. Section 4.2 then applies it to a middle size network. In section 4 the respective probabilities are equal on all links in the network (and in section 3 we proved a global distribution to exist for this special case). When transfer probabilities vary from transfer link to transfer link, a proof of the existence of such a useful distribution for our purpose is still missing. But at least the expert system shell SPIRIT helps to check the existence or non existence of useful distributions empirically. So in section 4.3 we show networks in SPIRIT and test whether they allow for such distributions or not. Section 5 then is a user-friendly survey of instruments for calculating indices and for characterizing actors in SPIRIT: diffusion and reception of an actor and his embeddedness in the net structure as well as actors' chances to receive messages from others. First advices are given how such indices can be calculated even for complex queries: What are diffusion, reception and embeddedness for certain groups of actors and how might these groups be composed. Section 6 is a resume of our work and points to future research.

2 Preliminaries

2.1 Syntax and knowledge processing on networks

Knowledge acquisition about a social network's probabilistic structure needs

- a set of n actors $a_1, \ldots, a_n$,
- a set of n binary variables $\{V_1, \ldots, V_n\}$ with values $V_i = v_i$ and $v_i = i/\bar{i}$
  - being $\boldsymbol{v} = (v_1, \ldots, v_n)$ respective configurations,
- a set of conditionals $V_l = l \mid V_k = k$ – or $l \mid k$ for short – for some $k \neq l \in \{1, \ldots, n\}$,
- a set of conditional probabilities p_{kl} for those $k \neq l \in \{1, \ldots, n\}$.

The semantics of $V_i = i/\bar{i}$ is that of a proposition. Actor a_i knows the message or has a certain attitude (i) or not ($\bar{i}$). The conditionals togehter with the probabilities shape links between actors: If actor a_k knows the message then actor a_l also does, with probability p_{kl}. The selected notation for propositions and conditionals – $v_i = i/\bar{i}$ instead of, e.g., $V_i = yes/no$ and $l \mid k$ instead of $V_l = yes \mid V_k = yes$ – permits a compact form of mathematical and logical developments throughout this paper.

Now we search for a distribution Q on $\mathcal{V} = \{\boldsymbol{v}\}$ which complies with the p_{kl} and for which $0 < Q(V_i = i) < 1$ for all $i = 1, \ldots, n$; such a Q we call useful. The second requirement prohibits any actor to be absolut ignorant or absolut certain about the message under Q. Why this demand is useful for our purpose, will be discussed later. The p_{kl} must be chosen carefully, as shows the following example. To make things transparent we sometimes use self-explanatory graphical representations.

Example 1

i) For two actors a_1, a_2 with the four configurations or states $1\ 2$, $1\ \bar{2}$, $\bar{1}\ 2$, $\bar{1}\ \bar{2}$ and for the desired conditional probabilities $Q(2 \mid 1) = p_{12} = 1$, $Q(1 \mid 2) = p_{21} = 0$, any Q on the states must comply $Q(1) = 0$ and hence is not useful.

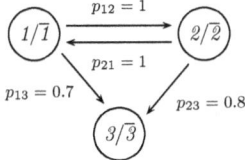

ii) For three actors a_1, a_2, a_3 with respective configurations or states $1\ 2\ 3$, $1\ 2\ \bar{3}$, ..., $\bar{1}\ \bar{2}\ \bar{3}$ and the desired conditional probabilities $Q(2 \mid 1) = p_{12} = 1$, $Q(1 \mid 2) = p_{21} = 1$, $Q(3 \mid 1) = p_{13} = 0.7$, $Q(3 \mid 2) = p_{23} = 0.8$ any Q on the states must comply $Q(1) = Q(2) = 0$ and again is not useful.

◊

The easy proofs of the statements in example 1 are left to the reader. The statements show the importance of a careful assignment of conditional probabilities or p-attenuation coefficients, respectively, for the network.

Finding a Q which respects all p_{kl} and also meets the above mentioned requirements is not sufficient for our purpose. Among all such Q we now determine the one with maximum entropy (MaxEnt). Important contributions are those of Shore and Johnson [16], Kern-Isberner [6] [7], Rödder et al. [13] [11]. MaxEnt distributions are a sophisticated tool for knowledge processing. In the context of AI, such Q is called a knowledge base.

Generalizing from the latter contribution we now allow for networks with a p-attenuated transfer structure and show respective equations for knowledge acquisition and processing. For this purpose solve

$$Q^* = \arg\max H(Q) = -\sum_v Q(v) \log_2 Q(v) \tag{2}$$

s.t. $Q(V_l = l \mid V_k = k) = p_{kl}$ for all links (k, l)

(2) respects the desired conditioned probabilities and does not create any not intended probabilistic dependencies on $\mathcal{V} = \{v\}$. For an axiomatic justification of (2) again cf. [6]. Once there exists Q^* and we have determined it solving (2), it represents unbiased knowledge about the Social Network's structure.

All knowledge so far is conditional: if-then. Whether an actor really has the message or attitude under consideration is not known so far. This kind of initialization of an actor's state is called evidentiation. For this let $E \triangleq V_i = i$ ($\overline{E} \triangleq V_i = \overline{i}$) be a fix (evident) proposition and let p_E ($p_{\overline{E}}$) be the probability of such proposition. Then evidentiation is the determination of distribution Q^{**} ($\overline{Q}^{**}$) on $\mathcal{V}$ which solves (3).

$$Q^{**}(\overline{Q}^{**}) = \arg\min R(Q, Q^*) = \sum_v Q(v) \log_2 \frac{Q(v)}{Q^*(v)} \quad \text{s.t.} \quad \begin{matrix} Q(E) = p_E \\ (Q(\overline{E}) = p_{\overline{E}}) \end{matrix} \quad (3)$$

Here R is the relative entropy or Kullback-Leibler divergence between Q^* and Q and $Q(E) = p_E$ ($Q(\overline{E}) = p_{\overline{E}}$) restricts distributions Q to those for which evidence holds. Q^{**} ($\overline{Q}^{**}$) is probabilistic structure on $\mathcal{V}$ under evidence. Once Q^{**} or $\overline{Q}^{**}$ is determined as in (3) it permits the evaluation of the network for given evidence. For any actor a_j, $Q^{**}(V_j = j)$ or $\overline{Q}^{**}(V_j = j)$ is the probability to have the message or attitude, if evidently E is p_E or $\overline{E}$ is $p_{\overline{E}}$. Evidencing E with a positive probability p_E requires a positive $Q^*(E)$. Evidencing is conditioning. Thus Q^{**} is the from Q^* inferred distribution given that E has probability p_E. (3) numerically is realized by so-called proportional scaling. Such scaling requires $Q^*(E) > 0$. Only then evidentiation of E is possible. Evidencing $\overline{E}$ with a positive $p_{\overline{E}}$ requires a positive $Q^*(\overline{E})$, by the same reasoning as before. A Q^* with this requirements for all possible E and $\overline{E}$ we call a useful Q^*. For more details on proportional scaling confer [10, pp. 104, 105].

2.2 Probability and information

Probability and information are strongly related. In this section we list some well-known facts about these relations, which the reader might find in any introductory textbook of information theory.

- If $A \subset \mathcal{V}$ is an arbitrary set of states on the network and $P(A)$ its probability, then $-\log_2 P(A)$ is the information we receive when we learn that A is true.
- $H(P) = -\sum_v P(v) \log_2 P(v)$ is the entropy in P. It is maximum for the uniform distribution P^0 on $\mathcal{V}$ and it is 0 for a P with $P(v) = 1$ for some v.
- $R(Q, P) = \sum_v Q(v) \log_2 \frac{Q(v)}{P(v)}$ for two appropriate distributions on $\mathcal{V}$ is the Kullback-Leibler divergence from P to Q. $H(P)$ measures the structural probabilistic independence in P and $R(Q, P)$ measures the change of structural independence from P to Q. Specially for the uniform P^0 we have $R(Q, P^0) = \log_2 |\{v\}| - H(Q) = H(P^0) - H(Q)$. In P^0 probabilistic independence is maximum, $R(Q, P^0)$ then yields dependency change against maximum independence.
- Let Q^{**} be the optimal solution of (3) for $p_E = 1$ and let $\overline{Q}^{**}$ be the respective solution for $p_{\overline{E}} = 1$. Then we have $R(Q^{**}, Q^*) = -\log_2 Q^*(V_i = i)$ and $R(\overline{Q}^{**}, Q^*) = -\log_2 Q^*(V_i = \overline{i})$. The authors in [11] call $-\log_2 Q^*(V_i = i)$ and $-\log_2 Q^*(V_i = \overline{i})$ actor a_i's diffusion and reception, respectively. Diffusion measures the penetration of evidence $p_E = 1$ throughout the net and reception measures the influence of evidence from others upon actor

a_i. For a deeper justification of these notions and for a mathematical proof see again [11], section 5.4.1.

Perhaps the last bullet-point needs an explication. If Q^* is the result of knowledge acquisition for a network with given attenuation coefficients p_{kl} like in (2), then the Kullback-Leibler divergence R between Q^* and Q^{**} ($\overline{Q}^{**}$) is available already in Q^*. Q^* "knows" the effect of evidencing with $p_E = 1$ or $p_{\overline{E}} = 1$ beforehand. There is no counting of arcs like in a graph theoretical model to predict such effects of evidence upon the whole net.

- For any two disjoint sets of variables $V_\mathcal{I}, V_\mathcal{J} \subset \{V_1, \ldots, V_n\}$ with $V_\mathcal{I} \cap V_\mathcal{J} = \emptyset$, $T(V_\mathcal{I}, V_\mathcal{J}; P) = \sum_{v_\mathcal{I} v_\mathcal{J}} P(v_\mathcal{I} v_\mathcal{J}) log_2 \frac{P(v_\mathcal{I} v_\mathcal{J})}{P(v_\mathcal{I})P(v_\mathcal{J})}$ is the transinformation, also called mutual information, between $V_\mathcal{I}$ and $V_\mathcal{J}$ under P. For further details cf. [14] and [18].
If $\mathcal{I} = \{l\}$ and $\mathcal{J} = \{1, \ldots, n\} \setminus \{l\}$, then $T(V_\mathcal{I}, V_\mathcal{J}; P)$ is called the embeddedness of actor l in the network, see [11, p. 7974]. Embeddedness of an actor is more than his diffusion or his reception. It measures the mutual information theoretical influence between himself and all the others.

Up to now we do not know whether equation (2) has a solution for all p_{kl} or not. As we said in the introduction, at least for $p_{kl} = p$ for all links (k, l) in the net there always exists a solution. The next section sketches the proof of this statement. The reader more interested in applications of the new concept might skip this issue and go directly to section 4.

3 Existence of Q^* for p-attenuation

Equation (2) always has a useful MaxEnt solution, when all links of the net have the same attenuation coefficient p. First, in lemma 1, we show the existence of a $\overline{Q}$ which respects all conditional postulations $\overline{Q}(V_l = l \mid V_k = k) = p$ for a (complete) clique, and then for an arbitrary net with n actors. This is not yet the distribution which maximizes entropy, however. Then, in theorem 1, for a complete clique in which all links (k, l) have attenuation $0 < p < 1$, we proof the existence of a useful MaxEnt Q^*. Later we generalize for arbitrary nets and finally we study the special cases $p = 1$ and $p = 0$. Resuming: For any $p \in [0, 1]$ and $p_{kl} = p$ all links (k, l) there exists a MaxEnt Q^* for (2) with $0 < Q^*(V_i = i) < 1$ for all $i = 1, \ldots, n$.

Lemma 1

i) For a clique with attenuation $p \in [0, 1]$ for each link, there exists a $\overline{Q}$ with $\overline{Q}(V_l = l \mid V_k = k) = p$ for all $l \neq k$.
ii) For an arbitrary net with attenuation $p \in [0, 1]$ on all links (k, l), $\overline{Q}$ from i) also yields $\overline{Q}(V_l = l \mid V_k = k) = p$ for all links (k, l).

Proof.

Ad i)

Make $\overline{Q}(1\ 2\ldots n) = 1 - \dfrac{n(1-p)}{p+n(1-p)}$

$\overline{Q}(1\ \overline{2}\ \overline{3}\ldots\overline{n}) = \overline{Q}(\overline{1}\ 2\ \overline{3}\ldots\overline{n}) = \overline{Q}(\overline{1}\ \overline{2}\ \overline{3}\ldots\overline{n-1}\ n) = \dfrac{1-p}{p+n(1-p)}$

and $\overline{Q}(v_1\ v_2\ldots v_n) = 0$, *otherwise.*

Thus, for all i

$\overline{Q}(i) = 1 - \dfrac{n(1-p)}{p+n(1-p)} + \dfrac{1-p}{p+n(1-p)} = \dfrac{1}{p+n(1-p)} > 0$

and $\overline{Q}(\overline{i}) = 1 - \dfrac{1}{p+n(1-p)}.$

Furthermore

$\overline{Q}(V_l = l \mid V_k = k) = \left(1 - \dfrac{n(1-p)}{p+n(1-p)}\right) \Big/ \dfrac{1}{p+n(1-p)} = p.$

Ad ii) Erase all missing links from the clique to get the desired net. Then obviously $\overline{Q}$ yields $\overline{Q}(V_l = l \mid V_k = k) = p$ for all links (k, l). □

In the following theorem 1 we again first consider a clique and furthermore restrict the p-attenuation coefficient p to $0 < p < 1$. Then we generalize to arbitrary networks and finally we abolish the restriction on p and study the cases $p = 0$ and $p = 1$.

Already in section 2.1 we claimed the existence of a distribution on $\mathcal{V}$ which makes all probabilities of $V_i = i$ positiv for all actors a_i. This demand is a necessary condition for later evidentiation, see again section 2.1. Unfortunately for a distribution Q the conditional $Q(V_l = l \mid V_k = k) = p$ or $Q(V_l = l, V_k = k) = p \cdot Q(V_k = k)$ is valid for $Q(V_k = k) = 0$. Such a Q might meet the conditional probability p but not allow evidentiation on actor a_k. The reader is invited to equally resume that for all i probabilities $V_i = \overline{i}$ must be positive. Otherwise unabeling evidentiation of $V_i = \overline{i}$ with $p_{\overline{E}}$. When now constructing a MaxEnt distribution Q^* in theorem 1 we must avoid such degenerated conditional probabilities.

Theorem 1

i) For a clique with attenuation $0 < p < 1$ for each link, the MaxEnt distribution $\tilde{Q}$ with $\tilde{Q}(V_l = l \mid V_k = k) = p$ for all $l \neq k$ yields $0 < \tilde{Q}(V_i = i) < 1$ for all $i = 1, \ldots, n$.

ii) For an arbitrary net with attenuation $0 < p < 1$ on all links (k, l), the MaxEnt distribution Q^* with $Q^*(V_l = l \mid V_k = k) = p$ is useful, i.e. $0 < Q^*(V_i = i) < 1$ for all $i = 1, \ldots, n$.

iii) For $p = 1$ and for $p = 0$ and an arbitrary net, a MaxEnt Q^* also exists and is useful.

Proof.

Ad i) For a n-clique with p-attenuation $0 < p < 1$, $\overline{\overline{Q}}$ with $\overline{\overline{Q}}(\bar{1}\,\bar{2}\ldots\bar{n}) = 1$ and 0 for all other configurations, the statement $\overline{\overline{Q}}(V_l = l \mid V_k = k) = p$ for all (k,l) is valid, see our remarks on degenerated conditional probabilities. $\overline{\overline{Q}}$ is not MaxEnt, however, as $H(\overline{\overline{Q}}) = 0$ and $H(\overline{Q})$ of lemma 1 has $H(\overline{Q}) > 0$. Let $\tilde{Q}$ with $H(\tilde{Q}) \geq H(\overline{Q})$ be the MaxEnt solution of (2). As $\overline{Q}$ is feasible for (2), $\tilde{Q}$ always exists.

Assume $\tilde{Q}(V_l = l) = 0$ for some l. Then any predecessor a_k of a_l also has $\tilde{Q}(V_k = k) = 0$, otherwise contradicting $\tilde{Q}(V_l = l \mid V_k = k) = p > 0$.

Assume $\tilde{Q}(V_l = l) = 1$ for some l. Then any predecessor a_k of a_l has $\tilde{Q}(V_k = k) = 0$, otherwise contradicting $\tilde{Q}(V_l = l \mid V_k = k) = p < 1$.

Running through all predecessors would make all actors' probability 0. This yields $\tilde{Q}(\bar{1}\,\bar{2}\ldots\bar{n}) = 1$ and hence $\tilde{Q}$ is not MaxEnt. Consequently $0 < \tilde{Q}(V_i = i) < 1$ for all i.

Ad ii) Let $\mathcal{R}_1 = \{(k,l) \mid (k,l)$ are links in the desired net$\}$.
Let $\mathcal{R}_2 = \{(i,j) \mid (i,j)$ are links which make the net complete$\}$.
Let $Q^*_{\mathcal{R}_1}$ and $Q^*_{\mathcal{R}_1 \cup \mathcal{R}_2}$ be optimal solutions of the equations

$$\min R(Q, P^0) \quad \text{s.t.} \quad Q(V_l = l \mid V_k = k) \quad (k,l) \in \mathcal{R}_1 \tag{4}$$

and

$$\min R(Q, P^0) \quad \text{s.t.} \quad Q(V_l = l \mid V_k = k) \quad (k,l) \in \mathcal{R}_1 \cup \mathcal{R}_2. \tag{5}$$

Then $Q^*_{\mathcal{R}_1 \cup \mathcal{R}_2}$ is also optimal solution of the equation

$$\min R(Q, Q^*_{\mathcal{R}}) \quad \text{s.t.} \quad Q(V_l = l \mid V_k = k) \quad (k,l) \in \mathcal{R}_1 \cup \mathcal{R}_2 \tag{6}$$

For the proof see [7], theorem 5.3 on page 86. If now $Q^*_{\mathcal{R}_1}(V_l = l) = 0$ for some l, then also $Q^*_{\mathcal{R}_1 \cup \mathcal{R}_2}(V_l = l) = 0$. And if $Q^*_{\mathcal{R}_1}(V_l = l) = 1$ for some l, then also $Q^*_{\mathcal{R}_1 \cup \mathcal{R}_2}(V_l = l) = 1$. So for all l $0 < Q^*_{\mathcal{R}_1}(V_l = l) < 1$, otherwise contradicting i).

Ad iii) $p = 1$ first.
$Q(V_l = l \mid V_k = k) = 1 \Rightarrow Q(V_l = \bar{l}, V_k = k) = 0$ for all links (k,l) and any Q. Make $Q^*(v) = 0$ for all v containing any $(k, \bar{l})$, and make all remaining probabilities equal under Q^*. Note that $1 > Q^*(V_l = l) > 0$ for all l.

$p = 0$ second.
$Q(V_l = l \mid V_k = k) = 0 \Rightarrow Q(V_l = l, V_k = k) = 0$ for all links (k,l) and any Q. Make $Q^*(v) = 0$ for all v containing any (k, l), and make all remaining probabilities equal under Q^*. Note that $1 > Q^*(V_l = l) > 0$ for all l. □

4 Network analysis under MaxEnt

4.1 Small nets and p-attenuation

For small nets with 2 or 3 actors we present marginal distributions, diffusion and reception under Q^* for varying p-attenuation values $p = 1.0, 0.9, 0.8, 0.7, 0.6,$

0.5. We add probabilities for inverse rules in some cases, so as to characterize modus tollens under MaxEnt.

In tables 1 to 4 we varied p-attenuation from 1.0 to 0.5. $Q^*(l \mid k) = 1.0$ means certain transfer of a message on such link, for $Q^*(l \mid k) = 0.5$ it is uncertain whether the message from a_k reaches a_l – 0.5 – or not – 0.5. We abstain from considering p-attenuation coeffcients underneath 0.5 in this paper.

Table 1 shows what we expected. The less p the less a_k's diffusion and the less a_l's reception. In table 2 we observe higher diffusions of a_k than in table 1. This is due to the fact that a_k now informs a_l and a_j instead of a_l, only. For the same reason the reception for a_l and a_j are less than for a_l like in table 1.

Table 1. Probabilities, diffusion and reception for $l \mid k$.

	Q^*				$-log_2 Q^*$			
p	k	$\bar{k}$	l	$\bar{l}$	k	$\bar{k}$	l	$\bar{l}$
1.0	0.33	0.67	0.67	0.33	1.59	0.59	0.59	1.59
0.9	0.41	0.59	0.66	0.34	1.29	0.76	0.59	1.57
0.8	0.45	0.55	0.64	0.36	1.15	0.87	0.65	1.46
0.7	0.48	0.52	0.60	0.40	1.06	0.94	0.75	1.31
0.6	0.49	0.51	0.55	0.45	1.02	0.99	0.86	1.15
0.5	0.50	0.50	0.50	0.50	1.00	1.00	1.00	1.00

Table 2. Probabilities, diffusion and reception for $j \mid k$ and $l \mid k$.

	Q^*						$-log_2 Q^*$					
p	j	$\bar{j}$	l	$\bar{l}$	k	$\bar{k}$	j	$\bar{j}$	l	$\bar{l}$	k	$\bar{k}$
1.0	0.60	0.40	0.60	0.40	0.20	0.80	0.74	1.32	0.74	1.32	2.32	0.32
0.9	0.63	0.37	0.63	0.37	0.32	0.68	0.67	1.43	0.67	1.43	1.63	0.57
0.8	0.62	0.38	0.62	0.38	0.41	0.59	0.69	1.40	0.69	1.40	1.31	0.75
0.7	0.59	0.41	0.59	0.41	0.46	0.54	0.76	1.29	0.76	1.29	1.12	0.89
0.6	0.55	0.45	0.55	0.45	0.49	0.51	0.87	1.15	0.87	1.15	1.03	0.97
0.5	0.50	0.50	0.50	0.50	0.50	0.50	1.00	1.00	1.00	1.00	1.00	1.00

Table 3 needs our special care. We observe higher reception values for a_l than in table 1, as actor a_l now receives messages from both, a_j and a_k. For $p = 1.0$ this reception is highest: 2.32. a_l receives from all other actors – here a_j and a_k – which do not have the message if a_l does not have it. $Q^*(l \mid k) = 1$ iff $Q^*(\bar{k} \mid \bar{l}) = 1$, e.g., that is modus tollens; modus tollens helps to find all actors in a net which send a message to a_l.

As soon as p becomes 0.9 instead of 1.0, the reasoning with modus tollens changes slightly: If now a_l would not know the message, then very likely (0.81) a_k does not know it either. Because if a_k would know it, very likely a_l knows it (0.9). This is a weak form of contradiction. With this reasoning, $-log_2 Q^*(\bar{l})$ is the correct measure for a_l's p-reception. A profound discussion of the case $p = 1.0$ you again find in [11, p. 7974].

Table 3. Probabilities, diffusion, reception and modus tollens for $l \mid j$ and $l \mid k$.

p	Q^*								$-\log_2 Q^*$					
	j	$\bar{j}$	k	$\bar{k}$	l	$\bar{l}$	$\bar{k}\mid\bar{l}$	j	$\bar{j}$	k	$\bar{k}$	l	$\bar{l}$	
1.0	0.40	0.60	0.40	0.60	0.80	0.20	1.00	1.32	0.74	1.32	0.74	0.32	2.32
0.9	0.46	0.54	0.46	0.54	0.76	0.24	0.81	1.14	0.88	1.14	0.88	0.40	2.05
0.8	0.48	0.52	0.48	0.52	0.70	0.30	0.69	1.07	0.94	1.07	0.94	0.52	1.72
0.7	0.49	0.51	0.49	0.51	0.63	0.37	0.60	1.03	0.97	1.03	0.97	0.66	1.44
0.6	0.50	0.50	0.5	0.50	0.57	0.43	0.54	1.01	0.99	1.01	0.99	0.82	1.21
0.5	0.50	0.50	0.50	0.50	0.50	0.50	0.50	1.00	1.00	1.00	1.00	1.00	1.00

Last not least we consider table 4. Interesting enough, actor a_j for $p = 1.0$ is neither a receiver nor a sender of a message. The reception from a_k is absorbed by the diffusion to a_l. This effect disappears for $0.5 < p < 1.0$. Anyhow, diffusion and reception in our context are relative measures: They show a joint balance between "in" and "out". How to isolate these parts from each other and how to develop this balance effect in mathematical equations is left for future research.

Table 4. Probabilities, diffusion and reception for $j \mid k$ and $l \mid j$.

p	Q^*						$-\log_2 Q^*$					
	j	$\bar{j}$	k	$\bar{k}$	l	$\bar{l}$	j	$\bar{j}$	k	$\bar{k}$	l	$\bar{l}$
1.0	0.50	0.50	0.25	0.75	0.75	0.25	1.00	1.00	2.00	0.42	0.42	2.00
0.9	0.59	0.41	0.37	0.63	0.74	0.26	0.76	1.29	1.43	0.67	0.44	1.92
0.8	0.60	0.40	0.44	0.56	0.68	0.32	0.73	1.34	1.20	0.83	0.55	1.65
0.7	0.58	0.42	0.48	0.52	0.62	0.38	0.78	1.27	1.07	0.93	0.70	1.38
0.6	0.55	0.45	0.50	0.50	0.55	0.45	0.87	1.14	1.01	0.98	0.85	1.17
0.5	0.50	0.50	0.50	0.50	0.50	0.50	1.00	1.00	1.00	1.00	1.00	1.00

4.2 Newcomb's Fraternity and p-attenuation

Newcomb's Fraternity is a famous example for any sociologist. That is why the authors in [11] picked it up and modified it slightly. As we want to compare attenuated and non-attenuated link transfers, it might be worthwhile to study the same example again, see figure 1. Originally the nodes showed 17 students at the University of Michigan in the years 1954 to 1956. The links represented friendship nominations amongst the members of the group. In the new context an isolated student no.18 is added and the links are interpreted as (non-attenuated) transfers of a message from actor to actor, rather than friendship nominations. If now p-attenuation is present p might mean some kind of effectiveness of such a link or the respective probability that the transfer realizes successfully. With 34 probabilistic rules and for varying p like in table 5 we generated respective distributions Q^* and provide all introduced indices such as diffusion (di), reception (re) and embeddedness (em) in table 6. For all p the results meet our intuition. A strong sender loses diffusion and a strong receiver loses reception with decreasing p. Embeddedness increases with growing p, being embeddedness zero for $p = 0.5$. $p = 0.5$ is the dead end of all message flow in the net.

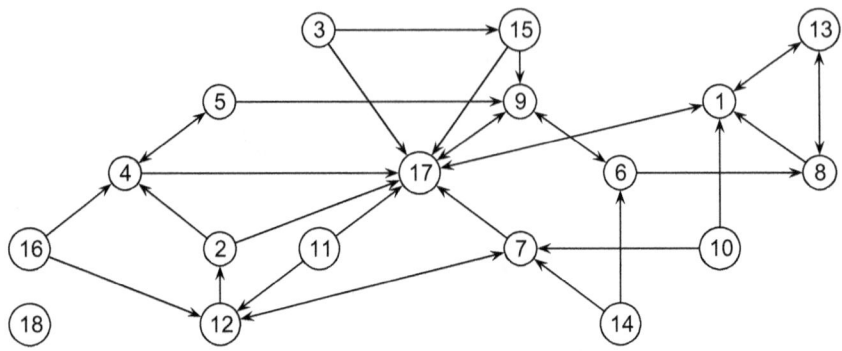

Fig. 1. Modified Newcomb Fraternity, cf. [11].

Table 5. p-attenuated set of conditionals for Newcomb's Fraternity.

no.	cond.	p-attenuation	no.	cond.	p-attenuation
1	(9 | 17)	[1.0], [0.9], [0.7], [0.5]	2	(1 | 17)	[1.0], [0.9], [0.7], [0.5]
3	(12 | 16)	[1.0], [0.9], [0.7], [0.5]	4	(4 | 16)	[1.0], [0.9], [0.7], [0.5]
5	(17 | 15)	[1.0], [0.9], [0.7], [0.5]	6	(9 | 15)	[1.0], [0.9], [0.7], [0.5]
7	(7 | 14)	[1.0], [0.9], [0.7], [0.5]	8	(6 | 14)	[1.0], [0.9], [0.7], [0.5]
9	(8 | 13)	[1.0], [0.9], [0.7], [0.5]	10	(1 | 13)	[1.0], [0.9], [0.7], [0.5]
11	(7 | 12)	[1.0], [0.9], [0.7], [0.5]	12	(2 | 12)	[1.0], [0.9], [0.7], [0.5]
13	(17 | 11)	[1.0], [0.9], [0.7], [0.5]	14	(12 | 11)	[1.0], [0.9], [0.7], [0.5]
15	(7 | 10)	[1.0], [0.9], [0.7], [0.5]	16	(1 | 10)	[1.0], [0.9], [0.7], [0.5]
17	(17 | 9)	[1.0], [0.9], [0.7], [0.5]	18	(6 | 9)	[1.0], [0.9], [0.7], [0.5]
19	(13 | 8)	[1.0], [0.9], [0.7], [0.5]	20	(1 | 8)	[1.0], [0.9], [0.7], [0.5]
21	(17 | 7)	[1.0], [0.9], [0.7], [0.5]	22	(12 | 7)	[1.0], [0.9], [0.7], [0.5]
23	(9 | 6)	[1.0], [0.9], [0.7], [0.5]	24	(8 | 6)	[1.0], [0.9], [0.7], [0.5]
25	(9 | 5)	[1.0], [0.9], [0.7], [0.5]	26	(4 | 5)	[1.0], [0.9], [0.7], [0.5]
27	(17 | 4)	[1.0], [0.9], [0.7], [0.5]	28	(5 | 4)	[1.0], [0.9], [0.7], [0.5]
29	(17 | 3)	[1.0], [0.9], [0.7], [0.5]	30	(15 | 3)	[1.0], [0.9], [0.7], [0.5]
31	(17 | 2)	[1.0], [0.9], [0.7], [0.5]	32	(4 | 2)	[1.0], [0.9], [0.7], [0.5]
33	(17 | 1)	[1.0], [0.9], [0.7], [0.5]	34	(13 | 1)	[1.0], [0.9], [0.7], [0.5]

Table 6. Diffusion, reception and embeddedness under p-attenuation.

act.	$p=1.0$ di	re	em	$p=0.9$ di	re	em	$p=0.7$ di	re	em	$p=0.5$ di	re	em
1	0.03	5.86	0.13	0.22	2.83	0.12	0.60	1.56	0.03	1.00	1.00	0.00
2	0.19	3.05	0.43	0.36	2.20	0.27	0.69	1.39	0.06	1.00	1.00	0.00
3	1.61	0.57	0.26	1.32	0.74	0.14	1.07	0.93	0.04	1.00	1.00	0.00
4	0.10	3.86	0.36	0.32	2.34	0.27	0.67	1.44	0.07	1.00	1.00	0.00
5	0.10	3.86	0.36	0.32	2.34	0.17	0.67	1.44	0.03	1.00	1.00	0.00
6	0.03	5.86	0.13	0.22	2.83	0.10	0.60	1.56	0.02	1.00	1.00	0.00
7	0.27	2.54	0.66	0.36	2.18	0.27	0.64	1.48	0.05	1.00	1.00	0.00
8	0.03	5.86	0.13	0.22	2.83	0.11	0.60	1.56	0.03	1.00	1.00	0.00
9	0.03	5.86	0.13	0.22	2.83	0.12	0.60	1.56	0.03	1.00	1.00	0.00
10	1.27	0.77	0.15	1.12	0.89	0.06	1.03	0.97	0.02	1.00	1.00	0.00
11	1.27	0.77	0.15	1.11	0.90	0.06	1.03	0.97	0.02	1.00	1.00	0.00
12	0.27	2.54	0.66	0.36	2.18	0.36	0.64	1.48	0.07	1.00	1.00	0.00
13	0.03	5.86	0.13	0.22	2.83	0.08	0.60	1.56	0.02	1.00	1.00	0.00
14	1.27	0.77	0.15	1.12	0.89	0.06	1.03	0.97	0.02	1.00	1.00	0.00
15	0.61	1.54	0.27	0.62	1.51	0.17	0.76	1.28	0.04	1.00	1.00	0.00
16	1.27	0.77	0.15	1.14	0.87	0.07	1.05	0.96	0.02	1.00	1.00	0.00
17	0.03	5.86	0.13	0.22	2.83	0.14	0.60	1.56	0.05	1.00	1.00	0.00
18	1.00	1.00	0.00	1.00	1.00	0.00	1.00	1.00	0.00	1.00	1.00	0.00

There is one important observation on p-attenuated networks which needs further research. The reader easily realizes from figure 1 that there are the following strongly connected subgraphs (SCS): $\{a_1, a_6, a_8, a_9, a_{13}, a_{17}\}, \{a_7, a_{12}\}$, $\{a_4, a_5\}$ and the isolated $\{a_{18}\}$. In such SCS every actor reaches every actor via a path of p-attenuated links. Under MaxEnt all actors of a SCS have something in common: They all have the same diffusion, the same reception and for $p = 1.0$ even the same embeddedness, confer table 6. In a certain way all actors of a SCS are similar, they have the same information theoretical characteristics: Their influence upon other actors – diffusion – is the same, the influence which other actors excert on them – reception – coincide. And even their embeddedness is equal when attenuation is absent. A proof of these facts is still missing.

4.3 Newcomb's Fraternity and p_{kl}-attenuation

As was outlined already in example 1 in section 1, attenuation coefficients cannot be chosen arbitrarily. Otherwise the resulting probability distribution on the network's states might not be useful. Consequently this statement remains valid for MaxEnt-distributions: An imprudent choice of p_{kl} might make Q^* a useless distribution on the network. Useless in that evidentiation of any actor becomes impossible.

Table 7 shows a set of useful and a set of useless link-probabilities for Newcomb's Fraternity. In the first case SPIRIT accepts the p_{kl} and builds a useful Q^*, in the second it doesn't. The reader notices that the p_{kl} differ only for one link.

For $l = 9$ and $k = 17$ in the useful case we have $p_{kl} = 0.5$ and in the useless one $p_{kl} = 0.9$, the rest is identical. The two sets of attenuation coefficients only blank-faced are close to each other, in their semantical meaning for the network they are not. The useless Q^* yields $Q^*(\overline{1}\ \overline{2}\ \overline{3}\ \ldots\ \overline{17}) = 1$ and $Q^*(\overline{18}) = 0.5$. Such Q^* does not permit useful evidentiation of any actor a_i like in equation (3).

Table 7. Useful and useless p_{kl}-attenuation coefficients.

no.	cond.	p_{kl} useful	useless	no.	cond.	p_{kl} useful	useless
1	(9 | 17)	[0.5]	[0.9]	2	(1 | 17)	[0.8]	[0.8]
3	(12 | 16)	[0.6]	[0.6]	4	(4 | 16)	[0.9]	[0.9]
5	(17 | 15)	[0.7]	[0.7]	6	(9 | 15)	[0.7]	[0.7]
7	(7 | 14)	[0.6]	[0.6]	8	(6 | 14)	[0.5]	[0.5]
9	(8 | 13)	[0.6]	[0.6]	10	(1 | 13)	[0.8]	[0.8]
11	(7 | 12)	[0.5]	[0.5]	12	(2 | 12)	[0.5]	[0.5]
13	(17 | 11)	[0.6]	[0.6]	14	(12 | 11)	[0.7]	[0.7]
15	(7 | 10)	[0.8]	[0.8]	16	(1 | 10)	[0.9]	[0.9]
17	(17 | 9)	[0.8]	[0.8]	18	(6 | 9)	[0.6]	[0.6]
19	(13 | 8)	[0.7]	[0.7]	20	(1 | 8)	[0.6]	[0.6]
21	(17 | 7)	[0.8]	[0.8]	22	(12 | 7)	[0.5]	[0.5]
23	(9 | 6)	[0.5]	[0.5]	24	(8 | 6)	[0.7]	[0.7]
25	(9 | 5)	[0.6]	[0.6]	26	(4 | 5)	[0.9]	[0.9]
27	(17 | 4)	[0.8]	[0.8]	28	(5 | 4)	[0.7]	[0.7]
29	(17 | 3)	[0.6]	[0.6]	30	(15 | 3)	[0.5]	[0.5]
31	(17 | 2)	[0.9]	[0.9]	32	(4 | 2)	[0.8]	[0.8]
33	(17 | 1)	[0.9]	[0.9]	34	(13 | 1)	[0.7]	[0.7]

Table 8 for the useful knowledge base provides diffusion, reception and embeddedness for all actors. Obviously, the observed characteristics of SCS in section 4.2 disappear with arbitrary p_{kl}. If there exists a mathematical characterization of useful p_{kl} for an arbitrary network remains an open question.

Table 8. Measures under the useful p_{kl}-attenuation coefficients.

actor	1	2	3	4	5	6	7	8	9	10	11	12	13	14	15	16	17	18
di	0.56	1.05	1.22	0.43	0.79	0.81	1.23	0.98	1.06	1.45	1.27	0.60	0.75	1.03	1.20	1.16	0.39	1.00
re	1.64	0.96	0.81	1.96	1.24	1.23	0.80	1.03	0.94	0.66	0.77	1.23	1.30	0.97	0.83	0.85	2.08	1.00
em	0.32	0.08	0.09	0.22	0.18	0.16	0.11	0.22	0.18	0.18	0.11	0.06	0.14	0.02	0.14	0.08	0.33	0.00

5 Survey of network analysis in SPIRIT

Once the expert system shell SPIRIT has learned all link-probabilities and built up a useful knowledge base Q^* about the network's structure it is ready for analyses. Figure 2 shows the Newcomb network in SPIRIT for useful p_{kl} like in table 7.

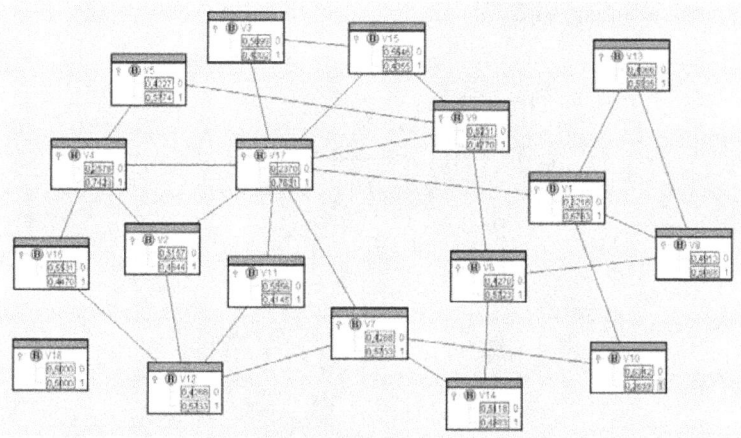

Fig. 2. Newcomb Network for useful p_{kl}.

The numbers in the bars for "1" show the marginal probabilities $Q^*(V_k = k)$, the numbers in the bars for "0" the probabilities $Q^*(V_k = \bar{k})$. Evidentiation $E \triangleq V_i = \text{i}$ with $p_E = 1$ like in equation (3) for actor 10 is shown in figure 3.

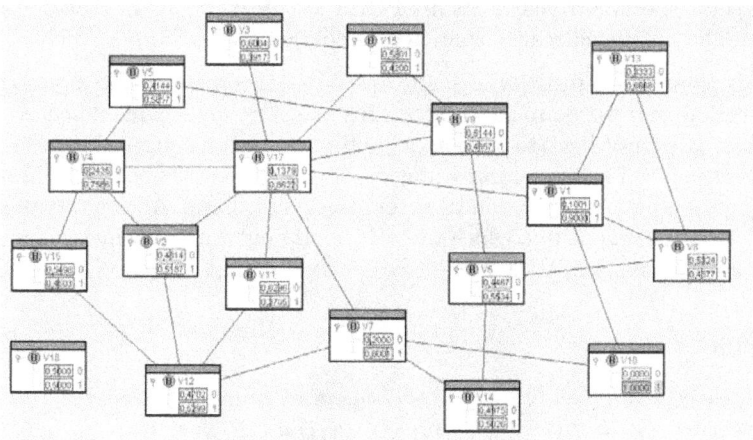

Fig. 3. Evidencing actor 10 for useful p_{kl}.

The numbers in the bars now are the conditioned probabilities given $p_E = 1.0$. The semantics of such numbers are conditioned probabilities that any actor receives the message given that a_i launches it. In attenuated networks these conditioned probabilities might not be very different from probabilities in the original knowledge base, cf. figures 2 and 3. Probabilities and conditioned probabilities might even be subject to second order uncertainty. More on that you find in [13], section 2.3.1.

SPIRIT allows for converting probabilities in information units. Figure 4 shows the respective converted Newcomb Network.

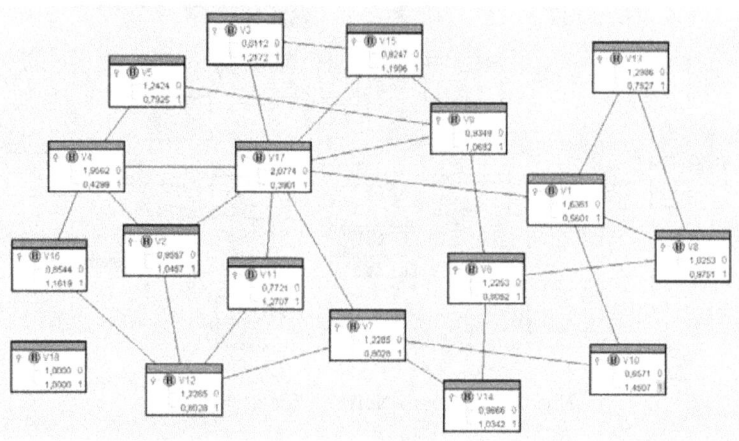

Fig. 4. Diffusion and reception for useful p_{kl}.

It provides diffusion and reception for all actors. Actor 3 has diffusion 1.22 and reception 0.81 as was outlined already in table 8. To calculate embeddedness needs some special arrangements in SPIRIT which are beyond the scope of this paper. The interested reader is invited to contact the authors.

That is what characteristics of single actors is concerned. In the remainder of this section we relate on analyzing groups of actors and their influence in the network. If we leaf back to section 2, we find elementary propositions $V_i = i/\bar{i}$ as constituents of network representations. Such elementary propositions might be composed by $^-$ (not), $\wedge$ (and), $\vee$ (or) and parentheses. Any such expression is a Composed Proposition which we call CP. If built with certain wariness such a CP can be true and false.

Example 2

i) $V_3 = 3$ and $V_{16} = 16$ are two elementary propositions in the Newcomb Network. $(V_3 = 3) \wedge (V_{16} = 16) = CP_1$ is a CP. It is true – $(CP_1 = 1)$ – for any interpretation of the net for which $V_3 = 3$ and $V_{16} = 16$, and false – $(CP_1 = \bar{1})$ – otherwise.

ii) $V_3 = 3$ and $V_{16} = 16$ are like in i). $(V_3 = 3) \vee (V_{16} = 16) = CP_2$ is a CP. It is true – $(CP_2 = 2)$ – for any interpretation for which $V_3 = 3$ or $V_{16} = 16$ or both, and false – $(CP_2 = \overline{2})$ – otherwise.

iii) $V_3 = 3$ and $\overline{V_{16} = 16}$ are two elementary propositions in the Newcomb Network. $(V_3 = 3) \wedge (\overline{V_{16} = 16}) = CP_3$ is a CP. It is true – $(CP_3 = 3)$ – for any interpretation for which $V_3 = 3$ and $\overline{V_{16} = 16}$, and false – $(CP_3 = \overline{3})$ – otherwise.

Even for CPs composed by many elementary propositions SPIRIT offers a comfortable handling. The reader might find technical details in [13].

For any CP_m the shell provides $Q^*(CP_m = m)$ and $Q^*(CP_m = \overline{m})$, respectively. An easy calculation of the negative logarithm informs the user about CP_m's diffusion and reception.

Furthermore, with $E \;\hat{=}\; CP_m = m$ or $\overline{E} \;\hat{=}\; CP_m = \overline{m}$ and equation (3) in section 2.1 an immediate evidentiation with corresponding probabilities p_E and $p_{\overline{E}}$ is possible. When doing so the user finds all actors' conditional probabilities under composed evidence in SPIRIT. The following example shows such evidentiation.

Example 3 We study Newcomb's Network without attenuation. Now evidencing $E \;\hat{=}\; CP_2 = 2$ from example 2 with $p_E = 1$ yields all conditional probabilities in the network. The following results give a deep insight in composed evidence. An isolated evidencing of $V_3 = 3$, only, would make all conditional probabilities of its successors equal to one, and hence for V_{15}. Dito for evidence $V_{16} = 16$ and its successors, and thus for V_4. Not so when evidencing $CP_2 = 2$. This time it is uncertain whether V_3 knows the message or V_{16}; at least one of them does. Neither V_{15} nor V_4 are certain, yet more probable than before. But V_{17} becomes true for $V_3 = 3$ or $V_{16} = 16$.

6 Summary and further research

Social Networks often are represented as graphs, being the nodes actors and the edges relations between the actors. There is an extensive literature on analyses of such graphs. The objective of these analyses is to find good characterizations of actors and their positions in the social fabric. In the present contribution we pick up the idea of a new form of network analysis. This new form identifies actors with binary variables as nodes and if-then rules as (directed) edges or links in the net. With this concept new results about each actor's position are available. What is the influence he exerts on others – diffusion –, how far is he influenced by others – reception –, and how is he cross-linked with others – embeddedness.

Sometimes the links represent weak relations, however. Not always an actor understands a message or accepts an immaterial good which others send to him. In this case attenuated contacts or probabilistic ties might be the loophole. For such probabilistic networks we relate on diffusion, reception and embeddedness again and show that the construction of a knowledge base under Maximum Entropy and manipulations of the net under Minimum Relative Entropy yield meaningful characterisitics of all actors' positions.

What if one actor *or* another one knows the message and dispatches it through the net, what if one actor knows it and the other denies it. What is the consequence of such ambiguity for diffusion, reception and embeddedness of actors. Can this idea be generalized for even more complex questions. We are at the beginning of such analyses.

Cohesive groups of actors are groups which are joined by common interests or attitudes. Are such affiliations identifiable in models under Maximum Entropy and Minimum Relative Entropy?

References

1. Ph. Bonacich. Power and centrality: A family of measures. *The American Journal of Sociology*, 92(5):1170–1182, 1987.
2. Ph. Bonacich and P. Lloyd. Eigenvector-like measures of centrality for asymmetric relations. *Social Networks*, 23(3):191–201, 2001.
3. I. Csiszár. I-divergence geometry of probability distributions and minimization problems. *The Annals of Probability*, 3(1):148–158, 1975.
4. Ê. Durkheim. *Die Regeln der soziologischen Methode*. Luchterhand, Neuwied/Berlin, 1961, first 1895.
5. L. Katz. A new status index derived from sociometric analysis. *Psychometrika*, 18(1):39–43, 1953.
6. G. Kern-Isberner. Characterizing the principle of minimum cross-entropy within a conditional-logical framework. *Artificial Intelligence*, 98(1-2):169–208, 1998.
7. G. Kern-Isberner. *Conditionals in nonmonotonic reasoning and belief revision: Considering conditionals as agents*. Springer-Verlag, Berlin, Heidelberg, 2001.
8. J.L. Moreno. *Who Shall Survive: A New Approach to the Problem of Human Interrelations*. Nervous and Mental Disease Publishing Co., Washington, DC, 1934.
9. M.E.J. Newman. *Networks: An introduction*. Oxford University Press, 2012.
10. W. Rödder. Conditional logic and the principle of entropy. *Artificial Intelligence*, 117(1):83–106, 2000.
11. W. Rödder, D. Brenner, and F. Kulmann. Entropy based evaluation of net structures deployed in social network analysis. *Expert Systems with Applications*, 41(17):7968–7979, 2014.
12. W. Rödder and F. Kulmann. Recall and reasoning - an information theoretical model of cognitive processes. *Information Sciences*, 176(17):2439–2466, 2006.
13. W. Rödder, E. Reucher, and F. Kulmann. Features of the expert-system-shell spirit. *Logic Journal of IGPL*, 14(3):483–500, 2006.
14. St. Roman. *Introduction to coding and information theory*. Undergraduate Texts in Mathematics. Springer, 1996.
15. J. Scott. *Social Network Analysis*. Sage Publications, London, 3rd edition, 2013.
16. J. Shore and R. Johnson. Axiomatic derivation of the principle of maximum entropy and the principle of minimum cross-entropy. *IEEE Transactions on Information Theory*, 26(1):26–37, 1980.
17. SPIRIT, 2011. Last accessed on 2015-09-15.
18. F. Topsœ. *Informationstheorie*. Teubner Studienbücher Mathematik, Stuttgart, 1974.

Towards a Behavioural Theory for Random Parallel Computing[1]

Klaus-Dieter Schewe[2], *Flavio Ferrarotti*[3],
Loredana Tec[4], *Qing Wang*[5]

Abstract. A *behavioural theory* comprises a set of postulates that characterise a particular class of algorithms, an abstract machine model that provably satisfies the postulates, and a proof that all algorithms stipulated by the postulates are captured by the abstract machine model. This article is dedicated to the development of a behavioural theory for random, parallel algorithms. For this first a theory for non-deterministic, parallel algorithms is developed, which captures the case of uniform distribution for the random choices made by the algorithms. This is followed by a discussion how the uniformity could be replaced by arbitrary distributions.

1 Introduction

The term "behavioural theory" was introduced to capture the gist of theories following the approach in the proof of Gurevich's celebrated *sequential ASM thesis* [11]. In a nutshell, a behavioural theory consists of three parts:

1. **Axiomatisation.** A set of *postulates* that characterise a particular class of algorithms (such as sequential algorithms) or systems in an intuitive and commonly accepted way.
2. **Plausibility.** An *abstract machine model* that provably satisfies the postulates.
3. **Characterisation.** A *proof* that all algorithms/systems stipulated by the postulates are faithfully captured by the abstract machine model.

While it seems to be a standard—though by no means trivial—mathematical task to provide a set of postulates and an abstract machine model that can be proven to be equivalent in the sense that they describe exactly the same systems, the challenge lies in the non-provable intuitive clarity of the postulates, by means of which the scientific community can be convinced. This is the case for the postulates characterising sequential algorithms:

---

[1] The research reported in this paper results from the project *Behavioural Theory and Logics for Distributed Adaptive Systems* supported by the **Austrian Science Fund (FWF): [P26452-N15]**. It was further supported by the Austrian Research Promotion Agency (FFG) through the COMET funding for the Software Competence Center Hagenberg.
[2] Software Competence Center Hagenberg, Austria, kd.schewe@scch.at
[3] Software Competence Center Hagenberg, Austria, flavio.ferrarotti@scch.at
[4] Software Competence Center Hagenberg, Austria, loredana.tec@scch.at
[5] The Australian National University, Australia, qing.wang@anu.edu.au

sequential time: Each sequential algorithm starts from some initial states and proceeds by means of transitions from states to successor states.
abstract state: Each state can be defined by means of a Tarski structure.
bounded exploration: The changes made by the algorithm when progressing from a state to its successor state must be captured in the finite description of the algorithm, i.e. that there is a finite set of ground terms such that coincidence of different states on these ground terms yields equality of the updates performed in the transition to the successor states.

There are many other behavioural theories, out of which we single out the *parallel ASM thesis* addressing the class of (synchronous) parallel algorithms, and the *DB-ASM thesis* addressing the class of non-deterministic database transformations.

Regarding the former one the behavioural theory by Blass and Gurevich [3, 4] was not convincing due to the lack of intuitive clarity of the postulates. We believe that our own *simplified parallel ASM thesis* [9] has overcome this problem, as it only requires two changes to the sequential ASM thesis: (1) in the bounded exploration postulate we replace ground terms by multiset comprehension terms, by means of which different computation "branches" or "proclets" become dependent on the state; (2) same as in the work of Blass and Gurevich we request an explicit *background postulate*, by means of which assumptions concerning structural elements such as pairs, multisets, truth values, etc. are made explicit. This also includes the use of meta-finite structures [10] to formally capture a reserve for the computation. The price to be paid for this simplification of the postulates is a severely more complicated characterisation proof, which has to exploit and generalise results from finite model theory [8]. Regarding the latter one the main contribution of the behavioural theory for database transformations by Schewe and Wang was the capture of non-determinism in a reduced form [13].

The problem we would like to address in this article is the development of a behavioural theory for random parallel algorithms, by means of which we see a potential to exploit the contributions by Gabriele Kern-Isberner to probabilistic reasoning [1, 2, 15] for the verification of random systems. According to Brassard and Bratley [7] a random algorithm is an algorithm that leaves some of its decisions to random choice. If no further assumptions are made, choices are usually assumed to underlie a uniform distribution. With this interpretation a random algorithm would be simply a non-deterministic, parallel algorithm. Therefore, we start our investigation by an approach to integrate the (simplified) parallel ASM thesis and the thesis for non-deterministic transformations. We will handle the development of a corresponding behavioural theory in Sections 2-4. We further discuss in Section 5 how the uniformity assumption can be removed, i.e. how arbitrary probability distributions could be exploited in a behavioural theory for random, parallel algorithms. This discussion is still preliminary in many respects. Finally, in Section 6 we outline further open problems such as the coupling with linguistic reflection, in which case the distributions in the choices would be subject to changes as well, and the randomisation of concurrent (i.e. asynchronous parallel) algorithms, for which a behavioural theory also exists [5]. Furthermore, we raise the question how to extend logics for parallel ASMs [12, 14] and DB-ASMs [16] in order to permit reasoning about random, parallel algorithms.

2 Postulates for Non-Deterministic Parallel Computing

In this section we modify and extend the postulates characterising parallel algorithms towards non-deterministic, parallel algorithms. For this we follow the presentation in the simplified parallel ASM thesis[6] [9].

2.1 Abstract States and Sequential Time

We start with the sequential time postulate. The main change is that instead of a one-step tranformation function τ we consider a binary relation: whenever $(\mathbf{S}, \mathbf{S}') \in \tau_A$ holds, $\mathbf{S}'$ will be a (possible) successor state of $\mathbf{S}$.

Postulate 1 (Non-Deterministic Sequential Time Postulate). A *non-deterministic, parallel algorithm* A is associated with a non-empty set $\mathcal{S}_A$ of *states*, a non-empty subset $\mathcal{I}_A \subseteq \mathcal{S}_A$ of *initial states*, and a relation $\tau_A \subseteq \mathcal{S}_A \times \mathcal{S}_A$ called the *one-step transformation relation* of A.

A *run* or a *computation* of a non-deterministic, parallel algorithm A is a finite or infinite sequence of states $\mathbf{S}_0, \mathbf{S}_1, \ldots$, where $\mathbf{S}_0$ is an initial state in $\mathcal{I}_A$ and $(\mathbf{S}_i, \mathbf{S}_{i+1}) \in \tau_A$ holds for every $i \geq 0$.

Postulate 2 (Abstract State Postulate). States of a non-deterministic parallel algorithm A are first-order structures. All states in $\mathcal{S}_A$ have the same vocabulary. A state and its successor states have the same base set. $\mathcal{S}_A$ and $\mathcal{I}_A$ are closed under isomorphisms. Any isomorphism between two states $\mathbf{S}_1$ and $\mathbf{S}_2$ is also an isomorphism between successor states $\mathbf{S}'_1$ and $\mathbf{S}'_2$.

Definition 1. Let $\mathbf{S}$ be the state of an algorithm of vocabulary Σ, let $f \in \Sigma$ be a function symbol of arity r and let $\bar{a}$ be an r-tuple in S^r. Then the pair $(f, \bar{a})$ is called a *location* in $\mathbf{S}$. The *content* of the location $(f, \bar{a})$ is the value $f^{\mathbf{S}}(\bar{a})$ in $\mathbf{S}$. If $\ell = (f, \bar{a})$ is a location in a state $\mathbf{S}$ and b is an element in the base set S, then the tuple (ℓ, b) is an *update* of $\mathbf{S}$. If $b = f^{\mathbf{S}}(\bar{a})$, then the update $((f, \bar{a}), b)$ is called a *trivial update*. A set of updates Δ is *consistent* if it has no clashing updates, i.e., if for every pair of updates (l_i, b_i) and (l_j, b_j) in Δ, we have that $l_i = l_j$ only if $b_i = b_j$.

The result of executing a consistent update set Δ in a state $\mathbf{S}$ is a new state $\mathbf{S} + \Delta$ such that for every location $l_i = (f_i, \bar{a}_i)$ of $\mathbf{S}$:

$$f_i^{\mathbf{S}+\Delta}(\bar{a}_i) = \begin{cases} b & \text{if } (l_i, b) \in \Delta; \\ f_i^{\mathbf{S}}(\bar{a}_i) & \text{if there is no } b \text{ with } (l_i, b) \in \Delta. \end{cases}$$

---

[6] As the extended CoRR/abs article is freely available we will avoid to repeat many of the tedious technical details here.

As for the sequential and parallel ASM theses it is clear that if $\mathbf{S}'$ is a successor state of $\mathbf{S}$, then there is a unique, consistent update set $\Delta = \Delta(\mathbf{S}, \mathbf{S}')$ such that applying Δ to $\mathbf{S}$ results in $\mathbf{S}'$, for which we use the notation $\mathbf{S}' = \mathbf{S} + \Delta(\mathbf{S}, \mathbf{S}')$. Thus, the transformation relation τ_A defines sets $\Delta(\mathbf{S})$ of update sets for all states $\mathbf{S}$ of A.

2.2 The Background of a Computation

The background of computation for non-deterministic, parallel algorithms is the same as for parallel algorithms. States are grounded in meta-finite structures, which in one side emphasises their intrinsic finiteness, but on the other side permits extensions by means of the so-called *reserve*, i.e. a set of values that do not appear in a state $\mathbf{S}$, but from which elements may be taken for the updates from $\mathbf{S}$ to some successor $\mathbf{S}'$. Furthermore, we have to distinguish between states of the algorithms and states of the *computation* of the algorithms, the latter ones including in addition the standard background needed for the computation.

Postulate 3 (Background Postulate). Let A be a non-deterministic, parallel algorithm with background class $\mathcal{K}$. The vocabulary $\Sigma_\mathcal{K}$ of $\mathcal{K}$ includes (at least) a binary *tuple constructor* and a *multiset constructor* of unbounded arity; and the vocabulary $\Sigma_\mathbf{B}$ of the background of the computation states of A includes (at least) the following *obligatory* function symbols:

- Nullary function (constants) symbols true, false, undef and $\oslash$.
- Unary function symbols reserve, atomic, first, second, Boole, $\neg$, $\{\cdot\}$, $\uplus$ and AsSet.
- Binary function symbols $=$, $\wedge$, $\vee$, $\rightarrow$, $\leftrightarrow$, $\uplus$ and $(,)$.

For details please refer to [9, Sect. 3].

2.3 Bounded Exploration

The fourth postulate for parallel algorithms is the bounded exploration postulate. It states that if two states coincide on a *bounded exploration witness*, i.e. a finite set of multiset comprehension terms, then these two states have the same update sets. This reflects the fact that all information that determines the updates must be included in the finite representation of the algorithms, in other words, what is read in a state determines the updates in that state. This remains the same for non-deterministic algorithms with the little change that what is read also determines the choices, i.e. if two states coincide on a bounded exploration witness, they should determine the same set of update sets.

We refer to the class of multiset comprehension terms of the form

$$\{t(x_1, \ldots, x_r) \mid \varphi(x_1, \ldots, x_r)\}$$

which have *no* free-variables and where t is an ordered tuple that represents (using some fixed encoding) a tuple $(t_0, \ldots, t_n)$ of terms with $\mathit{free}(t_0) \cup \cdots \cup \mathit{free}(t_n) = \{x_1, \ldots, x_r\}$, as *witness terms*, and denote them as

$$\{(t_0, \ldots, t_n) \mid \varphi(x_1, \ldots, x_r)\}$$

Two states of computation $\mathbf{S}_1$ and $\mathbf{S}_2$ of a same vocabulary Σ *coincide* over a set W of witness terms if for every $\alpha_i \in W$, we have that $\text{val}_{\mathbf{S}_1}(\alpha_i) = \text{val}_{\mathbf{S}_2}(\alpha_i)$.

Postulate 4 (Bounded Exploration Postulate). Let A be a parallel algorithm. Then there is a finite set W of witness terms, called *bounded exploration witness* of A, such that for every pair of states $\mathbf{S}_1$ and $\mathbf{S}_2$ of A, it holds that $\Delta(\mathbf{S}_1) = \Delta(\mathbf{S}_2)$ whenever the corresponding computation states coincide over W.

Note that bounded exploration implies that the set $\Delta(\mathbf{S})$ of update sets for a state $\mathbf{S}$ is finite, and that every update set $\Delta \in \Delta(\mathbf{S})$ is finite, too.

2.4 Slicing

So far we mainly argued that the postulates for deterministic, parallel algorithms would hold to a large extent also for non-deterministic, parallel algorithms. The main difference is that states may have several successor states, which implies that there is a set $\Delta(\mathbf{S})$ of update sets for each state $\mathbf{S}$, and that the evaluation of a bounded exploration witness W determines $\Delta(\mathbf{S})$.

Now let $\Phi = \{(t_1,\ldots,t_n) \mid \varphi(x_1,\ldots,x_r)\}$ be a witness term for some bounded exploration witness W. The evaluation of Φ in a state $\mathbf{S}$ determines a multiset of tuples $\bar{a} = (a_1,\ldots,a_r)$. For the deterministic case we argued that each such $\bar{a}$ corresponds to a branch of the parallel computation. For the non-deterministic case, however, this is no longer true, as Φ might as well determine the different choices the algorithm A can make in a state. That is, we will need a way to distinguish between the witness terms responsible for the parallelism and those responsible for non-deterministic choice.

Our key idea is the following. If Φ corresponds to non-deterministic choices, then its evaluation in a state $\mathbf{S}$ determines how many update sets are in $\Delta(\mathbf{S})$. In other words, if we "slice" Φ into several parts $\Phi_1,\ldots,\Phi_k$ by replacing φ by $\varphi_1,\ldots,\varphi_k$ (with $\varphi_1 \wedge \cdots \wedge \varphi_k \equiv \varphi$), then each Φ_i should give rise only to a subset $\Delta_i(\mathbf{S})$ of $\Delta(\mathbf{S})$.

Definition 2. Let $\Phi = \{(t_1,\ldots,t_n) \mid \varphi(x_1,\ldots,x_r)\} \in W$ be a witness term. For any formula ψ define

$$\Phi^+ = \{(t_1,\ldots,t_n) \mid \varphi(x_1,\ldots,x_r) \wedge \psi(x_1,\ldots,x_r)\} \text{ and}$$
$$\Phi^- = \{(t_1,\ldots,t_n) \mid \varphi(x_1,\ldots,x_r) \wedge \neg\psi(x_1,\ldots,x_r)\}$$

and let W^+ and W^- result from W by replacing Φ by Φ^+ or Φ^-, respectively. Φ *satisfies the slicing property* iff for every choice of ψ and every state $\mathbf{S}$ we obtain $\Delta(\mathbf{S}) = \Delta^+(\mathbf{S}) \cup \Delta^-(\mathbf{S})$, where $\Delta^+(\mathbf{S})$ and $\Delta^-(\mathbf{S})$ are the sets of update sets for $\mathbf{S}$ for the algorithms with bounded exploration witness W^+ and W^-, respectively.

Obviously, any bounded exploration witness W is the disjoint union of W_s and W_{ns}, where W_s contains all witness terms of W that satisfy the slicing property, and W_{ns} contains all other witness terms. As W_s is responsible for the non-determinism, we can formulate our last postulate for non-deterministic, parallel algorithms[7].

Postulate 5 (Slicing Postulate). If no witness term in the bounded exploration witness W of A satisfies the slicing property, then A is deterministic.

If A is deterministic, then $\Delta(\mathbf{S})$ contains only a single update set and thus A satisfies exactly the postulates for parallel algorithms [9]. In particular, there exists a parallel ASM that is behaviourally equivalent to A.

3 Non-Deterministic Abstract State Machines

We can assume that the concept of Abstract State Machines (ASMs) is well known [6]. Also non-deterministic choice has always been present in ASMs, though it has not been handled in the justifying behavioural theories, except for the research on non-deterministic database transformations [13, 16]. Here we briefly repeat the rules for non-deterministic ASMs. For further details on update sets yielded by ASMs see the elaborate presentation in [6].

Definition 3 (ASM Rules). The set $\mathcal{R}$ of ASM rules over a signature Σ is defined inductively by:

- If $f \in \Sigma$ is an n-ary function symbol and $t_0, t_1, \ldots, t_n$ are terms of vocabulary $\Sigma_\mathbf{S}$, then $f(t_1, \ldots, t_n) := t_0$ is an *assignment* rule in $\mathcal{R}$.
- If φ is a term of vocabulary $\Sigma_\mathbf{S}$ and r is a rule in $\mathcal{R}$, then **if** φ **then** r **endif** is a *conditional* rule in $\mathcal{R}$.
- If φ is a term of vocabulary $\Sigma_\mathbf{S}$ with $\mathit{free}(\varphi) \supseteq \{x_1, \ldots, x_k\}$ and r is a rule in $\mathcal{R}$, then **forall** $x_1, \ldots, x_k$ **with** φ **do** r **enddo** is a *forall* rule in $\mathcal{R}$.
- If φ is a term of vocabulary $\Sigma_\mathbf{S}$ with $\mathit{free}(\varphi) \supseteq \{x_1, \ldots, x_k\}$ and r is a rule in $\mathcal{R}$, then **choose** $x_1, \ldots, x_k$ **with** φ **do** r **enddo** is a *choice* rule in $\mathcal{R}$.

Theorem 1 (Plausibility). *Every non-deterministic, parallel ASM $\mathcal{M}$ defines a non-deterministic, parallel algorithm with the same vocabulary and background as $\mathcal{M}$.*

We omit the easy proof, as it is completely analogous to the plausibility proof for deterministic, parallel ASMs [9, Theorem 5.1]. Only the case of the choice rule has to be added.

4 The Characterisation Theorem

Our goal is now to prove the converse of Theorem 1, i.e. to show that every non-deterministic, parallel algorithm A as stipulated by our five postulates can be represented by a behaviourally equivalent non-deterministic, parallel ASM $\mathcal{M}$ with the same vocabulary and background. That is to generalise the correponding result for deterministic, parallel algorithms [9, Theorem 7.2].

---

[7] As this article describes research that is still in progress to some extent, we would like to state that we believe that the slicing postulate is too strong, i.e. it should be possible to simplify the theory using a weaker postulate. For the time being we do not have a proof for such a conjecture.

Theorem 2 (Characterisation). *For every non-deterministic, parallel algorithm A there is a behaviourally equivalent non-deterministic, parallel ASM $\mathcal{M}$.*

Proof (Sketch). Fix a bounded exploration witness W for A, and consider a state $\mathbf{S}$. Let $\Delta(\mathbf{S}) = \{\Delta_1, \ldots, \Delta_k\}$. For all $\Phi \in W_s$ we can replace the defining formula φ such that Δ_i is the sole update set in state $\mathbf{S}$ for the modified set W_i of witness terms. According to the slicing postulate the corresponding algorithm is deterministic. Therefore, using the other postulates we can exploit the Characterisation Theorem for deterministic, parallel algorithms [9, Theorem 7.2], which gives us an ASM rule r that yields the update set Δ_i in state $\mathbf{S}$. Consequently, the rule

choose $x_1, \ldots, x_k$ **with** $\varphi_1 \wedge \cdots \wedge \varphi_\ell$ **do** r **enddo**

(where the formulae φ_i correspond to the witness terms in W_s) defines a rule that yields $\Delta(\mathbf{S})$. Now proceed analogously to the proof for deterministic, parallel algorithms using Lemmata 7.9–7.11 in [9] to obtain the claimed result. □

5 From Non-Determinism to Randomness

Theorems 1 and 2 capture randomness in parallel computation, as long as we assume that all choices made by an algorithm are uniformly distributed, which is often assumed for random algorithms [7]. However, it would be desirable to extend the theory to algorithms, where arbitrary distributions could be exploited. We will concentrate here on the discrete case, for which we would only modify the sequential time postulate.

Postulate 6 (Random Sequential Time Postulate). A *random, parallel algorithm* A is associated with a non-empty set $\mathcal{S}_A$ of *states*, a non-empty subset $\mathcal{I}_A \subseteq \mathcal{S}_A$ of *initial states*, and probability distributions $d_\mathbf{S} : \mathcal{S}_A \to [0, 1]$. For $d_\mathbf{S}(\mathbf{S}') > 0$ we call $\mathbf{S}'$ a *possible successor state* of $\mathbf{S}$.

We leave all other postulates as they are. Then we can proceed as for the case of non-deterministic, parallel algorithms. In particular, for each state $\mathbf{S}$ we obtain a set $\Delta(\mathbf{S})$ of update sets corresponding to all its possible successor states, and $d_\mathbf{S}$ induces a probability distribution on $\Delta(\mathbf{S})$.

We modify the choice rule of ASMs to become

choose $x_1, \ldots, x_k$ **by** $d(x_1, \ldots, x_k)$ **with** φ **do** r **enddo**

which includes a (discrete) probability distribution d. In doing so, we believe that the proofs of the Plausibility Theorem 1 and of the Characterisation Theorem 2 can be generalised to the case of random, parallel algorithms. In particular, in the proofs the probability distributions $d_\mathbf{S}$ in Postulate 6 and d in the ASM rules should mutually determine each other. For the time being it is no more than a conjecture, and its proof may not be straightforward, as it was the case for the proof of the behavioural theory for synchronous parallel algorithms.

6 Concluding Remarks

There are many ways how the research sketched in this article could be taken further. For instance, we could look for a behavioural theory of asynchronous random, parallel algorithms, which would generalise the concurrent ASM thesis that was developed recently [5]. We could also incorporate linguistic reflection, i.e. the ability of an algorithm to modify itself. A behavioural theory for reflective, parallel algorithms is currently under development in our project "Behavioural Theory and Logic for Distributed Adaptive Systems".

However, the most exciting question is, how probabilistic logic with maximum entropy semantics, which is a major field of Gabriele Kern-Isberner's research [1, 2, 15], could be exploited in a logic for reasoning about random, parallel algorithms. For this probabilistic reasoning will have to be integrated into logics for variants of ASMs [12, 14, 16]. We leave this open problem as an open invitation for future research.

References

1. C. Beierle, M. Finthammer, and G. Kern-Isberner. Relational probabilistic conditionals and their instantiations under maximum entropy semantics for first-order knowledge bases. *Entropy*, 17(2):852–865, 2015.
2. C. Beierle and G. Kern-Isberner. Semantical investigations into nonmonotonic and probabilistic logics. *Annals of Mathematics and Artificial Intelligence*, 65(2-3):123–158, 2012.
3. A. Blass and Y. Gurevich. Abstract state machines capture parallel algorithms. *ACM Transactions on Computational Logic*, 4(4):578–651, 2003.
4. A. Blass and Y. Gurevich. Abstract state machines capture parallel algorithms: Correction and extension. *ACM Transactions on Computational Logic*, 9(3), 2008.
5. E. Börger and K.-D. Schewe. Concurrent Abstract State Machines. *Acta Informatica*, 2015. to appear.
6. E. Börger and R. F. Stärk. *Abstract State Machines. A Method for High-Level System Design and Analysis*. Springer, 2003.
7. G. Brassard and P. Bratley. *Algorithmics - theory and practice*. Prentice Hall, 1988.
8. H.-D. Ebbinghaus and J. Flum. *Finite Model Theory*. Perspectives in Mathematical Logic. Springer, Berlin Heidelberg New York, 2nd edition, 1999.
9. F. Ferrarotti, K.-D. Schewe, L. Tec, and Q. Wang. A new thesis concerning synchronised parallel computing – simplified parallel ASM thesis. *CoRR*, abs/1504.06203, 2015. available at http://arxiv.org/abs/1504.06203, submitted for publication.
10. E. Grädel and Y. Gurevich. Metafinite model theory. *Information and Computation*, 140(1):26–81, 1998.
11. Y. Gurevich. Sequential abstract-state machines capture sequential algorithms. *ACM Transactions on Computational Logic*, 1(1):77–111, 2000.
12. S. Nanchen and R. F. Stärk. A logic for secure memory access of abstract state machines. *Theoretical Computer Science*, 336(2-3):343–365, 2005.
13. K.-D. Schewe and Q. Wang. A customised ASM thesis for database transformations. *Acta Cybernetica*, 19(4):765–805, 2010.
14. R. F. Stärk and S. Nanchen. A logic for abstract state machines. *Journal of Universal Computer Science*, 7(11):980–1005, 2001.

15. M. Thimm and G. Kern-Isberner. On probabilistic inference in relational conditional logics. *Logic Journal of the IGPL*, 20(5):872–908, 2012.
16. Q. Wang. *Logical Foundations of Database Transformations for Complex-Value Databases*. Logos, 2010.

Part VI

Argumentation

On Argumentation-based Paraconsistent Logics

Leila Amgoud[1]

Abstract. *Argumentation* is an alternative approach for reasoning with inconsistent information. Starting from a knowledge base (a set of premises) encoded in a logical language, an argumentation-based logic defines *arguments* and *attacks* between them using the consequence operator associated with the language, then uses a *semantics* for evaluating the arguments. The plausible conclusions to be drawn from the knowledge base are those supported by "good" arguments.
In this paper, we discuss two families of such logics: the family of logics that uses extension semantics for the evaluation of arguments, and the one that uses ranking semantics. We discuss the outcomes of both families and compare them.

1 Introduction

An important problem in the management of knowledge-based systems is handling of inconsistency. Inconsistency may be present for mainly three reasons: i) A knowledge base may contain a default rule and strict rules encoding exceptions of the default rule [26]. The two kinds of rules may lead to opposite conclusions. ii) In model-based diagnosis [21], the description of the normal behavior of a system may be conflicting with the observations made on this system. iii) An inconsistent knowledge base may result from the union of several consistent knowledge bases pertaining to the same domain [12]. Moreover, in [18], Gabbay and Hunter claim that inconsistency in a database exists on purpose and may be useful if its presence triggers suitable actions that cope with it. They give the example of overbooking in airline booking systems.

Whatever the source of inconsistency, a paraconsistent logic is needed to deal with it. A paraconsistent logic consists of a language and a consequence operator which returns rational conclusions even from inconsistent sets of formulas.

There has been much work on constructing and investigating such logics. Two families can be distinguished: those that restore consistency (e.g., [10, 26, 27]) and those that tolerate inconsistency and cope with it (e.g., [6, 7, 13, 17]). One important instance of the first family computes the maximal (for set inclusion) consistent subbases of a knowledge base, then chooses the conclusions that follow from all those subbases ([27]). Regarding the second family, a prominent approach considers many-valued interpretations with the crucial particularity that they can be models of even inconsistent premises and thus, can be used to draw conclusions.

Since early nineties, due to its explanatory power, *argumentation* has become a promising approach for handling inconsistency. Like many-valued logics,

---

[1] IRIT – CNRS, 118, route de Narbonne, 31062, Toulouse Cedex 09, `amgoud@irit.fr`

it accepts inconsistency and copes with it. Starting from a knowledge base encoded in a particular logical language, an argumentation-based logic builds arguments and attack relations between them using a consequence operator associated with the language, then it evaluates the arguments using a semantics. Finally, it draws the conclusions that are supported by good arguments. Attacks generally refer to the inconsistency of the base.

In the argumentation literature, arguments are mainly evaluated using *extension*-based semantics (or extension semantics for short) as introduced by Dung in his seminal paper [14]. Extension semantics are functions transforming any argumentation graph into one or several subsets of arguments, called *extensions*, each of which representing a coherent point of view. Using the extensions, the set of arguments is partitioned into three disjoint categories: i) the arguments which are in all extensions (called *sceptically* accepted), ii) the arguments that are in some but not all extensions (called *credulously* accepted), and ii) the arguments which do not belong to any extension (called *rejected*). Examples of extension semantics are the well-known stable, preferred, complete, grounded, and admissible semantics proposed by Dung in [14], as well as their refinements like the recursive semantics [5] and ideal semantics [15].

More recently, another family of semantics, called *graded semantics*, is emerging. Those semantics do not compute extensions and are based on different principles. For instance, the number of attackers is taken into account while it does not play any role in extension semantics. Thus, the two families of semantics (extension semantics and graded semantics) may not provide the same evaluations of arguments.

Graded semantics assign to every argument a numerical value representing its *strength*. These values allow to rank-order the arguments from the most acceptable to the less acceptable ones. Ranking semantics [1], h-categoriser semantics [8, 25], value-based semantics [11], and game-theoretical semantics [22] are examples of graded semantics.

In [29], the authors presented a very interesting extension of Dung's semantics by introducing the notion of *stratified labellings*. Like graded semantics, the idea is to provide graded assessment of arguments by assigning a degree to each argument. However, the degree does not represent the strength of an argument but rather to what extent the argument is controversial. Thus, the corresponding semantics do not satisfy the mandatory properties of graded semantics. Interestingly enough, stratified labellings have close relationships with ranking functions like Z-ordering [23].

All the above semantics evaluate arguments solely on the basis of attacks and do not take into account the internal structure of arguments. Their input is a plain directed graph whose nodes and arrows represent abstract arguments and attacks.

Our aim in this paper is to discuss and compare the paraconsistent logics built on top of any Tarskian logic ([28]) and induced by each family of semantics. We consider in particular the most popular extension semantics (stable and preferred semantics introduced by Dung in [14]) and a *ranking semantics* defined more recently in [1]. We show that logics based on extension semantics

(ALES) return flat conclusions and restore consistency. Furthermore, they generalize to any Tarskian logic the paraconsistent logic defined by Rescher and Manor in [27] on top of propositional logic. The paraconsistent logics based on ranking semantics (ALRS) return ranked conclusions (from the most plausible to the less plausible ones) and tolerate inconsistency. Moreover, they are good candidates for measuring inconsistency in knowledge bases. Finally, we show that ALRS are more discriminating than ALES in that they solve inconsistency while ALES avoid it.

2 Argumentation-based logics

An argumentation-based logic is built on top of a logic. In this paper, we focus on Tarskian logics [28]. According to Tarski, a logic is a set of well-formed *formulae* and a *consequence operator* which returns the set of formulae that follow from another set of formulae. There are no requirements on the connectives used in the language. However, the consequence operator should satisfy some very basic properties.

Definition 1 (Logic). *A logic is a tuple $\langle \mathcal{F}, w, \mathtt{CN} \rangle$ where $\mathcal{F}$ is a set of well-formed formulae, w is a well-order[2] on $\mathcal{F}$, $\mathtt{CN}$ is a consequence operator, i.e., a function from $2^{\mathcal{F}}$ to $2^{\mathcal{F}}$ such that for $\Phi \subseteq \mathcal{F}$,*

- $\Phi \subseteq \mathtt{CN}(\Phi)$ *(Expansion)*
- $\mathtt{CN}(\mathtt{CN}(\Phi)) = \mathtt{CN}(\Phi)$ *(Idempotence)*
- $\mathtt{CN}(\Phi) = \bigcup_{\Psi \subseteq_f \Phi} \mathtt{CN}(\Psi)$ *(Compactness)*
- $\mathtt{CN}(\{\varphi\}) = \mathcal{F}$ *for some $\varphi \in \mathcal{F}$* *(Absurdity)*
- $\mathtt{CN}(\varnothing) \neq \mathcal{F}$ *(Coherence)*

Notation: $Y \subseteq_f X$ means that Y is a finite subset of X.

The well-ordering w enables to arbitrarily select a *representative* formula among equivalent ones. Its exact definition is not important for the purpose of the paper. Almost all well-known monotonic logics (classical logics, intuitionistic logics, modal logics, etc.) can be viewed as special cases of Tarski's notion of an abstract logic. AI introduced non-monotonic logics, which do not satisfy monotonicity [9].

The next definition introduces the concept of *adjunctive* logic.

Definition 2 (Adjunctiveness). *A logic $\langle \mathcal{F}, w, \mathtt{CN} \rangle$ is adjunctive iff for all φ and ψ in $\mathcal{F}$, there exists $\alpha \in \mathcal{F}$ such that $\mathtt{CN}(\{\alpha\}) = \mathtt{CN}(\{\varphi, \psi\})$.*

Intuitively, an adjunctive logic infers, from the union of two formulas $\{\varphi, \psi\}$, some formula(s) that can be inferred neither from φ alone nor from ψ alone

---

[2] A *well-order* on a set X is a relation with the property that every non-empty subset of X has a least element in this ordering.

(except, of course, when ψ ensues from φ or vice-versa). In fact, most well-known logics are adjunctive.[3] A logic which is not adjunctive could for instance fail to deny $\varphi \vee \psi$ from the premises $\{\neg\varphi, \neg\psi\}$.

The notion of *consistency* associated with such logics is defined as follows:

Definition 3 (Consistency). *A set $\Phi \subseteq \mathcal{L}$ is consistent wrt a logic $(\mathcal{L}, w, \mathtt{CN})$ iff $\mathtt{CN}(\Phi) \neq \mathcal{L}$. It is* inconsistent *otherwise.*

Before introducing the notion of argument, let us first define when pairs of formulas are equivalent.

Definition 4 (Equivalent formulas). *Let $\langle \mathcal{F}, w, \mathtt{CN} \rangle$ be a logic and $\varphi, \psi \in \mathcal{F}$. The formula φ is equivalent to ψ wrt logic $\langle \mathcal{F}, w, \mathtt{CN} \rangle$, denoted by $\varphi \equiv \psi$, iff $\mathtt{CN}(\{\varphi\}) = \mathtt{CN}(\{\psi\})$.*

The building block of argumentation-based logics is the notion of argument. An argument is a reason for concluding a formula. Thus, it has two main components: a *support* and a *conclusion*. In what follows, two arguments having the same supports and different yet equivalent conclusions are not distinguished, they are rather seen as the same argument. The reason is that those arguments are redundant and increase uselessly and misleadingly the argumentation graph both from a theoretical and computational point of view.

Definition 5 (Argument). *Let $\langle \mathcal{F}, w, \mathtt{CN} \rangle$ be a logic and $\Phi \subseteq_f \mathcal{F}$. An argument* built from Φ *is a pair (Ψ, ψ) such that:*

- *$\Psi \subseteq \Phi$ and Ψ is consistent,*
- *ψ is the w-smallest element of $\{\psi' \in \mathcal{F} \mid \psi' \equiv \psi\}$ such that $\psi \in \mathtt{CN}(\Psi)$,*
- *$\nexists \Psi' \subset \Psi$ such that $\psi \in \mathtt{CN}(\Psi')$.*

An argument (Ψ, ψ) is a sub-argument *of (Ψ', ψ') iff $\Psi \subseteq \Psi'$.*

Notations: Let $\langle \mathcal{F}, w, \mathtt{CN} \rangle$ be a logic and $\Phi \subseteq_f \mathcal{F}$. $\mathtt{Supp}$ and $\mathtt{Conc}$ denote respectively the *support* Ψ and the *conclusion* ψ of an argument (Ψ, ψ) built from Φ. $\mathtt{Arg}(\Phi)$ denotes the set of all arguments that can be built from Φ by means of Definition 5, $\mathtt{Sub}((\Psi, \psi))$ is a function that returns all the sub-arguments of argument (Ψ, ψ). For any $\mathcal{E} \subseteq \mathtt{Arg}(\Phi)$, $\mathtt{Concs}(\mathcal{E}) = \{\mathtt{Conc}(a) \mid a \in \mathcal{E}\}$ and $\mathtt{Base}(\mathcal{E}) = \bigcup_{a \in \mathcal{E}} \mathtt{Supp}(a)$. $\mathtt{Max}(\Phi)$ is the set of all maximal (for set inclusion) consistent subsets of Φ, i.e. for any $S \in \mathtt{Max}(\Phi)$, $S \subseteq \Phi$, S is consistent wrt logic $(\mathcal{F}, w, \mathtt{CN})$, and for any $\varphi \in \Phi \backslash S$, it holds that $S \cup \{\varphi\}$ is inconsistent. $\mathtt{MIC}(\Phi)$ denotes the set of all minimal (for set inclusion) inconsistent subsets of Φ, i.e., for any $S \in \mathtt{MIC}(\Phi)$, $S \subseteq \Phi$, S is inconsistent wrt logic $(\mathcal{F}, w, \mathtt{CN})$, and for any $\varphi \in S$, it holds that $S \backslash \{\varphi\}$ is consistent wrt logic $(\mathcal{F}, w, \mathtt{CN})$. Finally, $\mathtt{Free}(\Phi) = \{\varphi \in \Phi \mid \forall S \in \mathtt{MIC}(\Phi), \varphi \notin S\}$, i.e., $\mathtt{Free}(\Phi)$ is the set of formulae of Φ that are not involved in any minimal (for set inclusion) inconsistent subset of Φ.

---

[3] Some fragments of well-known logics fail to be adjunctive, e.g., the pure implicational fragment of classical logic as it is negationless, disjunctionless, and, of course, conjunctionless.

Since information may be inconsistent, arguments may attack each other. In what follows, such attacks are captured by a binary relation, denoted by $\mathcal{R}$. For two arguments a, b, $(a,b) \in \mathcal{R}$ (or $a\mathcal{R}b$) means that a *attacks* b. For the sake of generality, $\mathcal{R}$ is left *unspecified*. It can thus be instantiated in various ways (see [19] for examples of instantiations of $\mathcal{R}$). However, we assume that it is based on inconsistency.

Definition 6 (Conflict-dependency). Let $\langle \mathcal{F}, w, \text{CN} \rangle$ be a logic and $\Phi \subseteq_f \mathcal{F}$. An attack relation $\mathcal{R} \subseteq \text{Arg}(\Phi) \times \text{Arg}(\Phi)$ is conflict-dependent *iff for all* $a, b \in \text{Arg}(\Phi)$, if $(a,b) \in \mathcal{R}$ then $\text{Supp}(a) \cup \text{Supp}(b)$ *is inconsistent*.

All existing attack relations are conflict-dependent with a notable exception, *undercutting* [24], which prevents the application of defaults in case of logics built on top of rule-based languages like ASPIC [4].

As said before, an argumentation-based logic defines from each set of formula a directed graph whose nodes are arguments and arrows are attacks between them.

Definition 7 (Argumentation function). *An* argumentation function $\mathcal{G}$ *on a logic* $\langle \mathcal{F}, w, \text{CN} \rangle$ *transforms any set* $\Phi \subseteq_f \mathcal{F}$ *into a finite directed graph*[4] $\langle \text{Arg}(\Phi), \mathcal{R} \rangle$ *where* $\mathcal{R} \subseteq \text{Arg}(\Phi) \times \text{Arg}(\Phi)$ *is a conflict-dependent attack relation.*

We are now ready to introduce *argumentation-based logics* (AL). An AL is a logic (in the sense of Definition 1) which is defined upon a *base logic*. The latter is supposed to behave in a rational way when information is consistent but exhibits an irrational behaviour in presence of inconsistency. Propositional logic is an example of such logic. AL restricts thus the base logic's inference power. An AL proceeds as follows: For any set Φ of formulas in the base logic, it defines its corresponding argumentation graph. The conclusions to be drawn from Φ using the consequence operator of the AL are the formulae that are supported by good arguments according to a given semantics $\mathcal{S}$.

Definition 8 (AL). *An argumentation-based logic (AL) is a logic* $\mathcal{L} = \langle \mathcal{F}, w, \text{CN}' \rangle$ *which is based on base logic* $\langle \mathcal{F}, w, \text{CN} \rangle$, *argumentation function* $\mathcal{G}$ *on* $\langle \mathcal{F}, w, \text{CN} \rangle$ *and semantics* $\mathcal{S}$, *where for any* $\Phi \subseteq_f \mathcal{F}$, $\text{CN}'(\Phi) \subseteq \{\varphi \in \mathcal{F} \mid \exists a \in \text{Arg}(\Phi)$ *with* $\mathcal{G}(\Phi) = \langle \text{Arg}(\Phi), \mathcal{R} \rangle$ *and* $\text{Conc}(a) \equiv \varphi$ *wrt logic* $\langle \mathcal{F}, w, \text{CN} \rangle\}$. *If* Φ *is consistent wrt* $\langle \mathcal{F}, w, \text{CN} \rangle$, *then* $\text{CN}'(\Phi) = \text{CN}(\Phi)$.

In the next two sections, we define more precisely the consequence operators CN' of the logics induced by extensions semantics and ranking ones.

3 Logics induced by extension semantics

The most popular semantics were proposed by Dung in his seminal paper [14]. Those semantics as well as their refinements (e.g. in [5, 15]) partition the powerset of the set of arguments into two classes: *extensions* and *non-extensions*.

---

[4] In the literature, the pair $\langle \text{Arg}(\Phi), \mathcal{R} \rangle$ is also called *argumentation system*.

Every extension represents a coherent point of view. We illustrate the kind of paraconsistent logics induced by such semantics, namely naive, stable and preferred. Before giving the formal definitions of the three semantics, we first introduce two key concepts on which they are based.

Definition 9 (Conflict-freeness–Defence). Let $\mathcal{T} = \langle \mathcal{A}, \mathcal{R} \rangle$ be an argumentation graph, $\mathcal{E} \subseteq \mathcal{A}$ and $a \in \mathcal{A}$.

- $\mathcal{E}$ is conflict-free iff $\nexists a, b \in \mathcal{E}$ such that $a\mathcal{R}b$.
- $\mathcal{E}$ defends an argument a iff $\forall b \in \mathcal{A}$ such that $b\mathcal{R}a$, $\exists c \in \mathcal{E}$ such that $c\mathcal{R}b$.

Definition 10 (Semantics). Let $\mathcal{T} = \langle \mathcal{A}, \mathcal{R} \rangle$ be an argumentation graph and $\mathcal{E} \subseteq \mathcal{A}$.

- $\mathcal{E}$ is a naive extension iff it is a maximal (w.r.t. set $\subseteq$) conflict-free set.
- $\mathcal{E}$ is an admissible set iff it defends all its elements.
- $\mathcal{E}$ is a complete extension iff it is an admissible set that contains any argument it defends.
- $\mathcal{E}$ is a preferred extension iff it is a maximal (w.r.t. set $\subseteq$) set that is conflict-free and defends its elements.
- $\mathcal{E}$ is a stable extension iff it is conflict-free and attacks any argument in $\mathcal{A} \backslash \mathcal{E}$.
- $\mathcal{E}$ is a grounded extension iff it is a minimal (w.r.t. set $\subseteq$) complete extension.
- $\mathcal{E}$ is an ideal extension iff it is a maximal (w.r.t. set $\subseteq$) admissible set contained in every preferred extension.

Notations: $\text{Ext}_x(\mathcal{T})$ denotes the set of all extensions of $\mathcal{T}$ under semantics x where $x \in \{n, p, s\}$ and n (resp. p and s) stands for naive (respectively preferred and stable). When we do not need to refer to a particular semantics, we write $\text{Ext}(\mathcal{T})$ for short. Since any argumentation framework $\mathcal{T}$ has a single grounded and a single ideal extension, they will be denoted respectively by $\text{GE}(\mathcal{T})$ and $\text{IE}(\mathcal{T})$.

Example 1. The argumentation graph depicted below

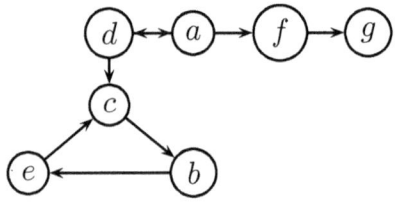

has five naive extensions: $\mathcal{E}_1 = \{a, c, g\}$, $\mathcal{E}_2 = \{d, e, f\}$, $\mathcal{E}_3 = \{b, d, f\}$, $\mathcal{E}_4 = \{a, e, g\}$, $\mathcal{E}_5 = \{a, b, g\}$; one stable $\mathcal{E}_3$ and two preferred extensions $\mathcal{E}_3$ and $\mathcal{E}_6 = \{a, g\}$.

It is worth recalling that stable extensions are naive (respectively preferred) extensions but the converses are not always true. Moreover, an argumentation framework may have no stable extensions.

Let us now define the plausible conclusions that may be drawn from a set of formulae Φ by an argumentation-based logic. The idea is to infer a formula φ from Φ iff it is the conclusion of at least one argument in every extension of the argumentation graph built from Φ.

Definition 11 (ALES). *An argumentation-based logic induced from extension semantics (ALES) is a logic $\mathcal{L} = \langle \mathcal{F}, w, \mathtt{CN}' \rangle$ based on base logic $\langle \mathcal{F}, w, \mathtt{CN} \rangle$, argumentation function $\mathcal{G}$, and semantics $x \in \{n, p, s\}$, where*

$$\text{for all } \Phi \subseteq_f \mathcal{F}, \text{ for all } \varphi \in \mathcal{F},$$
$\varphi \in \mathtt{CN}'(\Phi)$ *iff* $\forall \mathcal{E} \in \mathtt{Ext}_x(\mathcal{G}(\Phi)), \exists a \in \mathcal{E}$ *s.t.* $\mathtt{Conc}(a) \equiv x$ *wrt logic* $\langle \mathcal{F}, w, \mathtt{CN} \rangle$.

In [3] a comprehensive study has been made on the family of logics described in this section. It has been shown that when the argumentation graph built over a set of formulae satisfies two key properties, then there is a full correspondence between the naive extensions of the graph and the maximal consistent subsets of the set of formulae. Before presenting the formal result, let us first recall the two properties.

Postulates (Closure under sub-arguments – Consistency) Let $\mathcal{G}$ be an argumentation function on logic $\langle \mathcal{F}, w, \mathtt{CN} \rangle$ and $\Phi \subseteq_f \mathcal{F}$. For all $\mathcal{E} \in \mathtt{Ext}(\mathcal{G}(\Phi))$,

- if $a \in \mathcal{E}$, then $\mathtt{Sub}(a) \subseteq \mathcal{E}$. We say that $\mathcal{G}(\Phi)$ is closed under sub-arguments.
- $\mathtt{Concs}(\mathcal{E})$ is consistent. We say that $\mathcal{G}(\Phi)$ satisfies consistency.

The following result shows that there is a one-to-one correspondence between the naive extensions of an argumentation graph and the maximal (for set inclusion) subsets of the set of formulae over which the graph is built. Indeed, each maximal consistent subset gives birth to a naive extension using the function $\mathtt{Arg}$ and each naive extension returns a maximal consistent set of formulae using the function $\mathtt{Base}$.

Theorem 1. *[3] Let $\mathcal{G}$ be an argumentation function on an adjunctive logic $\langle \mathcal{F}, w, \mathtt{CN} \rangle$, and let $\Phi \subseteq_f \mathcal{F}$. If $\mathcal{G}(\Phi)$ satisfies consistency and is closed under sub-arguments (under naive semantics), then:*

- *For all $\mathcal{E} \in \mathtt{Ext}_n(\mathcal{G}(\Phi))$, $\mathtt{Base}(\mathcal{E}) \in \mathtt{Max}(\Phi)$.*
- *For all $\mathcal{E}_i, \mathcal{E}_j \in \mathtt{Ext}_n(\mathcal{G}(\Phi))$, if $\mathtt{Base}(\mathcal{E}_i) = \mathtt{Base}(\mathcal{E}_j)$ then $\mathcal{E}_i = \mathcal{E}_j$.*
- *For all $\mathcal{E} \in \mathtt{Ext}_n(\mathcal{G}(\Phi))$, $\mathcal{E} = \mathtt{Arg}(\mathtt{Base}(\mathcal{E}))$.*
- *For all $\mathcal{S} \in \mathtt{Max}(\Phi)$, $\mathtt{Arg}(\mathcal{S}) \in \mathtt{Ext}_n(\mathcal{G}(\Phi))$.*

Let us now characterize the set of inferences that may be drawn from a set of formulae Φ by any argumentation-based logic under naive semantics. It coincides with the set of inferences that are drawn from the maximal consistent subsets of Φ.

Theorem 2. *[3] Let $\mathcal{L} = \langle \mathcal{F}, w, \mathtt{CN}' \rangle$ be an ALES based on adjunctive logic $\langle \mathcal{F}, w, \mathtt{CN} \rangle$, argumentation function $\mathcal{G}$, and naive semantics. For all $\Phi \subseteq_f \mathcal{F}$, if $\mathcal{G}(\Phi)$ satisfies consistency and is closed under sub-arguments, then*

$$\mathtt{CN}'(\Phi) = \bigcap_{\mathcal{S}_i \in \mathtt{Max}(\Phi)} \mathtt{CN}(\mathcal{S}_i).$$

It is worth noticing that the paraconsistent logics ALES restore consistency and return flat consequences. Moreover, they generalize to any Tarskian logic the universal logic defined by Rescher and Manor in [27].

In both previous theorems, the base logic is considered adjunctive. An important question is what about the case where the base logic is not adjunctive? It was shown in [3] that in that case the argumentation-based logic may choose arbitrarily *some* maximal consistent subsets of a knowledge base, leading thus to counter-intuitive outcomes.

A similar study has been conducted for stable and preferred semantics. It has been shown that there are two families of attack relations. The first family leads to coherent argumentation graphs (i.e., their stable extensions coincide with their preferred ones). Furthermore, stable extensions coincide with the naive ones. Such graphs coincide then with the above discussed ones. The ideal extension of these graphs coincide with the grounded extension. It also coincides with the intersection of the preferred (thus naive, stable) extensions. Furthermore, it is exactly the set $\mathtt{Arg}(\mathtt{Free}(\Phi))$, i.e., the set of arguments built over the set of free formulae of Φ. Finally, the conclusions drawn under ideal and grounded semantics are the formulae that follow using the base logic from the free formulae of a given set of formulae.

The second family of attack relations allows choosing only *some* maximal consistent subsets of a knowledge base. Thus, the corresponding argumentation-based logics infer counter-intuitive conclusions under stable and preferred semantics. The grounded and ideal extensions of such graphs may lead to counter-intuitive results as well.

Let us now illustrate this approach with propositional logic, an instance of Tarskian logic. We assume that there is a *finite number of variables* in the language. This assumption, very common in the literature, ensures the finiteness condition of Definition 1. The attack relation between arguments is *assumption-attack* introduced for the first time in [16].

Definition 12 (Assumption attack). *Let $\langle \mathcal{F}, w, \mathtt{CN} \rangle$ be propositional logic. An argument $\langle \Psi, \psi \rangle$ attacks an argument $\langle \Psi', \psi' \rangle$, denoted by $\langle \Psi, \psi \rangle \mathcal{R}_{as} \langle \Psi', \psi' \rangle$, iff $\exists \varphi \in \Psi'$ s.t. $\psi \equiv \neg \varphi$.*

Assumption attack is among the attacks relation that lead to rational argumentation-based logics.

Theorem 3. *[3] Let $\mathcal{G}$ be an argumentation function on propositional logic $\mathcal{L} = \langle \mathcal{F}, w, \mathtt{CN} \rangle$ such that for all $\Phi \subseteq_f \mathcal{F}$, $\mathcal{G}(\Phi) = \langle \mathtt{Arg}(\Phi), \mathcal{R}_{as} \rangle$. It holds that*

$$\mathtt{Ext}_n(\mathcal{G}(\Phi)) = \mathtt{Ext}_s(\mathcal{G}(\Phi)) = \mathtt{Ext}_p(\mathcal{G}(\Phi)).$$

$$\text{IE}(\mathcal{G}(\varPhi)) = \text{GE}(\mathcal{G}(\varPhi)) = \text{Arg}(\text{Free}(\varPhi)).$$

Let us now consider the following example.

Example 2. Let $\varPhi = \{p, \neg p, q, p \to \neg q\}$ be a propositional knowledge base. This base has three maximal (for set inclusion) consistent subbases:

- $\varPhi_1 = \{p, q\}$,
- $\varPhi_2 = \{p, p \to \neg q\}$,
- $\varPhi_3 = \{\neg p, q, p \to \neg q\}$.

An argumentation-based logic induced from naive, stable, or preferred semantics will draw from $\varPhi$ the tautologies since they are the only common consequences of the three subbases.

Note that none of the two conflicts $\{p, \neg p\}$ and $\{p, q, p \to \neg q\}$ is solved. Such output may seem unsatisfactory in general and in multi-agent systems where one needs an efficient way for solving conflicts between agents. Let us now have a closer look at the knowledge base $\varPhi$. The four formulae in $\varPhi$ do not have the same responsibility for inconsistency. For instance, the degree of blame of p is higher than the one of q since it is involved in more conflicts. Moreover, p is frontally opposed while q is opposed in an indirect way. Similarly, $\neg q$ is more to blame than q since it follows from the controversial formula p.

To sum up, *rational* argumentation-based logics induced from extension semantics generalize, to any Tarskian logic, the paraconsistent logic defined by Rescher and Manor in [27] on top of propositional logic. Such logics coincide with their base logic in case of a consistent knowledge base. When the latter is inconsistent, they only draw the formulae that follow logically (using the base logic) from the set of formulae which are not involved in inconsistency. This means that they leave conflicts unsolved as shown in Example 2.

4 Logics induced by ranking semantics

Ranking semantics have been introduced in [1] as an alternative approach for evaluating arguments. Their basic idea is to rank arguments from the most to the less acceptable ones, instead of computing extensions. They should satisfy axioms (i.e., desirable properties), some of them are mandatory while others are optional. In what follows, we investigate the argumentation-based logics induced from *burden-based semantics* (Bbs), a ranking semantics introduced in [1]. Bbs assigns a *burden number* to every argument. The heavier the burden of an argument, the weaker its attacks.

Definition 13 (Burden numbers). Let $\mathcal{T} = \langle \mathcal{A}, \mathcal{R} \rangle$ be a finite argumentation graph, $i \in \{0, 1, \ldots\}$, and $a \in \mathcal{A}$. We denote by $\text{Bur}_i(a)$ the burden number of a in the ith step:

$$\text{Bur}_i(a) = \begin{cases} 1 & \text{if } i = 0; \\ 1 + \sum_{b \in \text{Att}(a)} \frac{1}{\text{Bur}_{i-1}(b)} & \text{otherwise.} \end{cases}$$

where $\text{Att}(a) = \{b \in \mathcal{A} \mid (b, a) \in \mathcal{R}\}$.

By convention, if $\text{Att}(a) = \emptyset$, then

$$\sum_{b \in \text{Att}(a)} \frac{1}{\text{Bur}_{i-1}(b)} = 0.$$

Let us illustrate this function in the following example.

Example 3. stable Consider the argumentation graph depicted in Example 1. The burden numbers of each argument are summarized in the table below.

Step i	a	b	c	d	e	f	g
0	1	1	1	1	1	1	1
1	2	2	3	2	2	2	2
2	1.5	1.33	2	1.5	1.5	1.5	1.5
3	1.66	1.5	2.33	1.66	1.75	1.66	1.66
4	1.60	1.42	2.17	1.60	1.66	1.60	1.60
⋮	⋮	⋮	⋮	⋮	⋮	⋮	⋮

It is worth pointing out that the function Bur converges, and thus each argument a has a single burden number $\text{Bur}(a) = \lim_{i \to \infty} \text{Bur}_i(a)$.

There are different ways of comparing pairs of arguments, each of which leads to a new semantics. For instance, one may compare the final burden numbers of arguments (i.e., the ones got by the limit). The corresponding semantics satisfies all the mandatory axioms defined in [1]. Moreover, it allows *compensation*. Indeed, it considers that two weak attacks are equivalent to a strong one. Another alternative consists of comparing arguments lexicographically as follows:

Definition 14 (Bbs). *The* burden-based semantics Bbs *transforms any argumentation graph* $\mathcal{T} = \langle \mathcal{A}, \mathcal{R} \rangle$ *into the ranking* $\text{Bbs}(\mathcal{T})$ *on* $\mathcal{A}$ *such that* $\forall a, b \in \mathcal{A}$, $\langle a, b \rangle \in \text{Bbs}(\mathcal{T})$ *iff one of the two following cases holds:*

$\forall i \in \{0, 1, \ldots\}$, $\text{Bur}_i(a) = \text{Bur}_i(b)$;
$\exists i \in \{0, 1, \ldots\}$, $\text{Bur}_i(a) < \text{Bur}_i(b)$ *and* $\forall j \in \{0, 1, \ldots, i-1\}$, $\text{Bur}_j(a) = \text{Bur}_j(b)$.

Intuitively, $\langle a, b \rangle \in \text{Bbs}(\mathcal{T})$ means that a is *at least as acceptable as* b. Let us see in an example how the semantics works.

Example 4. stable According to Bbs, the argument b is strictly more acceptable than a, d, f, and g which are themselves equally acceptable and strictly more acceptable than e. Finally, e is more acceptable than c.

It is worth noticing that Bbs considers finite argumentation graphs. However, from a finite set of formulae, Definition 5 may generate an infinite number of arguments. Of course, most of them are redundant. In order to avoid such useless arguments, we assume that a logic satisfies the following additional condition:

- $\{\text{CN}(\{\varphi\}) \mid \varphi \in \mathcal{F}\}$ is finite (Finiteness)

Finiteness ensures a finite number of non-equivalent formulae. This condition, not considered by Tarski, will avoid redundant arguments. It is worth recalling that classical logic satisfies finiteness when the number of propositional variables is finite, which is a quite common assumption in the literature.

Property 1. For all $\Phi \subseteq_f \mathcal{F}$, $\text{Arg}(\Phi)$ is finite.

The plausible conclusions of an argumentation-based logic that uses ranking semantics are simply those supported by at least one argument. Note that a formula and its negation may both be plausible. This means that the approach tolerates inconsistency. More importantly, the conclusions are ranked from the most to the least plausible ones. A formula is ranked higher than another formula if it is supported by an argument which is more acceptable than any argument supporting the second formula. The notation $\varphi \geq \psi$ means that φ is at least as plausible as ψ.

Definition 15 (ALRS). *An argumentation-based logic induced from ranking semantics (ALRS) is a logic $\mathcal{L} = \langle \mathcal{F}, w, \text{CN}' \rangle$ based on base logic $\langle \mathcal{F}, w, \text{CN} \rangle$, argumentation function $\mathcal{G}$, and semantics Bbs, where for all $\Phi \subseteq_f \mathcal{F}$, $\mathcal{G}(\Phi) = \langle \text{Arg}(\Phi), \mathcal{R} \rangle$, and*

- $\text{CN}'(\Phi) = \{\varphi \in \mathcal{F} \mid \exists a \in \text{Arg}(\Phi) \text{ and } \text{Conc}(a) \equiv \varphi \text{ wrt logic } \langle \mathcal{F}, w, \text{CN} \rangle\}$.
- *for all $\varphi, \psi \in \text{CN}'(\Phi)$, $\varphi \geq \psi$ iff $\exists a \in \text{Arg}(\Phi)$ such that $\text{Conc}(a) \equiv \varphi$ and $\forall b \in \text{Arg}(\Phi)$ such that $\text{Conc}(b) \equiv \psi, \langle a, b \rangle \in \text{Bbs}(\mathcal{G}(\Phi))$.*

Unlike certain well-known inconsistency-tolerating logics (like the 3- and 4-valued ones [6, 13]), the above logics satisfy the following crucial property: if the premises are consistent, the conclusions coincide with those of CN. They satisfy other important properties like ranking free formulae above non-free ones. Recall that free formulae are those that are not involved in any minimal (for set inclusion) inconsistent subset of a knowledge base. They also consider that any formula is at most as plausible as its logical consequences and thus equivalent formula are equally plausible (see [2] for a complete study of these logics).

Let us now illustrate this family of logics by considering propositional logic as base logic and assumption-attack as attack relation.

Example 5. illust Let $\Phi = \{p, \neg p, q, p \rightarrow \neg q\}$ be a propositional knowledge base. Assume that the non-equivalent formulae selected by the well-ordering w are as follows:

$p \wedge \neg p$	$\neg p \wedge \neg q$	$\neg p$	$p \rightarrow \neg q$	$p \vee \neg p$
	$\neg p \wedge q$	$\neg q$	$p \rightarrow q$	
	$p \wedge \neg q$	$p \leftrightarrow q$	$q \rightarrow p$	
	$p \wedge q$	$p \leftrightarrow \neg q$	$\neg p \rightarrow q$	
		q		
		p		

The set $\text{Arg}(\Phi)$ contains the 21 following arguments.

a	$\langle\{\neg p,q\},\neg p\wedge q\rangle$	k	$\langle\{q,p\to\neg q\},p\leftrightarrow\neg q\rangle$
b	$\langle\{q,p\to\neg q\},\neg p\wedge q\rangle$	l	$\langle\{q\},q\rangle$
c	$\langle\{p,p\to\neg q\},p\wedge\neg q\rangle$	m	$\langle\{p\},p\rangle$
d	$\langle\{p,q\},p\wedge q\rangle$	n	$\langle\{\neg p\},p\to\neg q\rangle$
e	$\langle\{\neg p\},\neg p\rangle$	o	$\langle\{p\to\neg q\},p\to\neg q\rangle$
f	$\langle\{q,p\to\neg q\},\neg p\rangle$	y	$\langle\{\neg p\},p\to q\rangle$
g	$\langle\{p,p\to\neg q\},\neg q\rangle$	z	$\langle\{q\},p\to q\rangle$
h	$\langle\{p,q\},p\leftrightarrow q\rangle$	r	$\langle\{p\},q\to p\rangle$
i	$\langle\{\neg p,q\},p\leftrightarrow\neg q\rangle$	s	$\langle\{p\},\neg p\to q\rangle$
j	$\langle\{p,p\to\neg q\},p\leftrightarrow\neg q\rangle$	t	$\langle\{q\},\neg p\to q\rangle$
		u	$\langle\emptyset,p\vee\neg p\rangle$

The set $\mathtt{CN}'(\Phi)$ contains the conclusions of the arguments and their equivalent formulae. Due to the large number of attacks, we do not give them here. From the argumentation graph, the following burden numbers are computed (table on the left). The ranking on $\mathtt{Arg}(\Phi)$ is as shown in the table on the right.

	$i=0$	$i=1$	$i=2$	$i=3$
u	1	1	1	1
m,s,r	1	3	1.83	2.41
e,n,y	1	2	1.33	1.54
l,z,t	1	2	1.25	1.48
o	1	2	1.25	1.48
d,h	1	4	2.05	2.89
b,f,k	1	3	1.50	1.96
c,g,j	1	4	2.05	2.89
a,i	1	3	1.58	2.02

u
o,l,z,t
e,n,y
b,f,k
a,i
m,r,s
c,d,g,h,j

The conclusions of the arguments are ranked as follows:

$p\vee\neg p$
$p\to\neg q,\ q,\ p\to q,\ \neg p\to q$
$\neg p$
$\neg p\wedge q,\ p\leftrightarrow\neg q$
$p,\ q\to p$
$p\wedge\neg q,\ \neg q,\ p\wedge q,\ p\leftrightarrow q$

Recall that equivalent formula are equally plausible. For instance, $p\wedge p$ is as plausible as p. Note that $\neg p$ is more plausible than p, and q is more plausible than $\neg q$. Thus, unlike ALES, logics that use ranking semantics solve both conflicts of the base Φ.

The ranking of formulae produced by the previous logic is not arbitrary. It not only satisfies some rationality postulates discussed in [2], it also captures in some cases a well-known *inconsistency measure* [20]. The latter assigns a degree of blame to each formula of a knowledge base. This degree is the number of minimal inconsistent susbsets of the base (called conflicts) in which the formula is involved. It was shown in [2] that if a formula φ of a knowledge base is involved in more conflicts than another formula ψ of the base, then ψ is more plausible than φ. This result is only true in case each formula in the base cannot be inferred from another consistent subset of the base.

Theorem 4. *[2] Let* $\mathcal{L} = \langle \mathcal{F}, w, \text{CN}' \rangle$ *be an ALRS based on propositional logic* $\langle \mathcal{F}, w, \text{CN} \rangle$, *argumentation function* $\mathcal{G}$ *such that for all* $\Phi \subseteq_f \mathcal{F}$, $\mathcal{G}(\Phi) = \langle \text{Arg}(\Phi), \mathcal{R}_{as} \rangle$, *and semantics* Bbs. *Let* $\Phi \subseteq \mathcal{F}$. *If for all* $\varphi \in \Phi$, $\nexists \Psi \subseteq \Phi \setminus \{\varphi\}$ *such that* Ψ *is consistent and* $\varphi \in \text{CN}(\Psi)$, *then for all* $\varphi, \psi \in \text{CN}'(\Phi) \cap \Phi$, *if* $|\{\Psi \in \text{MIC}(\Phi) \mid \varphi \in \Psi\}| > |\{\Psi' \in \text{MIC}(\Phi) \mid \psi \in \Psi'\}|$, *then* $\psi > \varphi$ *(i.e.,* $\psi \succeq \varphi$ *and* $\varphi \not\succeq \psi$*).*

Works on inconsistency measures focus only on the formulae of the base and completely neglect their logical consequences. ALRS focus on both. That's why the two approaches may not find the same results in the general case. Indeed, it may be the case that a formula φ of a base is involved in more conflicts than another formula ψ of the same base, but φ follows logically from a subset of the base and this subset constitutes a more acceptable argument than the one supporting ψ. Thus, ALRS will rank φ higher than ψ while the inconsistency measure will prefer ψ.

To sum up, argumentation-based logics induced from ranking semantics tolerate inconsistency in that they may infer inconsistent conclusions. Furthermore, they rank-order the conclusions with regard to plausibility. Finally, unlike ALES, they solve inconsistency.

5 Conclusion

Argumentation is a natural approach for handling inconsistency. It is more akin to the way humans deal with inconsistency in everyday life. Indeed, it constructs arguments pro and arguments con claims, then it evaluates the arguments before concluding.

This paper discussed two families of argumentation-based paraconsistent logics: the family of logics that use naive (respectively stable, preferred, grounded and ideal) semantics, and the family of logics that use ranking semantics, namely Bbs. We have shown that argumentation logics are efficient since, under extension semantics, they generalize well-known logics and, under ranking semantics, they outperform them.

References

1. L. Amgoud and J. Ben-Naim. Ranking-based semantics for argumentation frameworks. In *SUM*, pages 134–147, 2013.
2. L. Amgoud and J. Ben-Naim. Argumentation-based ranking logics. In *Proceedings of the 2015 International Conference on Autonomous Agents and Multiagent Systems, (AAMAS'2015)*, pages 1511–1519, 2015.
3. L. Amgoud and Ph. Besnard. Logical limits of abstract argumentation frameworks. *Journal of Applied Non-Classical Logics*, 23(3):229–267, 2013.
4. Leila Amgoud, Martin Caminada, Claudette Cayrol, Marie-Christine Lagasquie, and Henry Prakken. Towards a consensual formal model: inference part. *Deliverable of ASPIC project*, 2004.
5. P. Baroni, M. Giacomin, and G. Guida. Scc-recursiveness: a general schema for argumentation semantics. *Artificial Intelligence Journal*, 168:162–210, 2005.

6. N. D. Belnap. A Useful Four-Valued Logic. In J.M. Dunn and G. Epstein, editors, *Modern Uses of Multiple-Valued Logic*, pages 7–37. Oriel Press, 1977.
7. S. Benferhat, D. Dubois, and H. Prade. How to infer from inconsistent beliefs without revising? In *14th International Joint Conference on Artificial Intelligence (IJCAI'95)*, pages 1449–1455, 1995.
8. Philippe Besnard and Anthony Hunter. A logic-based theory of deductive arguments. *Artificial Intelligence*, 128(1-2):203–235, 2001.
9. D. G. Bobrow. Special issue on non-monotonic reasoning. *Artificial Intelligence Journal*, 13 (1-2), 1980.
10. G. Brewka. Preferred subtheories: an extended logical framework for default reasoning. In *Proceedings of the 11th International Joint Conference on Artificial Intelligence, (IJCAI'89)*, pages 1043–1048, 1989.
11. C. Cayrol and M.-C. Lagasquie-Schiex. Graduality in Argumentation. *Journal of Artificial Intelligence Research (JAIR)*, 23:245–297, 2005.
12. L. Cholvy. Automated reasoning with merged contradictory information whose reliability depends on topics. In *European Conference on Symbolic and Quantitative Approaches to Reasoning under Uncertainty, ECSQARU'95*, pages 125–132, 1995.
13. I.M.L. D'Ottaviano and N.C.A. da Costa. Sur un problème de Jaśkowski. In *Comptes Rendus de l'Académie des Sciences de Paris*, volume 270, pages 1349–1353, 1970.
14. P. M. Dung. On the Acceptability of Arguments and its Fundamental Role in Non-Monotonic Reasoning, Logic Programming and n-Person Games. *AIJ*, 77:321–357, 1995.
15. P.M. Dung, P. Mancarella, and F. Toni. Computing ideal skeptical argumentation. *Artificial Intelligence Journal*, 171:642–674, 2007.
16. Morten Elvang-Gøransson, John Fox, and Paul Krause. Acceptability of arguments as 'logical uncertainty'. In *2nd European Conference on Symbolic and Quantitative Approaches to Reasoning and Uncertainty (ECSQARU'93)*, volume 747 of *Lecture Notes in Computer Science*, pages 85–90, Granada, Spain, November 1993. Springer.
17. M. Fitting. Kleene's three-valued logics and their children. *Fundamenta Informaticae*, 20:113–131, 1994.
18. D. Gabbay and A. Hunter. Making inconsistency respectable: a logical framework for inconsistency in reasoning. *Fundamentals of Artificial Intelligence Research*, LNAI 535:19–32, 1991.
19. N. Gorogiannis and A. Hunter. Instantiating abstract argumentation with classical logic arguments: Postulates and properties. *Artificial Intelligence*, 175(9-10):1479–1497, 2011.
20. Anthony Hunter and Sébastien Konieczny. On the measure of conflicts: Shapley inconsistency values. *Artificial Intelligence*, 174(14):1007–1026, 2010.
21. J. De Kleer. Using crude probability estimates to guide diagnosis. *Artificial Intelligence*, 45:381–391, 1990.
22. P-A Matt and F. Toni. A game-theoretic measure of argument strength for abstract argumentation. In *Proceedings of 11th European Conference on Logics in Artificial Intelligence, JELIA'08*, volume 5293 of *Lecture Notes in Computer Science*, pages 285–297. Springer, 2008.
23. J. Pearl. Qualitative probabilities for default reasoning, belief revision, and causal modeling. *Artifitial Intelligence*, 84(1-2):57–112, 1996.
24. John Pollock. How to reason defeasibly. *Artificial Intelligence J.*, 57(1):1–42, 1992.

25. F. Pu, J. Luo, Y. Zhang, and G. Luo. Argument ranking with categoriser function. In *Knowledge Science, Engineering and Management - 7th International Conference, KSEM 2014, Proceedings*, pages 290–301, 2014.
26. R. Reiter. A logic for default reasoning. *Artificial Intelligence*, 13(1-2):81–132, 1980.
27. N. Rescher and R. Manor. On inference from inconsistent premises. *Journal of Theory and decision*, 1:179–219, 1970.
28. A. Tarski. *Logic, Semantics, Metamathematics (E. H. Woodger, editor))*, chapter On Some Fundamental Concepts of Metamathematics. Oxford Uni. Press, 1956.
29. M. Thimm and G. Kern-Isberner. On controversiality of arguments and stratified labelings. In *Proceedings of Computational Models of Argument, (COMMA'2014)*, pages 413–420, 2014.

Further Applications of the Gabbay-Rodrigues Iteration Schema in Argumentation and Revision Theories

Dov Gabbay[1], Odinaldo Rodrigues[2]

Abstract. In [8], we proposed an iteration schema which operated on an extended argumentation framework whose nodes were assigned initial values in $[0, 1]$, coming from some application area, e.g., revision theory. We showed that the schema generated a new set of node values at each iteration and that after a finite number of steps no new values in the open interval $(0, 1)$ were generated. Any remaining nodes with values in the set $\{0, 1\}$ retain those values during all future iterations. The sequence eventually converges by turning as few values in $(0, 1)$ into $\{0, 1\}$ as necessary in order to yield a complete extension in the traditional sense (interpreting the value 1 as "in", the value 0 as "out", and any other value as "undecided". This traditional extension is the best at accommodating the $\{0, 1\}$-part of the initial set of values. Although the iteration schema operates on values in $[0, 1]$, in this work, we show a simplified form operating on the set $\{0, \frac{1}{2}, 1\}$ which is more suitable for use in a practical implementation.

1 Introduction

An *abstract argumentation framework* is a tuple $\langle S, R \rangle$, where S is a set of arguments and R is an attack relation on $S \times S$. Semantics of argumentation frameworks can be defined in terms of *extensions*, which are subsets of the set of arguments S with special properties.

In [8], we proposed the Gabbay-Rodrigues Iteration Schema. The schema is an iterative method that can be used for calculating extensions in the traditional Dung sense mentioned above. The schema takes an assignment of initial values $V_0 : S \longmapsto [0, 1]$ and produces a new assignment V_{i+1} for each iteration $i \geqslant 0$. The values in the schema eventually converge and in the limit of the sequence we can construct a complete extension by taking the nodes with value 1.

One disadvantage of the schema is the need to calculate the limit values of the sequence. This can be approximated by iterating sufficiently many times until the difference between the values of the nodes of two consecutive iterations falls below a certain threshold ε. In GRIS [9], Rodrigues set the threshold

---

[1] Department of Informatics, King's College London, Bar Ilan University, Ramat Gan, Israel, and University of Luxembourg, Luxembourg; dov.gabbay@kcl.ac.uk, WWW home page: http:///www.inf.kcl.ac.uk/staff/dg

[2] Department of Informatics, King's College London, The Strand, London, WC2R 2LS, UK, odinaldo.rodrigues@kcl.ac.uk, WWW home page: http://www.inf.kcl.ac.uk/staff/odinaldo

ε as the upper bound of the relative error introduced due to the rounding in the calculations of the target machine.[3] In experimental tests, this required around 100 iterations. Since in GRIS the schema is used to "ground" the strongly connected components of an argumentation framework and hence invoked often, if we were able to obtain the same results without the need to do any approximation, this would increase the overall performance of the computation of the semantics significantly. This paper proposes and discusses a discrete version of the method called the *Discrete Gabbay-Rodrigues Iteration Schema* and its application in argumentation and revision theories. The computational steps of this method are developed in the main body of this paper and are summarised in Section 6. The reader might wish to refer to the summary to get a more complete overview of the paper.

The following example motivates how an argumentation framework can have requested values coming from revision theory. The interested reader may wish to refer to [6, 7] for a more in-depth analysis of the relationship between revision and argumentation theories.

Example 1 (Numerical Revision Under Constraints). John and Mary are young students and they are getting married. As part of the planning of the wedding, John and Mary, as well as both sets of parents, put forward names of people they want to invite to the wedding.

This can be modelled as a database Δ of names, where $x \in \Delta$ means "invite x to the wedding". There are however constraints on Δ. It may be the case that x and y are not in good terms and if x is invited, then y should not be invited. So let $\mathcal{C}$ be a set of contraints of the form

$$x \to \neg y.$$

The system $\Delta \cup \mathcal{C}$ should not be regarded as a classical theory with negation, in the sense that if it is inconsistent it proves everything. We do not mean that if both x and y are invited then we should invite everybody. We would rather regard $\Delta \cup \mathcal{C}$ as a set violating the constraints and that Δ should be revised to a smaller $\Delta' \subsetneq \Delta$ which satisfies the constraints.

The organisers decided to ask everyone who put a name forward to rank them by a number, indicating how they feel about x being invited. These number were aggregated into a single number $h(x)$ from $\{0, \frac{1}{2}, 1\}$, for every x, where $h(x) = 0$ means "do not invite x" and $h(x) = 1$ means "absolutely must invite x" and $h(x) = \frac{1}{2}$ means "indifferent with respect to x's invitation".

In order now to revise the theory Δ, we present the constraints in a relational form, where xRy means "if you invite x, do not invite y". This turns $\langle \Delta, R \rangle$ into an abstract argumentation framework and the function h into a numerical assignment to the elements of Δ. Our task is to seek a function h', as near h as possible, such that

$$\Delta' = \{x \mid h'(x) = 1\}$$

---

[3] When the threshold is reached we can no longer reliably tell if the difference in value is a legitimate approximation of the limit value or was introduced due to a rounding error in the arithmetic unit of the target machine.

If the wishes of everyone can be accomodated without conflicts, then we want $\Delta' = \Delta$. Otherwise, we want to invite as many people as possible subject to the constraint that "everyone whose attendance is not objected by anyone invited is also invited."

To solve this problem and other problems like it, we turn to the *discrete* version of the Gabbay-Rodrigues Iteration Schema [8].

The rest of the paper is structured as follows. In Section 2 we provide some background material from argumentation theory. In Section 3, we re-introduce the full-fledged version of the Gabbay-Rodrigues Iteration Schema. This is followed by the presentation of its discrete version in Section 4. We show some examples in Section 5 and conclude with a discussion in Section 6.

2 Background

This section provides a very brief overview of the key concepts from argumentation theory that will be needed in the remainder of this paper. For a more comprehensive introduction to these concepts the reader is referred to [5, 2, 3, 1].

We mentioned in the introduction that the semantics of an argumentation framework can be given in terms of extensions. In [2] Caminada has shown that the semantics can be alternatively presented in terms of labelling functions. Essentially, a labelling function λ is an assignment $\lambda : S \longmapsto \{\mathbf{in}, \mathbf{und}, \mathbf{out}\}$. There is a direct correspondence between the labelling semantics and the notion of extensions, by taking an extension to be the set of arguments with label **in** in labelling functions with special properties (defined below).

As numerical assignments will also give values to all arguments, it will be easier for us to think in terms of labellings instead of extensions. The correspondence between the numerical values and the values in $\{\mathbf{in}, \mathbf{und}, \mathbf{out}\}$ is given later in Definition 5. We start by defining the status of the label of an argument with respect to the labels of its attackers.

Definition 1 (Illegal labelling of an argument [4]). *Let $\langle S, R \rangle$ be an argumentation framework and λ a labelling function for S.*

1. *An argument $X \in S$ is illegally labelled* **in** *by λ if $\lambda(X) = \mathbf{in}$ and there exists $Y \in Att(X)$ such that $\lambda(Y) \neq \mathbf{out}$.*
2. *An argument $X \in S$ is illegally labelled* **out** *by λ if $\lambda(X) = \mathbf{out}$ and there is no $Y \in Att(X)$ such that $\lambda(Y) = \mathbf{in}$.*
3. *An argument $X \in S$ is illegally labelled* **und** *by λ if $\lambda(X) = \mathbf{und}$ and either for all $Y \in Att(X)$, $\lambda(Y) = \mathbf{out}$ or there exists $Y \in Att(X)$, such that $\lambda(Y) = \mathbf{in}$.*

The conditions above are used to define labelling functions associated with admissible sets, grounded and complete extensions. The sets are defined in terms of the arguments labelled **in** by the corresponding labelling functions.

Definition 2 (Admissible labelling function [4, Definition 8]). *An admissible labelling function is a labelling function without arguments that are illegally labelled* **in** *and without arguments that are illegally labelled* **out**.

It is easy to see that any illegal labelling function λ_0 can be turned into an admissible labelling function λ_{da} by successively turning each node illegally labelled **in** or **out** into **und**.[4] This process was called a *contraction sequence* in [4]. The labelling function λ_{da} obtained in this way is the largest element of the set of all admissible labellings that are smaller or equal to λ.[5]

Definition 3 (Complete labelling function [4, Definition 9]). *A complete labelling function is a labelling function without any illegally labelled arguments.*

After a contraction sequence, by successively assigning the correct legal label to each illegally labelled undecided node in λ_{da} we eventually end up with a labelling function λ_{CP} without any illegally labelled nodes. By Definition 3, λ_{CP} is a complete labelling function. This corrective process was called an *expansion sequence* in [4]. Caminada and Pigozzi have further shown that λ_{CP} is the smallest element of the set of all complete labellings that are larger or equal to λ_{da} (in the sense of Footnote 5).

If we start with $\lambda_{da}(X) = $ **und**, for all $X \in S$, then obviously λ_{CP} will correspond to the smallest complete labelling function of all, which is the grounded labelling function.

Definition 4 (Grounded labelling function [4, Definition 10]). *Let λ be a complete labelling function. λ is a grounded labelling function if and only if the set $\{X \in S \mid \lambda(X) = \mathbf{in}\}$ is minimal with respect to set inclusion among all complete labelling functions.*

These results can be understood in our numerical setting through the following two-way translation mechanism.

Definition 5 (Caminada-Pigozzi/Gabbay-Rodrigues Translation). *A labelling function λ and a valuation function V can be inter-defined according to the table below.*

$\lambda(X) \to V_\lambda(X)$		$V(X) \to \lambda_V(X)$	
in	$\to$ 1	1	$\to$ in
out	$\to$ 0	0	$\to$ out
und	$\to$ $\frac{1}{2}$	$(0,1)$	$\to$ und

What this gives us is that it is possible to turn labellings that are not complete (i.e., are not associated with a complete extension in Dung's sense) via a contraction sequence followed by an expansion sequence. The Gabbay-Rodrigues Iteration Schema can be used in a numerical context and in the limit of the sequence, the computed values will correspond to the same complete labelling. This is explained in more detail in the next section.

---

[4] Here "da" reminds us that λ_{da} is the "down-admissible" labelling function resulting from λ_0.

[5] λ_1 is smaller or equal to λ_2 if $in(\lambda_1) \subseteq in(\lambda_2)$ and $out(\lambda_2) \subseteq out(\lambda_2)$. Conversely, λ_1 is larger or equal to λ_2 if $in(\lambda_1) \supseteq in(\lambda_2)$ and $out(\lambda_2) \supseteq out(\lambda_2)$.

3 The Gabbay-Rodrigues Iteration Schema

In [8], we proposed the Gabbay-Rodrigues Iteration Schema. The schema is an iterative method that can be used for calculating extensions in the traditional Dung sense. The schema takes an assignment of initial values $V_0 : S \longmapsto [0, 1]$ and produces a new assignment V_{i+1} for each iteration $i \geqslant 0$. We will use U to denote the unit interval $[0, 1]$. The schema is defined as follows.

Definition 6. *Let $\mathcal{N} = \langle S, R \rangle$ be an argumentation framework and V_0 be an assignment of values to the nodes in S. The* Gabbay-Rodrigues Iteration Schema *is defined by the following system of equations $\boldsymbol{T}$, where for each node $X \in S$, the value $V_{i+1}(X)$ is defined in terms of the values of the nodes in $\{X\} \cup Att(X)$ in iteration V_i as follows:*

$$V_{i+1}(X) = (1 - V_i(X)) \cdot \min \left\{ \tfrac{1}{2}, 1 - \max_{Y \in Att(X)} V_i(Y) \right\} + \\ V_i(X) \cdot \max \left\{ \tfrac{1}{2}, 1 - \max_{Y \in Att(X)} V_i(Y) \right\} \quad \text{(T)}$$

The schema guarantees that all node values generated remain in the unit interval U:

Proposition 1 ([8]). *Let $\mathcal{N} = \langle S, R \rangle$ be an argumentation framework and $V_0 : S \longmapsto U$ an assignment of initial values to the nodes in S. Let each assignment V_i, $i > 0$, be calculated by the Gabbay-Rodrigues Iteration Schema for $\mathcal{N}$. It follows that $V_i(X) \in U$, for all $i \geqslant 0$ and all $X \in S$.*

In order to understand what the schema computes based on the initial assignment V_0, it will prove useful to introduce some terminology first.

Definition 7. *For any assignment of values $V : S \longmapsto U$ define the sets $in(V) = \{X \in dom\ V \mid V(X) = 1\}$ and $out(V) = \{X \in dom\ V \mid V(X) = 0\}$.*

Definition 8. *Let $V : S \longmapsto U$ be an assignment of values to the nodes in S. The set of crisp values with respect to V, C_V is the defined as the set $in(V) \cup out(V)$*

By abuse of notation, we will also use $in(\lambda)$ (resp. $out(\lambda)$) to designate the nodes labelled **in** (resp. **out**) by λ and C_λ to denote $in(\lambda) \cup out(\lambda)$.

As in a contraction followed by an expansion sequence, the schema also operates in two phases. In the first phase, all nodes X whose initial values are crisp and whose equation

$$V(X) = 1 - \max_{Y \in Att(X)} \{V(Y)\} \quad (1)$$

is not satisfied have their values re-assigned to a value in the open interval $(0, 1)$. Intuitively, what this does is to turn illegal crisp values into the undecided range, but unlike in a contraction sequence, more than one illegal crisp value can be turned into undecided in the same iteration. The schema does not turn values in $(0, 1)$ into $\{0, 1\}$, as can be seen by the theorem below, although as many values as required will approximate their correct "legal" $\{0, 1\}$ values so as to yield a complete extension.[6]

---

[6] The correct "legal" values are obtained in the *limit* of the sequence $\lim_{i \to \infty} V_i$.

Theorem 1 ([8]). *Let $\mathcal{N} = \langle S, R \rangle$ be an argumentation framework, $\boldsymbol{T}$ a system of equations for $\mathcal{N}$ using the Gabbay-Rodrigues Iteration Schema, and $V_0 : S \longmapsto U$ an assignment of initial values to the nodes in S. Let V_0, V_1, V_2, ... be a sequence of value assignments where each V_i, $i > 0$, is generated by $\boldsymbol{T}$. Then the following properties hold for all $X \in S$ and for all $k \geqslant 0$*

1. *If $V_k(X) = 0$, then $V_{k+1}(X) \neq 1$.*
2. *If $V_k(X) = 1$, then $V_{k+1}(X) \neq 0$.*
3. *If $0 < V_k(X) < 1$, then $0 < V_{k+1}(X) < 1$.*

We say that a sequence becomes *stable* if no new undecided nodes are generated between iterations.[7]

Definition 9. *Let $\mathcal{N} = \langle S, R \rangle$ be an argumentation framework and $V_0 : S \longmapsto U$ an assignment of initial values to the nodes in S. A sequence of assignments $V_i : S \longmapsto U$ where each $i > 0$ is said to be stable at iteration k, if for all nodes $X \in S$ we have that*

1. *If $V_k(X) \in \{0, 1\}$, then $V_{k+1}(X) = V_k(X)$; and*
2. *k is the smallest value for which the condition above holds.*

The theorem below shows that once all crisp values stabilise between two iterations, they remain unchanged throughout the rest of the sequence.

Theorem 2 ([8]). *Let $\mathcal{N} = \langle S, R \rangle$ be an argumentation framework, $\boldsymbol{T}$ its GR system of equations, and V_0 an initial assignment of values to the nodes in S. Let V_0, V_1, V_2, ... be a sequence of value assignments where each V_i, $i > 0$, is generated by $\boldsymbol{T}$. Assume that for some iteration i and all nodes $X \in S$ such that $V_i(X) \in \{0, 1\}$, we have that $V_{i+1}(X) = V_i(X)$, then for all $j \geqslant 1$, $V_{i+j}(X) = V_i(X)$.*

Since no new crisp values are generated and S is finite, then at some iteration k the last set of illegal crisp values is changed into the undecided range.

Corollary 1 ([8]). *Let $\mathcal{N} = \langle S, R \rangle$ be an argumentation framework, $V_0 : S \longmapsto U$ an assignment of initial values to the nodes in S and $\boldsymbol{T}$ its GR system of equations. The following hold:*

1. *If the sequence of value assignments is not stable at iteration k, then there exists $X \in S$, such that $V_k(X) \in \{0, 1\}$ and $V_{k+1}(X) \in (0, 1)$.*
2. *Let $|S| = n$. Then, the sequence is stable for some $k \leqslant n$.*

Corollary 1 shows that for some value $0 \leqslant k \leqslant |S|$, the sequence of value assignments $V_0(X), V_1(X), V_2(X), \ldots$ eventually becomes stable. That is, there exists $k \geqslant 0$, such that for all $j \geqslant 0$ and all nodes X

- if $V_k(X) = 0$, then $V_{k+j}(X) = 0$;
- if $V_k(X) = 1$, then $V_{k+j}(X) = 1$; and

---

[7] Stability here for us just means that the schema has corrected all incorrectly assigned crisp values and hence no new values that were not already in the $(0, 1)$ interval will be generated.

- if $V_k(X) \in (0,1)$, then $V_{k+j}(X) \in (0,1)$.

Obviously, if the initial assignment already gives a value in $(0,1)$ to all nodes, then the schema is already stable at the outset:

Proposition 2. *Consider the assignment V_0 such that $V_0(X) = \frac{1}{2}$ for all $X \in S$. Then the Gabbay-Rodrigues Iteration Schema is stable at iteration 0.*

Proof. The initial assignment has no nodes in $\{0,1\}$, so condition 1. for stability of Definition 9 is vacuously satisfied and obviously 0 is the smallest iteration value for which this holds.

We mean "stable" in the sense that all crisp values at the stable point remain unchanged. Stability in the sequence does not guarantee that all values are legal, and hence it does not guarantee that the corresponding labelling is complete. What remains to be done is to "correct" as many illegal undecided nodes so as to yield an extension. We have shown in [8], that this is achieved in the limit of the sequence:

Theorem 3 ([8]). *Let $\langle S, R \rangle$ be an argumentation framework; V_0 be an initial assignment of values to the nodes in S; λ_0 an initial labelling of these nodes; and V_0 and λ_0 faithful to each other according to Definition 5. Let λ_{da} be the labelling at the end of a contraction sequence from λ_0 and λ_{CP} the labelling at the end of an expansion sequence after λ_{da}. Let k be the point at which the sequence $V_0, V_1, \ldots$ becomes stable and $V_e(X)$ the value of the node X in the limit of the sequence of values calculated through the Gabbay-Rodrigues Iteration Schema. Then λ_{CP} and V_e agree with each other according to Definition 5.*

The above theorem therefore establishes a correspondence between the results obtained in the limit of the sequence of the Gabbay-Rodrigues Iteration Schema and those obtained after a contraction and expansion sequence (in Caminada-Pigozzi's sense). However, by using the Gabbay-Rodrigues Iteration Schema in a numerical context, we can also use the values at the stable point as the admissible values closest to the initial assignment of values (in Dung's sense). The values at the stable point are not used in this paper.

We now present a simplified version of the schema which only operates on the set of values $\{0, \frac{1}{2}, 1\}$.

4 The Discrete Gabbay-Rodrigues Iteration Schema

We would like to simplify the calculations of the Gabbay-Rodrigues Iteration Schema by avoiding having to approximate the limit values of the sequence and yet keeping the same final results. This is the objective of the discrete schema we present next.

Definition 10. *Let $\mathcal{N} = \langle S, R \rangle$ be an argumentation framework and V_0 be an assignment of values from $\{0, \frac{1}{2}, 1\}$ to the nodes in S. The Discrete Gabbay-Rodrigues Iteration Schema is defined by the following system of equations*

(T_d), where the value $V_{i+1}(X)$ of each node in iteration $i + 1$ is defined in terms of the values of the nodes in iteration i as follows:

$$V_{i+1}(X) = 1 - \max_{Y \in Att(X)} \{V_i(Y)\} \qquad (T_d)$$

The theorem below shows that, as for the full-fledged Gabbay-Rodrigues Iteration Schema, initial assignments corresponding to complete labelling functions are preserved.

Theorem 4. *Let $\langle S, R \rangle$ be an argumentation framework and V_i an assignment of values from $\{0, \frac{1}{2}, 1\}$ to the nodes in S. λ_{V_i} is a complete labelling function if and only if $V_{i+1}(X) = V_i(X)$, for all $X \in S$.*

Proof. ($\Rightarrow$) Take any $X \in S$ and assume λ_{V_i} is a complete labelling function.

1. If $V_i(X) = 1$, then $\lambda_{V_i}(X) =$ **in**. Since λ_{V_i} is a complete labelling function, then for all $Y \in Att(X)$, $\lambda_{V_i}(Y) =$ **out**. Therefore, $\max_{Y \in Att(X)}\{V_i(Y)\} = 0$, and hence $V_{i+1}(X) = 1 - 0 = 1$.
2. If $V_i(X) = 0$, then $\lambda_{V_i}(X) =$ **out**. Since λ_{V_i} is a complete labelling function, then there exists $Y \in Att(X)$, such that $\lambda_{V_i}(Y) =$ **in**. Therefore, $\max_{Y \in Att(X)}\{V_i(Y)\} = 1$, and hence $V_{i+1}(X) = 1 - 1 = 0$.
3. If $V_i(X) = \frac{1}{2}$, then $\lambda_{V_i}(X) =$ **und**. Since λ_{V_i} is a complete labelling function, then there exists $Y \in Att(X)$, such that $\lambda_{V_i}(Y) =$ **und** and for no $Y \in Att(X)$ do we have that $\lambda_{V_i}(Y) =$ **in**. Therefore, $\max_{Y \in Att(X)}\{V_i(Y)\} = \frac{1}{2}$, and hence $V_{i+1}(X) = 1 - \frac{1}{2} = \frac{1}{2}$.

Therefore, for all nodes $X \in S$, if λ_{V_i} is a complete labelling function, then $V_{i+1}(X) = V_i(X)$.

($\Leftarrow$) Take any $X \in S$, and assume that $V_{i+1}(X) = V_i(X)$.

1. If $V_{i+1}(X) = 1$, then $\max_{Y \in Att(X)}\{V_i(Y)\} = 0$, and hence for all $Y \in Att(X)$, $V_i(Y) = 0$. It follows that λ_{V_i} labels all attackers Y of X **out**. Since $V_i(X) = V_{i+1}(X)$, then $\lambda_{V_i}(X) =$ **in** and therefore λ_{V_i} legally labels X.
2. If $V_{i+1}(X) = 0$, then there exists $Y \in Att(X)$, such that $V_i(Y) = 1$. Therefore, one of the attackers of X is labelled **in** by λ_{V_i}. Since $V_i(X) = V_{i+1}(X)$, $\lambda_{V_i}(X) =$ **out** and hence λ_{V_i} legally labels X.
3. If $V_{i+1}(X) = \frac{1}{2}$, then $\max_{Y \in Att(X)}\{V_i(Y)\} = \frac{1}{2}$. Therefore, there exists $Y \in Att(X)$, such that $V_i(Y) = \frac{1}{2}$ and for no $Y \in Att(X)$ do we have that $V_i(Y) = 1$. This means one attacker of X is labelled **und** by λ_{V_i}, but no attacker of X is labelled **in** by it. Since $V_i(X) = V_{i+1}(X)$, $\lambda_{V_i}(X) =$ **und**. Therefore, X is legally labelled by λ_{V_i}.

It follows that if for all nodes $X \in S$, $V_{i+1}(X) = V_i(X)$, then all nodes $X \in S$ are legally labelled by λ_{V_i} and hence λ_{V_i} is a complete labelling function.

Corollary 2. *Let $\langle S, R \rangle$ be an argumentation framework and take the assignment of values $V_i : S \longmapsto \{0, \frac{1}{2}, 1\}$. If λ_{V_i} is a complete labelling function, then $V_{i+j}(X) = V_i(X)$, for all $X \in S$ and all $j \geq 0$.*

Proof. Theorem 4 shows that if λ_{V_i} is a complete labelling function, then $V_{i+1}(X) = V_i(X)$ for all $X \in S$. Since each iteration only depends on the values of the nodes of the previous iteration and $V_{i+1}(X) = V_i(X)$ for all $X \in S$, then all subsequent iterations will produce the exact same values.

The above corollary shows that the sequence of values generated by the Discrete Gabbay-Rodrigues Iteration Schema does not change when the values of an iteration correspond to a complete labelling function. In practice what this means is that the discrete schema can be used to check whether an initial labelling function is complete. If the initial assignment corresponding to the labelling function is complete, then the values calculated at the second iteration will remain the same for all nodes.

It is easy to see that the complexity of this check is not higher than that of using labelling functions on $\{\textbf{out}, \textbf{in}, \textbf{und}\}$ and that the check can be implemented in a single loop such as the one in Algorithm 1.

Algorithm 1 Checking whether a labelling is complete

1: **procedure** ISCOMPLETE($\langle S, R, V \rangle$)
2:    **for** all nodes $X \in S$ **do**
3:       **if** $V(X) \neq 1 - \max_{Y \in Att(X)}\{V(Y)\}$ **then return false**   ▷ The value of the node X is illegal
      return true   ▷ The values of all nodes are legal

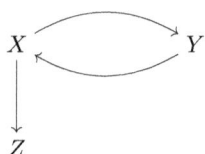

Fig. 1. An argumentation framework with multiple complete extensions.

Example 2. Consider the argumentation framework of Fig. 1 and the complete labellings $\lambda_0 = \{X = \textbf{und}, Y = \textbf{und}, Z = \textbf{und}\}$; $\lambda_1 = \{X = \textbf{in}, Y = \textbf{out}, Z = \textbf{out}\}$ and $\lambda_2 = \{X = \textbf{out}, Y = \textbf{in}, Z = \textbf{in}\}$. The associated translations are $V_{\lambda_0} = \{X = \frac{1}{2}, Y = \frac{1}{2}, Z = \frac{1}{2}\}$; $V_{\lambda_1} = \{X = 1, Y = 0, Z = 0\}$; and $V_{\lambda_2} = \{X = 0, Y = 1, Z = 1\}$. The table below shows how the Discrete Gabbay-Rodrigues Iteration Schema evolves using V_{λ_i} as the initial values for V_0^i.

$$V_0^0(X) = \tfrac{1}{2} \qquad V_1^0(X) = 1 - \max\{\tfrac{1}{2}\} = \tfrac{1}{2}$$
$$V_0^0(Y) = \tfrac{1}{2} \qquad V_1^0(Y) = 1 - \max\{\tfrac{1}{2}\} = \tfrac{1}{2}$$
$$V_0^0(Z) = \tfrac{1}{2} \qquad V_1^0(Z) = 1 - \max\{\tfrac{1}{2}\} = \tfrac{1}{2}$$

$$V_0^1(X) = 1 \qquad V_1^1(X) = 1 - \max\{0\} = 1$$
$$V_0^1(Y) = 0 \qquad V_1^1(Y) = 1 - \max\{1\} = 0$$
$$V_0^1(Z) = 0 \qquad V_1^1(Z) = 1 - \max\{1\} = 0$$

$$V_0^2(X) = 0 \qquad V_1^2(X) = 1 - \max\{1\} = 0$$
$$V_0^2(Y) = 1 \qquad V_1^2(Y) = 1 - \max\{0\} = 1$$
$$V_0^2(Z) = 1 \qquad V_1^2(Z) = 1 - \max\{0\} = 1$$

Unlike the full-fledged Gabbay-Rodrigues Iteration Schema, its discrete version *does not* correct all possible illegal initial assignments. For instance, if the assignment $V_0 = $ all-**in** is given to the Discrete Gabbay-Rodrigues Iteration Schema, then the sequence of values will not converge, as shown in Example 3. The sequence in the full-fledged Gabbay-Rodrigues Iteration Schema will however converge to the values $X = \tfrac{1}{2}, Y = \tfrac{1}{2}, Z = \tfrac{1}{2}$, which do correspond to the empty complete extension.

Example 3. Consider the argumentation framework of Fig. 1 and the initial assignment $V_0 = \{X = 1, Y = 1, Z = 1\}$. The table below shows how the Discrete Gabbay-Rodrigues Iteration Schema evolves using V_0 as initial values.

$$V_0(X) = 1 \quad V_1(X) = 1 - \max\{1\} = 0 \quad V_2(X) = 1 - \max\{0\} = 1 \ldots$$
$$V_0(Y) = 1 \quad V_1(Y) = 1 - \max\{1\} = 0 \quad V_2(Y) = 1 - \max\{0\} = 1 \ldots$$
$$V_0(Z) = 1 \quad V_1(Z) = 1 - \max\{1\} = 0 \quad V_2(Z) = 1 - \max\{0\} = 1 \ldots$$

We will see that under the particular initial assignment all-**und**, the discrete version of the schema *will* always converge to values whose corresponding labelling function is complete.

This means that given the all-**und** initial assignment for an argumentation framework $\langle S, R \rangle$, the Discrete Gabbay-Rodrigues Iteration Schema will compute its grounded extension.

Theorem 5. *Let $V_k(X)$ be the values of the Gabbay-Rodrigues Iteration Schema at the stable point. Then the labelling λ_{V_k} is admissible.*

Proof. We only need to show that if $V_k(X) = 1$ then X is legally labelled **in** by λ_{V_k} and that if $V_k(X) = 0$ then X is legally labelled **out** by λ_{V_k}.

So suppose $V_k(X) = 1$. Since the sequence is stable at k, $V_{k+1}(X) = V_k(X) = 1$. Hence,

$$1 = 0 \cdot \min\{\tfrac{1}{2}, 1 - \max_{Y \in Att(X)}\{V_i(Y)\}\} + 1 \cdot \max\{\tfrac{1}{2}, 1 - \max_{Y \in Att(X)}\{V_i(Y)\}\}$$
$$= \max\{\tfrac{1}{2}, 1 - \max_{Y \in Att(X)}\{V_i(Y)\}\}$$

Therefore, $\max_{Y \in Att(X)}\{V_i(Y)\} = 0$, and hence for all $Y \in Att(X)$, $V_k(Y) = 0$. Therefore, λ_{V_k} labels all attackers of X **out**, and hence X is legally labelled **in** by λ_{V_k}.

On the other hand, if $V_k(X) = 0$, since the sequence is stable at k, $V_{k+1}(X) = V_k(X) = 0$. Hence,

$$0 = 1 \cdot \min\{\frac{1}{2}, 1 - \max_{Y \in Att(X)}\{V_i(Y)\} + 0 \cdot \max\{\frac{1}{2}, 1 - \max_{Y \in Att(X)}\{V_i(Y)\}$$

$$= \min\{\frac{1}{2}, 1 - \max_{Y \in Att(X)}\{V_i(Y)\}$$

Therefore, $\max_{Y \in Att(X)}\{V_i(Y)\} = 1$, and hence there exists $Y \in Att(X)$ such that $V_k(Y) = 1$. Therefore, λ_{V_k} labels some attacker of X **in**, and hence X is legally labelled **out** by λ_{V_k}.

Proposition 3. *Let λ be an admissible labelling function, then no expansion sequence can change the labels of the nodes in C_λ.*

Proof. 1. Take $\lambda(X) = $ **in**. Since λ is admissible, then all attackers of X must be labelled **out** by λ, and hence none of them is labelled **und**.
The significance of this is that no attacker of X can have its label changed to **in** in any expansion sequence, and hence X cannot itself be changed from **in** to **out** or to **und**. So the label of X remains **in** in all expansion sequences.
2. Take $\lambda(X) = $ **out**. Since V is admissible, then for at least one attacker Y of X do we have that $\lambda(Y) = $ **in**, and therefore, by 1. above, $\lambda(Y)$ cannot change from **in**. Furthermore, any change in the values of any other attacker of X cannot affect X's label, since it is already **out**, and hence it remains **out**.

The counterpart for the above proposition using the schema (T_d) is shown below.

Proposition 4. *Let λ be an admissible labelling and let V_0 be V_λ according to Definition 5. Now consider the schema (T_d). If $V_0(X) \in \{0,1\}$, then $V_1(X) = V_0(X)$.*

Proof. 1. Suppose $V_0(X) = 1$. Since λ is admissible, then all attackers Y of X are labelled **out**, and hence for all such attackers $V_0(Y) = 0$. Therefore,

$$V_1(X) = 1 - \max_{Y \in att(X)}\{V_0(Y)\} = 1 - 0 = 1 = V_0(X).$$

2. Suppose $V_0(X) = 0$. Since λ is admissible, then there exists one attacker Y of X which is labelled **in**, and hence $V_0(Y) = 1$. Therefore,

$$V_1(X) = 1 - \max_{Y \in att(X)}\{V_0(Y)\} = 1 - 0 = 1 = V_0(X).$$

So it is easy to see that given an admissible assignment of values V_0, (T_d) will only change the values of the nodes X such that $V_0(X) \in (0,1)$. But what nodes can change? Suppose $V_0(X) = \frac{1}{2}$. We have three cases

1. Either $max_{Y \in Att(X)}\{V_0(Y)\} = 1$, and then $V_1(X) = 0$. Notice that this change cannot alter the value of any node Z attacked by X with $V_0(Z) \in \{0, 1\}$. If V_0 is admissible and X attacks Z, then certainly $V_0(Z) \neq 1$. If $V_0(Z) = 0$, then again since V_0 is admissible, Z would have been attacked by another node W such that $V_0(W) = 1$, so the change of X to 1 is irrelevant to Z.
2. Or $max_{Y \in Att(X)}\{V_0(Y)\} = 0$, and then $V_1(X) = 1$. Notice that this change cannot again alter the value of any node Z attacked by X with $V_0(Z) \in \{0, 1\}$. If V_0 is admissible and X attacks Z, then if $V_0(Z) = 0$, it would have been attacked by some node W with $V_0(W) = 1$, and so $V_1(Z)$ will remain 0. We cannot have that $V_0(Z) = 1$, since V_0 is admissible and $V_0(X) = \frac{1}{2}$, so the change in the value of X to 1 is irrelevant to all nodes attacked by it.
3. Or $max_{Y \in Att(X)}\{V_0(Y)\} = \frac{1}{2}$, and then $V_1(X) = \frac{1}{2}$. Therefore all nodes attacked by X will remain unaffected.

Conjecture 1. Let $V_e(X)$ be the equilibrium value of the node X calculated according to the Gabbay-Rodrigues Iteration Schema and $V_e^d(X)$ its value calculated according to the discrete version of the schema, where

$$V_0^d(X) = \begin{cases} V_k(X), & \text{if } V_k(X) \in \{0,1\} \\ \frac{1}{2}, & \text{otherwise} \end{cases} \quad (2)$$

Then for all nodes X, $V_e(X) = V_e^d(X)$.

Sketch of proof.

1. Use the full-fledged Gabbay-Rodrigues Iteration Schema. By Theorem 5, the values in V_k correspond to an admissible labelling.
2. Proposition 1 shows that the sequence of values becomes stable at some iteration k. Definition 9 says that crisp values do not change after the stable point. Turn the remaining values in $(0, 1)$ into $\frac{1}{2}$, generating V_0^d. Notice that V_0^d is still admissible, since no nodes with values in $\{0, 1\}$ were changed.
3. Run the discrete version of the Schema using V_0^d as initial values. Proposition 4 guarantees that old crisp values remain the same throughout.
4. All nodes for which the sum of the value of the attackers is 0 will turn into 1. This may then change the value of some nodes attacked by them and so forth. Any change of values will not affect the original crisp values in V_k, because of Proposition 4.
   Proceeding in this way will generate the minimal complete labelling including the initial admissible labelling.

5 Worked examples

This section presents some examples and discusses both the differences between the two versions of the schema and how they can be combined.

1. We have seen in Fig. 1, that the Discrete Gabbay-Rodrigues Iteration Schema is not guaranteed to converge if given an arbitrary illegal initial assignment. So for the argumentation framework of Fig. 1, and initial assignment $V_0^d(X) = V_0^d(Y) = V_0^d(Z) = 1$, we get the values of the nodes in odd steps of the iteration as being 0 and in the even steps as being 1, without convergence. The full-fledged Gabbay-Rodrigues Iteration Schema does not suffer from this. The values calculated will be as follows:

Discrete

	V_0^d	V_1^d	V_2^d
X	1	0	1
Y	1	0	1
Z	1	0	1

Full-fledged

	V_0	V_k	V_e
X	1	$\frac{1}{2}$	$\frac{1}{2}$
Y	1	$\frac{1}{2}$	$\frac{1}{2}$
Z	1	$\frac{1}{2}$	$\frac{1}{2}$

Our suggestion is to let the discrete version take over from iteration k, giving (in the discrete version):

	V_0^d	$V_i^d = V_0^d$
X	$\frac{1}{2}$	$\frac{1}{2}$
Y	$\frac{1}{2}$	$\frac{1}{2}$
Z	$\frac{1}{2}$	$\frac{1}{2}$

Which does converge in a finite number of steps (without the need to calculate the limit of the sequence).

2. Consider the argumentation framework of Fig. 2. It has the grounded labelling $\lambda_1 = \{X = \textbf{in}, Y = \textbf{out}, W = Z = \textbf{und}\}$ corresponding to the grounded extension $E_1 = \{X\}$.
Given the all-$\frac{1}{2}$ initial assignment V_0 below, the Discrete Gabbay-Rodrigues Iteration Schema will compute the subsequent values in the sequence as follows.

	V_0^d	V_1^d	V_2^d	V_3^d	
X	$\frac{1}{2}$	1	1	...	
Y	$\frac{1}{2}$	$\frac{1}{2}$	0	...	$\Big\}\lambda_1$
W	$\frac{1}{2}$	$\frac{1}{2}$	$\frac{1}{2}$	...	
Z	$\frac{1}{2}$	$\frac{1}{2}$	$\frac{1}{2}$	...	

The values converge at iteration 2, and the corresponding labelling function λ_{V_2} is the same as λ_1, so $in(V_2) = E_1$.
Notice that the complete labellings $\lambda_2 = \{X = \textbf{in}, Y = \textbf{out}, W = \textbf{in}, Z = \textbf{out}\}$ and $\lambda_3 = \{X = \textbf{in}, Y = \textbf{out}, W = \textbf{out}, Z = \textbf{in}\}$ cannot be obtained using the Discrete Gabbay-Rodrigues Iteration Schema directly using $\frac{1}{2}$ as initial values. This is because λ_1 and λ_2 are complete but not grounded.

However, if the initial assignments $V_0^{d_1}(X) = 1$, $V_0^{d_1}(Y) = 0$, $V_0^{d_1}(W) = 1$, $V_0^{d_1}(Z) = 0$ and $V_0^{d_2}(X) = 1$, $V_0^{d_2}(Y) = 0$, $V_0^{d_2}(W) = 0$, $V_0^{d_2}(Z) = 1$ are given to the discrete schema, the values will immediately stabilise as they correspond to complete extensions:

	$V_0^{d_1}$	$V_1^{d_1}$
X	1	1
Y	0	0
W	1	1
Z	0	0

	$V_0^{d_2}$	$V_1^{d_2}$
X	1	1
Y	0	0
W	0	0
Z	1	1

$$X \longrightarrow Y \longrightarrow W \longleftarrow Z$$

Fig. 2. A sample argumentation framework.

6 Conclusions and Discussion

In [8] we put forward the Gabbay-Rodrigues Iteration Schema, which given any initial assignment of values to the nodes of an argumentation framework, will successively turn each node with an illegal value 1 or 0 into the undecided range (i.e., in the open interval $(0,1)$). The schema will then, *in the limit* of the sequence of values, turn each illegal undecided values into 0 or 1 so as to yield a complete extension.

The disadvantage of this is that we need some means of computing the values in the limit of the sequence. Every computer has an upper bound of the relative error introduced due to the rounding in the arithmetic calculations. When the maximum variation in node values between two successive iterations becomes smaller than or equal to this value, we can no longer be certain if the variation is genuine or due to rounding errors. In [9], Rodrigues used this as the halting point of the approximation.

In this paper we proposed a simplified version of the schema, which we called the *Discrete Gabbay-Rodrigues Iteration Schema*.

Using the simplified schema, we can only give two guarantees: 1) if the initial assignment corresponds to a complete labelling (i.e., yields a complete extension), then the values of the schema will remain the same; 2) If the initial values are all $\frac{1}{2}$, then the schema will converge to values corresponding to the grounded labelling of the argumentation framework. However, there is no guarantee of correction of any other initial illegal values other than $\frac{1}{2}$.

Our suggestion is to combine the two schema as follows.

1. Start with the Gabbay-Rodrigues Iteration Schema and iterate until no new nodes with undecided values are generated (say, iteration k). This is the stable point for the schema.

2. At the stable point k, Corollary 1 and Theorem 1 guarantee that all crisp values are stable (you can read this as "they are legal"). This is also the largest possible set of such values and nodes cannot swap values within $\{0,1\}$.

   The limit theorem of the Gabbay-Rodrigues Iteration Schema gives us that all values will eventually converge to one of $\{0, \frac{1}{2}, 1\}$. This means that the remaining values in $(0,1)$ will all converge to one of $\{0, \frac{1}{2}, 1\}$ **and** this convergence will not affect the previously calculated crisp values.

   Instead of approximating the limit, let us take the discrete version of the iteration schema and use the following assignment V_0^d as the initial assignment for the discrete schema:

$$V_0^d(X) = \begin{cases} V_k(X), & \text{if } V_k(X) \in \{0,1\} \\ \frac{1}{2}, & \text{otherwise} \end{cases} \qquad (3)$$

3. Now apply the discrete schema with the initial values V_0^d.

$$V_{i+1}^d(X) = 1 - \max_{Y \in Att(X)} \{V_i^d(Y)\} \qquad (\mathbf{T_d})$$

   The initial crisp values in V_0^d are all legal, since $V_0^d = V_k$ is admissible, so by Proposition 4, they will not change.

   The illegal $\frac{1}{2}$ values will change, but only so as to yield a complete extension. We conjecture that this extension has to be the same as the one calculated by the full-fledged Gabbay-Rodrigues Iteration Schema, since that extension is the minimal extension including the crisp values calculated at the stable point.

References

1. P. Baroni, M. Caminada, and M. Giacomin. An introduction to argumentation semantics. *The Knowledge Engineering Review*, 26:365–410, 12 2011.
2. M. Caminada. On the issue of reinstatement in argumentation. In *Proceedings of the 10th European Conference on Logics in Artificial Intelligence*, JELIA'06, pages 111–123, Berlin, Heidelberg, 2006. Springer-Verlag.
3. M. Caminada and D. M. Gabbay. A logical account of formal argumentation. *Studia Logica*, 93(2-3):109–145, 2009.
4. M. Caminada and G. Pigozzi. On judgment aggregation in abstract argumentation. *Autonomous Agents and Multi-Agent Systems*, 22(1):64–102, 2011.
5. P. M. Dung. On the acceptability of arguments and its fundamental role in non-monotonic reasoning, logic programming and n-person games. *Artificial Intelligence*, 77:321–357, 1995.
6. M. A. Falappa, A. J. García, G. Kern-Isberner, and G. R. Simari. On the evolving relation between belief revision and argumentation. *The Knowledge Engineering Review*, 26:35–43, 2 2011.
7. M. A. Falappa, A. J. García, G. Kern-Isberner, and G. R. Simari. Stratified belief bases revision with argumentative inference. *Journal of Philosophical Logic*, 42(1):161–193, 2013.
8. D. M. Gabbay and O. Rodrigues. Equilibrium states in numerical argumentation networks. *Logica Universalis*, pages 1–63, 2015.

9. O. Rodrigues. Gris system description. In Matthias Thimm and Serena Villata, editors, *System Descriptions of the First International Competition on Computational Models of Argumentation (ICCMA'15)*, pages 37–40. Cornell University Library, 2015. http://arxiv.org/abs/1510.05373.

Author Index

Abraham, Michael, 195
Ajroud, Amen, 210
Amgoud, Leila, 377

Beierle, Christoph, 1, 297
Belfer, Israel, 195
Benferhat, Salem, 210
Biskup, Joachim, 133
Brewka, Gerhard, 1

Delgrande, James, 73
Dellnitz, Andreas, 348
Douven, Igor, 265
Dubois, Didier, 280

Eiter, Thomas, 150

Falappa, Marcelo A., 223
Fermé, Eduardo, 243
Ferrarotti, Flavio, 365
Fine, Benjamin, 23
Finthammer, Marc, 297

Gabbay, Dov, 195, 392
Garcia, Alejandro, 223
Gonçalves, Sara, 243

Hansson, Sven Ove, 84
Howarth, Elizabeth, 316

Kulmann, Friedhelm, 348

Lukasiewicz, Thomas, 175

Martinez, Maria Vanina, 175
Moldenhauer, Anja, 31
Molinaro, Cristian, 175

Paris, Jeff, 316
Potyka, Nico, 332
Prade, Henri, 280
Predoiu, Livia, 175

Rödder, Wilhelm, 348
Ragni, Marco, 98
Redl, Christoph, 150
Renne, Bryan, 73
Rodrigues, Odinaldo, 392
Rosenberger, Gerhard, 23, 31

Schüller, Peter, 150
Schewe, Klaus-Dieter, 365
Schild, Uri, 195
Simari, Gerardo I., 175
Simari, Guillermo, 223
Spohn, Wolfgang, 112

Tadros, Cornelia, 133
Tec, Loredana, 365
Thalheim, Bernhard, 52
Thimm, Matthias, 1
Tropmann-Frick, Marina, 52

Wang, Qing, 365

www.ingramcontent.com/pod-product-compliance
Lightning Source LLC
Chambersburg PA
CBHW071327190426
43193CB00041B/912